Microplastics in Aquatic Environments

Microplastics in Aquatic Environments

Sources, Impacts, and Sustainable Management Strategies

Written by

Melvin S. Samuel

Marquette University, Milwaukee, WI, United States

ELSEVIER

Elsevier
Radarweg 29, PO Box 211, 1000 AE Amsterdam, Netherlands
125 London Wall, London EC2Y 5AS, United Kingdom
50 Hampshire Street, 5th Floor, Cambridge, MA 02139, United States

ISBN: 978-0-443-45619-0

For information on all Elsevier publications visit our website at
https://www.elsevier.com/books-and-journals

Publisher: Peter Llewellyn
Acquisitions Editor: Maria Elekidou
Editorial Project Manager: Naman Singh
Production Project Manager: Aishwarya R.
Cover Designer: Greg Harris

Typeset by TNQ Tech

Dedication

This book represents the culmination of sustained scientific inquiry, interdisciplinary collaboration, and solution-driven research led by ***Dr. Melvin S. Samuel****, whose work over the past decade has significantly contributed to advancing the understanding of microplastics, per- and polyfluoroalkyl substances (PFAS), and emerging contaminants in aquatic environments.* ***Dr. Melvin's*** *expertise lies at the intersection of environmental chemistry, aquatic toxicology, materials science, and regulatory science, with a strong emphasis on translating complex datasets into actionable strategies for environmental protection and sustainable management.*

Dr. Melvin's *contributions to microplastics research have focused on elucidating the sources, transport pathways, environmental fate, and bioavailability of microplastic particles across freshwater, estuarine, and marine systems. His work has addressed critical knowledge gaps related to microplastic size distributions, polymer-specific behavior, aging and weathering processes, and interactions with sediments, dissolved organic matter, and biological systems. Through integrated laboratory studies, field-based monitoring, and comparative regional analyses, his research highlights how microplastics act not merely as inert particles, but as dynamic environmental vectors capable of influencing chemical transport, microbial colonization, and ecosystem functioning.*

A defining aspect of ***Dr. Melvin's*** *research is his focus on the coupled behavior of microplastics and cooccurring contaminants, particularly PFAS. Recognizing PFAS as a class of highly persistent, mobile, and toxic compounds, his work has explored how microplastics can serve as sorptive surfaces and transport media for PFAS and other organic pollutants. This integrated perspective provides critical insight into combined exposure risks, pollutant bioaccumulation, and the limitations of conventional water treatment systems in addressing complex contaminant mixtures. His contributions emphasize the need for holistic contaminant management strategies that account for interactions among physical, chemical, and biological stressors.*

Dr. Melvin *has also made substantial contributions to understanding the ecological and health implications of microplastic pollution. His work examines impacts on aquatic organisms across trophic levels, including ingestion, physiological stress responses, and implications for food-web transfer. Extending beyond ecological systems, his research evaluates human exposure pathways through drinking water, seafood, and airborne microplastics, framing these risks within toxicological and* *regulatory contexts. By integrating toxicology, exposure science, and environmental monitoring, his work supports evidence-based risk assessment and public health decision-making.*

A core strength of this book lies in its reliance on intensive data collection, standardized analytical approaches, and global case studies, reflecting ***Dr. Melvin's*** *commitment to robust, reproducible science. The volume draws upon datasets and case examples from diverse geographic regions, including Asia, Europe, North America, Africa, and other rapidly developing and industrialized areas. These regional studies capture variability in pollution sources, hydrological conditions, waste management practices, and regulatory frameworks, enabling comparative analysis and globally relevant conclusions. Such an approach reinforces the importance of localized data while supporting the development of harmonized global assessment frameworks.*

Dr. Melvin's *expertise extends to advanced analytical and monitoring technologies, including spectroscopy- and microscopy-based identification of microplastics, high-resolution chemical characterization, and emerging digital and AI-assisted tools for detection and classification. His work critically evaluates the limitations of current methodologies, emphasizing the urgent need for standardization, inter-laboratory comparability, and scalable monitoring solutions. These insights are particularly relevant for policymakers, regulators, and practitioners seeking reliable metrics for environmental assessment and compliance.*

Beyond diagnosis, ***Dr. Melvin's*** *research is strongly solution-oriented. His contributions to this book highlight technological innovations for microplastic and PFAS remediation, including filtration systems, material substitutions, green chemistry approaches, and nature-based solutions. He advocates for integrated strategies that combine engineering controls, policy interventions, circular economy principles, and public engagement, recognizing that technological solutions alone are insufficient without systemic change. His work aligns scientific innovation with sustainability goals, corporate responsibility, and environmental justice.*

This book is dedicated not only to advancing scientific knowledge but also to supporting environmental professionals, regulators, industry leaders, and communities who are actively working to mitigate microplastic and PFAS pollution. It reflects ***Dr. Melvin's*** *belief that interdisciplinary collaboration, transparent data sharing, and evidence-based governance are essential for protecting aquatic ecosystems, safeguarding food security, and ensuring public health resilience.*

Ultimately, this volume stands as a reflection of ***Dr. Melvin's*** *commitment to bridging science, policy, and technology. It is dedicated to present and future generations of researchers and practitioners who seek to transform environmental challenges into opportunities for sustainable innovation and to build cleaner, safer, and more resilient aquatic systems worldwide.*

Contents

About this book

Microplastic pollution has emerged as one of the most pervasive and underestimated environmental threats of modern times. Aquatic ecosystems, including rivers, lakes, estuaries, groundwater systems, and oceans, serve as primary sinks for microplastics generated through the degradation of larger plastic debris and the direct release of micro-sized particles from consumer, industrial, and urban activities. These ecosystems are fundamental to biodiversity, drinking water security, food production, and climate regulation, making microplastic contamination a critical issue with direct implications for environmental sustainability and human health.

This book is important because it presents a comprehensive and integrated synthesis of microplastic pollution across aquatic environments, moving beyond isolated studies to examine the full continuum from sources and transport to ecological, toxicological, and societal impacts. It systematically addresses how microplastics originate, accumulate, and interact with both biotic and abiotic components of aquatic systems, providing readers with a holistic understanding of their environmental behavior and consequences.

The novelty of this book lies in its systems-level and interdisciplinary approach. Rather than treating microplastics as inert particles, this book emphasizes their role as active environmental vectors that facilitate biofilm formation, complex with cooccurring pollutants, and influence biogeochemical processes, including greenhouse gas emissions. By integrating ecological impacts with analytical advancements, human risks, regulatory frameworks, and remediation technologies, this book bridges gaps between environmental science, public health, policy, and sustainability research.

The book holds significant value for a diverse readership. Researchers and students will benefit from a structured overview that identifies emerging research gaps and methodological challenges. Policymakers and regulators will find evidence-based insights to support governance frameworks and environmental standards. Environmental professionals and industry stakeholders will gain practical perspectives on technological solutions and sustainable strategies for microplastic reduction and remediation.

The objective of this book is to enhance scientific literacy and decision-making capacity across disciplines. Readers will gain a clear understanding of microplastic pathways in aquatic ecosystems, their impacts on biodiversity and food webs, potential toxicological implications for humans, and the effectiveness of current detection, monitoring, and treatment approaches. Importantly, the book encourages the translation of scientific knowledge into actionable strategies at local, regional, and global levels.

Microplastic pollution is deeply connected to day-to-day human life. Everyday activities such as consuming bottled water, washing synthetic textiles, using personal care products, and disposing of plastic waste contribute directly to microplastic loading in aquatic environments. By tracing these pathways, the book enables readers to recognize how routine behavior influences environmental and human exposure, fostering awareness and responsibility at the individual and community levels.

This book uniquely integrates scientific evidence, country-specific case studies, and regulatory analysis into a single global framework for understanding and managing microplastic pollution. It brings together findings from diverse geographic regions, including Malaysia, the United States (southern coastlines, California, Alaska, and the Gulf of Mexico), Canada (British Columbia and Arctic freshwater systems), Mexico, Pakistan, and India, to demonstrate how microplastic occurrence, transport, and biological uptake vary across marine, freshwater, Arctic, and estuarine environments. Beyond documenting contamination patterns, the book critically examines international treaties, global and regional policies, national regulations, and industry standards, while also addressing human-rights considerations and governance challenges associated with microplastic exposure. By linking empirical field studies with policy instruments such as plastic management strategies, microbead controls, Clean Water Act mechanisms, global health initiatives, and emerging international frameworks, the book provides a comprehensive, evidence-based foundation for strengthening regulation, guiding technology development and mitigation strategies, and advancing coordinated global action on microplastic pollution in a changing climate.

From a societal perspective, reducing microplastic contamination supports public health protection, food security, and ecosystem resilience. Cleaner water sources reduce exposure risks, healthier aquatic ecosystems sustain fisheries and livelihoods, and informed regulatory actions contribute to long-term environmental stability. Addressing microplastic pollution is, therefore, not only an environmental necessity but also a life-preserving and preventive public health measure.

Ultimately, this book underscores that microplastics are indicators of broader sustainability challenges related to production, consumption, and water management. By integrating scientific evidence with practical solutions and societal relevance, the book offers readers both knowledge and purpose, encouraging informed choices, responsible policies, and collective action toward healthier aquatic ecosystems and a safer future.

"The presence of microplastics in water is evidence not of technological failure, but of delayed responsibility."

Preface

Dr. Melvin S. Samuel is an accomplished researcher and interdisciplinary scientist whose academic and professional contributions span analytical instrumentation, water treatment, microbiology, polymer science, and emerging contaminant remediation. With over 15+ years of cumulative research and industry experience, his work has consistently focused on addressing complex environmental and public-health challenges through rigorous scientific inquiry, technological innovation, and translational research.

Melvin S. Samuel, PhD, Environmental Scientist - Aquatic Pollution, Microplastics, PFAS Remediation & Policy Solutions.

Dr. Melvin earned his PhD from Vellore Institute of Technology (VIT), India, and subsequently advanced his research career through prestigious academic and applied research appointments. He served as a Scientist at the Indian Institute of Technology (IIT) Kharagpur, India, followed by roles as a Research Associate at the University of Wisconsin Milwaukee (UWM), USA, and a Senior Research Analyst at Marquette University, USA. Across these positions, Dr. Melvin contributed to interdisciplinary research spanning analytical instrumentation, material chemistry, water treatment, and emerging contaminants. In addition to academic excellence, Dr. Melvin has collaborated extensively with industry organizations, including Sloan, Badger Meter, Watts, Kohler, and the US Army Corps of Engineers, supporting applied research and technology translation. Dr. Melvin serves as a Subject Matter Expert for the startup of AegisGreen Technology and Innovation, providing sustainable-driven solutions and emerging environmental technologies. Over the course of his career, he has authored more than 125+ peer-reviewed scientific publications, accumulating over 7500+ citations, with an h-index of 42, underscoring the impact and relevance of his scholarly work within the global scientific community.

A defining aspect of Dr. Melvin's contributions lies in his work on emerging contaminants, particularly microplastics, nanoplastics, per- and polyfluoroalkyl substances (PFAS), heavy metals, biofilm, and persistent organic pollutants (POPs). His research has explored contaminant fate, transport, transformation, and removal mechanisms across aquatic, sedimentary, and biological matrices. In parallel, Dr. Melvin has played a pivotal role in the development and evaluation of advanced treatment technologies, including batch systems, biological degradation, photocatalysis, electrochemical processes, membrane technologies, nanostructured materials, and hybrid remediation approaches. His work bridges fundamental science with real-world applications, contributing to scalable solutions for drinking water treatment, industrial wastewater management, and environmental remediation. Notably, he is a named inventor on a granted U.S. patent (US 2023/0363392 A1), reflecting his commitment to innovation and technology translation. Beyond technical research, Dr. Melvin has actively contributed to regulatory science, environmental compliance, and sustainability governance, engaging in international standards such as ISO 14001, ISO 45001, and environmental risk assessment frameworks.

The conceptual motivation for the book entitled ***Microplastics in Aquatic Ecosystems: Sources, Impacts, and Sustainable Management Strategies*** was inspired not only by scientific evidence but also by cultural narratives that amplify environmental awareness. The ***Sardar*** film emphasizes how exposure of plastic bottles to sunlight accelerates the release of harmful chemicals such as bisphenol A (BPA), polyethylene terephthalate (PET), and microplastic pollution-related compounds, leading to severe health consequences for consumers, all of which are central to this book. While the ***Sardar*** film dramatizes the health outcomes, the underlying mechanism, chemical migration from plastics into drinking water, is well established. This aligns directly with this book's focus on abiotic interactions, pollutant complexation, and toxicological implications of plastic-derived contaminants, particularly in aquatic environments where plastics fragment into micro and nanoplastics. The film also portrays water commodification, public exposure to plastic-derived chemicals, and long-term health consequences, which mirror real-world challenges documented in environmental and toxicological literature. Dr. Melvin extends sincere appreciation to Karthi (Actor) and P. S. Mithran (Director) for their impactful storytelling and performances.

This book reflects Dr. Melvin's cumulative scientific experience, scholarly rigor, and commitment to addressing global environmental challenges through both technical innovation and societal engagement. By integrating robust scientific evidence with broader cultural and ethical perspectives, the book aims to contribute not only to academic scholarship but also to informed policymaking and public understanding of microplastic pollution in aquatic ecosystems.

Acknowledgments

Dr. Melvin S. Samuel gratefully acknowledges the academic mentorship, professional guidance, and collaborative support that have shaped his interdisciplinary research career and contributed to the development of this book. He extends his sincere appreciation to **Dr. Patrick McNamara, PhD, PE**, and **Dr. Brooke Mayer, PhD, PE**, Professors in the Department of Civil, Construction, and Environmental Engineering at Marquette University, for their sustained mentorship, technical insights, and encouragement in advancing research on microplastics, emerging contaminants, and environmental systems. Their leadership in water quality, environmental engineering, and applied sustainability has significantly influenced the scientific rigor and practical orientation of this work.

Dr. Melvin also expresses his deep gratitude to **Dr. N. D. Pradeep Singh**, Professor in the Department of Chemistry at the Indian Institute of Technology Kharagpur, India, whose mentorship and expertise in analytical chemistry and chemical engineering were instrumental in strengthening the author's research foundation, analytical depth, and critical thinking. Special appreciation is extended to **Dr. Sekar Viswanathan**, Vice President, VIT University, for providing the opportunity to pursue both the MS.and PhD programs, which played a pivotal role in shaping Dr. Melvin's scholarly independence and long-term commitment to chemical engineering and materials science research. The author also sincerely acknowledges **Dr. Ramalingam Chidambaram**, Senior Professor at VIT University, for his mentorship during the PhD program; his guidance in scientific rigor, translational research, and technology development laid a strong foundation for advancing emerging contaminant solutions from academia to industry.

The author further acknowledges **Dr. Shama P. Mirza**, Professor of Chemistry & Biochemistry and Director of the SAILARC Laboratory at the University of Wisconsin–Milwaukee, for providing sustained access to advanced Shimadzu analytical instrumentation over a period of 7 years. This support enabled company-driven research on PFAS and microplastics, facilitating analytical validation, technology scale-up, and peer-reviewed PFAS publications. The work directly contributed to the development of destruction technologies capable of transforming persistent PFAS into non-toxic end products, strengthening the pathway from laboratory innovation to deployable environmental solutions.

This book is grounded in Dr. Melvin's extensive research experience spanning microplastics, PFAS, emerging contaminants, environmental chemistry, advanced materials, and water and wastewater treatment systems, as reflected in his peer-reviewed publications, patents, and interdisciplinary collaborations. He is also thankful to his collaborators and research peers whose discussions and contributions enriched the perspectives presented in this volume. Finally, Dr. Melvin expresses his heartfelt gratitude to family and friends for their unwavering support, encouragement, and understanding throughout his academic and professional journey.

Introduction

Dr. Melvin S. Samuel has carefully written this book in a clear, systematic, and reader-friendly manner to make complex concepts related to microplastic pollution accessible to researchers, students, policymakers, environmental professionals, and interdisciplinary audiences. Each chapter reflects extensive in-depth research, critical scientific analysis, and comprehensive evaluation of global studies, regulatory frameworks, environmental impacts, analytical techniques, transport mechanisms, toxicological effects, and sustainable remediation strategies associated with microplastics. Through detailed case studies, advanced scientific discussions, and globally relevant perspectives, this book provides a thorough and authoritative understanding of microplastic pollution while bridging the gap between environmental science, policy, engineering, and public health.

This book helps readers understand how microplastics have become an invisible but widespread part of daily life through drinking water, seafood, food packaging, air, household products, and aquatic environments. Dr. Melvin presents the subject in a scientifically detailed yet accessible manner, enabling both scientific and general audiences to recognize the sources, risks, environmental impacts, and long-term consequences of microplastic contamination. The book increases public awareness, supports informed lifestyle and environmental decisions, and provides valuable knowledge on how individuals, communities, industries, and policymakers can contribute toward reducing plastic pollution and protecting environmental and human health for future generations.

Microplastics (MPs) have emerged as one of the most persistent and complex environmental contaminants of the 21st century, attracting increasing global attention due to their widespread occurrence, environmental persistence, ecological toxicity, and potential implications for human health. These microscopic plastic particles originate from both primary and secondary sources and are continuously introduced into aquatic ecosystems through wastewater discharge, industrial activities, urban runoff, atmospheric deposition, aquaculture, agricultural practices, and degradation of larger plastic debris. Once released into the environment, MPs undergo extensive transport, transformation, accumulation, and interaction processes across freshwater, marine, estuarine, and terrestrial ecosystems, thereby creating multidimensional environmental challenges. This book provides a comprehensive and interdisciplinary overview of the sources, transport mechanisms, environmental fate, ecological impacts, detection strategies, toxicological implications, regulatory perspectives, and sustainable mitigation approaches associated with microplastic contamination in aquatic systems.

Chapter 1 introduces the fundamental concepts associated with the sources, transport, and accumulation of microplastics in aquatic environments. The chapter discusses the formation of primary and secondary MPs, freshwater contamination pathways, environmental and anthropogenic factors governing their distribution, and the role of hydrodynamics, stormwater systems, meteorological events, and biological carriers in their dispersion. Particular emphasis is placed on accumulation patterns in marine and freshwater systems and the ecological consequences associated with MP persistence. This chapter establishes the scientific foundation required to understand the environmental behavior of MPs and highlights the complexity of their transport across interconnected ecosystems.

Chapter 2 focuses on the interactions of microplastics with biotic and abiotic components of the ecosystem. The chapter comprehensively explains contaminant adsorption and desorption mechanisms, including hydrophobic interactions, electrostatic interactions, pore filling, hydrogen bonding, and van der Waals forces. It further discusses biofilm formation on microplastic surfaces and their role in PFAS adsorption, along with the influence of salinity, organic matter, polymer morphology, ionic strength, and environmental conditions on contaminant behavior. The chapter provides critical insight into how MPs function as vectors for toxic contaminants and influence water quality, sediment chemistry, and cryosphere environments.

Chapter 3 examines the impact of microplastics on biodiversity, food webs, and ecosystem services. The chapter highlights the transformation, persistence, and toxicity of MPs and their role as carriers of microbes and co-contaminants in aquatic systems. It discusses the direct and indirect effects of MPs on aquatic organisms, trophic transfer, bioaccumulation, biomagnification, nutrient cycling, and carbon and phosphorus dynamics. Additionally, the chapter explores socio-economic consequences, climate-related implications, and the importance of circular economy principles and life cycle assessment in plastic management strategies. The major takeaway from this chapter is the recognition that MPs influence not only individual organisms but also ecosystem stability, food security, and climate resilience.

Chapter 4 addresses the emerging role of microplastics in biofilm formation, pollutant complexation, and greenhouse gas emissions. The chapter discusses microbial colonization on MP surfaces, the development of plastisphere communities, antibiotic resistance dissemination, and pollutant enrichment on biofilm-coated MPs. It further explains the influence of MP-associated biofilms on greenhouse gas emissions and climate impacts, while also discussing advanced characterization methods for MP-associated microbial communities. This chapter introduces novel perspectives on the interactions between MPs, microorganisms, and environmental pollutants, highlighting their role in shaping ecosystem-level biochemical processes and climate-related effects.

Chapter 5 presents a detailed overview of detection, monitoring, and analytical techniques for microplastic assessment. It discusses advanced spectroscopy- and microscopy-based tools, including FTIR spectroscopy, Raman spectroscopy, fluorescence microscopy, thermal analysis, pyrolysis-GC/MS, SEM, ICP-MS, laser-induced breakdown spectroscopy, and hyperspectral imaging. The chapter also evaluates field sampling methodologies, laboratory analysis protocols, standardization challenges, and future analytical strategies for accurate MP quantification and identification. This chapter is particularly significant for researchers and environmental laboratories seeking robust methodologies for monitoring MPs in diverse environmental matrices.

Chapter 6 explores the potential health risks and toxicological implications of microplastics in humans and animals. The chapter comprehensively discusses exposure pathways, including ingestion, inhalation, and dermal contact, followed by their toxicological impacts on the gastrointestinal, respiratory, reproductive, nervous, immune, and renal systems. Mechanisms such as oxidative stress, inflammation, apoptosis, autophagy, microbial imbalance, and pulmonary fibrosis are also discussed in detail. The chapter integrates in vitro and in vivo studies to provide critical understanding of the long-term biological consequences of MP exposure and highlights important knowledge gaps requiring future investigation.

Chapter 7 investigates microplastic contamination in fisheries, seafood, agricultural products, processed foods, beverages, and drinking water, emphasizing implications for food security and human health. The chapter discusses contamination pathways, bioaccumulation, retention in aquatic organisms, and ecotoxicological risks associated with seafood consumption. Furthermore, it examines analytical methods for isolating MPs from food products and evaluates the economic consequences of MP contamination in aquaculture and fisheries industries. This chapter provides valuable insight into how MPs enter the human food chain and affect global nutritional sustainability.

Chapter 8 provides extensive regional case studies describing microplastic contamination across diverse aquatic environments worldwide. The chapter includes investigations from Malaysia, North America, California, Arctic Sea ice systems, British Columbia estuaries, Great Slave Lake, the Gulf of Mexico, Baja California, Pakistan, and Indian freshwater ecosystems. These case studies collectively demonstrate the spatial variability, seasonal distribution, polymer composition, ecological implications, and regional management challenges associated with MPs. The chapter offers real-world evidence supporting the global nature of MP pollution and highlights the importance of region-specific monitoring and mitigation strategies.

Chapter 9 discusses regulatory frameworks and global perspectives on microplastic pollution. It reviews international treaties, national policies, industry standards, WHO initiatives, Clean Water Act provisions, and governance strategies related to plastic pollution prevention and microplastic management. The chapter also evaluates policy gaps, climate change influences, and the environmental fate of MPs under changing climatic conditions. The integration of regulatory science with environmental management makes this chapter particularly valuable for policymakers, environmental agencies, and sustainability professionals.

Chapter 10 highlights technological solutions for microplastic remediation, including filtration systems, coagulation-flocculation methods, geopolymer technologies, biodegradable materials, microalgae-assisted remediation, and wastewater treatment approaches. The chapter evaluates emerging technologies for MP detection, separation, degradation, and prevention, while discussing challenges associated with field implementation and scalability. Special attention is given to sustainable remediation systems and life cycle assessment-based treatment strategies. This chapter demonstrates how scientific innovation can contribute toward practical and scalable solutions for MP pollution control.

Finally, Chapter 11 presents sustainable strategies for reducing microplastic contamination in aquatic systems. The chapter emphasizes circular economy approaches, bio-circular green technologies, public awareness initiatives, international regulatory frameworks, and artificial intelligence-assisted detection and mitigation techniques. Advanced AI applications including machine learning, FTIR-assisted identification, Raman-based image processing, neural networks, and deep learning for MP quantification are extensively discussed. The chapter concludes by highlighting future perspectives on integrating sustainability, technological innovation, environmental policy, and community engagement for long-term microplastic management.

A major strength of this book is its comprehensive integration of global case studies, regional assessments, and international regulatory perspectives related to microplastic contamination in aquatic ecosystems. The book systematically examines microplastic occurrence, distribution, transport pathways, ecological risks, and mitigation strategies across geographically diverse environments, including Malaysia, North America, California, Alaska, British Columbia, the Gulf of Mexico, Baja California, Pakistan, and Indian freshwater systems. These regional investigations provide valuable comparative insights into how climatic conditions, hydrodynamics, wastewater discharge, urbanization, fisheries, industrialization, and anthropogenic activities influence the spatial and temporal distribution of microplastics across marine, estuarine, freshwater, coastal, and Arctic ecosystems. The inclusion of region-specific studies on oysters, mussels, benthic fish, crabs, sediments, surface waters, and estuarine biota further strengthens the environmental and ecological relevance of this work.

The book also provides a detailed evaluation of international regulatory frameworks, governance mechanisms, and policy strategies aimed at addressing microplastic pollution at both national and global scales. Dedicated discussions include international treaties, global plastic management policies, microbead regulations, World Health Organization initiatives, Clean Water Act provisions, regional environmental policies, and emerging frameworks for sustainable plastic governance. Emphasis is placed on policy gaps, environmental justice considerations, global production trends, and the urgent need for harmonized international monitoring standards and targeted mitigation policies. By integrating environmental science with governance, sustainability, and policy development, this book serves as a globally relevant scientific resource for researchers, environmental agencies, policymakers, industry stakeholders, and academic institutions working toward long-term management and mitigation of microplastic pollution.

Relevance and Importance to the United States of America: This book is highly relevant to the United States of America due to the increasing concerns surrounding microplastic contamination in US freshwater, marine, estuarine, coastal, and wastewater systems. The book provides scientifically grounded insights into microplastic occurrence, transport pathways, ecological impacts, human exposure risks, and regulatory challenges across several North American regions, including California, the Gulf of Mexico, the St. Lawrence Estuary, British Columbia coastal systems, and urban aquatic environments. These case studies offer important environmental and public health perspectives that are directly applicable to US water quality management, fisheries, wastewater treatment infrastructure, and coastal ecosystem protection. The book further supports ongoing US efforts related to environmental sustainability, plastic pollution reduction, Clean Water Act implementation, emerging contaminant regulation, and climate resilience planning. By integrating advanced analytical techniques, risk assessment strategies, regulatory frameworks, and sustainable remediation technologies, this work provides valuable guidance for environmental agencies, policymakers, researchers, water utilities, and academic institutions across the United States. The inclusion of global perspectives alongside US-focused environmental challenges strengthens the book's contribution toward developing evidence-based strategies for long-term microplastic monitoring, pollution prevention, and ecosystem protection.

Overall, this book offers a multidisciplinary and globally relevant perspective on microplastic pollution by integrating environmental science, analytical chemistry, toxicology, environmental engineering, climate science, regulatory policy, and sustainable remediation technologies. One of the major novelties of this volume lies in its ability to connect fundamental mechanisms of MP generation and transport with emerging themes such as biofilm-mediated pollutant interactions, greenhouse gas emissions, artificial intelligence-driven analytical approaches, circular economy strategies, and region-specific environmental case studies. The book serves as a valuable scientific resource for researchers, environmental professionals, policymakers, graduate students, analytical scientists, and regulatory agencies seeking comprehensive understanding and advanced insights into the environmental fate, ecological significance, and mitigation of microplastics in aquatic systems.

Concluding Perspective*:* Through this book, Dr. Melvin envisions a future where aquatic ecosystems, drinking water resources, marine biodiversity, and environmental systems are protected from the growing threats posed by microplastics and emerging contaminants. The author believes that clean water and healthy aquatic environments are fundamental to human survival, ecological balance, food security, and sustainable development. By combining extensive scientific research, global case studies, environmental risk assessments, regulatory perspectives, and sustainable remediation approaches, this book seeks to increase awareness, inspire responsible environmental action, and support future innovations for pollution prevention and ecosystem restoration. A world with cleaner rivers, lakes, oceans, and groundwater systems represents not only environmental recovery, but also improved public health, safer food systems, healthier biodiversity, and a more sustainable future for generations to come. Dr. Melvin hopes this book will serve as both a scientific resource and a call for collective global action toward reducing plastic pollution and protecting the planet from the long-term consequences of emerging contaminants.

> *A cleaner water system today creates a healthier environment, safer ecosystems, and a sustainable future for generations to come. Reducing microplastics and emerging contaminants begins with awareness, responsibility, and collective action* — ***Dr. Melvin S. Samuel***

CHAPTER

1 Sources, transport, and accumulation of microplastics in aquatic environments

Melvin S. Samuel

Department of Civil, Construction, and Environmental Engineering, Marquette University, Milwaukee, WI, United States

Highlights

- Meteorological events drive Microplastic mobilization, transport, and redistribution in aquatic systems.
- Hydrodynamics and atmospheric processes control Microplastic dispersion and accumulation pathways.
- Environmental metrics such as land use and watershed characteristics indicate Microplastic loading.
- Anthropogenic pressures increase Microplastic inputs into freshwater and coastal environments.
- Material properties and seasonal dynamics govern Microplastic fate in freshwater systems.

1.1 Introduction

The widespread utilization of plastic polymers in contemporary society is attributed to their exceptional mechanical properties, chemical stability, and economic viability. Since their industrial inception in the mid-20th century, the global output of plastic materials has increased hugely, with production volumes reaching approximately 367 million metric tons by 2020 (Zhang et al., 2020). Between 2020 and 2022, global plastic output increased from 380.4 million metric tons (Mt) to 400.3 Mt, reflecting sustained industrial and consumer demand (Naqash et al., 2024). While these attributes have facilitated extensive applications across diverse sectors, the inherent resistance of plastics to environmental degradation has precipitated significant ecological challenges. One of the most critical concerns is the fragmentation of large MP debris into smaller MPs, a process driven by physical, chemical, and biological weathering mechanisms. MPs are defined as synthetic polymer particles with diameters less than 5 mm and are now recognized as persistent contaminants in terrestrial and aquatic ecosystems. Based on their origin, MPs are classified into two principal categories: primary and secondary (Ashrafy et al., 2023; Barrows et al., 2018).

Primary MPs are engineered at microscopic scales for targeted application in industrial and consumer products. These include microbeads incorporated into cosmetic formulations, abrasive agents used in manufacturing processes, and polymeric pellets commonly composed of polyethylene (PE), polypropylene (PP), and polystyrene (PS) utilized as raw materials in plastic fabrication (Conkle et al., 2018). Pathways for environmental entry of these particles include domestic wastewater effluents, accidental release during industrial handling, and failures experienced during transportation and storage operations (De Falco et al., 2019). The pervasive dependence of plastic materials, coupled with

Microplastics in Aquatic Environments. https://doi.org/10.1016/B978-0-443-45619-0.00001-7

systemic inefficiencies in waste management infrastructure, has triggered extensive contamination across global ecosystems. A pivotal investigation quantified the annual marine influx of mismanaged plastic waste from costal zones to range between 4.8 and 12.7 million metric tons, highlighting the magnitude of anthropogenic contributions to oceanic pollution (Akdogan and Guven, 2019). Such contamination exerts deleterious effects on aquatic biota, spanning trophic levels from planktonic organisms to apex marine predators. A comprehensive global assessment estimated that 19–23 million metric tons of plastic waste constituting nearly 11% of total annual plastic waste entered aquatic systems within a single year. This influx is not solely attributable to surface runoff or direct disposal; landfill leachate has also been identified as a significant vector for MP transport. Rainwater infiltration through municipal solid waste layers generates leachate enriched with MP particles, contributing to downstream contamination (Borrelle et al., 2020).

Empirical evidence consistently demonstrates the predominance of primary MPs in diverse freshwater and marine environments. These particles have been extensively documented in oceanic water, lacustrine systems, and fluvial networks (McCormick et al., 2020), underscoring their ubiquity and persistence. Of particular concern is the role of municipal wastewater treatment plants as conductive MPs distribution. Despite technological advancements in filtration and sedimentation processes, it is estimated that approximately 65 million MP particles are discharged daily into receiving water bodies via treated effluent streams (Conley et al., 2019). These findings accentuate the critical for enhanced waste management protocols, including upstream interventions, advanced filtration technologies, and policy driven mitigation strategies aimed at curbing MP pollution and safeguarding aquatic ecosystem integrity.

Spatial heterogeneity in MP concentrations is evident across global soil matrices. A recent comparative analysis reported MP contamination in agricultural soils (items/kg) as follows: Germany (2), Pakistan (2545), Chile (306), Spain (3607), Iran (48), Turkey (16.5), Japan (48), Canada (541), and Tunisia (161). In contrast, soil proximal to landfill site exhibited markedly elevated MP burdens, with reported values including Iran (544), Portugal (1,500,000), India (1780) and the United Kingdom (12,300) (Shi et al., 2024). Aquatic environments similarly display regional disparities in MP concentrations. Elevated levels have been documented in the Mediterranean Sea, English Channel, West Baltic Sea, and the Northeast Pacific Ocean, whereas comparatively lower concentrations have been observed in the Coral Sea, off the Australian coast, the Scottish shoreline, and Polar Regions such as Antarctica (Bohdan, 2022) (Table 1.1). Nations such as India, Pakistan, and China have been identified as critical hotspots for MP pollution, driven by high population densities, rapid industrial expansion, and insufficient waste treatment infrastructure (Chen et al., 2024).

Meta analytical evaluation of MP concentration across the five major oceanic bodies revealed values ranging from 0.002 to 62.5 items/m^3, with a global average of 2.76 items/m^3. The Atlantic Ocean exhibited the highest mean concentration at 4.98 item/m^3 (Mutuku et al., 2024). Regional case studies from China further illustrate the variability in MP pollution. In Dongguan City, surface water such as the Hanxi River, Xiaohai River, and Dongguan Canal recorded MP concentrations ranging from 351.09 to 1646.22 items/kg, predominantly composed of polyvinyl chloride (PVC) and polyethylene (PE) polymers (Zhou et al., 2024). Investigations along with the Yangtze River, a major fluvial system reported MP concentrations ranging from 5.13 items/L to 113.9 items/kg, with fibrous MPs derived from PE and polypropylene (PP) being the most prevalent, and particle sizes predominantly below 1 mm (Xia et al., 2024). Collectively, these findings underscore the extensive and heterogeneous nature of MP contamination and reinforce the imperative for globally coordinated monitoring framework, policy interventions, and technological innovations aimed at mitigations aimed at mitigating the environmental and ecological impacts of MPs.

Table 1.1 Summary of microplastic abundance and detection reported in global studies, highlighting the extent of contamination across diverse aquatic environments.

				Results		
Country (site)	**Sampling**	**Pretreatment**	**Analysis**	**Size**	**Concentration**	**References**
Antarctic (Byers Peninsula)	Two nylon drifting nets (333 and 100 μm)	33% H_2O_2, hypersaline solution (NaCl)	μ-FT–IR (micro-Fourier transform infrared spectroscopy)	Fibers: 400–3546 μm (average 1118 μm), films: 10–1026 μm (average 199 μm)	0.47–1.43 items/1000 m^3 (0.95 items/1000 m^3)	Gonzalez-Pleiter et al. (2020)
Atlantic Ocean	A flow pump system and a towed planktonic manta net	–	FT–IR	Net >5 mm (30%), Pump<2.5 mm	~60 items/m^3 (pump) ~1829 items/m^3 (net)	Silvestrova and Stepanova (2021)
Canada (Baynes Sound, Vancouver Island)	Jar (1 L glass jars) and Bucket (12 L stainless bucket) samples	Potassium Hydroxide (KOH) solution	FT–IR	–	0.69 MP/L (1 L samples and 0.12 MP/L (10 L samples)	Covernton et al. (2019)
China (The North Yellow sea)	Niskin hydrophore with a 30 μm steel sleeve	30% H_2O_2 in the presence of a Fe(II) catalyst, NaCl solution (1.2 g/cm^3), sodium iodide (NaI) solution (1.6 g/cm^3) * surface seawater: only NaCl	μ-FT–IR	<0.5 mm (35.7%–83.5%) surface waters	545 ± 282 items/m^3	Zhu et al. (2018)
Italy, Slovenia	Manta nets (net opening of 330 μm)	H_2O_2 digestion	ATR–FT–IR	330 μm – 1 mm (SMP, 34%), 1–5 mm (LMP, 64%, >5 mm (Meso-plastics 2%)	315,009 ± 568,578 items/kg^2 (217 ± 575 g/km^2)	Zeri et al. (2018)

Continued

Table 1.1 Summary of microplastic abundance and detection reported in global studies, highlighting the extent of contamination across diverse aquatic environments.—cont'd

Country (site)	Sampling	Pretreatment	Analysis	Results		References
				Size	Concentration	
United States (Chesapeake Bay)	The manta trawl (330 μm mesh net)	Wet peroxide oxidation (WPO) method with a Fe (II) catalyst, NaCl	FT–IR	0.10–10 mm (average: 4 mm)	32,000 particles/km^2 (0.160 particles/m^3)	Bikker et al. (2020)
United States (lower Hudson river estuary)	333 μm Neuston net with a 1 L plastic bottle attachment	Fe (II) WPO, ~5 M NaCl	Stereomicroscope	–	2016, 243,772 particles/km^2, 2017: 143,204 particles/km^2, 2018: 830,762 particles/km^2. 2019:244,142 particles/km^2	Polanco et al. (2020)

FIGURE 1.1

Microplastics in the world oceans based on studies.

At the global scale, clear spatial patterns in MP abundance across lakes have not yet emerged, likely due to heterogeneous local pollution sources and limited sampling coverage. Although large scale lake studies remain scarce, existing research highlights significant MP burdens. MP in the world oceans based on studies (Fig. 1.1). Eriksen et al. (2014) estimated approximately 43,000 particles/km^2 in Lake Superior, while Baldwin et al. (2020) reported 466,000 particles/km^2 in tributaries of the Great Lakes. Mason et al. (2016) identified about 17,000 particles/km^2 in the surface waters of Lake Michigan. In Europe, reported concentrations include 2.68–3.36 particles/m^3 in Lake Chiusi, Italy (Hale et al., 2020), and much lower values in Swiss lakes, ranging from 0.06 particles/m^3 in Lake Zurich to 1.2 particles/m^3 in Lake Maggiore (Fischer et al., 2016).

1.2 Primary and secondary sources of microplastics

MP contamination in the environment can be broadly classified into two principal categories—primary and secondary MPs based on their origin and formation mechanisms. Differentiation between these sources is critical for designating targeted mitigation strategies and informing regulatory frameworks.

1.2.1 Primary microplastics

Primary MPs are deliberately engineered to be microscopic in scale, typically less than 5 mm in diameter. These particles are incorporated into a wide array of consumer products and industrial applications. Representative sources include: (i) Tire and road wear particles (TRWPs): Generated through mechanical abrasion between vehicle tires and road surfaces (Fig. 1.2). (ii) Synthetic textile fibers: Released during laundering of garments composed of polymers such as polyester, nylon, and acrylic. (iii) Personal care formulations: Containing microbeads used for exfoliation in facial scrubs, toothpaste, and cosmetic products. (iv) Plastic pellets: Utilized as feedstock in polymer manufacturing process. (v) Urban particulates: Including city dust, artificial turf fragments, and marine coatings that shed MPs via weathering and frictional forces (Horton et al., 2017). Wastewater treatment plants (WWTPs), stormwater runoff and atmospheric dispersion discharge these particles into the environment. The large plastic debris, finer MP particles are subsequently released into aquatic environments due to the evading conventional filtration system.

In coastal region such as China, a large portion of WWTPs is located adjacent to water bodies, assisting direct entry of MPs into marine and freshwater systems (Liu et al., 2019). Agricultural practices also contribute to MP pollution through the application of sewage sludge as fertilizer and the use of polymeric materials. Over time, these materials degrade and release MPs into the soil, which mobilize into nearby watershed via precipitation or irrigation runoff (Corradini et al., 2019). In response to the escalating ecological risks posed by MPs, several jurisdictions have enacted legislative measures to restrict their use. The European Union, for instance, has implemented comprehensive regulations targeting both intentionally added MPs and those released inadvertently from sources such as tires, textiles, paints and cigarette filters (European Commission, 2019).

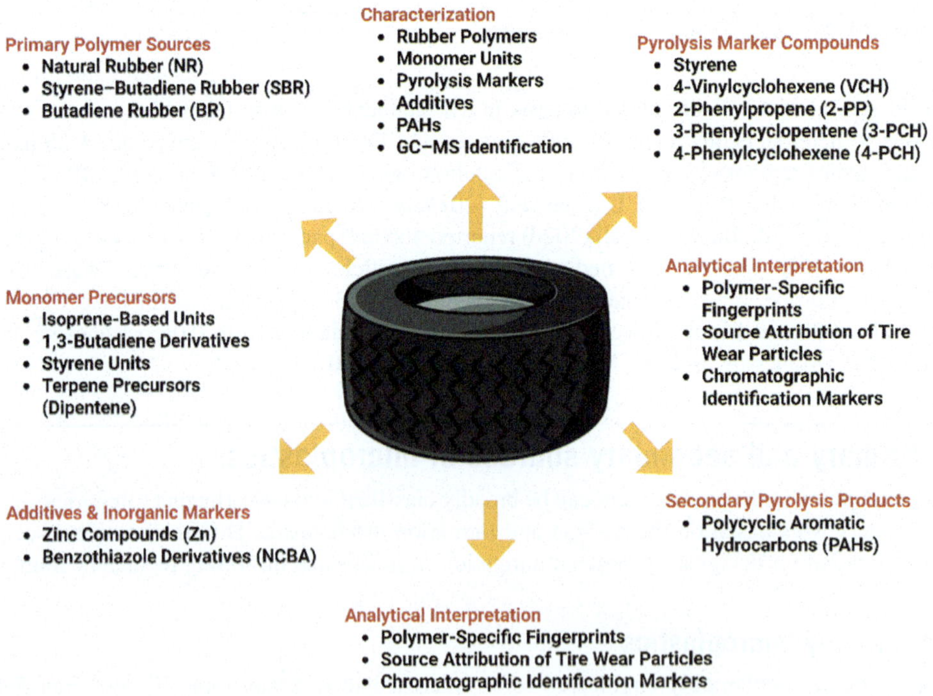

FIGURE 1.2

Wide array of consumer products and industrial applications- tire and road wear particles (TRWPs).

1.2.2 Secondary microplastics

Secondary MPs originate from the environmental fragmentation of larger plastic debris. This transformation is driven by a combination of physical, chemical, and biological degradation processes, which gradually reduce MP items such as bottles, bags, and packaging materials into micro sized particles. Key degradation mechanisms include: (i) Photodegradation: ultraviolet (UV-B) radiation includes polymer chain scission, leading to fragmentation. (ii) Mechanical abrasion: Physical forces from waves, sediments, and coastal substrates erode plastic surfaces. (iii) Thermal and Oxidative degradation: Exposure to heat and oxygen promotes cracking and embrittling. (iv) Biodegradation: Microbial enzymatic activity contributes to polymer breakdown. (v) Hydrolysis: Chemical reactions with water molecules facilities depolymerization, particularly in marine settings (Gewert et al., 2015; Covernton et al., 2019).

Environmental factors such as temperature, oxygen availability, and UV intensity modulate the rate of degradation. Certain polymers possess intrinsic properties that promote self-catalyzed degradation forming reactive oxygen containing moieties that further facilitate fragmentation. Nonetheless, the overall decomposition rate remains insufficient to offset the accumulation of plastic waste, primarily due to the limited metabolic capabilities of microbial consortia and suboptimal environmental conditions (e.g., pH, temperature). Floating plastic debris undergoes dynamic vertical movement within the water column through a process of fouling and defouling. Marine organisms colonize the plastic surface, increasing its density and causing it to sink. Subsequent microbial degradation of the biofilm reduces density, allowing the plastic to resurface. This cyclical behavior impedes consistent degradation and prolongs the environmental persistence of MPs (Fazey and Ryan., 2016). Morphological alterations in MPs particles such as surface pitting, fissuring, and discoloration serve as indicators of their degradation history. Analyzing these features across diverse ecosystems provides valuable insights into the transport dynamics, environmental fate, and longevity of MPs (Blettler et al., 2018).

1.2.3 Migration of microplastics from aquatic systems to terrestrial environments

In the interface between aquatic and terrestrial ecosystems, in the context of MP pollution, Wastewater treatment plants (WWTPs) serve as critical nodes. During the treatment process, MPs present in influent wastewater are predominantly sequestered within sewage sludge. The subsequent application of this sludge as fertilizer on agricultural land constitutes a direct and significant pathway for the translocation of MPs from aquatic systems to soil environments. Empirical data from China indicate that MP concentrations in sewage sludge range from 1.60×10^3 to 56.4×10^3 particles per kilograms. Extrapolations based on these values suggest that approximately 1.56×10^{14} MP particles may be introduced into agricultural soils annually through sludge amendment practices. Notably, WWTPs retain an estimated 88% of MPs within sludge matrices, while only 12% persist in treated effluent, underscoring the role of sludge as a major reservoir for MPs. Comparative analyses reveal that soils treated with sewage sludge exhibit MP concentrations up to 256% higher than those untreated soils, reinforcing the significance of WWTPs in facilitating terrestrial MP concentration (Li et al., 2018).

Despite these insights, methodological challenges persist in accurately quantifying MPs in soil. The absence of standardized natural blank samples and the potential for contamination during sampling and pretreatment procedures can result in inflated concentration estimates (Corradini et al.,

2019). Moreover, MPs may leach from laboratory materials or procedural artifacts, further complicating analytical precision. The reuse of sewage sludge establishes a tangible conduit between aquatic and terrestrial domains. Wastewater is inherently rich in nutrients, fostering microbial proliferation. MPs due to their surface properties and persistence, can act as substrate for microbial colonization and serve as vectors for both commensal and pathogenic microorganisms. Elevated MP concentrations in downstream environments have been documented, highlighting the cumulative nature of this pollution pathway. To comprehensively elucidate the environmental fate of MPs, future research must extend beyond wastewater effluents to encompass sediments, sludge deposits, and shoreline interfaces. A multidisciplinary approach integrating environmental monitoring, microbial ecology, and regulatory oversight is imperative to curtail MP dissemination and mitigate their long-term ecological impacts across interconnected ecosystems.

1.2.4 Abundance of microplastics in freshwater

The prevalence and ecological implications of MPs in marine environments, investigations into their occurrence, distribution, and behavior within freshwater systems remain comparatively underexplored. Given the factor that freshwater serves as primary sources for drinking water and irrigation, understanding the MP dynamics in these ecosystems is essential for protecting both environmental integrity and public health protection (Table 1.2).

1.2.4.1 Influence of anthropogenic activities

Freshwater systems that are located close to industrial, agricultural, or urban development tend to show higher level of MP concentrations. For instance, Asia's largest textile manufacturing region in China, MP level recorded at 13.3 particles per liter higher than the industrial areas (Deng et al., 2020). The elevated concentration was due to the industrial production and commercial activities. Similarly in Nigeria's Ox Bow Lake, over 73% of identified MPs which as spherical beads, originating from local artisanal bead. In Taihu Lake in China a major contributor to national GDP through industrial and agricultural output exhibited MP concentrations ranging from 3.4 to 25.6 particles per liter. In Australia's Gold cost, which is artificial wetland constructed to manage urban runoff found to contain MPs in both water and sediment matrices. These MPs is derived from vehicular tire wear. Hungary's Carpathian Basin conducted research on MP level and found that the pond outlet contained lower MP concentrations compared to inlets, revealing some water bodies function as retention zones for MPs (Bordos et al., 2019).

1.2.4.2 Source Attribution via particle morphology

The morphological characteristics of MPs such as shape and textile offer valuable insights into their origin and transport pathways. Common morphological types include beads, flakes, fibers, and fragments, each associated with unique anthropogenic sources (Fig. 1.3). During the preproduction of plastic materials used in manufacturing pellet shaped MP originate. Synthetic textiles, fishing gear and clothing release fibrous MP this is due to the mechanical degradation of larger plastic items. Italy's Ofanto River, black polyethylene flakes were detected in the surrounding area (Campanale et al., 2020). In India Netravathi River, MP fibers dominated the water system, found that the fibers were more prevalent in benthic sediments. These findings imply ineffective waste disposal practices and redundant wastewater treatment infrastructure as this significantly contributes MP pollution in freshwater.

Table 1.2 Results of microplastic analysis in freshwater environments.

				Results		
Country (site)	**Sampling**	**Pretreatment**	**Analysis**	**Composition**	**Concentration**	**References**
Argentina (Rio de la Plata estuary)	Bucket, plankton net (36 μm)	Fe (II) 0.05 M and 30% H_2O_2, NaCl (5M) 30% H_2O_2	Stereomicroscope	–	139 MP/m^3	Pazos et al. (2018)
Australia (Cooks river)	14 L metal bucket through 37 μm plankton net	NaCl solution (1.2 kg/L), 30% KOH:NaClO, Nile red Sodium hydroxide (NaOH)	Microscope	–	400–17,383 $particles/m^3$	Hitchcock (2020)
Canada (lake Ontario)	Stainless steel bucket with PP rope	Saturated calcium chloride ($CaCl_2$, $\rho = 1.4$ g/L)	Stereomicroscope, μ-Raman	Cellulose, unknown, Anthropogenic (Dye-Based, unknown) PET, PE, PVC, and PVC additive	13.3 ± 15.5 particles L^{-1} in wastewater effluent, 0.9 ± 1.3 particles L^{-1} in agricultural runoff, 0.8 ± 0.7 particles in the lake	Grbić et al. (2020)
China (Yellow river)	Stainless-steel bucket	NaCl solution (1.3 g/mL), H_2O_2 (30%)	Optical electron microscope. ATR–FT–IR	PE, PP, PS	Wet season: 497 (380–583) items/L dry season: 930 (623–1392) items/L	Han et al. (2020)
India (Netravathi river)	Stainless steel bucket of 10-L volume	H_2O_2 (30%) in the presence of Ferrous solution (Fe (II); 0.05 M), $ZnCl_2$	Stereozoom microscope ATR–FT–IR	PE (58.33%) and PET (28.57%), PVC	288 $pieces/m^3$ (56–2328 $pieces/m^3$)	Amrutha and Warrier (2020)
United Kingdom (Trent catchment)	The metal bucket on a telescopic pole	30% H_2O_2	Dissection microscope, ATR–FT–IR	PE (12), PP (3), PS (2), PVA (2), UA (1)	178 plastic particles	Stanton, T et al. (2020)
United States (Gallatin river watershed)	Grab sample (average of 1.3L)	–	Stereomicroscope, μFT–IR	Semi synthetic cellulose (30%), PES (20%), PET (13%), cotton (5%), PP (4%), PA (4%), Urethane (4%), etc.	1.2 microplastics L^{-1} (0–67.5 microplastics L^{-1})	Barrows et al. (2018)

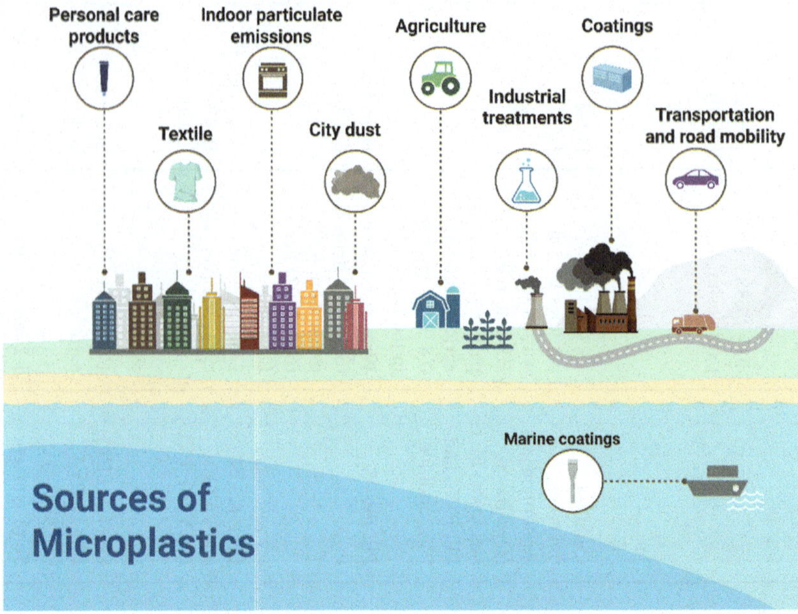

FIGURE 1.3

Schematic illustration of the major factors governing microplastic sources, transport pathways, and accumulation in water environments.

1.2.4.3 Hydrological drivers of microplastic distribution

Hydrological processes exert a substantial influence on the spatial distribution and transport dynamics of MPs within freshwater systems (Fig. 1.4). MP concentrations often increase downstream, particularly in urbanized and estuarine regions. Studies conducted on China's Suzhou and Huangpu rivers revealed elevated MP levels in downstream segments, a pattern similarly observed in Korea's Nakdong River and India's Netravathi River (Amrutha and Warrier, 2020), Conversely, smaller urban waterways may exhibit a reduction in MP fiber concentrations as they approach marine outflows, potentially due to dilution effects or sedimentation processes.

These observations underscore the role of urbanization and anthropogenic activity in shaping MP contamination profiles in freshwater environments. Hydrological parameters including flow velocity, precipitation patterns, and water retention time modulate the accumulation, dispersion, and ecological impact of MPs across the systems.

1.2.5 Factors influencing the distribution of microplastic in freshwater environments

Physiochemical Characteristic of MP. The spatial heterogeneity and concentration gradients of MPs in freshwater systems are governed by a complex interplay between their physicochemical properties and ambient environmental conditions. Parameters such as particle size, polymer density, surface

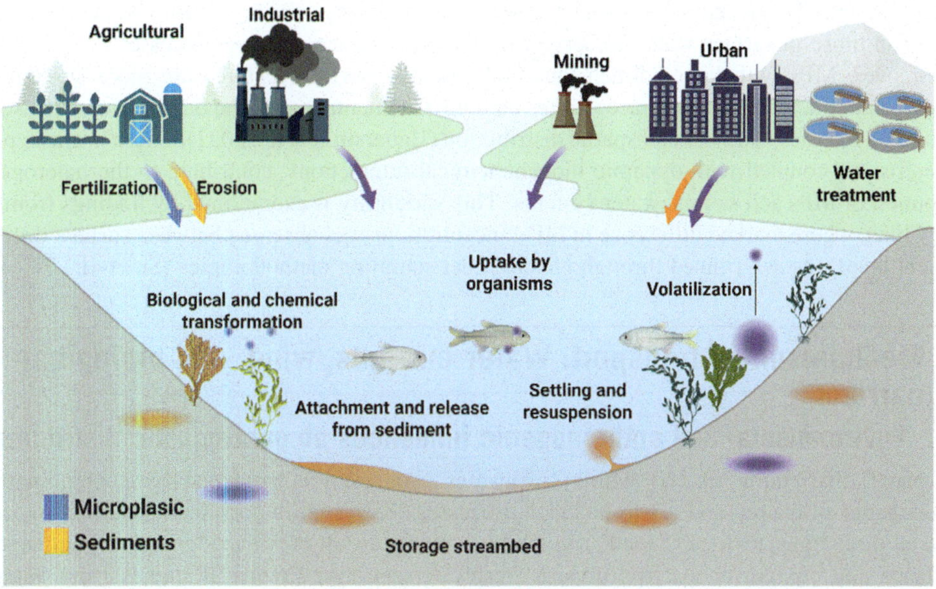

FIGURE 1.4

Environmental, hydrological, industrial, and urban-related factors collectively influence the transport, distribution, and accumulation of microplastics in aquatic environments.

morphology, and chemical composition critically influence their transport, deposition, and persistence within aquatic matrices. Empirical investigations, including those conducted in stormwater retention ponds in Denmark have consistently demonstrated a predominance of smaller sized MPs relative to larger counterparts. This phenomenon aligns with hydrodynamic principles described by Stokes law, which suggests that the terminal settling velocity of a particle is proportional to the square of its diameter. As a result, larger MPs exhibit enhanced sedimentation potential or may remain suspended at the water air interface depending on their density and shape.

Field studies from the Ciwalengke River in Indonesia revealed a satisfied distribution of MPs, with smaller particles predominantly suspended in the water column and larger particles concentrated benthic sediments. The combined influence of polymer density and particle size governs these settling dynamics, facilitating the accumulation of MPs in sedimentary stones. Similar patterns have been observed in the Rhine and Meuse Rivers in the Netherlands, where both floating MPs and substantial sedimentary deposits have been documented, particularly in estuarine regions (Leslie et al., 2017). Although high density polymers such as polyvinyl chloride (PVC) and polyethylene terephthalate (PET) exhibit inherent sedimentation tendencies, low density polymers (e.g., polyethylene and polypropylene) may also accumulate in sediments over time. This counterintuitive behavior is attributed to physicochemical interactions including aggregation with dissolved organic matter, microbial colonization (biofouling), and incorporation into fecal pellets, all of which reduce particle buoyancy and enhance settling efficiency (Porter et al., 2018). In the downstream reaches of South

Korea's Nakdong River, sediment associated MP concentrations were reported to be nearly 2800-fold higher than those in surface water, underscoring the significance of these processes.

Moreover, MPs with larger dimensions and irregular surface textures are more susceptible to biofilm formation, which not only accelerates their environmental aging and degradation but also alters their hydrodynamics behavior and spatial distribution (Haberstroh et al., 2021). These intrinsic particle characteristics, coupled with dynamic biogeochemical interactions, contribute to the heterogeneous distribution of MPs across freshwater systems. This variability is exemplified by findings from Taihu Lake, China, where marked difference in MP concentrations was observed between surface waters and sediment layers, as determined through plankton net sampling methodologies (Su et al., 2016).

1.3 Mechanisms of transport: Water currents, wind, and biological carriers

1.3.1 Environmental and anthropogenic influences on microplastic distribution

The spatial distribution of MP within freshwater ecosystem is governed by a combination of anthropogenic pressures and environmental processes. Key contributors include urban expansion, industrial discharges, textile effluents, and land use practices, all of which significantly influence MP abundance and composition across diverse freshwater settings. Empirical studies from Patagonia, Argentina, have highlighted the role of urban settlements and fishing activities in elevating MP concentrations in regional lakes, underscoring the influence of localized human behavior on aquatic contamination profiles. Similarly, research conducted in Lake Ontario, Canada, revealed a strong spatial correlation between MP abundance and proximity to urban infrastructure, identifying metropolitan zones as critical hotspots for plastic pollution (Alfonso et al., 2020)

Chemical analyses of MPs frequently detected polymers such as polyethylene (PE), polyethylene terephthalate (PET), polyvinyl chloride (PVC), and cellulose based compounds. These materials are commonly derived from industrial effluents, roadway runoff, mismanaged solid waste, and municipal wastewater. Notably, cellulose acetate, a constituent of cigarette filters, has been proposed as a reliable indicator of urban derived MP pollution due to its widespread use and persistence in the environment. Comparable findings have been reported in the Suzhou and Huangpu Rivers in China, where elevated MP concentrations were observed in urban and estuarine segments, further reinforcing the link between urbanization and MP loading (Luo et al., 2019).

In the United States, Lake Mead exhibited increased MP levels in areas subjected to recreational and industrial activities, suggesting that direct human interaction with water bodies contributes to localized contamination (Baldwin et al., 2020). However, the role of wastewater treatment plants (WWTPs) as MP sources appears to be context dependent. Studies in the Trent catchment, the Untied Kingdom, reported no significant increase in MP concentrations downstream of WWTP effluent discharge points, indicating variability in treatment efficacy and influent composition (Stanton et al., 2020). Moreover, land use patterns play a pivotal role in shaping MP emissions. Commercial and industrial zones have been shown to release higher quantities of MPs compared to residential or roadside environment. On the Tibetan Plateau, agricultural practices particularly the application of plastic films have been implicated in MP contamination of adjacent freshwater systems (Feng et al., 2020). These observations emphasize the necessity of incorporating land use typologies into MP monitoring frameworks to accurately assess and manage pollution sources. Collectively, these findings highlight the multifactorial nature of MP distribution in freshwater environments and underscore the importance of integrating environmental variables and anthropogenic activities into comprehensive pollution assessment models.

1.3.2 Environmental metrics and anthropogenic pressure in microplastic distribution

Recent advancements in spatial analysis have introduced the night light index as a novel proxy quantifying anthropogenic pressure across geographic regions. This metric integrates both resident and transient population activities, offering a refined approach to assessing human induced environmental stressors (Feng et al., 2020). Application of this index in the Rio de la Plata estuary, Argentina, revealed elevated MP concentration in density urbanized zones characterized by substantial sewage discharge and high turbidity levels. These findings underscore the dominant role of urban infrastructure and wastewater outflows in shaping MP pollution gradients. Interestingly, statistical analyses indicated that conventional environmental parameters such as temperature, pH, turbidity, conductivity, salinity, dissolved oxygen (DO), total dissolved solids (TDS), and wind velocity did not exhibit significant correlations with MP abundance in this region (Pazos et al., 2018). This suggests that anthropogenic factors exert a more pronounced influence on MPs distribution than ambient physicochemical conditions in certain contexts.

Conversely, research conducted in the Fengshan River, Taiwan, demonstrated robust associations between MP concentrations and several water quality indicators, including chemical oxygen demand (COD), suspended solids (SS), and the river pollution index (Tien et al., 2020). These metrics reflect organic and particulate loading in aquatic systems, implying that localized water quality parameters can substantially affect MP transport, aggregation, and deposition dynamics. Given the variability in findings across geographic and hydrological contest, further interdisciplinary research is warranted to elucidate the complex interactions between MPs and aquatic physicochemical conditions. Such investigations should aim to integrate spatial modeling polymer specific behavior, and real time water quality monitoring to enhance predictive capabilities and inform targeted mitigation strategies.

1.3.3 Influence of meteorological events and transport mechanisms on microplastic dispersion

Meteorological phenomena, particularly precipitation events, exert a substantial influence on the mobilization and spatial redistribution of MP within freshwater systems. Episodic weather events can trigger abrupt fluctuations in MP concentrations, particularly through enhanced hydrodynamic activity and terrestrial runoff. A study conducted on the Crooks River in Australia documented a dramatic 40-fold increase in MP abundance following a major storm event. Peak concentrations were observed within 48 h post precipitation, followed by a gradual decline over the subsequent 5-day period (Hitchcock, 2020). This is attributed to the influx of land-based MPs and the resuspension of previously settled particles due to elevated flow velocities and turbulence (Hurley et al., 2018)

Rainfall also influences the compositional profile of MPs. In the Trent region of the United Kingdom, stormwater samples revealed the approximately 22% of MPs originated from tire and road wear, highlighting the role of surface runoff in transferring urban pollutants into aquatic environments (Grbić et al., 2020). Similarly, investigations along South Korea's Nakdong River indicated that up to

80% of the annual MP load occurred during the rainy season, with over 90% of these particles remaining suspended in the water column, facilitating downstream transport (Eo et al., 2019). These findings underscore the necessity of integrating stormwater dynamics into MP transport models, particularly given their role in bridging terrestrial and marine systems. In agricultural landscapes such as Tibetan Plateau, irrigation return flows have been identified as vectors for MP migration from riverine systems into adjacent soils, further complicating the environmental fate of these particles (Feng, S et al., 2020).

Hydrological parameters including tidal mixing, flow mixing, flow regimes, and retention time also modulate MP distribution across water and sediment compartments. These processes influence particle deposition, suspension, and vertical migration, contributing to spatial heterogeneity in MP concentrations (Tien et al., 2020). Atmospheric mechanisms further extend the dispersal range of MPs. Wind and precipitation can facilitate long range transport, introducing MPs into remote or elevated regions. In Turkey's urban Crater Lake, situated at an altitude of 2380 m, small MP fragments were detected in areas devoid of direct agricultural activity, implicating Aeolian transport as a contributing factor (Çomaklı et al., 2020). Additionally, tire derived MP have been identified in ponds adjacent to highways, suggesting atmospheric deposition as a plausible pathway for localized contamination (Liu et al., 2019). Collectively, these observations highlight the multifaceted nature of MP transport and emphasize the importance of incorporating meteorological and hydrological variables into comprehensive environmental monitoring and risk assessment frameworks.

1.3.4 Role of stormwater dynamics and hydrological conditions in microplastic mobilization

Effective stormwater management has emerged as a critical invention for mitigating MP influx into aquatic ecosystems, particularly considering the ecological risks posed to biota exposed to transported MPs (Fig. 1.5). Hydrodynamics conditions, especially those induced by meteorological events, significantly influence the transport, redistribution, and bioavailability of MPs. Observations from the Hillsborough River in Florida revealed that under quiescent flow regimes, advection was the dominant mechanism governing MP flux. In contrast, turbulent conditions such as those generated during storm events enhanced both vertical and lateral transport of MP by order of magnitude (Haberstroh et al., 2021). Such high energy hydrological disturbance can resuspend previously settled MP into the water column, thereby altering their vertical stratification independent of polymer density or particle morphology.

The magnitude of MP concentration shifts during these events is modulated by the size and hydrological characteristics of the receiving water body. In India's Netravathi River, MP concentrations exhibited a strong positive correlation with rising water levels, suggesting that increased discharge volumes facilitate the mobilization of land-based MP into fluvial systems (Amrutha and Warrier, 2020). Weather induced hydrological dynamics exert cascading effects on MP migration pathways, ecological exposure, and contamination bioavailability. In the Fenghan River Taiwan, MP uptake in fish populations was found to be associated with fluctuations in suspended solid (SS), pH, and electrical conductivity, indicating that these physicochemical parameters influence both particle dispersion and biological assimilation (Tien et al., 2020). Furthermore, MPs may act as vectors for hydrophobic organic pollutants such as polycyclic aromatic hydrocarbons (PAHs), which can be adsorbed onto particle surfaces and subsequently bioaccumulated by aquatic organisms. This compound specific

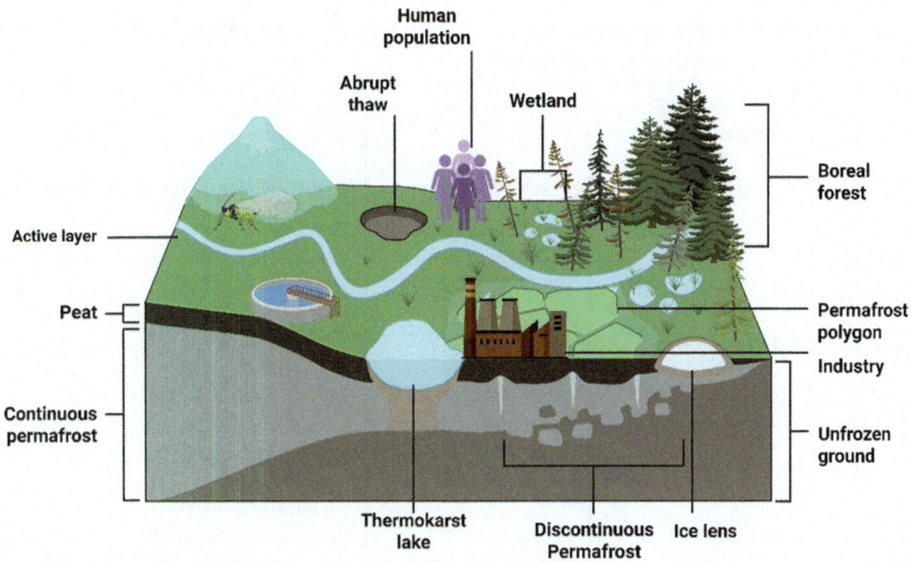

FIGURE 1.5

Schematic illustration of the major factors governing microplastic sources, transport pathways, and accumulation in water environments.

interaction amplifies the ecological threat posed by MPs, transforming them from inert particles into carries of toxic substances. These findings underscore the importance of integrating stormwater hydrology, particle behavior, and contaminant dynamics into comprehensive MP risk assessment frameworks. Such approaches are essential for developing targeting mitigation strategies and safeguarding aquatic biodiversity against the compounded impacts of MP pollution.

1.4 Accumulation patterns in freshwater and marine environments

1.4.1 Fate and accumulation

MP traverse intricate environmental pathways, originating predominantly from urban centers, industrial zones, and manufacturing hubs. These particles are subsequently mobilized through fluvial networks rivers and lakes before entering marine systems, where they disperse across surface waters, coastal margins, and abyssal ocean basins (Table 1.3). The generation of plastic waste is multifactorial, driven by regional consumption behaviors, the efficacy of waste management infrastructure, and the scale of industrial production activities. Post consumption, plastics may undergo various end-of-life trajectories, including reuse, mechanical recycling, incineration, landfilling, or inadvertent environmental release. While circular economy practices such as reuse and recycling offer sustainable alternatives by retaining plastics within closed loop systems, most plastic waste ultimately escapes containment due to infrastructural deficiencies and improper disposal practices.

Table 1.3 Overview of microplastic sources in soils, aquatic environments, and food chains, along with reported remediation techniques.

Source/ Medium	MPs Produced/ Concentration (pieces/kg or pieces/L)	Types of MPs	Remedial strategies	References
Soil (agricultural)	0.1–7 pieces/kg	Fragmented plastic fibers, pellets and films	Utilizing biodegradable plastics, soil washing, and phytoremediation	Horton et al. (2017)
Freshwater	0.1–10 pieces/L	Fragmented plastics, fibers, beads and films	Water filtration, wetland restoration, and phytoremediation	Auta et al. (2017)
Marine water	0.3–10 pieces/L	Pellets, beads and fragments	Marine plastic clean-up, beach cleaning programs, and chemical treatments	Thiele et al. (2021)
Industrial effluents	Varies (up to 100 pieces/L)	Fragmented plastics, pellets, and fibers)	Recycling plastic waste and advanced filtration systems	Shen et al. (2022)
Fish	4000 pieces/kg of fish tissue	Fragments & fibers of polystyrene, polyamide, polyethylene, polyester	Filtration coupled with wastewater treatment plans	Srisiri et al. (2024)
Beverages	2.2–9.8 pieces/L	Polyacrylonitrile, polyethylene, polyethylene terephthalate	Wastewater treatment combined with membrane filtration, electrocoagulation	Basaran et al. (2024)

The environmental dissemination of MPs is modulated by geographic features, meteorological conditions, and the physicochemical attributes of the particles themselves such as polymer density, particle size, and buoyancy. Computational modeling efforts have sought to predict MP dispersal patterns, consistently indicating that terrestrial sources are the primary contributors to marine plastic pollution (Akdogan and Guven, 2019; Bikker, 2020). It is estimated that global riverine systems transport between 1.15 and 2.41 million metric tons of plastic annually into the oceans, accounting for approximately 80%–94% of total marine plastic input.

1.4.2 Interaction with marine biota

MPs are now prevalent across marine habitats, including pelagic zones, benthic sediments, and deep-sea environments. Over 1400 marine species have been documented to interact with plastic debris, either through ingestion or physical entanglement. Due to their diminutive size and visual resemblance to pray items, MPs are frequently mistaken for food by marine organisms. Ingestion has been observed across a wide taxonomic range, including corals, zooplankton, phytoplankton, crustaceans, teleost fish, and apex predators such as cetaceans, sea turtles, and pinnipeds. A major ecological concern is the trophic transfer of MPs within marine food webs. Lower trophic organisms ingest MPs, which are then bioaccumulated and biomagnified through successive trophic levels, ultimately reaching human consumers. This phenomenon is particularly critical for species reliant on phytoplankton, as MP readily from aggregates with autotrophic microorganisms thereby increasing the likelihood of ingestion and disrupting primary productivity.

1.4.3 Transport, threats of microplastic and appropriate countermeasures

MPs suspended in the water column or buoyant at the surface are subjected to long range transport via oceanic circulation systems. Ocean gyres large scale, wind driven circulatory systems influenced by the Coriolis effect Gyre, are estimated to contain approximately 35% of the global oceanic plastic burden (Welden and Lusher, 2017). Additional oceanographic processes, including thermohaline circulation, Coriolis induced deflection, and wind driven surface drift, further modulate the spatial redistribution of MPs. These mechanisms facilitate the movement of plastic particles between pelagic zones and coastal environments, contributing to the dynamic and persistent nature of marine plastic pollution. The escalating presence of MPs in the environment has become critical for intensified research in recent years. Continuous plastic production, combined with inadequate waste collection and disposal particles, has resulted in large volumes of plastic debris entering natural ecosystems. Once released, these materials undergo physical, chemical, and biological degradation, forming MPs that persist and circulate within environmental compartments.

MPs readily enter aquatic food webs, beginning with ingestion by microorganisms and lower trophic organisms. Through trophic transfer, these particles move upward in the food chain, ultimately reaching humans. Exposure occurs primarily through ingestion of contaminated food and water, as well as inhalation of airborne MPs. Particles at the micro- and nanoscale have been detected in human respiratory and digestive systems, indicating that MPs pose risks that cannot be overlooked. Particle size plays a critical role in determining biological interactions. MPs smaller than approximately 1.5 μm are of particular concern, as they may cross cellular membranes and directly interact with tissues. Detecting such small particles remains technically challenging, as conventional techniques become ineffective. Advanced analytical approaches, such as pyrolysis-based methods, are therefore increasingly required. Equally important is the strict control of sampling and pretreatment environments, as airborne contamination can significantly distort analytical results. As particle sizes decrease, the need for controlled environments and in situ analytical techniques becomes more critical to ensue data reliability.

Plastic debris already present in the environment will continue to fragment over time, posing long-term threats to ecosystems and human health. To prevent further escalation, comprehensive and enforceable waste management strategies must be implemented to reduce plastic discharge at the source. From a research perspective, standardized protocols for sample collection, pretreatment, and analysis are urgently needed to improve data comparability. In parallel, advanced identification techniques capable of detecting nano-sized plastics should be developed. Finally, it is essential to investigate MP transport pathways into the human body and identify strategies to reduce exposure. In case where complete elimination is unrealistic, defining safe exposure thresholds may become necessary to guide regulatory decision-making.

1.5 Conclusion

This chapter synthesizes current knowledge on the sources, transport mechanisms, and accumulation behavior of MPs in aquatic environments. The influence of environmental factors on MP distribution and transformation, as well as their potential impacts, has been critically evaluated. Despite substantial progress, significant gaps remain particularly regarding MPs smaller than 1.5 μm, whose detection and characterization are limited by current analytical capabilities and contamination risk during sampling and processing. The existing research has successfully identified key environmental drivers influencing MP behavior and provided valuable direction for future investigations. Advancing this field requires influencing MP behavior and provides valuable direction for future investigations. Advancing this field requires the harmonization of sampling and analytical protocols, alongside the development of robust techniques for micro- and nanoscale plastic detection.

Research must increasingly focus on aged MPs and their interactions with coexisting environmental contaminants, as these represent more realistic exposure scenarios. Given the dynamic nature of MPs, long-term monitoring strategies are more informative than single-point sampling approaches. Ultimately, meaningful mitigation will depend not only on scientific advancements but also on improved plastic waste management, upgraded wastewater treatment infrastructure, and coordinated policy interventions. Addressing MP pollution is a shared global responsibility that demands integrated scientific, technological, and regulatory solutions.

References

Akdogan, Z., Guven, B., 2019. Microplastics in the environment: a critical review of current understanding and identification of future research needs. Environ. Pollut. 254, 113011.

Alfonso, M.B., et al., 2020. First evidence of microplastics in nine lakes across Patagonia (South America). Sci. Total Environ. 733, 139385.

Amrutha, K., Warrier, A.K., 2020. The first report on the source-to-sink characterization of microplastic pollution from a riverine environment in tropical India. Sci. Total Environ. 739, 140377.

Ashrafy, S.M., et al., 2023. Microplastics: classification, sources, toxicological effects, and removal technologies. Chemosphere 313, 137421.

Auta, H.S., Emenike, C.U., Fauziah, S.H., 2017. Distribution and importance of microplastics in the marine environment: a review of the sources, fate, effects, and potential solutions. Environ. Int. 102, 165–176. https://doi.org/10.1016/j.envint.2017.02.013.

Baldwin, A.K., et al., 2020. Microplastics in Lake Mead national recreation area, USA: occurrence and biological uptake. PLoS One 15 (5), e0228896.

Barrows, A.P.W., et al., 2018. A watershed-scale, citizen science approach to quantifying microplastic concentration in a mixed land-use river. Water Res. 147, 382–392.

Basaran, B., Aytan, Ü., Sentürk, Y., Ozçifçi, Z., Akçay, H.T., 2024. Microplastic contamination in some beverages marketed in Türkiye: characteristics, dietary exposure and risk assessment. Food Chem. Toxicol. 189, 114730. https://doi.org/10.1016/j.fct.2024.114730.

Bikker, J., et al., 2020. Microplastics and other anthropogenic particles in the surface waters of the Chesapeake Bay. Mar. Pollut. Bull. 156, 111257.

Blettler, M.C.M., et al., 2018. Freshwater plastic pollution: recognizing research biases and identifying knowledge gaps. Sci. Total Environ. 612, 635–643.

Bohdan, A., 2022. Global distribution patterns and sources of microplastics in marine environments. Sci. Total Environ. 833, 155212.
Bordos, G., et al., 2019. Identification of microplastics in fish ponds and natural freshwater environments of the Carpathian Basin, Europe. Chemosphere 216, 110–116.
Borrelle, S.B., et al., 2020. Predicted growth in plastic waste exceeds efforts to mitigate plastic pollution. Science 369 (6510), 1515–1518.
Campanale, C., et al., 2020. Microplastics and their possible sources: the example of the Ofanto River in southeast Italy. Environ. Pollut. 258, 113284.
Chen, J., et al., 2024. Geographic distribution and trends of microplastic pollution in Asia: a comprehensive review. Environ. Res. 232, 116391.
Çomaklı, E., et al., 2020. Assessment of microplastic pollution in a high-altitude crater lake: a case study of an urban crater lake in Erzurum, Turkey. Water, Air, Soil Pollut. 231 (6), 275.
Conkle, J.L., et al., 2018. Assessment of microplastic concentrations in human-consumed seafood. Environ. Pollut. 238, 999–1006.
Conley, K., et al., 2019. Wastewater treatment plants as a source of microplastics to an urban estuary: removal efficiencies and loading per capita over one year. Water Res. 3, 100030.
Corradini, F., et al., 2019. Evidence of microplastic accumulation in agricultural soils from sewage sludge disposal. Sci. Total Environ. 671, 411–420.
Covernton, G.A., et al., 2019. Size and shape matter: a preliminary analysis of microplastic sampling technique in seawater studies with implications for ecological risk assessment. Sci. Total Environ. 667, 124–132.
De Falco, F., et al., 2019. The contribution of washing processes of synthetic clothes to microplastic pollution. Sci. Rep. 9, 6633.
Deng, H., et al., 2020. Microplastic pollution in water and sediment in a textile industrial area. Environ. Pollut. 258, 113658.
Eo, S., et al., 2019. Spatiotemporal distribution and annual load of microplastics in the Nakdong River, South Korea. Water Res. 160, 228–237.
Eriksen, M., et al., 2014. Plastic pollution in the world's oceans. PLoS One 9, e111913.
European Commission, 2019. Restriction of Intentionally Added Microplastics: Proposal for REACH Annex XVII. European Commission, Brussels.
Fazey, F.M.C., Ryan, P.G., 2016. Biofouling on buoyant marine plastics: an experimental study into the effect of size on surface longevity. Environ. Pollut. 210, 354–360.
Feng, S., et al., 2020. Microplastics in a remote region of the Tibetan Plateau. Sci. Total Environ. 739, 140087.
Fischer, E.K., et al., 2016. Microplastic pollution in lakes and lake shoreline sediments: a case study on Lake Bolsena and Lake Chiusi. Environ. Pollut. 213, 648–657.
Gewert, B., Plassmann, M.M., MacLeod, M., 2015. Pathways for degradation of plastic polymers floating in the marine environment. Environ. Sci. Process. Impacts 17, 1513–1521.
Gonzalez-Pleiter, M., et al., 2020. First detection of microplastics in the freshwater of an Antarctic specially protected area. Mar. Pollut. Bull. 161, 111811.
Grbić, J., et al., 2020. Microplastics entering northwestern Lake Ontario are diverse and linked to urban sources. Water Res. 174, 115623.
Haberstroh, C.J., et al., 2021. Effects of hydrodynamics on the cross-sectional distribution and transport of plastic in an urban coastal river. Water Environ. Res. 93 (2), 186–200.
Hale, R.C., et al., 2020. A global perspective on microplastics. J. Geophys. Res., Oceans 125.
Han, M., et al., 2020. Distribution of microplastics in surface water of the lower Yellow River near the estuary. Sci. Total Environ. 707, 135601.
Hitchcock, J.N., 2020. Storm events as key moments of microplastic contamination in aquatic ecosystems. Sci. Total Environ. 734, 139436.

Horton, A.A., Walton, A., Spurgeon, D.J., Lahive, E., Svendsen, C., 2017. Microplastics in freshwater and terrestrial environments: evaluating the current understanding to identify the knowledge gaps and future research priorities. Sci. Total Environ. 586, 127–141. https://doi.org/10.1016/j.scitotenv.2017.01.190.

Hurley, R., Woodward, J., Rothwell, J.J., 2018. Microplastic contamination of river beds reduced by catchment-wide flooding. Nat. Geosci. 11 (4), 251–257.

Leslie, H.A., et al., 2017. Microplastics en route: field measurements in the Dutch river delta and Amsterdam canals, wastewater treatment plants, North Sea sediments and biota. Environ. Int. 101, 133–142.

Li, X., et al., 2018. Microplastics in sewage sludge from wastewater treatment plants in China. Water Res. 142, 75–85.

Liu, F., et al., 2019. Microplastics in urban and highway stormwater retention ponds. Sci. Total Environ. 671, 992–1000.

Luo, W., et al., 2019. Comparison of microplastic pollution from urban creeks to coastal waters. Environ. Pollut. 246, 174–182.

Mason, S.A., et al., 2016. Pelagic plastic pollution within the surface waters of Lake Michigan, USA. J. Great Lake. Res. 42, 753–759.

McCormick, A.R., et al., 2020. Microplastic is an abundant and distinct microbial habitat in an urban river. Environ. Sci. Technol. 54 (3), 1616–1626.

Mutuku, J.N., et al., 2024. A global review of microplastics in the oceans: abundance, sources, and ecological risks. Mar. Pollut. Bull. 195, 115320.

Naqash, N., et al., 2024. Global trends in plastic production and their environmental implications. Sustain. Mater. Technol. 34, e00642.

Pazos, R.S., et al., 2018. Microplastics integrating the coastal planktonic community in the Río de la Plata estuary. Environ. Pollut. 243, 134–142.

Polanco, H., et al., 2020. Microplastics in surface water of the lower hudson river estuary. Mar. Pollut. Bull. 161, 111702.

Porter, A., et al., 2018. Role of marine snow in microplastic fate and bioavailability. Environ. Sci. Technol. 52 (12), 7111–7119.

Shen, M., et al., 2022. Microplastics in landfill and leachate: occurrence, environmental behavior and removal strategies. Chemosphere 305, 135325. https://doi.org/10.1016/j.chemosphere.2022.135325.

Shi, L., et al., 2024. Quantifying microplastic contamination in soils: an international comparative study. Environ. Monit. Assess. 196 (1), 18.

Silvestrova, K., Stepanova, N., 2021. Distribution of microplastics in the Atlantic Ocean surface layer. Mar. Pollut. Bull. 162, 111836.

Srisiri, S., Haetrakul, T., Dunbar, S.G., Chansue, N., 2024. Microplastic contamination in edible marine fishes from the upper Gulf of Thailand. Mar. Pollut. Bull. 198, 115785. https://doi.org/10.1016/j.marpolbul.2023.115785.

Stanton, T., et al., 2020. Freshwater microplastic concentrations vary through space and time. Environ. Pollut. 263, 114481.

Su, L., et al., 2016. Microplastics in Taihu Lake, China. Environ. Pollut. 216, 711–719.

Thiele, C.J., Hudson, M.D., Russell, A.E., Saluveer, M., Sidaoui-Haddad, G., 2021. Microplastics in fish and fishmeal: an emerging environmental challenge? Sci. Rep. 11 (1), 1–12. https://doi.org/10.1038/s41598-021-81499-8.

Tien, C.J., Wang, Z.X., Chen, C.S., 2020. Microplastics in water, sediment and fish from the Fengshan River system: relationship to aquatic factors and accumulation of polycyclic aromatic hydrocarbons by fish. Environ. Pollut. 265, 114962.

Welden, N.A.C., Lusher, A.L., 2017. Integrative assessment of microplastics. Integrated Environ. Assess. Manag. 13, 483–487.

Xia, R., et al., 2024. Monitoring microplastic pollution in the Yangtze River. Sci. Total Environ. 889, 164261.

Zeri, C., et al., 2018. Floating plastics in adriatic waters: from macro- to micro-scale. Mar. Pollut. Bull. 136, 341–350.

Zhang, Y., et al., 2020. Importance of atmospheric transport for microplastics deposited in remote areas. Environ. Pollut. 255, 113133.

Zhou, Y., et al., 2024. Microplastic pollution in agricultural soils of Guizhou Province, China. Environ. Sci. Pollut. Control Ser. 31 (12), 18321–18335.

Zhu, L., et al., 2018. Microplastic pollution in the north yellow sea, China. Sci. Total Environ. 636, 20–29.

CHAPTER

2 Interactions of microplastics with biotic and abiotic components of the ecosystem

Melvin S. Samuel

Department of Civil, Construction, and Environmental Engineering, Marquette University, Milwaukee, WI, United States

Highlights

- Microplastics absorb contaminants through hydrophobic, electrostatic, and π-π interactions, pore filing, and hydrogen bonding.
- Biofilm formation on Microplastics enhances PFAS retention and persistence in aquatic systems.
- Environmental conditions (Organic matter, salinity, ionic strength, divalent cations) strongly control adsorption behavior.
- Polymer type, surface roughness, and porosity govern contaminants' sorption efficiency.
- Microplastics degrade water quality and sediments, impacting cryosphere systems and coral reef health.

2.1 Introduction

The global surge in plastic production has outpaced the development of effective waste management systems, rendering the safe disposal of plastic contaminants increasingly unmanageable. This imbalance has led to widespread environmental contamination and increased concerns regarding public health risks. A significant proportion of plastic waste ultimately enters aquatic ecosystems, where it persists and accumulates over time. MPs defined as plastic less than 5 mm in diameter constitute a major fraction of this pollution and have been detected in diverse aquatic environments, including oceans, rivers, and freshwater lakes. MPs and per and polyfluoroalkyl substances (PFAS) enter aquatic ecosystems through multiple anthropogenic pathways. Key contributors include agricultural runoff, tourism related waste, transportation activities, and effluent discharges from wastewater treatment plants (WWTPs). Inadequate land-based waste management further exacerbates MP contamination, as rainwater runoff facilities their entry into freshwater systems. WWTPs are widely recognized as significant point sources of MP in aquatic environments (Liu et al., 2021). Despite high removal efficiencies, the sheer volume of treated wastewater results in substantial MP release. Once discharged, MPs are transported via rivers and lakes, eventually reaching marine system. Global estimates suggest that rivers contribute approximately 500,000 metric tons of plastic waste to the oceans annually (Strokal et al., 2023).

Similarly, PFAS contamination in aquatic environments is closely linked to human activities. Major sources include the use of aqueous film forming foams (AFFF) in firefighting, leachate from

Microplastics in Aquatic Environments. https://doi.org/10.1016/B978-0-443-45619-0.00002-9

landfills containing PFAS laden consumer products (e.g., textiles, packaging, paints), and industrial and domestic wastewater discharges. PFAS concentrations vary widely, with domestic wastewater typically exceeding 100 ng/L, while industrial sources such as textile and electroplating effluents can reach concentrations of 4268 ng/L and 78 ng/L, respectively (Jia et al., 2023).

The pervasive distribution of MP in marine and freshwater systems is largely governed by their physicochemical characteristics. Properties such as polymer density, particle size, surface charge, color, and aggregation behavior influence their buoyancy, transport, and fate in the environment. Additionally, abiotic factors (e.g., photodegradation, oxidation) and biotic interactions (e.g., microbial colonization or biofouling) further modify these properties, affecting how MP are distributed across water columns and sediment layers. MPs are broadly classified into two types based on their origin. (a) Primary MPs are intentionally manufactured at microscopic scales for use on commercial and industrial applications. These include microbeads in personal care products, plastic pellets (nurdles), and synthetic fibers from textiles. (b) Secondary MPs are generated through the fragmentation of larger plastic debris due to mechanical, chemical, or biological degradation processes (Fig. 2.1). Although primary MPs are initially produced within a size range of approximately 0.1–5000 μm, they can undergo further fragmentation, resulting in the formation of nanoparticles typically ranging from 1 to 100 nm. The emergence of nanoparticles presents a significant analytical challenge due to their small size, which complicates detection, quantification, and risk assessment, particularly in food matrices and biological tissues.

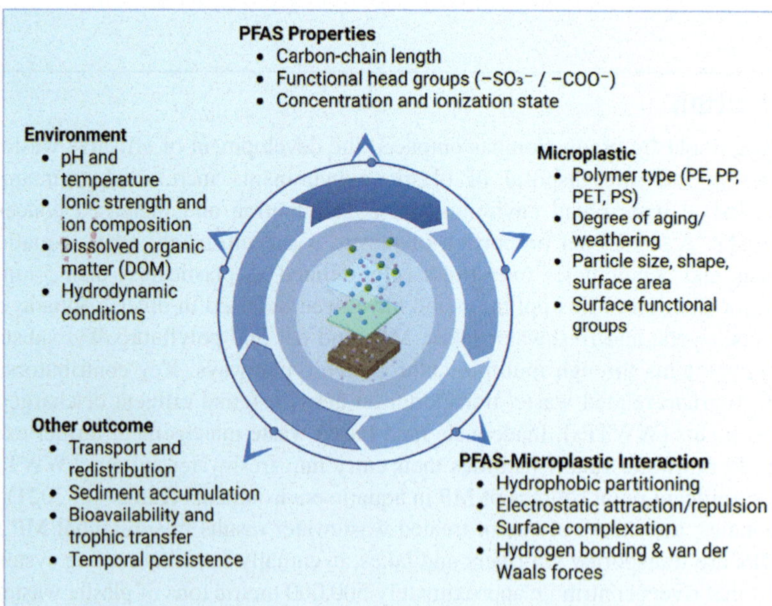

FIGURE 2.1

Factors affecting the interaction of PFAS and microplastics.

MP originates from a variety of sources. These include the breakdown of discarded plastic products in the environment and the direct release of microbeads from cosmetic and personal care products. Wastewater treatment plants (WWTPs) serve as critical nodes in the transport of MPs, particularly in urban areas. Although WWTPs are designed to remove solid waste, their efficiency in retaining MPs varies with particle size and treatment technology. A study by Kalcíkova et al. (2017) using a laboratory scale sequencing batch reactor demonstrated that approximately 50% of microbeads were retained in activated sludge. Notably, smaller particles (60–70 μm) were more likely to be captured compared to larger ones, while a significant fraction still escaped into the effluent stream. These findings highlight the limitation of current wastewater treatment processes in fully mitigating MP release and underscore the need for advanced filtration technologies and regular oversight.

2.2 Adsorption and desorption of contaminants on microplastics

The treatment of MP contaminated water and wastewater presents a multifaceted challenge, particularly when MP acts as a carrier of hazardous contaminants. The adsorption and desorption of MP is influenced by several interrelated factors, including the mechanisms of contaminant adsorption, the physicochemical properties of MPs, their fragmentation behavior, and the treatment processes. Understanding these mechanisms is essential for developing advanced remediation strategies, especially for MP that have absorbed persistent and emerging pollutants such as per and polyfluoroalkyl substances (PFAS) and polychlorinated biphenyls (PCBs) (Fig. 2.2).

2.2.1 Mechanism of contaminant adsorption on microplastics

MPs have emerged as significant vectors for a wide range of environmental contaminants, particularly persistent organic pollutants (POPs) and emerging substances of concern. Among these, polycyclic aromatic hydrocarbons (PAHs), PCBs, and PFAS are frequently detected in association with MPs in aquatic environments. PFAS, have gotten increasing attention due to their widespread occurrence in surface water, drinking water supplies, and wastewater effluents, as well as their exception chemical stability and resistance to degradation (Sun et al., 2019). PFAS compounds, including perfluorooctane sulfonate (PFOS) and perfluorooctanoic acid (PFOA), are of particular concern due their adverse effects on human and ecological health. Their amphiphilic structure comprising a hydrophobic fluorinated carbon chain and a negatively charged functional group facilitates strong interaction with MP surfaces. MPs exhibit a high affinity for hydrophobic and semi polar contaminants such as PBDEs, EDCs and PPCPs. This is largely attributed to their large surface area, hydrophobic polymer matrix, and surface roughness, which collectively enhances sorption capacity. Once adsorbed, these contaminants can be transported across environmental compartments, increasing their bioavailability and ecological risk. The adsorption and desorption behavior of contaminants of MPs is influenced by a complex interplay of factors, including:

- MP characteristics: Polymer type (e.g., Polyethylene (PE), Polypropylene (PP), Polystyrene (PS), Polyvinyl chloride (PVC)), surface area, porosity, crystallinity, and degree of weathering.
- Environmental conditions: pH, temperature, salinity, ionic strength, and redox potential.
- Contaminant properties: Solubility, molecular structure, polarity, and charge distribution (Wang et al., 2020a,b).

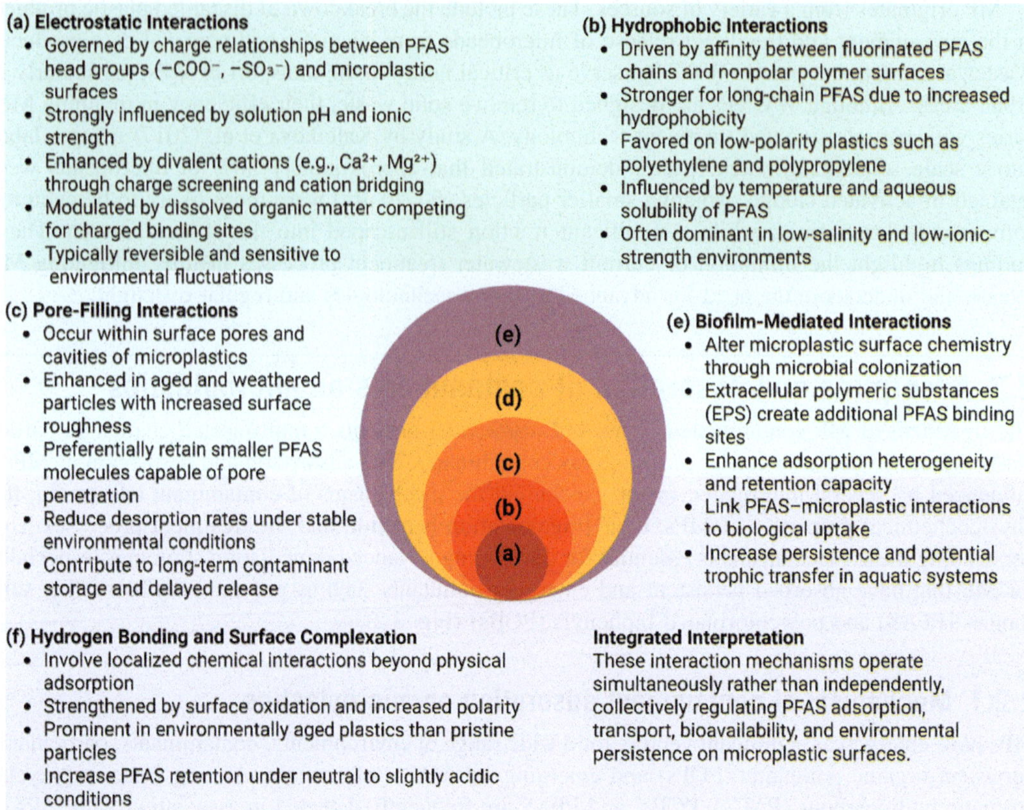

FIGURE 2.2

Potential mechanisms governing PFAS adsorption onto microplastics.

MPs and PFAS are highly persistent contaminants that resist conventional treatment due to strong polymeric backbones and C–F bonds. Recent advances emphasize destructive and transformational technologies, including photocatalytic oxidation, electrochemical treatment, and thermal processes, to enable permanent removal rather than phase transfer Samuel et al. (2022), McNamara et al. (2023), Samuel et al. (2022), (2023) demonstrated high-efficiency degradation, defluorination, and mineralization using advanced oxidation and electrochemical treatments. Collectively, these works support scalable, technology-ready solutions that bridge laboratory innovation with deployable PFAS and microplastic remediation systems.

Benzo(a)pyrene adsorbed onto PVC MP exhibited time and concertation dependent kinetics, with toxicity levels exceeding those of the contaminant or MP alone. This highlights the potential for synergistic effects and the role of MPs as carriers of toxic substances in sedimentary environments (Caruso, 2019).

The mechanisms underlying contaminant adsorption on MP include:

- Hydrophobic interactions: dominant for nonpolar compounds and long chain PFAS.
- Electrostatic interactions: Influenced by the surface charge of MPs and the ionic nature of the contaminant.
- Hydrogen bonding and covalent bonding: Particularly relevant for PFAS and other polar compounds.
- Pore blockage and site competition: Especially in complex matrices with multiple co-contaminants (Enfrin et al., 2020).

2.2.2 Hydrophobic interactions

Hydrophobic interactions are among the most influential mechanisms governing the adsorption of contaminants onto MPs in aquatic environments. These interactions arise from the thermodynamically unfavorable contact between hydrophobic molecular groups and water. As a result, hydrophobic moieties tend to aggregate, minimizing their exposure to the aqueous phase and leading to the clustering of nonpolar substances (Torres et al., 2021). Given that most MP found in the environment are composed of hydrophobic polymers such as PE, PP and PS hydrophobic interactions dominate their sorption behavior. Numerous studies have confirmed that these interactions are the primary mechanism facilitating the adsorption of PFAS onto MPs (Brahana et al., 2023). The strength of hydrophobic interactions between MPs and contaminants is influenced by several physicochemical properties, including polymer composition, surface energy, and environmental aging. A commonly used metric to quantify hydrophobicity is the water contact angle (θ). MPs with contact angles greater than 90 degrees are classified as hydrophobic (Zhuang and Wang, 2023). Pristine PS pellets exhibited a contact angle of 127.75 degrees, which decreased to 98.81 degrees after 9 days of exposure to Fenton's reagent, indicating a reduction in hydrophobicity due to oxidative aging (Yang et al., 2020). Aging processes such as UV exposure, chemical oxidation, and mechanical abrasion can introduce polar functional groups onto MP surfaces, thereby reducing their hydrophobic character and altering their sorption capacity. This transformation may significantly impact the environmental fate of MPs and their role as vectors for hydrophobic contaminants like PFAS. The hydrophobicity of MPs, often quantified by the octanol-water partition coefficient and their degree of environmental aging or weathering, further modulate adsorption behavior (Wang et al., 2019). Among common polymers, PE especially in its rubbery form has demonstrated higher sorption capacities compared to other MP types.

In the case of PFAS, adsorption mechanisms are governed by both electrostatic and hydrophobic interactions. The negatively charged head group of PFAS interacts with positively charged MP surfaces, while the hydrophobic fluorinated tail promoted aggregation on nonpolar surface (Gagliano et al., 2020). Short chain PFAS tend to favor electrostatic interactions, whereas long chain PFAS exhibits stronger hydrophobic binding, often forming molecular aggregates on MP surfaces. These adsorption mechanisms are not only dependent on the physicochemical properties of the contaminants but also on the specific characteristics of the MPs themselves. Understanding this interaction is critical for assessing the environmental fate of contaminant laden MPs and for designing effective removal strategies in water and wastewater treatment systems. Understanding the dynamics of hydrophobic interactions is essential for predicting contaminant behavior in aquatic systems and for developing effective strategies to mitigate the risks associated with MP pollution.

2.2.3 Electrostatic interactions

Electrostatic interactions arise from the attraction between oppositely charged species or repulsion between similarly charged surfaces. These interactions are highly sensitive to environmental conditions such as pH and ionic strength. The pH dependent nature of adsorption is commonly used to investigate electrostatic interactions between MPs and charged contaminants, including PFAS (Zhuang and Wang, 2023). Wang et al. (2015) reported that the adsorption of PFOS on PE and PS increased under acidic conditions. As pH decreases, the surfaces of PE and PS become protonated, acquiring a positive charge that enhances the attraction of anionic PFOS molecules. Additionally, the presence of cations such as Na^{+} and Ca^{2+} in the aqueous environment can further enhance PFAS adsorption by compressing the electrical double layer and promoting electrostatic attraction. Observed a significant increase in PFOS adsorption on PE in the presence of elevated concentrations of these ions.

2.2.4 Pore filling mechanism

The pore filling mechanism involves the diffusion of contaminants into nano and microspores within polymer matrices, particularly glassy polymers such as polyamide (PA), PVC, and PS (Huffer and Hofmann, 2016). Adsorption via pore filling is influenced by the molecular size of the contaminant and the pore size of the MPs. Larger pores are occupied first, and adsorption slows as smaller pores are filled (Atugoda et al., 2021). Weathering processes such as UV exposure and chemical degradation can increase surface porosity, enhancing the potential for pore filling interactions (Zhang et al., 2011). While this mechanism has been demonstrated for contaminants like ciprofloxacin (Liu et al., 2019).

2.2.5 Van der Waals and π-π interactions

Van der Waals forces, although common in MP antibiotic interactions, are considered negligible in PFAS adsorption due to the low polarizability and small molecular size of PFAS compounds (Llorca et al., 2018). These wear intermolecular forces may contribute to overall adsorption but are unlikely to dominate PFAS-MP interactions. Van der Waals and π-π interactions, on the other hand, involve electron coupling between aromatic rings. Aromatic polymers such as PS and PFAS compounds containing benzene rings (e.g., sodium perfluorononyloxybenzenesulfonate (OBS) may engage in π-π stacking. While this mechanism has been proposed, it has not yet been experimentally confirmed and warrants further investigation (Dai et al., 2022).

2.2.6 Hydrogen bonding

Hydrogen bonding occurs between donor and acceptor functional group and is influenced by the chemical structure of both the MP and the contaminant. Although the strong C–F bond in PFAS limits hydrogen bonding potential, the oxygen atoms in PFAS functional groups can participate in hydrogen bonding with suitable MP surfaces (Dai et al., 2022). MPs, enriched with oxygen containing groups such as–OH and –COOH, are more likely to form hydrogen bonds with PFAS (Li et al., 2013). Ateia et al. (2020) found that PA66 (nylon) exhibited the highest PFAS removal efficiency, attributed to hydrogen bonding interactions. Bhagwat et al. (2021) similarly observed enhanced PFOS adsorption on PA fibers, likely due to polarized N–H bonds in the polymer chain. Recent studies have highlighted complex hydrogen bonding between amide groups in PA (proton donors) and carboxyl or sulfonyl groups in PFAS (proton acceptors), further supporting the role of hydrogen bonding in PFAS-MP interactions (Mejias et al., 2023).

2.2.7 Biofilm formation on microplastics and its role in PFAS adsorption

MPs in natural environments are subjected to continuous physical, chemical, and biological weathering (Fig. 2.3). These processes facilitate the colonization of MPs surfaces by microbial communities, leading to the formation of biofilms complex assemblages of microorganisms embedded in extracellular polymeric substances (Zheng et al., 2023). Biofilm development transforms the physicochemical properties of MPs, including their hydrophobicity, surface charge, density, and functional group composition, thereby altering their environmental behavior such as transport, degradation, and contamination adsorption (Zhang et al., 2022a,b). Biofilms can significantly influence the sorption dynamics of MPs. Wang et al. (2020a,b) demonstrated that biofilm coated MPs exhibited altered

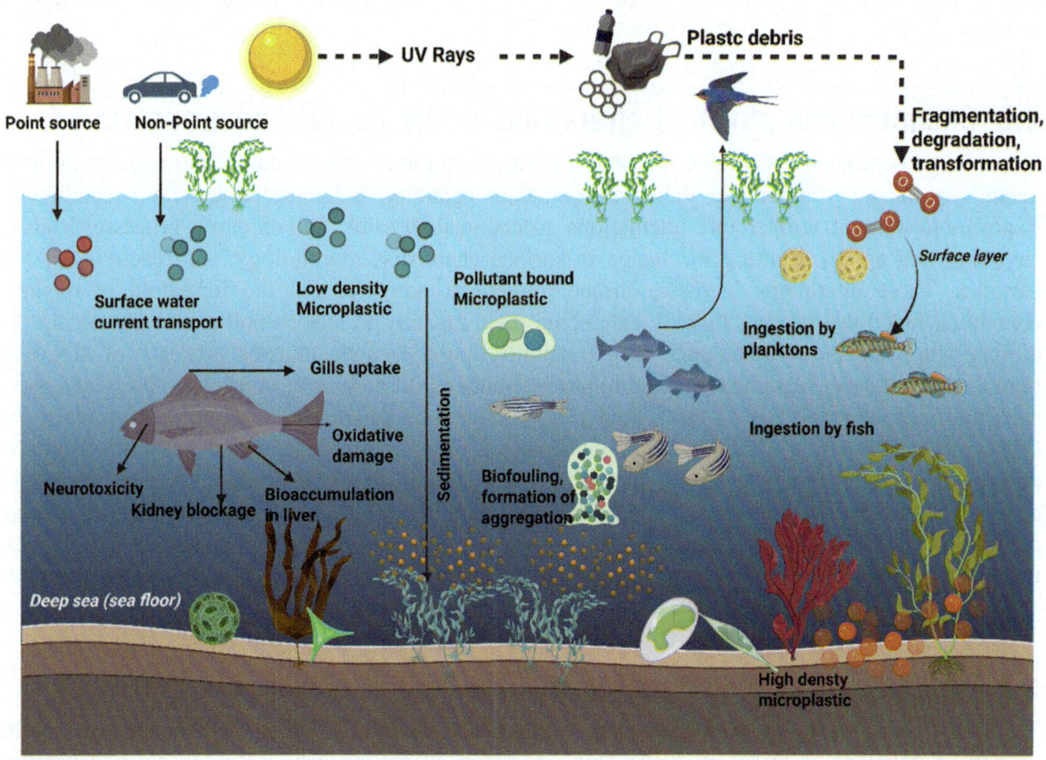

FIGURE 2.3

Illustration of the primary sources and formation pathways of microplastics in aquatic environments. Plastic materials degrade under ultraviolet radiation into microplastics that accumulate in surface waters and interact with aquatic biota, exemplified here by fish.

adsorption behavior for contaminants such as tetracycline and copper ions (Cu^{2+}), affecting their mobility in aquatic systems. Moreover, biofilms themselves possess the capacity to adsorb and PFAS, acting as secondary vectors for these persistent pollutants (Penland et al., 2020). In a study of the Seine River, Munoz et al. (2018) observed spatial and temporal variations in PFAS concentrations and suggested that the composition and structure of biofilm communities were key factors influencing PFAS bioaccumulation. Bhagwat et al. (2021) found that biofilm covered MPs had a significantly higher capacity to adsorb PFOS compared to pristine MPs, highlighting the synergistic role of biofilms in enhancing contaminant retention. Beyond adsorption, biofilm may also offer potential sites for the biodegradation of PFAS. The microbial consortia within biofilms could facilitate PFAS breakdown. However, initial experimental results have shown limited success, likely due to the extreme chemical stability and resistance to degradation characteristic of PFAS compounds (Zhang et al., 2022a,b). This underscores the need for further research into microbial pathways and enzymatic mechanisms capable of degrading PFAS in biofilm associated environments. In summary, biofilm not only modifies the surface properties and sorption behavior of MPs but may also play a role in the environmental fate of PFAS through adsorption and potential biodegradation. Further studies should focus on elucidating the microbial ecology of MP associated with biofilms and their functional roles in PFAS transformation and transport.

2.3 Chemical and physical alterations in the aquatic environment

MP undergo continuous chemical and physical transformations once released into aquatic environments, significantly influencing their environmental behavior and interaction with contaminants. Exposure to sunlight, temperature fluctuations, mechanical abrasion, and oxidative processes leads to environmental aging, resulting in change in surface chemistry, morphology, and reactivity of MP particles. These alterations increase surface roughness, porosity, and the formation of oxygen-containing functional groups, thereby enhancing their capacity to adsorb pollutants. Chemical conditions within aquatic systems such as pH, salinity, ionic strength, and the presence of dissolved organic matter further regulate MP contaminant interactions. Variations in salinity and divalent cations can modify electrostatic forces and bridging effects, influencing sorption dynamics, particularly for PFAS and other ionic pollutants. Simultaneously, organic matter can either promote or inhibit adsorption by competing for binding sites or forming surface coatings on MPs. Physical alterations also affect the distribution and fate of MPs, determining whether particles remain suspended, aggregate with natural particulates, or settle into sediments. Together, these chemical and physical transforming govern the mobility, persistence, and ecological impact of MPs, underscoring their role as dynamic and evolving pollutants within aquatic environments.

2.3.1 Influence of organic matter on PFAS adsorption to microplastics

The presence of organic matter in aquatic environments plays a pivotal role in modulating the adsorption behavior of PFAS on MPs. This interaction is governed by complex physicochemical mechanisms, including co-sorption and molecular complexation between PFAS and OM (Ateia et al., 2020). In systems where MPs are present, PFAS may adsorb onto OM coated MP surfaces through a combination of electrostatic and hydrophobic interactions. Organic matter, predominantly composed of anionic species, can absorb onto MP via repulsive electrostatic forces. Simultaneously, PFAS molecules characterized by hydrophobic fluorinated tails can bind to organic matter coated surfaces through hydrophobic interactions, facilitating their retention in the system (Gagliano et al., 2020). This dual interaction mechanism suggests that organic matter may act as a mediator in PFAS-MP binding dynamics.

Experimental studies have demonstrated that the type of organic matter significantly influences PFAS adsorption efficiency. For instance, hydrophobic organic matter has been shown to enhance PFAS retention on granular activated carbon (GAC), whereas hydrophilic organic matter exhibits negligible impact on PFAS adsorption onto both GAC and anion exchange (AE) resins. However, in other contexts, organic matter may inhibit PFAS sorption. Wang et al. (2015) reported that increasing organic matter concentrations (up to 500 mg/L) suppressed PFOA adsorption on activated carbon fibers, likely due to competitive sorption and pore blockage. Further investigations revealed a 40% reduction in adsorption equilibrium for perfluorinated carboxylates (PFCAs) in the presence of organic matter, reinforcing the notion of competitive site occupation (Du et al., 2014). Proteins within organic matter have also been implicated in reducing PFAS retention through combined hydrophobic and electrostatic interactions (Pramanik et al., 2020). Similarly, the adsorption of PFCAs on bamboo derived activated carbon (BdAC) was limited by competition with co-existing organic matter for active binding sites (Du et al., 2014). These findings underscore the complexity of PFAS adsorption mechanisms in MP rich environments influenced by organic matter. While organic matter can enhance or inhibit PFAS retention depending on its composition and concentration, the underlying conditions governing these interactions remain poorly understood. Further research is essential to elucidate the specific roles of organic matter in contaminant MP dynamics and to inform the development of effective strategies for the removal of contaminant laden MP from aquatic systems.

2.3.2 Factors related to environmental conditions and polymer characteristics

MPs exhibit a wide range of physical and chemical properties that influence their environmental behavior and interactions with contaminants (Fig. 2.4). These properties include particle morphology (e.g., fibers, fragments, spheres), polymer composition (e.g., polyethylene (PE), low density polyethylene (LDPE), polyacrylate (Pas), particle size, and surface characteristics (Jiang, 2018). The efficiency of MP removal from aquatic systems is closely linked to these attributes, as well as to environmental parameters such as pH, ionic strength, salinity, and the presence of natural organic matter. MPs those subjected to environmental degradation processes such as photooxidation, seawater corrosion, and mechanical abrasion often exhibit enhanced sorption capacity. This is likely due to changes in polymer structure, including reduced molecular weight and increased surface roughness, which promote contaminant binding (Wang et al., 2018).

Influence of pH and Ionic strength. Solution pH and ionic composition are critical factors affecting the adsorption behavior of contamination on MPs. The presence of co-existing inorganic anions and variations in pH can alter the surface charge of MP and the ionization state of contaminants, thereby influencing sorption dynamics (Du et al., 2014). Studies have shown that the adsorption of perfluorocarboxylate anions decreases with increasing pH, consistent with trends observed for tother anionic pollutants (Deng et al., 2012). Sorption of perfluorinated compounds onto carbon nanotubes

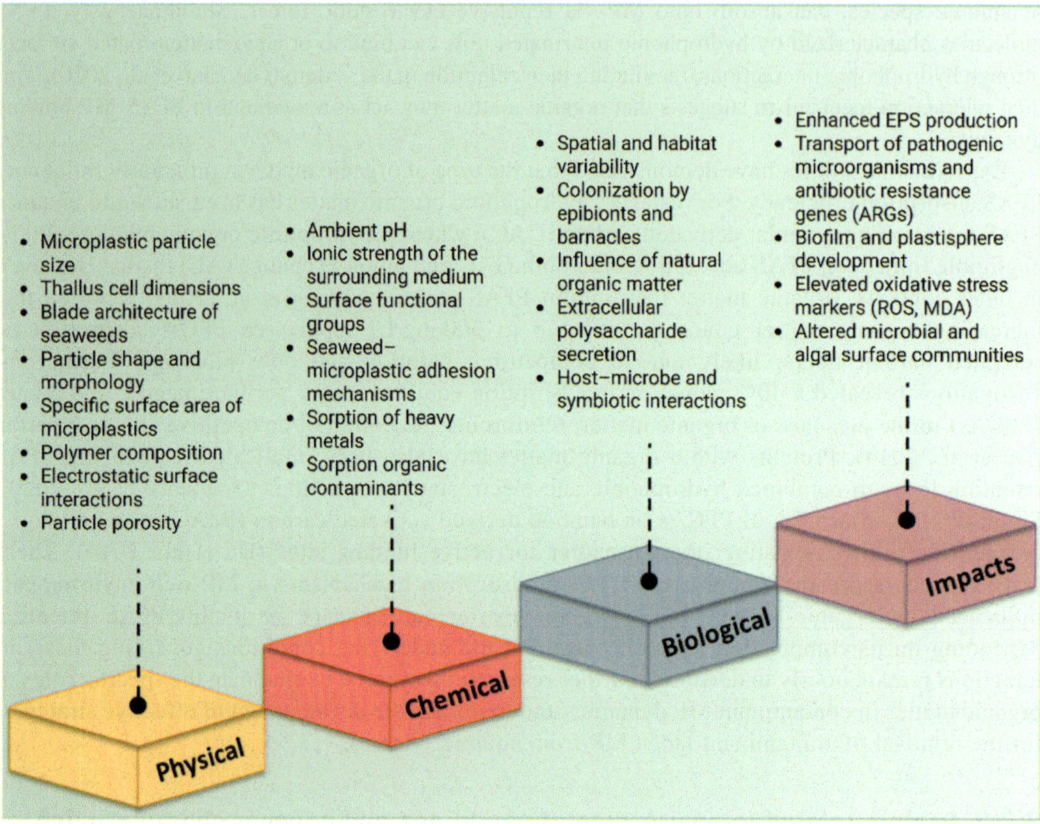

FIGURE 2.4

Overview of the major physical, chemical, and biological drivers influencing microplastic including adsorption mechanisms and resulting ecological consequences.
ARGs, Antibiotic resistance genes; *EPS*, Extracellular polymeric substances; *HM*, Heavy metals; *MP*, Microplastics; *ROS*, Reactive oxygen species.

(CNTs) was significantly reduced at higher pH levels due to the development of negative zeta potentials, which repel similarly charged anions. Conversely, PFOS adsorption onto PE and PS MPs increased under acidic conditions, although this effect was not observed for all PFAS compounds (e.g., FOSA showed no significant pH dependent sorption) (Wang et al., 2018). The sorption behavior is also influenced by MP particle size and polymer type. A study by Zhan et al. (2016) demonstrated that smaller MPs exhibited higher sorption capacity for PCBs, particularly in synthetic seawater compared to ultrapure water or organic solvents. The affinity of MPs for organic compounds followed the order: polyamide (PA) < polyethylene (PE) < polyvinyl chloride (PVC) < polystyrene (PS), although this ranking may vary depending on particle dispersion and specific molecular interactions such as π-π stacking (Huffer and Hofmann, 2016).

2.3.3 Role of divalent cations and ionic strength

The presence of divalent cations such as calcium (Ca^{2+}) and Magnesium (Mg^{2+}) can enhance contaminant adsorption even at elevated pH levels. PFOS sorption in sediment was significantly higher at pH 8.0 than at pH 7.0 when $CaCl_2$ concentrations were increased from 0.005 to 0.5 mol/L, indicating a sixfold increase in sorption at the higher pH (You et al., 2010). This phenomenon is attributed to the formation of cation bridges between the adsorbent surface and PFAS molecules, facilitated by the availability of basic binding sites (Du et al., 2014). Additionally, increased ionic strength particularly in the presence of monovalent cations line sodium (Na^{+}) can weaken the electrical double layer surrounding MPs, thereby enhancing PFAS adsorption. Xiao et al. (2011) reported improved PFAS removal under conditions of elevated ionic strength and reduced pH, suggesting that manipulating these parameters may mitigate contaminant adsorption on MPs and improve treatment outcomes.

2.3.4 Influence of salinity, surface morphology, and polymer type on contaminant sorption to microplastic

MPs serve as dynamic sorbent for a wide range of environmental contaminants, with their sorption behavior influenced by external conditions such as salinity, surface roughness, porosity, and polymer composition. These factors collectively determine the extent and efficiency of contaminant adsorption, particularly for POPs and PFASs.

2.3.4.1 Salinity effects on PFAS adsorption

Salinity plays a critical role in modulating the adsorption of PFAS onto MPs. Elevated salinity levels enhance PFAS sorption through hydrophobic interactions and the salting out effect, which reduces solubility and promotes partitioning onto nonpolar surfaces. Experimental studies have shown that PFOS exhibits increased adsorption on PE and PS in saline environments, indicating favorable sorption in seawater conditions (Gagliano et al., 2020). Fluoropolymers with extended carbon chain (10–14 carbons) also demonstrate high adsorption capacities in saline media. However, sorption behavior varies across polymer types. PS and polystyrene carboxylate (PS–COOH) exhibit greater affinity for PFAS compared to high density polyethylene (HDPE). Interestingly, Perfluorohexanoic Acid (PFHxA) adsorption on HDPE was significantly higher in freshwater (70%) than in seawater (20%), suggesting that salinity can suppress sorption depending on the polymer matrix (Llorca et al., 2018). These findings imply that salinity manipulation may serve as a strategy to mitigate PFAS retention on MPs, although sorption kinetics remain polymer dependent.

2.3.4.2 Surface roughness and porosity

Surface morphology including roughness and porosity substantially influences the sorption capacity of MPs. Rough surfaces and porous structures provide more active sites for contaminant binding, enhancing uptake efficiency. PFOS removal using polymer resins in industrial wastewater was found to depend on the resin's matrix and porosity (Deng et al., 2010). HDPE MPs, characterized by a granular shape and limited intraparticle diffusion, exhibited the lowest PFAS adsorption capacity. Hydrophobic interactions between MPs and PFAS tend to increase with longer chain compounds, and PS generally shows higher sorption affinity than PE and PVC in freshwater environments (Llorca et al., 2018). MPs collected from coastal regions in China were found to adsorb POPs such as PCBs, PAHs, and DDTs, indicating their role as environmental vectors for organic pollutants (Zhang et al., 2015).

2.3.4.3 Polymer specific sorption behavior

The sorption capacity of MPs varies significantly with polymer type and contaminant chemistry. Studies have reported high sorption of DDT on PVC and PE in freshwater, and equilibrium adsorption capacities for pesticide such as hexaconazole (HEX: 41.56 μg/g), myclobutanil (MYC: 18.94 μg/g) and triadimenol (TRI:10.48 μg/g) were found to correlate with their respective log K_{ow} (Fang et al., 2019). Tetracycline exhibited greater sorption on PS MPs than on PE and PP, especially in the presence, likely due to π-π interactions between aromatic rings in tetracycline and PS polymers (Xu et al., 2018). Fulvic acid, a component of natural organic matter, was shown to reduce tetracycline sorption on MPs by competing for binding sites through hydrogen bonding and complexation.

Different sorption mechanisms have been identified for various contaminants:

- Hydrophobic interaction: Dominant for bisphenol analogs on PVC MPs.
- Electrostatic repulsion: Observed for ionized bisphenol F (BPF) in neutral solutions, leading to reduced sorption.
- Noncovalent bonding: Hydrogen and halogen bonds enhance bisphenol adsorption on PVC MPs.

Adsorption efficiency increased with MP dosage until equilibrium was reached, with Bisphenol AF (BPAF) showing he highest affinity among bisphenols. Polyamide (PA) demonstrated strong sorption of 17β-estradiol (Beta) and antibiotics via hydrogen bonding (Liu et al., 2019). Sorption of aromatic compounds followed the order: PA < PE < PVC < PS, consistent with pi-pi interaction potential (Huffer and Hofmann, 2016). These insights underscore the importance of polymer selection and environmental conditions in determining the sorption behavior of MPs. Understanding these interactions is essential for predicting contaminant fate and designing effective remediation strategies in aquatic systems.

2.3.4.4 Other environmental factors influencing adsorption behavior

Beyond the well-established influences of pH, salinity, and polymer type, additional environmental parameters play a critical role in modulating the adsorption behavior of MPs and associated contaminants. Among these, temperature and the presence of co-occurring particulate matter are emerging as important yet underexplored factors.Temperature can significantly impact the adsorption kinetics and thermodynamics of contaminants onto MPs. Xue et al. (2021) investigated the sorption behavior of copper ions (Cu^{2+}) and oxytetracycline (OTC) on pristine thermoplastic polyurethane (TPU) MPs under both single and co-contaminant conditions, including light aging scenarios. Their findings revealed that the adsorption process was spontaneous and thermodynamically favorable, with positive enthalpy change (ΔH) values indicating endothermic adsorption driven by heat absorption. Similarly, Khumalo et al. (2022) applied an extended Flory Huggins model to examine the adsorption of PFAS compounds specially PFOA, PFOS, and sulfamethazine (SMT) onto PE and polypropylene (PP) MPs. Their results demonstrated that elevated temperatures enhanced PFAS adsorption, suggesting that thermal energy facilities strengthen interactions between MPs and contaminants. However, excessively high temperatures may also alter the physicochemical properties of both MPs and PFAS, potentially affecting their structural integrity and sorption behavior.

2.4 Impact on water quality and sediment contamination

2.4.1 Influence of co-occurring particulate matter

Natural aquatic environments are complex matrices containing various suspended solids, which may interfere with MPs-contaminant interactions (Table 2.1). Particulate matter such as activated carbon, natural clays, and mineral oxides can compete with MPs for contaminant binding or alter the partitioning dynamics of MPs in water (Atugoda et al., 2021). These particles may act as alternative sorbents, modify the surface chemistry of MPs, or influence aggregation and sedimentation processes. Despite their potential significance, the role of co-existing particulate matter in modulating MP adsorption behavior remains poorly understood. No comprehensive studies have yet quantified the competitive or synergistic effects of these particles on contaminant uptake by MPs in natural settings. This knowledge gap highlights the need for further research to explore the interactions between MPs and environmental particulates under realistic conditions.

2.4.2 Impact of MP on crysophere environments

Rapid warming in polar and high-altitude regions has accelerated the melting and retreat pf cryospheric components, including glaciers, ice sheets, sea ice and permafrost. These changes have led to widespread permafrost degradation, reduction in sea-ice extent, and enhanced glacier mass loss, with cascading impacts on hydrological systems and downstream ecosystems. As climate change progresses, the cryosphere is increasingly recognized as both a temporary sink and an emerging secondary source of MPs. Rising global temperatures intensify glacier, icesheet, and sea-ice melt, resulting in the mobilization of MPs that were previously trapped within ice matrices. The scale of MPs release depends on their concentration within snow and ice and on melt dynamics, both of which are highly

Table 2.1 Types and mechanisms of microplastic impacts on marine organisms and ecosystems.

Types of impacts	Species	References
Reduced food consumption and feeding habits	Sea cucumbers, sea stars, lobsters, crabs & fish	Hamilton et al. (2021); Pantos (2022)
Effects on health and toxicity	Cnidarians, sponges, molluscs, echinoderms, and fish	Iwalaye et al. (2020); Kim et al. (2023)
Ecological spillover	Benthic communities, diatoms, zooplanktons, and microbiological communities	Devereux et al. (2021); Mojiri et al. (2024)
Reduced ecosystem performance	Mangroves, Salt marshes, seagrass meadows, and coral reefs	John et al. (2022); Fallon and Freeman (2021); Miller et al. (2023)
Reduction in species diversity	Marine flora and fauna	Bhusare et al. (2022)

sensitive to climatic conditions. This process creates a positive feedback loop: climate driven melting release MPs into surrounding environments, while increased MP accumulation on snow and ice surfaces may further accelerate warning by modifying surface energy balance.

Previous studies have shown that dark particulates such as black carbon and mineral dust reduce surface albedo and enhance ice melt. Similarly, MPs of various colors can function as heat-absorbing particles, increasing solar radiation absorption and further lowering albedo. Experimental and field observations suggest that MP accumulation on ice surfaces can amplify melt processes. In Arctic Sea ice, MP increase ice permeability and solar energy absorption, leading to an estimated reduction on surface albedo of approximately 11% and promoting further ice loss. Atmospheric MPs have also been shown to contribute to radiative forcing. Recent estimates indicate an effective radiative forcing 0.044 ± 0.399 mWm^{-2} assuming uniform vertical distribution up to 10 km, and -0.746 ± 0.5543 mWm^{-2}, when MPs are restricted to the atmospheric boundary layer (Revell et al., 2021). Although these values remain small compared with total aerosol radiative forcing, continued growth in plastic production suggests that the climatic influence of atmospheric MPs may increase in the future. In addition, MPs may act as cloud ice nuclei, potentially influencing cloud formation processes and contributing incrementally to global radiative forcing.

Beyond their role in cryosphere-climate feedback, MPs act as vectors for POPs. During melting events, MPs released from ice can transport adsorbed contaminants into downstream aquatic systems. The interaction between MPs and POPs is governed by hydrophobic interactions, electrostatic forces, and other non-covalent mechanisms such as halogen bonding, π- π interactions, and hydrogen bonding. Sorption behavior is influenced by MP properties including polymer polarity, particle size, crystallinity, aging state, and surface area as well as environmental factors such as pH, temperature, and ionic strength. Pollutant-specific characteristics, including hydrophobicity and ionization state, further regulate adsorption and desorption dynamics. Experimental studies have demonstrated strong sorption of perfluorooctanesulfonate (PFOS) onto polyethylene MPs, with smaller particles exhibiting higher adsorption capacities. Under simulated fish gut conditions, up to 70%–80% of PFOS was shown to desorb, highlighting the potential for trophic transfer and contaminant mobilization across food webs.

The presence of MPs in remote snow and ice environments also poses risks to fragile crysopheric ecosystems. MPs can be ingested by microbial communities, invertebrates, and other low-trophic level organisms, potentially altering trophic interactions and introducing toxic substances into food webs. Antarctic amphipods have been reported to ingest between 20 and 45 MP particles per individual within a 48-hour exposure period, demonstrating the potential for rapid bioaccumulation. The widespread occurrence of MPs in Arctic and Antarctic regions affects both terrestrial and aquatic ecosystems. Ingestion of MPs has been documented in juvenile polar cod, a keystone species in Arctic food webs, indicating the potential for trophic transfer. MPs have also been detected at Arctic beaches, suggesting multiple exposure pathways for wildlife. Observed biological effects vary depending on species, particle size, and polymer type, but commonly include physical damage, digestive tract blockage, reduced feeding efficiency and impaired nutrient uptake. Bioaccumulation of MPs and associated contaminants poses risks for apex predators.

The combined effects of MP exposure and other climate related stressors such as ocean warming and acidification are likely to intensify negative impacts on ecosystem resilience. MP accumulation hotspots have been identified in the Southern Ocean, where interactions between plastic pollution and climate stressors are especially pronounced. Seasonal entrapment of MPs with ice and their subsequent release during melt periods prolong environmental bioavailability and increase ingestion risks for marine organisms. Recent evidence also suggests that some marine species can fragment ingested MPs into nanoplastics during digestion, further enhancing the potential for nano-scale remobilization in Antarctic marine ecosystems. MPs have been detected across a range of Antarctic environmental matrices and biota, including sediments, benthic organisms, krill, and fish. Antarctic krill (Euphausia superba) collected from the South Shetland and South Orkney Islands were found to contain approximately 0.20–0.29 MP particles per individual, with more than 90% of particles smaller than 150 μm and predominantly fibrous shape (Bargagli and Rota, 2022). Polyethylene, polypropylene, and polyethylene terephthalate were the most common polymers identified. Despite these findings, confirmed evidence of MPs transfer to higher trophic levels such as seabirds and marine mammals in cryospheric regions remains limited.

Crysopheric invertebrates including amphipods, copepods, chironomids, and other benthic organisms play critical roles in cold region food webs. MPs can accumulate in these organisms through ingestion, direct physical contact, or environmental uptake from surrounding snow, ice, and sediments. Exposure has been associated with oxidative stress, cellular damage, impaired physiological functions, inflammation, and potential reproductive toxicity. The relatively low metabolic rates characteristic of many cryospheric invertebrates may influence their sensitivity to micro and nanoplastic toxicity. However, current knowledge remains insufficient to fully evaluate organism-level and ecosystem-scale consequence of MP exposure in cryospheric systems. Fish inhabiting cryospheric environments, such as Arctic char, icefish, and other cold-water species, are integral to polar food webs and serve as important food resources for human populations. Micro and nanopastics can accumulate fish tissues through direct uptake from surrounding waters or via consumption of contaminated MPs. The extent of trophic transfer and biomagnification of micro and nanoplastics within cryospheric food webs remains poorly understood and demands further investigation. Accumulation of these particles in edible fish tissues also raises concerns regarding potential human exposure through seafood consumption. Although research on MPs in the cryosphere is still emerging, available evidence highlights the urgency of understanding their long-term fate and ecological impacts. Future studies should focus on interactions between MPs and co-occurring pollutants, biogeochemical cycles, and ecosystem processes. Biofouling the colonization of MP surfaces by microorganisms requires particular attention, as it can alter particle density, transport behavior, and ecological effects. Additionally, the potential for MP degradation under cryospheric conditions and the ecosystem level consequences of such transformations remain poorly explored. Addressing these knowledge gaps is essential for assessing the long-term implications of MP pollution for cryospheric biodiversity, trophic dynamics, and ecosystem functioning.

2.4.3 Impact of MPs on corals

A substantial proportion of published studies on MPs in coral reef environments has been conducted in Asia, accounting for approximately 54.9% ($n = 39$) of the total literature. In contrast, only a single study has been reported from Africa, revealing a pronounced knowledge gap, particularly with respect to East African coral reef ecosystems. Most existing investigations are concentrated in the South China Sea, Bay of Bengal, and Indonesian Sea, regions where comparatively high MP abundances have also been documented (Slynkova et al., 2026).

Reported MP concentrations vary considerably across Large Marine Ecosystems (LMEs). These differences are largely influenced by sampling methodologies, analytical protocols, and site-specific characteristics. Areas with intense tourism, dense human populations, or significant coastal development generally exhibit higher MP loas. However, meaningful comparisons across studies employing the most reported units were considered when comparing MP abundance across LMEs. In the water column, the highest MP concentration was reported from the South China Sea, with values reaching 389,280 particles m^{-3} for MPs smaller than 350 μm. In contrast, the lowest abundance was recorded in the Insular Pacific-Hawaiian LME, with only 0.046 particles m^{-3} for MPs smaller than 300 μm. MP abundance also varies with sampling depth, with surface waters generally exhibiting higher concentrations than subsurface layers (Slynkova et al., 2026).

Sediment associated MP showed a wide concentration range, rom 3.35 particles kg^{-1} in the Pacific Central American Coastal LME to 682 particles kg^{-1} in the South China Sea (Lin et al., 2024). In coral tissues, both the lowest and highest MP concentrations were reported from the Indonesian Sea, ranging from 0.0122 particles g^{-1} to 14.18 particles g^{-1} (Wijayanti et al., 2024). MP ingestion by corals is strongly influenced by poly size, feeding behavior, and target capture mechanisms, which determine species-specific exposure and uptake rates. Passive adhesion is considered the primary mechanism by which corals capture MPs from the water column. MPs can adhere to coral mucus layers and tissues surfaces without active ingestion. In addition to passive adhesion, corals may also interact with MPs through active feeding, followed by partial or complete extension. Longterm exposure experiments have demonstrated that corals can ingest MPs alongside natural food particles. Prolonged exposure of multiple coral species to bio fouled PE fragments showed no clear feeding preference, indicating that corals may ingest MPs opportunistically during normal feeding activities. However, other studies report contrasting results, suggesting that corals preferentially consume natural prey and that biofilm development on MPs does not necessarily enhance ingestion rates.

Feeding selectively also varies among coral species and MP types. Experiments exposing *Pocillopora verrucosa* and *Stylophora pi*stillata to a mixture of MPs revealed species specific ingestion patterns. *P. verrucosa* preferentially ingested PE fragments, with partial egestion, whereas *S. pistillata* primarily ingested MP fibers and subsequently egested all particles (Reichert et al., 2024). Similar selective feeding behavior has been observed in other coral species exposed to mixed polymer types, indicating that corals can, to some extent, distinguish MPs from natural food items based on particles characteristics. Although corals are capable of egesting ingested MPs, this process is energetically demanding and may impose physiological stress, Ony a small fraction of ingested MPs is ultimately retained within coral tissues, yet repeated ingestion and egestion cycles may contribute to sub-lethal effects over time.

2.5 Research gaps and future directions

Despite growing evidence of MP-PFAS interactions, comprehensive studies remain limited due to challenges in extraction and analytical methodologies. WWTPs and, marine environments serve as both sources and sinks for these pollutants, yet integrated research addressing their combined fate and transport is scarce. Future investigations should prioritize the development of standardized protocols for simultaneous detections of MPs and PFAS. Expanding monitoring efforts cross WWTPs, coastal zones, and inland water bodies will be essential for understanding their environmental partitioning cand informing mitigation strategies. While individual effects of MPs, PFAS, and other contaminants have been widely studied, their combined and interactive toxicological effects are not well understood.

MPs can act as vectors for multiple pollutants, yet dose response relationships and synergistic effects under environmentally relevant concentrations remain largely unexplored. Further research should emphasized mixture toxicity, chronic exposure scenarios, and multi-stress interactions, particularly in ecosystems already affected by climate change, eutrophication, and acidification.

Despite growing scientific evidence, translation of research findings into policy and management frameworks remains limited. There is a need for studies that evaluate the effectiveness of mitigation strategies, including wastewater treatment upgrades, plastic product redesign, and circular-economy interventions. Future research should bridge science and policy by quantifying environmental benefits, economic trade-offs, and feasibility of large-scale interventions aimed at reducing MP inputs and long-term ecological degradation.

2.6 Conclusion

MP represents more than inert debris in aquatic systems; they are chemically and biologically active particles that reshape contaminant behavior, ecosystem processes, and exposure pathways. Through adsorption and desorption mechanisms, MP interact with heavy metals, organic pollutants, ad PFAS, altering their persistence, mobility, and combined toxicity. These interactions are strongly influenced by MP aging, polymer composition, surface oxidation, and biofilm colonization, which together determine whether MPs act as contaminant sinks or secondary sources. The coupling of MPs with water chemistry and sediments intensifies contamination in benthic environments, promotes pollutant retention, and facilitates transfer through aquatic food webs. Seaweed emerge as particularly important mediators in this process, serving simultaneously as bio accumulators of metals and MP and as direct dietary components for humans and marine organisms. This dual role amplifies ecological and human health risks while posing economic challenges to the rapidly expanding global seaweed industry, especially in regions with which MP pollution loads.

Despite growing research attention, substantial gaps remain in understanding environmentally realistic exposure scenarios, long-term aging effects, and the toxicological consequences of combined MP-contaminant exposure, particularly for micro- and nanoplastics. Addressing these gaps requires standardized analytical methods, integrated multi-stress experiments, and closer alignment between environmental science, food safety, and policy frameworks. Ultimately, mitigating long-term aquatic degradation will depend on coordinated strategies that reduce plastic inputs at the source, promote circular-economy practices, and incorporate MP-contaminant interactions into global water quality and ecosystem management policies.

References

Ateia, M., Zheng, T., Calace, S., Tharayil, N., Pilla, S., Karanfil, T., 2020. Sorption behavior of real microplastics (MPs): insights for organic micropollutants adsorption on a large set of well-characterized MPs. Sci. Total Environ. 720.

Atugoda, T., Vithanage, M., Wijesekara, H., Bolan, N., Sarmah, A.K., Bank, M.S., et al., 2021. Interactions between microplastics, pharmaceuticals and personal care products: implications for vector transport. Environ. Int. 149.

Bargagli, R., Rota, E., 2022. Microplastic interactions and possible combined biological effects in Antarctic marine ecosystems. Animals 13, 162.

Bhagwat, G., Tran, T.K.A., Lamb, D., Senathirajah, K., Grainge, I., O'Connor, W., et al., 2021. Biofilms enhance the adsorption of toxic contaminants on plastic microfibers under environmentally relevant conditions. Environ. Sci. Technol. 55, 8877–8887.

Bhusare, S.R., Satheesh, S., Anil, A.C., Sujith, P.P., 2022. Impact of microplastics on marine species diversity and abundance: a critical review. Mar. Pollut. Bull. 176, 113446. https://doi.org/10.1016/j.marpolbul.2022.11344.

Brahana, P.J., Al Harraq, A., Saab, L.E., Roberg, R., Valsaraj, K.T., Bharti, B., 2023. Uptake and release of perfluoroalkyl carboxylic acids (PFCAs) from macro and microplastics. Environ. Sci. Process. Imp. 25, 1519–1531.

Caruso, G., 2019. Microplastics as vectors of contaminants. Mar. Pollut. Bull. 146, 921–924.

Dai, Y., Zhao, J., Sun, C., Li, D., Liu, X., Wang, Z., et al., 2022. Interaction and combined toxicity of microplastics and per- and polyfluoroalkyl substances in aquatic environment. Front. Environ. Sci. Eng. 16, 136.

Deng, S., Yu, Q., Huang, J., Yu, G., 2010. Removal of perfluorooctane sulfonate from wastewater by anion exchange resins: effects of resin properties and solution chemistry. Water Res. 44, 5188–5195.

Deng, S., Zhang, Q., Nie, Y., Wei, H., Wang, B., Huang, J., Yu, G., Xing, B., 2012. Sorption mechanisms of perfluorinated compounds on carbon nanotubes. Environ. Pollut. 168, 138–144.

Devereux, R., Hartl, M.G., Bell, M., Capper, A., 2021. The abundance of microplastics in Cnidaria and ctenophora in the north sea. Mar. Pollut. Bull. 173, 112992. https://doi.org/10.1016/j.marpolbul.2021.112992.

Du, Z., Deng, S., Bei, Y., Huang, Q., Wang, B., Huang, J., Yu, G., 2014. Adsorption behavior and mechanism of perfluorinated compounds on various adsorbents—a review. J. Hazard. Mater. 274, 443–454.

Enfrin, M., Lee, J., Le-Clech, P., Dumée, L.F., 2020. Kinetic and mechanistic aspects of ultrafiltration membrane fouling by nano- and microplastics. J. Membr. Sci. 601, 117890.

Fallon, B.R., Freeman, C.J., 2021. Plastics in Porifera: the occurrence of potential microplastics in marine sponges and seawater from Bocas del Toro, Panama. PeerJ 9, e11638. https://doi.org/10.7717/peerj.11638.

Fang, S., Yu, W., Li, C., Liu, Y., Qiu, J., Kong, F., 2019. Adsorption behavior of three triazole fungicides on polystyrene microplastics. Sci. Total Environ. 691, 1119–1126.

Gagliano, E., Sgroi, M., Falciglia, P.P., Vagliasindi, F.G.A., Roccaro, P., 2020. Removal of poly- and perfluoroalkyl substances (PFAS) from water by adsorption: role of PFAS.

Hamilton, B.M., Bourdages, M.P., Geoffroy, C., Vermaire, J.C., Mallory, M.L., Rochman, C.M., Provencher, J.F., 2021. Microplastics around an Arctic seabird colony: particle community composition varies across environmental matrices. Sci. Total Environ. 773, 145536. https://doi.org/10.1016/j.scitotenv.2021.145536.

Huffer, T., Hofmann, T., 2016. Sorption of non-polar organic compounds by micro-sized plastic particles in aqueous solution. Environ. Pollut. 214, 194–201.

Iwalaye, O.A., Moodley, G.K., Robertson-Andersson, D.V., 2020. The possible routes of microplastics uptake in sea cucumber Holothuria cinerascens (Brandt, 1835). Environ. Pollut. 264, 114644. https://doi.org/10.1016/j.envpol.2020.114644.

Jia, Y.Q., Shan, C., Fu, W.Y., Wei, S., Pan, B.C., 2023. Occurrences and fates of per- and polyfluoralkyl substances in textile dyeing wastewater along full-scale treatment processes. Water Res. 242.

Jiang, J.-Q., 2018. Occurrence of microplastics and its pollution in the environment: a review. Sustain. Prod. Consum. 13, 16–23.

John, J., Nandhini, A.R., Velayudhaperumal Chellam, P., Sillanpää, M., 2022. Microplastics in mangroves and coral reef ecosystems: a review. Environ. Chem. Lett. 20 (1), 397–416. https://doi.org/10.1007/s10311-021-01326-4.

Kalcíkova, G., Alic, B., Skalar, T., Bundschuh, M., Zgajnar Gotvajn, A., 2017. Wastewater treatment plant effluents as source of cosmetic polyethylene microbeads to freshwater. Chemosphere 188, 25–31. https://doi.org/10.1016/j.chemosphere.2017.08.131.

Khumalo, S.M., Lasich, M., Bakare, B.F., Rathilal, S., 2022. Sorption of perfluorinated and pharmaceutical compounds in plastics: a molecular simulation study. Water 14.

Kim, J.A., Kim, J.H., Park, Y.S., Kang, C.K., Choi, C.Y., 2023. Micro-polystyrene plastic and benzo[α]pyrene exposure affects the endocrine system and causes physiological stress in *Carassius auratus*. Comp. Biochem. Physiol. C Toxicol. Pharmacol. 271, 109695. https://doi.org/10.1016/j.cbpc.2023.109695.

Li, X.Y., Pignatello, J.J., Wang, Y.Q., Xing, B.S., 2013. New insight into adsorption mechanism of ionizable compounds on carbon nanotubes. Environ. Sci. Technol. 47, 8334–8341.

Lin, J., Zhao, Y.-M., Zhan, Z.-G., Zheng, J.-Y., Zhou, Q.-Z., Peng, J., Li, Y., Xiao, X., Wang, J.-H., 2024. Microplastics in remote coral reef environments of the Xisha Islands in the South China Sea: source, accumulation and potential risk. J. Hazard. Mater. 469, 133872.

Liu, X.M., Xu, J., Zhao, Y.P., Shi, H.H., Huang, C.H., 2019. Hydrophobic sorption behaviors of 17β-estradiol on environmental microplastics. Chemosphere 226, 726–735.

Liu, W.Y., Zhang, J.L., Liu, H., Guo, X.N., Zhang, X.Y., Yao, X.L., et al., 2021. A review of the removal of microplastics in global wastewater treatment plants: characteristics and mechanisms. Environ. Int. 146.

Llorca, M., Schirinzi, G., Martínez, M., Barcelo, D., Farre, M., 2018. Adsorption of perfluoroalkyl substances on microplastics under environmental conditions. Environ. Pollut. 235, 680–691.

McNamara, P.J., Samuel, M.S., Sathyamoorthy, S., Moss, L., Valtierra, D., Cortes Lopez, H., Nigro, N., Somerville, S., Liu, Z., 2023. Pyrolysis transports and transforms PFAS from biosolids to py-liquid. Environ. Sci.: Water Res. Technol. 9, 386–395. https://doi.org/10.1039/D2EW00677D.

Mejias, C., Martin, J., Santos, J.L., Aparicio, I., Alonso, E., 2023. Adsorption of perfluoroalkyl substances on polyamide microplastics: effect of sorbent and influence of environmental factors. Environ. Res. 216.

Miller, M.E., Motti, C.A., Hamann, M., Kroon, F.J., 2023. Assessment of microplastic bioconcentration, bioaccumulation and biomagnification in a simple coral reef food web. Sci. Total Environ. 858, 159615. https://doi.org/10.1016/j.scitotenv.2022.159615.

Mojiri, A., Vishkaei, M.N., Zhou, J.L., Trzcinski, A.P., Lou, Z., Kasmuri, N., Rezania, S., Gholami, A., Vakili, M., Kazeroon, R.A., 2024. Impact of polystyrene microplastics on the growth and photosynthetic efficiency of diatom Chaetoceros neogracile. Mar. Environ. Res. 194, 106343. https://doi.org/10.1016/j.marenvres.2024.106343.

Munoz, G., Fechner, L.C., Geneste, E., Pardon, P., Budzinski, H., Labadie, P., 2018. Spatio-temporal dynamics of per and polyfluoroalkyl substances (PFASs) and transfer to periphytic biofilm in an urban river: case-study on the River Seine. Environ. Sci. Pollut. Res. 25, 23574–23582.

Pantos, O., 2022. Microplastics: impacts on corals and other reef organisms. Emerg. Top. Life Sci. 6 (1), 81–93. https://doi.org/10.1042/etls20210236.

Penland, T.N., Cope, W.G., Kwak, T.J., Strynar, M.J., Grieshaber, C.A., Heise, R.J., et al., 2020. Trophodynamics of per- and polyfluoroalkyl substances in the food web of a large Atlantic slope river. Environ. Sci. Technol. 54, 6800–6811.

Pramanik, B.K., Roychand, R., Monira, S., Bhuiyan, M., Jegatheesan, V., 2020. Fate of road-dust associated microplastics and per- and polyfluorinated substances in stormwater. Process Saf. Environ. Prot. 144, 236–241.

Reichert, J., Tirpitz, V., Plaza, K., Worner, E., Bosser, L., Kühn, S., Primpke, S., Schubert, P., Ziegler, M., Wilke, T., 2024. Common types of microdebris affect the physiology of reef-building corals. Sci. Total Environ. 912, 169276.

Revell, L.E., Kuma, P., Le Ru, E.C., Somerville, W.R., Gaw, S., 2021. Direct radiative effects of airborne microplastics. Nature 598, 462–467.

Samuel, M.S., Shang, M., Niu, J., 2022. Photocatalytic degradation of perfluoroalkyl substances in water using a duo-functional tri-metallic-oxide hybrid catalyst. Chemosphere 293, 133568. https://doi.org/10.1016/j.chemosphere.2022.133568.

Samuel, M.S., Govindan, K., Mayer, B.K., McNamara, P.J., 2023. Electrochemical degradation of 6:2 fluorotelomer sulfonic acid using boron-doped diamond electrodes. ACS Fall Meeting, Division of Biochemical Technology, United States.

Slynkova, N., Leusch, F.D.L., Pitt, K.A., Hoogenboom, M.O., Ziajahromi, S., 2026. A systematic review of microplastics in coral reef ecosystems: abundance, distribution, toxicity, and future research directions. Mar. Pollut. Bull. 223 (2026), 119010.

Strokal, M., Vriend, P., Bak, M.P., Kroeze, C., van Wijnen, J., van Emmerik, T., 2023. River export of macro- and microplastics to seas by sources worldwide. Nat. Commun. 14, 4842.

Sun, J., Dai, X., Wang, Q., Van Loosdrecht, M.C.M., Ni, B.-J., 2019. Microplastics in wastewater treatment plants: detection, occurrence and removal. Water Res. 152, 21–37.

Torres, F.G., Dioses-Salinas, D.C., Pizarro-Ortega, C.I., De-la-Torre, G.E., 2021. Sorption of chemical contaminants on degradable and non-degradable microplastics: recent progress and research trends. Sci. Total Environ. 757.

Wang, F., Shih, K.M., Li, X.Y., 2015. The partition behavior of perfluorooctanesulfonate (PFOS) and perfluorooctanesulfonamide (FOSA) on microplastics. Chemosphere 119, 841–847.

Wang, F., Wong, C.S., Chen, D., Lu, X., Wang, F., Zeng, E.Y., 2018. Interaction of toxic chemicals with microplastics: a critical review. Water Res. 139, 208–219.

Wang, J., Liu, X.H., Liu, G.N., Zhang, Z.X., Wu, H., Cui, B.S., et al., 2019. Size effect of polystyrene microplastics on sorption of phenanthrene and nitrobenzene. Ecotoxicol. Environ. Saf. 173, 331–338.

Wang, F., Zhang, M., Sha, W., Wang, Y., Hao, H., Dou, Y., Li, Y., 2020a. Sorption behavior and mechanisms of organic contaminants to nano and microplastics. Molecules 25, 1827.

Wang, X., Zheng, H., Zhao, J., Luo, X.X., Zhenyu, W., Xing, B.S., 2020b. Photodegradation elevated the toxicity of polystyrene microplastics to grouper (*Epinephelus moara*) through disrupting hepatic lipid homeostasis. Environ. Sci. Technol. 54, 6202–6212.

Wijayanti, D.P., Indrayanti, E., Haryanti, D., Sibero, M.T., Widyarizia, M.E., Zahra, A.Q.A., Bachtiar, M., 2024. Characteristic and histological evidence of microplastic in scleractinian corals of Java Sea. Egypt. J. Aquat. Res. 50 (1), 95–102.

Xiao, F., Zhang, X., Penn, L., Gulliver, J.S., Simcik, M.F., 2011. Effects of monovalent cations on the competitive adsorption of perfluoroalkyl acids by kaolinite: experimental studies and modeling. Environ. Sci. Technol. 45, 10028–10035.

Xu, B., Liu, F., Brookes, P.C., Xu, J., 2018. Microplastics play a minor role in tetracycline sorption in the presence of dissolved organic matter. Environ. Pollut. 240, 87–94.

Xue, X.D., Fang, C.R., Zhuang, H.F., 2021. Adsorption behaviors of the pristine and aged thermoplastic polyurethane microplastics in Cu(II)-OTC coexisting system. J. Hazard. Mater. 407.

Yang, H.L., Lai, H., Huang, J., Sun, L.W., Mennigen, J.A., Wang, Q.Y., et al., 2020. Polystyrene microplastics decrease F–53B bioaccumulation but induce inflammatory stress in larval zebrafish. Chemosphere 255, 127040.

You, C., Jia, C., Pan, G., 2010. Effect of salinity and sediment characteristics on the sorption and desorption of perfluorooctane sulfonate at sediment-water interface. Environ.Pollut. 158, 1343–1347.

Zhan, Z., Wang, J., Peng, J., Xie, Q., Huang, Y., Gao, Y., 2016. Sorption of 3,3′,4,4′-tetrachlorobiphenyl by microplastics: a case study of polypropylene. Mar. Pollut. Bull. 110, 559–563.

Zhang, Q.Y., Deng, S.B., Yu, G., Huang, J., 2011. Removal of perfluorooctane sulfonate from aqueous solution by crosslinked chitosan beads: sorption kinetics and uptake mechanism. Bioresour. Technol. 102, 2265–2271.

Zhang, W., Ma, X., Zhang, Z., Wang, Y., Wang, J., Wang, J., Ma, D., 2015. Persistent organic pollutants carried on plastic resin pellets from two beaches in China. Mar. Pollut. Bull. 99, 28–34.

Zhang, G.S., Chen, J.P., Li, W.Y., 2022a. Conjugative antibiotic-resistant plasmids promote bacterial colonization of microplastics in water environments. J. Hazard. Mater. 430.

Zhang, S.F., Zhang, W.W., Ju, M.W., Qu, L., Chu, X.T., Huo, C., et al., 2022b. Distribution characteristics of microplastics in surface and subsurface Antarctic seawater. Sci. Total Environ. 838.

Zheng, Z.J., Huang, Y., Liu, L.A., Wang, L., Tang, J.C., 2023. Interaction between microplastic biofilm formation and antibiotics: effect of microplastic biofilm and its driving mechanisms on antibiotic resistance gene. J. Hazard. Mater. 459.

Zhuang, S.T., Wang, J.L., 2023. Interaction between antibiotics and microplastics: recent advances and perspective. Sci. Total Environ. 897.

Further reading

Chain length. Effect of organic matter and challenges in adsorbent regeneration. Water Res. 171.

CHAPTER

Impact of microplastics on biodiversity, food webs, and ecosystem services

3

Melvin S. Samuel

Department of Civil, Construction, and Environmental Engineering, Marquette University, Milwaukee, WI, United States

Highlights

- Microplastic affect aquatic organisms directly and indirectly across all trophic levels
- Food-web interaction and nutrient cycling are disrupted by microplastic accumulation
- Long-term exposure leads to biodiversity loss and ecosystem imbalance
- Essential ecosystem services such as water purification and habitat stability are threatened.
- Understanding these impacts is critical for global biodiversity conversation strategies.

3.1 Introduction

Global plastic production has escalated dramatically over the past decades, rising from just 1.5 million metric tons in 1950 to approximately 442.5 million metric tons in 2023, with projections indicating it could reach 1.2 billion metric tons by in 2060 (Jaganmohan, 2024). This surge in production has placed immense pressure on waste management systems, which are struggling to cope with the volume of discarded plastic. By 2025, it is estimated that over 250 million metric tons of plastic waste will be present in the environment, worsening pollution and ecological degradation (Miller et al., 2023). Alarming, only 9% of plastic waste is currently recycled or upcycled, while 22% is improperly managed and discarded as litter. The remaining waste is either incinerated (19%) or deposited in landfills (60%), contributing to long-term environmental contamination (Nakano et al., 2024).

A significant portion of this mismanaged waste finds its way into aquatic and terrestrial ecosystems. Approximately 22% of the total plastic waste equivalent to 10% of the global plastic trash is estimated to enter marine environment, where it accumulates and persists over time. If current trends continue, researchers warn that by 2050, the oceans may contain more plastic than fish. Microplastics (MPs), defined as plastic particles smaller than 5 mm by the U.S. National Oceanic and Atmospheric Administration (NOAA) and the European Chemicals Agency (ECHA), are particularly concerning due to their widespread presence and persistence (Ghosh et al., 2023). These MPs are categorized into two types: primary MPs, which are manufactured at micro-scale for use in products like microbeads, glitter, and synthetic fibers, and secondary MPs, which result from the breakdown of larger plastic items due to environment stressors such as UV radiation and mechanical abrasion (Ghosh et al., 2023). Both primary and secondary MPs are persistent pollutants, especially in marine ecosystems, where

Microplastics in Aquatic Environments. https://doi.org/10.1016/B978-0-443-45619-0.01003-7

they contribute significantly to water pollution. Notably, over 35% of oceanic MPs originate from synthetic textiles, such as polyester, acyclic, and nylon, which shed microfibers during washing (UNEP, 2023). These particles are frequently detected in marine environments, including deep-sea habitats, shorelines, and coral reefs, with rivers, floods, and wind acting as major transport mechanisms. Coastal and marine human activities also directly release large quantities of plastic waste into the ocean.

3.2 Microplastics chemical and physical transformations in the environment

3.2.1 Toxicity, persistence, and ecological impact

As microplastics age in natural environments, their surface chemistry evolves due to exposure to reactive species such as hydroxyl radicals, oxygen, and nitrogen oxides. These interactions, driven by photochemical and oxidative degradation, lead to the formation of the new functional groups on the plastic surface, including hydroxyl, carboxyl, and carbonyl groups (Guo and Wang, 2019). This chemical alteration causes surface cracking, which not only facilitates further degradation but also increases the likelihood of the particle that may prolong retention times in the digestive tracts of the organisms and intensify toxic effects upon ingestion (Ding et al., 2024). Microbial colonization is also influenced by these surface modifications. The newly formed functional groups enhance microbial adhesion and biofilm formation, potentially altering the ecological interactions between MPs and biota. Although these surface role in determining toxicity under realistic environmental conditions remains uncertain. MPs are predominantly hydrophobic, which affects their dispersion in aquatic systems and their affinity for hydrophobic organic pollutants. However, environmental aging and surface treatments can modify their hydrophobicity and surface charge, influencing how they interact with other particles and microorganisms (Zhou et al., 2024).

Degradation of MPs in marine and coastal environments occurs through various weathering processes, including photodegradation, thermal degradation, biodegradation, and thermos-oxidative degradation. These processes involve chemical reactions that break down the polymer chains, crystallinity, particle size, and density. Despite these transformations, MPs remain highly stable and can persist in ecosystems for decades (Xiang et al., 2022). The lifecycle of the MPs begins with their release into terrestrial or aquatic environments, followed by transport through water systems. Once in aquatic ecosystems, MPs bioaccumulate in organisms such as zooplankton, small fish, and larger marine species, eventually entering the human food chain. Studies have shown that ingestion of MPs can impair immune and digestive systems in marine life and may pose direct health risks to human due to their ability to carry toxic substances and absorb pollutants like antibiotics (Miller et al., 2020).

Ocean-based sources contribute approximately 10%–20% of the total MPs found in aquatic environments. These sources include maritime activities such as commercial fishing, offshore industrial operations, coastal tourism, and marine transportation (UNEP, 2023). A major contributor is discarded or lost fishing gear-particularly nylon nets and plastic monofilament lines-which can persist at various depths in the ocean. Each year, over 600,000 tons of fishing equipment are abandoned in marine waters, significantly adding to the microplastic load. Additionally, military and commercial vessels often release microplastic waste directly into the sea, further intensifying the problem (Peng et al., 2023). Offshore industries, especially those involved in petrochemical production, also discharge substantial amounts of plastic debris into marine ecosystems. Although ocean-based sources are less dominant than land-based ones, they still play a critical role in marine microplastic pollution, necessitating targeted mitigation strategies.

Atmospheric deposition is another significant pathway through which MPs enter marine environments. All airborne MPs originate from human activities and can be classified into industrial, agricultural, and domestic sources. Industrial emissions primarily stem from plastic incineration processes (Zhang et al., 2024). In agriculture, plastic mulching-using polymeric films to regulate soil temperature and moisture is a notable contributor (Nizzetto et al., 2016). Domestic sources include synthetic clothing, which sheds microfibers during wear and washing, and mechanical drying processes that release plastic fragments into the air. Additional contributors include tire wear, traffic emission, nail salon activities, wind erosion of larger plastic debris, and vented drying of synthetic materials. Microplastics can also be transferred from the ocean to the atmosphere through mechanisms such as bubble trust aerosolization and wave action during turbulent sea conditions (Allen et al., 2020). Once airborne, MPs can remain suspended for extended periods and travel long distances via wind currents, eventually settling in coastal or marine environments through dry deposition. This process occurs when wind velocity decreases, allowing MPs to fall directly onto the surface without precipitation. These particles then enter marine ecosystem, where they pose risks to aquatic organisms through ingestion and bioaccumulation. Wet deposition is another atmospheric pathway, where MPs become embedded in raindrops or snowflakes during cold formation. As precipitation occurs, these microplastic-laden droplets fall into oceans or coastal waters, contributing to direct contamination. In nearby regions, rainfall can wash MPs into rivers and streams, which ultimately transport them to the sea (Ryan et al., 2023). Sea spray also plays a role in this cycle; wind and wave action generate droplets of seawater containing MPs, which can be carried by air currents and later redeposited into marine environments. This continuous exchange between the atmosphere and ocean underscores the complexity and persistence of microplastic pollution in aquatic ecosystems.

3.2.2 Microplastic particles as carriers of microbes and co-contaminants in aquatic systems

Microplastic provide durable, hydrophobic surfaces that rapidly become colonized by microorganisms once they enter water bodies. This colonized layer often referred to as the plastisphere behaves differently from surrounding free-living microbial communities and from biofilms formed on natural particles. Biofilm development on microplastics can increase microbial biomass and alter community composition, creating micro-habitats that support survival, persistence, and dissemination of microbes and associated genetic traits, including antibiotic resistance genes (ARGs) (Fig. 3.1).

Entry of Microplastics and initial conditioning in water: After introduction into aquatic environments (Via litter, runoff, wastewater effluent, or fragmented debris), microplastic surfaces rapidly acquires an organic matter (e.g., protein, polysaccharides). This conditioning layer changes surface charge and wettability and acts as a primer of microbial attachment, making microplastics suitable platforms for microbial colonization and subsequent biofilm growth.

Chemotaxis and microbial attachment to Microplastic surfaces: Microbes can be chemotactically attracted toward nutrient-rich conditioned microplastic surfaces and early colonizers. Once attached, they form microcolonies, initiate extracellular polymeric substances (EPS) production, and develop structured biofilms.

Biofilm maturation and enrichment of ARGs: As the biofilm matures, microplastics may become hotspots for ARGs and mobile genetic elements (MGEs) due to.

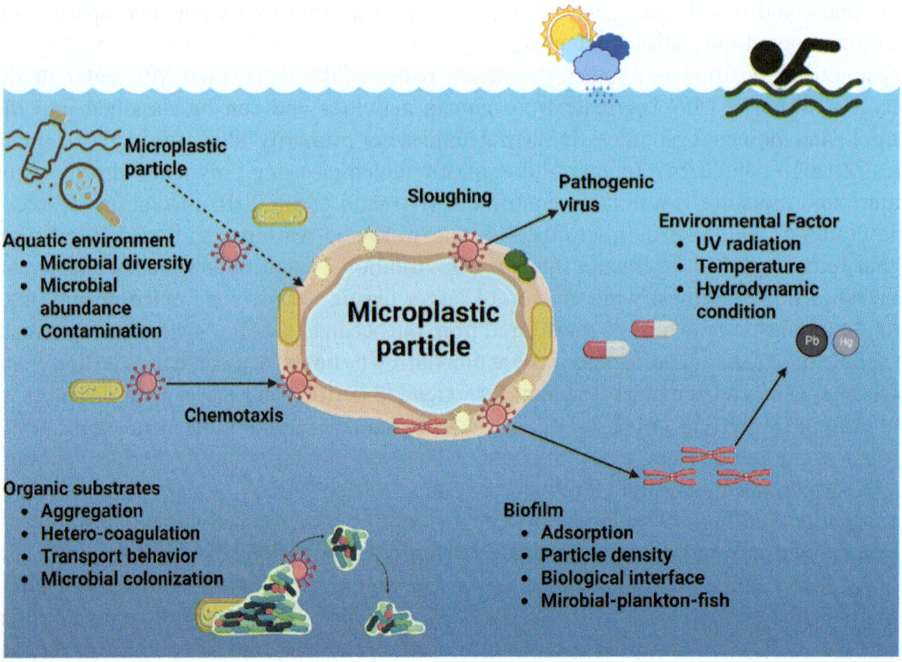

FIGURE 3.1

Schematic representation of pathways linking microplastics with microbial contamination in aquatic environments. Microplastics serve as vectors for biofilm-associated microorganisms, ARGs, and metals, with accumulation and release governed by environmental conditions such as UV exposure and temperature. The transport of pathogens and contaminants via microplastics presents potential risks to both ecosystems and human health.

- High cell density and proximity
- EPS that can retain extracellular DNA, and
- Co-occurrence of selective chemicals (e.g., antibiotics, metals)

Co-transport of antibiotics, heavy metals, and pathogens: Microplastics can carry not only bacteria but also antibiotics and heavy metals, which can act as a selective platform that favor resistant organisms and resistance gene maintenance. Wastewater related studies report plastisphere associated ARGs and virulence factors in discharged waters, reinforcing the role of MPs, as transport vectors from engineered systems to receiving waters.

Environmental drivers shaping plastisphere composition (UV, temperature, etc.): Fig. 3.1, highlights environmental factors (e.g., UV and temperature). These stressors can reshape microbial communities, influence EPS production, and affect the balance between colonization and detachment. Biofilm structure and function are dynamic and can change with seasonality and environmental conditions, impacting how strongly microplastics retain microbes and contaminants over time.

Sloughing/detachment and dispersal to the wider environment: A key step shown is sloughing biofilm fragments and attached microbes detaching from microplastic surfaces. Sloughing enables wider dissemination of bacteria, pathogens, and gene-bearing cells, potentially spreading ARGs across habitats. Detached fragments can also seed new surfaces, amplifying dispersal at ecosystem scales.

3.3 Microplastic pollution is widespread across major marine habitats

Each year, millions of metric tons of plastic waste are transported from terrestrial sources into marine environments. The distribution and concentration of MPs in these habitats are shaped by a combination of anthropogenic activities-such as urbanization, tourism, and industrial discharge-and environmental dynamics like river flow, wind patterns, ocean currents, tides, and cyclonic systems (UNEP, 2023). Regions with intense hydrodynamic activity tend to exhibit higher concentrations of MPs due to enhanced transport and dispersion mechanisms.

3.3.1 Environmental and human factors influencing its distribution and persistence

In **Asia,** extensive studies have documented a wide range of microplastic types including fragments, films, foams, pellets, and microbeads, with secondary MPs-formed from the breakdown of larger plastics being the most prevalent. China ranks among top global producers of the plastic waste, contributing significantly to marine microplastic pollution in the region (Curren et al., 2021).

Across **Europe**, MPs have been detected in freshwater systems, marine sediments, and aquatic organisms. The concentration of MPs correlates strongly with population density and human activity. Urban centers and tourist destinations exhibit elevated levels of microplastic contamination due to increased waste generation and environmental stress.

In **South Africa**, beach sediments contain between 340.7 and 4757 items per square meter, while offshore waters near **Namibia** show lower concentrations, ranging from 204.5 to 1491.7 items per cubic meter in the water column (Kanhai et al., 2020). These variations highlight the influence of both coastal activity and oceanic conditions on microplastic distribution.

North America, flanked by the Atlantic and Pacific Oceans, experiences dynamic marine conditions that facilitates widespread microplastic dispersal. In **Greenland,** the average density of MPs in the water column rose from 0.99 to 2.38 items/meter cubic, between 2005 and 2014 (Shahul Hamid et al., 2018). The **northern Gulf of Mexico**, particularly along the **Louisiana coast**, exhibits some of the highest global concentrations of MPs, surpassing even semi-enclosed seas like the Mediterranean.

The **North Atlantic Ocean** features distinct microplastic accumulation zones within oceanic eddies. Cyclonic eddies, which rotate counterclockwise and promote upwelling, contains around 20,000 items/kilometer square. These eddies, rotating clockwise with downwelling, hold up to 170,000 items/kilometer square. These eddies are rich in MPs, Zooplankton, and nutrients, indicating their role in concentrating and retaining microplastics (OCED, 2024).

In the **Arctic Ocean,** MPs have been found embedded in ice cores at concentrations ranging from 38 to 234 items/meter cube, underscoring the long-term persistence of these particles in polar environments. Common polymers identified include polyamide (PA), polypropylene (PP), polystyrene (PS), and polyethylene (PE).

Finally, **deep-sea-habitats** are not exempt from microplastic pollution. Sediments and corals collected from depts of 3500 m in the northeast Atlantic, southwest Indian Ocean, and Mediterranean basin revealed 13.4 items per 50 mL of material, Highlighting the extensive reach of microplastic contamination into the planets most remote marine ecosystems (Woodall et al., 2014).

3.3.2 Ocean-to-atmosphere microplastic transport

Recent studies have revealed that over 30% of airborne particles found above ocean surfaces are composed of plastic, indicating that marine environments play a prominent role in the aerosolization of MPs. This hypothesis is supported by the observed similarities in morphology, concentration, and polymer composition between MPs found in the atmosphere and those present in surface seawater. Aerosolized MPs, typically measuring around 20–13 μm, are released from the ocean surface and may account for up to 11% of point-source emissions of MPs into the atmosphere. Once airborne, these particles can be transported by wind, contributing to the formation of suspended particulate matter, cloud droplets, and changes in hydrophobic behavior. The interaction between deposition, atmospheric transport, and re-aerosolization creates a cyclic exchange of MPs between the ocean and the atmosphere (Allen et al., 2020) estimated that approximately 136,000 tons of marine plastics are blown ashore annually, highlighting the recirculation of MPs from marine to terrestrial environments.

Laboratory experiments have demonstrated that the air-water interface facilitates the release of nano plastics-particles smaller than 0.1 μm through bubble-bursting mechanisms. The efficiency of this aerosolization process is influenced by the surface properties of plastics, particularly their hydrophobicity, which affects their tendency to agglomerate near the ocean surface. Masry et al. (2021) found that factors such as ultraviolet (UV) degradation, the age of plastic (especially polyethylene), and smaller polystyrene particles with diameter of 600 nm or cluster coated with sulfur atoms can be emitted into the atmosphere during bubble rupture events. Despite these findings, further research is needed to fully understand the dynamics of the air-water interface and the transfer efficiency of different polymer types. The presence of nano plastic in ambient air samples has yet to be conclusively verified. Moreover, similar patterns in MP concentrations and compositions have been observed in terrestrial aquatic systems and atmospheric deposition, suggesting that other water bodies-such as rivers and lakes-may also contributes to airborne MP pollution through entrainment in moving water (Huang et al., 2022).

3.4 Direct and indirect effects on aquatic organisms

3.4.1 Impacts of microplastics on marine organisms and food webs

Microplastics infiltrate marine environments through various pathways, including land-based sources like urban runoff and wastewater discharge, marine activities such as fishing, aquaculture, and natural process like fragmentation and littering. Once in the ocean, MPs are dispersed by currents and winds, accumulating in gyres, coastlines, or sinking to the seabed. These particles act as carriers for harmful substance-such as heavy metals, persistent organic pollutants (POPs), and pathogenic microbes-which can be absorbed from surrounding waters or leached from the plastic itself. When ingested by marine organisms, MPs can impair growth, reproduction, and immune function, and trigger oxidants stress and endocrine disruption (Fang, 2023).

***Phytoplankton*:** The foundational producers in marine ecosystems, are particularly vulnerable. Exposure to MPs has been shown to inhibit cell growth, reduce photosynthetic efficiency, and alter gene expression. Studies on porphyridium purpureum revealed that concentrations of 50–100 mg/L of polystyrene MPs significantly hindered algal development, reducing chlorophyll content and increasing stress enzyme activity (Huang et al., 2022). Molecular docking simulations further demonstrated that PSMPs interact with photosystem II proteins in diatoms, reducing protein and chlorophyll levels by up to 35.5% and 64.4% respectively. Biodegradable MPs like micro-polylactic acid (mPLA) and micro-polybutylene succinate (mPBS) also suppressed growth in chlorella which vary in size, charge, and aging state, may have even more complex effects than those observed in laboratory settings (Arat, 2024).

Zooplankton: The key consumers in marine food webs, are also adversely affected. Experiments with brine shrimp (*Artemia parthenogenetica*) exposed to polystyrene (PS) and polyethylene (PE). MPs showed reduced growth, survival, and intestinal health, along with oxidative and immune stress (Li et al., 2021). These disruptions compromise lipid production and energy transfer to higher trophic levels. Additionally, MPs ingested by zooplankton are expelled in fecal pellets, which serve as food for other marine species. These pellets, altered in density and sinking rate, facilitate the spread of MPs throughout the ecosystem (Cole et al., 2016).

Impact on Corals: Coral reefs, which support diverse marine ecosystems, are increasingly contaminated by MPs. Studies have shown that long-term exposure to MPs affects coral physiology in species such *as Acropora muricata*, *Pocilloporid verrucosa*, *Porites Lutea*, and *Heliopora coerulea*. While some species exhibit minimal physiological disruption, others show species-specific responses, suggesting potential compensatory mechanisms. *Turbinaria mesenterina* exposed to PS microparticles exhibited significant microplastic accumulation and heightened immune activity. This was linked to altered symbiotic communication between coral host cells and algae, driven by elevated membrane lipid levels and increased susceptibility to lipid peroxidation, which ultimately led to cell death and physiological dysfunction (Tang et al., 2024). Short-term exposure to PE, MPs has also been shown to reduce polyp length, impair adaptability, disrupt zooxanthellae density, and damage the anthozoan-algae symbiotic relationship. These effects include excessive lipid peroxide production, tissue necrosis, reduced skeletal growth, and coral bleaching (Chen et al., 2022). The severity of these impacts often correlates with MPs concentration and polymer type, underscoring the vulnerability of coral ecosystems to plastic pollution.

Impact on Fish: Fish are highly susceptible to MP contamination, both through direct ingestion from seawater and indirectly via consumption of contaminated zooplankton. Research by Hawke et al. (2024), revealed that petroleum-based polymers like PE (300–600 μm, 2.5%of food weight) impair aerobic metabolism, swimming behavior, and escape responses in *Forsterygion capito*. Biopolymers had a lesser impact, suggesting that biodegradable alternatives may be less harmful to marine life. Biochemical studies have shown that MPs disrupt enzyme activity and blood chemistry in fish. For instance, *Cyrinus carpio* exposed to PE-MPs exhibited elevated levels of creatine phosphokinase and alkaline phosphatase, alongside reduced concentrations of globulin, albumin, triglycerides, total protein, and cholesterol (Saha et al., 2023).

Similarly, *Orechromics niloticus* exposed to varying MP concentrations showed increases levels of creatine, uric acid, glucose, and liver enzymes, indicating organ stress and dysfunction. MPs also interfere with neurotransmission by inhibiting acetylcholinesterase (AChE), a key enzyme in neural signaling. Suppression of AChE can lead to receptor overstimulation and neurotoxicity. Oxidative stress makers were significantly elevated in rainbow trout exposed to plastic additives, further confirming the neurotoxic potential of MPs. A study of 1337 fish from Turkish waters found MPs in the stomach of 34%, intestines of 41% and both organs in 58% of specimens, with most particles being blue and fiber-shaped (Guven et al., 2017).

MPs also promote lipid peroxidation (LPO), which damages cell membranes and disrupt neurotransmitter balance, contributing to neurological disorders in fish (Pastorino et al., 2023). They can destroy dopamine-producing neurons and harm cholinergic and GABAergic cells. Nematodes lacking a blood-brain barrier, are particularly vulnerable to MP-induced neurotoxicity. Behavioral abnormalities have also been observed, including gastrointestinal obstruction due to MP accumulation in the colon. Juvenile fish exposed to MPs show reduced body size, mass, and survival rates (Naidoo and Glassom, 2019). These physiological and developmental impairments threaten fish populations and pose ecological and economic risks, especially for commercially valuable species.

The marine environment is one of the most extensively studied ecosystems for MPs contamination, particularly due to its role in supplying seafood to human populations. Numerous studies have quantified MPs in marine species using metrics such as particles per gram or per individual. Marine organisms can ingest MPs passively through gill filtration, actively by mistaking them for prey, or indirectly by consuming contaminated prey (Ratnasari et al., 2024). Bivalves, crustaceans, and various fish species are commonly found to contain MPs. For instance, globally significant commercial fish such as European hake (*Merluccius*), red mullet (*Mullus Barbatus*), European pilchard (*Sardina pilchardus*), and Atlantic cod (*Gadus morhua*) have been shown to accumulate MPs in their digestive tracts (Barboza et al., 2018). Similar findings were reported for edible fish species from Brazil's eastern coast, including *Rhizoprionodon Landii* and *Scomberomorus cavalla* (De a Miranda & De Carvalho-Souza, 2016). In Portuguese coastal waters, species like *Dicentrarchus labrax*, *Trachurus*, and *Scomber Colias* exhibited varying levels of contamination, with *S. Colias* showing the highest gastrointestinal MP concentration (42%), likely due to differences in filtration rates, elimination efficiency, and proximity to shorelines (Barboza et al., 2018).

In the Persian Gulf, MPs were detected in the gills and muscles of over 128 benthic and pelagic species, raising concerns about toxicological effects and trophic transfer of contaminants (Akhbarizadeh et al., 2019). Mediterranean clupeid fish such as Sardina pilchardus and Engraulis *Encrasicolus* were found to contain microplastic fibers both internally and externally, which is troubling given their role as primary food sources for other marine species (Savoca et al., 2020). MPs can also be ingested by predators through contaminated prey like fish larvae, mollusks, polychaetas, and small crustaceans, facilitating their movement through marine food webs. Even though the gastrointestinal tract is typically removed during seafood processing, the presence of MPs in edible marine species remains a global concern due to potential human exposure (Savoca et al., 2019). In Bangladesh, economically important shellfish such as brown shrimp (*Metapenaeus monoceros*) and tiger shrimp (*Penaeus monodon*) were found to contain 3.40 ± 1.23 and 3.87 ± 1.05 particles/g of MPs, respectively. FTIR analysis confirmed the presence of polyamide-6 and rayon-materials commonly used in marine gear like ropes and nets (Hossain et al., 2020). The accumulation of MPs in aquatic environments reduces energy availability for growth and reproduction, compromising the health of marine organisms. MPs can leach toxic chemicals that bioaccumulate and biomagnifying through food webs, leading to a range of adverse effects: reduced fertilization and larval abnormalities, oxidative and intestinal damage, lowered metabolic rates and body mass, neurotoxicity, impaired growth and predatory behavior (Farrell and Nelson, 2013), diminished swimming performance, altered behavior, and even mortality.

Marine mammals are exposed to MPs through two primary pathways, direct ingestion from the water column while feeding and indirect ingestion via trophic transfer-consuming prey that has already ingested plastic particles. Baleen whales, which filter large volumes of seawater to feed on krill and plankton, are particularly vulnerable. Studies estimate that species like the blue whale (*Balenoptera musculus*) may ingest up to 10 million MP particles daily equivalent to approximately 43.6 kg of plastic mostly through contaminated prey rather than direct water intake (Kahane-Rapport et al., 2022). In contrast, humpback whales that feed primarily on fish may ingest around 200,000 MPs per day, while those feeding on krill may consume up to four million particles.

The health implications of such high ingestion rates remain largely unknown. However, concerns include the potential for MPs to fragment into nano plastics (NPs) within the digestive systems of prey and predators, leading to further environmental release through excretion. Additionally, the chemical composition of MPs including additives and adsorbed pollutants may pose toxicological risks. The energetic cost of processing indigestible plastic material could also strain the metabolic systems of these large mammals.

Stranding studies provide further evidence of MPs contamination in marine mammals. An investigation along the British coast found MPs in the gastrointestinal tracts of 50 stranded animals, including dolphins, porpoises, seals, and a pygmy sperm whale, with an average of 5.5 particles per animal. The polymers identified included nylon, polyester, rayon, PET, PE, PP, phenoxy resin, and LDPE (Nelms et al., 2019). Gray seals (*Halichoerus grypus*) were found to contain one to four MPs per individual. In the North Sea, nine stranded sperm whales were found to have ingested 322 plastic particles of various sizes, while humpback whales showed the presence of 45 MPs, including PE, PP, PVC, PA, and PET (Besseling et al., 2015).

Coastal dolphins, particularly those that forage near the seabed, appear to ingest more MPs than surface-feeding species, likely due to higher sediment contamination (Di Beneditto and Ramos, 2014). Pereira et al. (2023) analyzed the stomach contents of stranded Guiana dolphins (Sotalia guianensis) in the Southwestern Tropical Atlantic and found widespread contamination, with polyurethane (PU) and PET being the most common polymers. The study also noted a seasonal trend, with smaller particles more frequently detected during rainy periods, and a strong correlation between MPs mass and particle count, suggesting trophic transfer as a major contamination route.

While these studies confirm the presence of MPs in marine mammals, the long lifespans and large body sizes of these species make it difficult to assess the full extent of physiological and ecological impacts. Nonetheless, the potential for both acute and chronic toxicity, digestive obstruction, and energy depletion underscores the urgency of further research. Understanding how MPs affect these sentinel's species may also offer insights into the broader implications for human health, given our shared exposure to marine food webs.

3.4.2 Marine-derived microplastics and human health risks

Microplastics (MPs) and nano plastics (NPs) originating from marine environments are increasingly recognized as a public health concern due to their presence in seafood, sea salt, drinking water, and even the air (Fig. 3.2). These particles accumulate in commercial aquatic species such as fish, shellfish, crabs, and mussels, which are widely consumed by humans. This poses a particular risk to coastal

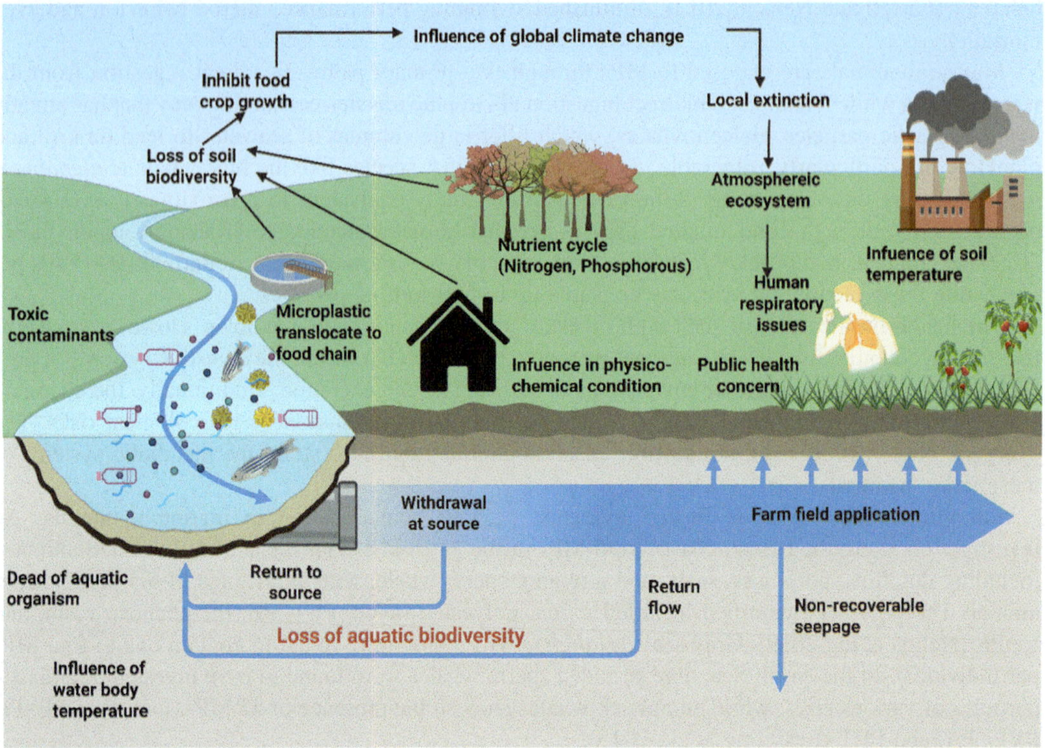

FIGURE 3.2

Overview of microplastic and nanoplastic effects on ecosystem services and climate processes across multiple environmental compartments.

communities, artisanal fishers, aquaculture workers, and indigenous populations who rely heavily on traditional seafood diets (Yu et al., 2020).

Humans are exposed to MPs through multiple pathways:

- **Ingestion**: Eating contaminated seafood is the most direct route. MPs have been found in the gastrointestinal tracts and tissues of various marine organisms, and although some parts are removed during processing, residual contamination remains a concern.
- **Inhalation**: MPs suspended in air, especially in coastal and urban areas, can be inhaled. Studies suggest that inhalation may lead to 3–15 times higher exposure than ingestion (Catarino et al., 2018).
- **Drinking water**: Coastal water sources may contain MPs, which can enter the human body through consumption.

Once inside the body, MPs can trigger a range of adverse biochemical effects. These include:

- **Immune system disruption**: inflammation, and genotoxicity (Wright and Kelly, 2017).

- **Oxidative stress**: Exposure to polyethylene (PE) and polystyrene (PS) MPs increases reactive oxygen species (ROS) in various human cell lines, including brain, intestinal, and lung epithelial cells.
- **Mitochondrial and lysosomal dysfunction**: apoptosis, and impaired cell viability have been observed in studies using PS particles on human and animal models (Frohlich et al., 2014).

Inhalation studies using human lung epithelial A549 cells-a model for alveolar type II cells-have shown that MPs can:

- Activate innate immunity by altering gene and protein expression.
- Disrupt membrane transporters like P-glycoprotein/MDR1, affecting cellular uptake of substances (Salomon and Ehrhardt, 2011).
- Interfere with apoptosis, cell cycle progression, and transcriptional regulation.
- Exhibit combined toxicity when interacting with other pollutants like phthalate esters.

Ingestion studies using intestinal cell models have confirmed that PS-NPs can penetrate the epithelial barrier, raising concerns about long-term effects. Cellular uptake is influenced by particle size, surface charge, and exposure duration, with carboxylate PS NPs (40–200 nm) entering cells via energy-dependent mechanisms (Santos et al., 2017). A recent study by Hu et al. (2024), detected MPs in human testicular tissue, with average concentrations of 328.44 μg/g. The predominant polymer was PE, and the presence of MPs correlated with changes in testicular and epididymal weights, sperm count, and other reproductive parameters-suggesting a potential link to male fertility issues. Overall, chronic exposure to MPs may lead to systemic inflammation, immune dysregulation, and reproductive toxicity. These findings highlight the urgent need for further research into the long-term health effects of marine-derived microplastics and the development of strategies to mitigate human exposure.

3.5 Disruptions in trophic transfer and nutrient cycling

3.5.1 Bioaccumulation and biomagnification of microplastics across trophic levels

Microplastics (MPs), typically defined as plastic particles smaller than 5 mm, are pervasive in aquatic ecosystems and pose growing concerns regarding their movement through food webs. Although data on the trophic transfer, bioaccumulation, and biomagnification of MPs especially in humans remain limited, emerging research provides insight into how these particles travel across trophic levels and accumulate in organisms. Due to their high surface area and chemical properties, MPs readily adhere to biological surfaces, including phytoplankton. Positively and neutrally charged MPs show greater affinity for attachment compared to negatively charged ones (Nolte et al., 2017). These phytoplankton, which serve as primary producers, are consumed in large quantities by herbivorous fish such as *Dorosoma cepedianum* and *Pimephales promelas*, introducing MPs into the aquatic food chain (Bhatt and Chauhan, 2023).

Once ingested, MPs tend to accumulate in fish due to low egestion rates, and secondary consumers such as omnivorous and carnivorous fish further ingest these contaminated organisms. Studies have shown that predatory fish contain higher MP loads (average 6.09 particles per fish) compared to omnivores (5.85) and herbivores (1.88), demonstrating clear biomagnification across trophic levels

(Bhatt and Chauhan, 2024). These contaminated fish eventually enter the human food supply through commercial fishing and seafood markets. Curren et al. (2021) found significant MPs contamination in seafood samples from Singapore supermarkets, highlighting the unintentional ingestion of MPs by consumers. Beyond food, MPs can enter the human body through drinking water, airborne particles, and even table salt. Smith et al. (2018) reported that while approximately 90% of ingested MPs are excreted, the remaining 10% may persist and accumulate in human tissues. MPs have been detected in the lungs, placenta, bloodstream, and other organs (Hu et al., 2024), raising concerns about their long-term health effects. As xenobiotic substances, MPs can act as carriers for toxic metals and harmful additives such as bisphenol A and polybrominated diphenyl ethers. Due to their hydrophobic nature, MPs can adsorb and transport hazardous chemicals, functioning as vectors that amplify toxicity a phenomenon often referred to as the "Trojan Horse" effect. This mechanism not only affects aquatic life but may also pose risks to human health, including inflammation, oxidative stress, and endocrine disruption (Fig. 3.3). Although comprehensive studies on the health impacts of MPs in humans are still lacking, the evidence underscores the need for further investigation into their biological behavior and toxicological consequences.

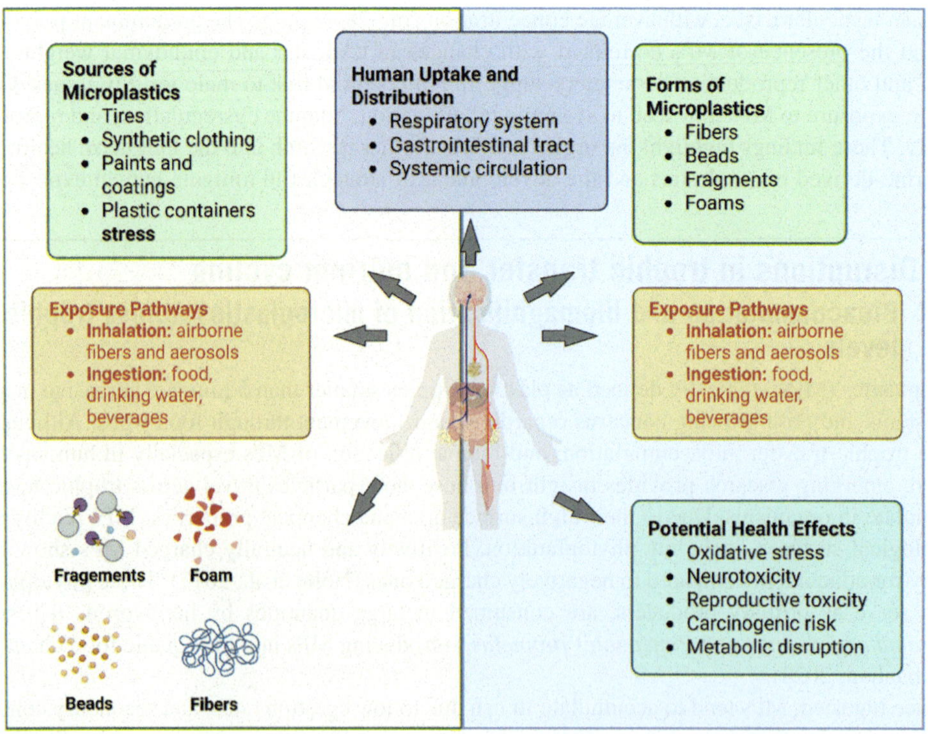

FIGURE 3.3

Human exposure to microplastics occurs via inhalation and ingestion, leading to systemic distribution and potential adverse health outcomes.

3.5.2 Impact of microplastics on nutrient cycling in terrestrial and aquatic ecosystems

Nutrient cycling is a vital ecological process that governs the movement and transformation of essential elements such as nitrogen, carbon, and phosphorus-between living organisms and their surrounding environment. However, human activities have increasingly disrupted these natural cycles, and MPs are emerging as a significant pollutant influencing nutrient dynamics across ecosystems. In terrestrial environments, MPs interfere with nitrogen cycling by altering soil microbial communities and enzyme activities. Nitrogen cycling involves complex microbial-mediated processes like nitrification and denitrification, which are sensitive to environmental changes. The presence of MPs in soil can reduce microbial diversity and disrupt enzymatic functions critical to nitrogen transformations. Biodegradable MPs like polylactic acid (PLA) can serve as a carbon source, enhancing microbial activity and promoting nitrogen fixation by stimulating bacteria such as Cyanobacteria, Bradyrhizobium, and Frankiale. This is often accompanied by increased activity of the nitrogen-fixing gene nifH in biofilms formed on agricultural plastic residues.

MPs can also suppress key enzymes involved in nitrogen cycling. Studies have shown that exposure to MPs can reduce the activity of amoA and nirK, which encode ammonia monooxygenase and nitrite reductase, respectively. The effects vary depending on the type of plastic and its chemical additives. For instance, low-density polyethylene (LDPE) has been associated with increased abundance of nitrogen cycling genes (nifH, AOB, amoA, nirK), while polyvinyl chloride (PVC) may inhibit ammonium oxidation but enhance denitrification. Polystyrene (PS) MPs have been linked to reduced microbial biomass and decreased leucine aminopeptidase activity, further impairing soil nutrient turnover. In aquatic systems, MPs contribute to nutrient cycling disruptions through biofilm formation on their surfaces. These biofilms increase the porosity of MPs and facilitate microbial colonization, accelerating the oxidation of ammonium (NH_4^+) and nitrite (NO_2^-), as well as enhancing denitrification. The presence of MPs in water bodies can thus alter nitrogen fluxes and microbial interactions, with implications for water quality and ecosystem health. Overall, the influence of MPs on nutrient cycling is complex and context-dependent, shaped by factors such as polymer type, concentration, environmental conditions, and microbial community composition. Continued research is essential to fully understand the long-term ecological consequences of microplastic pollution on biogeochemical processes.

3.5.3 Impact of microplastics on the carbon cycle

The carbon cycle is a critical global process that governs the transformation and movement of carbon among the atmosphere, oceans, soil, and living organisms. With rising concerns over climate change and greenhouse gas emissions-particularly carbon dioxide (CO_2) understanding how pollutants like microplastics (MPs) influence this cycle has become increasingly important (Fig. 3.2). MPs, composed primarily of carbon-rich polymers such as polyethylene (PE) and polystyrene (PS), can significantly alter carbon cycling in both terrestrial and aquatic ecosystems. In soils, MPs disrupt the carbon cycle by modifying physicochemical properties, microbial diversity, plant growth, and litter decomposition. Their hydrophobic nature reduces soil moisture and creates aerated microenvironments, which can accelerate the mineralization of soil organic matter (SOM). This process increases the release of carbon dioxide, potentially intensifying greenhouse gas emissions.

Moreover, MPs can act as non-plant derived carbon sources in the soil. Their presence can increase dissolved organic carbon (DOC) levels, as observed in studies where river water mixed with nano plastics showed higher total organic carbon (21.21 mg/L) compared to unpolluted water (9.78 mg/L). This suggests that MPs may enhance carbon accumulation in aquatic environments, although the long-term implications remain unclear. In plant-soil systems, MPs influence the allocation of photosynthetically fixed carbon. For example, low concentrations (1%) of PVC MPs have been shown to increase carbon allocation to shoots while slightly altering the root-to-shoot ratio. However, this can also lead to phytotoxic effects that suppress root development. At higher concentrations (1%–5%), MPs improve soil aeration, water retention, and reduce bulk density, which can mitigate root stress and enhance carbon allocation to roots. This root-associated carbon, known as rhizodeposition, is vital for supporting symbiotic fungi that contribute to carbon sequestration and soil health. The type, size, and concentration of MPs, along with soil characteristics and environmental conditions, all influence how MPs affect carbon cycling. For instance, while some MPs may stimulate microbial activity and carbon turnover, others may inhibit microbial decomposition or alter root-fungal interactions, which are essential for long-term carbon storage.

In summary, MPs can both disrupt and enhance aspects of the carbon cycle, depending on their properties and environmental context. Their influence on soil structure, microbial function, and plant physiology underscores the need for further research to fully understand their role in global carbon dynamics and ecosystem stability.

3.5.4 Effect of microplastics on the phosphorus cycle

Phosphorus is a vital nutrient for all living organisms, playing a central role in energy transfer, cellular function, and ecosystem productivity. However, the intrusion of MPs into soil environments has been shown to disrupt phosphorus cycling in complex ways:

- **Reduction in Total Phosphorus**: Studies have found that MPs particularly those derived from polyvinyl chloride (PVC) and polystyrene (PS) can reduce total phosphorus levels in soil. This may be due to their interference with microbial activity and nutrient binding processes.
- **Enhanced Phosphorus Availability in Acidic Soils**: Interestingly, PVC MPs containing plasticizers have been shown to increase phosphorus accessibility in acidic red soils. This enhancement is linked to increased inorganic phosphorus solubilization and organic phosphorus mineralization, which boosts the pool of bioavailable phosphorus.
- **Microbial Mediation**: The presence of MPs can alter microbial communities and enzyme activities that regulate phosphorus transformations. These changes may either inhibit or promote phosphorus cycling depending on the type, concentration, and chemical composition of the MPs, as well as local soil conditions.

3.5.5 Strategic management and future recommendations

Microplastic pollution is a global environmental challenge that requires coordinated scientific, political, and societal responses. Effective management strategies should focus on prevention, remediation, and innovation.

Elimination of MPs from Personal Care Product's: Microbeads in products like toothpaste, facial scrubs, and body washes are a direct source of MPs. Countries such as Canada, Australia, and the United States have initiated bans on these products, with the U.S. implementing a full ban in 2017. Consumer awareness and regulatory enforcement are key to reducing this source.

Adoption of Biodegradable Plastics: Bioplastics, which degrade into harmless compounds through microbial and enzymatic action, offer a promising alternative to conventional plastics. These materials can reduce the persistence of MPs in the environment and support circular economy models.

Enhanced Recycling and Wastewater Treatment: Improving solid waste management infrastructure-including recycling, energy recovery, and synthetic crude production-can significantly reduce plastic leakage into aquatic systems. Upgrading wastewater treatment plants with advanced filtration technologies is essential to intercept MPs before they reach rivers and oceans.

Development of Cleanup and Bioremediation Technologies: Recent research has identified microbial strains capable of degrading common plastics:

- *Enterobacter asburiae* and *Exiguobacterium* (from insect guts) show promise in breaking down PE and PS.
- *Ideonella sakaiensis* 201-F6 produces enzymes that convert PET into environmentally benign monomers-terephthalic acid and ethylene glycol.

These biological solutions could revolutionize plastic waste management by enabling eco-friendly degradation pathways.

3.6 Impact of micro- and nano plastics on ecosystem services and climate change

The widespread use of plastics-evidenced by the purchase of over one million plastic bottles every minute-has escalated into a global environmental crisis with significant implications for climate change. Microplastics (MPs) and nano plastics (NPs) are increasingly recognized as pollutants that disrupt ecosystem services across terrestrial, aquatic, and atmospheric domains. In polar regions, particularly the Southern Ocean, MPs have been found embedded in sea ice, where they correlate positively with chlorophyll levels, suggesting that biological activity contributes to their deposition. These particles are seasonally trapped and released with ice formation and melting, making them more accessible to marine organisms and increasing the likelihood of ingestion.

MPs and NPs interfere with ecosystem functions through physical and biological pathways. In terrestrial ecosystems, they alter soil structure, reduce microbial diversity, and impair nutrient cycling. One of the most critical impacts is on carbon sequestration-a key ecosystem service-where MPs, composed of approximately 80% carbon, complicate the distinction between natural soil organic carbon and plastic-derived carbon. This interference forms a slow decomposing "MPs cycle" that affects soil carbon storage and contributes to global change. Additionally, MPs disrupt nitrogen and phosphorus cycles, leading to oxidative stress and reduced crop productivity, as observed in wheat cultivation. These disruptions threaten food security and may contribute to future agricultural crises. In aquatic environments, MPs accelerate eutrophication and greenhouse gas (GHG) emissions by promoting microbial processes such as mineralization, nitrification, and denitrification, which release CO_2, CH_4, and N_2O. The accumulation of MPs also stimulates algal growth and organic contamination, further destabilizing aquatic food webs. The petrochemical industry's massive plastic output exacerbates climate instability, challenging global efforts to limit temperature rise to 1.5–2°C as outlined in the Paris Agreement.

Marine ecosystems are particularly vulnerable, as MPs affect phytoplankton photosynthesis, zooplankton development, and the marine carbon pump. These disruptions compromise the ocean's role as a carbon sink, hindering its ability to sequester CO_2 and regulate climate. MPs also influence solar radiation dynamics in ocean surface layers, triggering climate feedback mechanisms. The "plastisphere"-microbial communities living on plastic surfaces-contributes to GHG emissions through microbial metabolism, yet the extent of its impact on nutrient cycling and microbial diversity remains poorly understood.

Overall, MPs and NPs pose multifaceted threats to ecosystem services and climate regulation. Their persistence and interaction with biogeochemical processes demand urgent research to quantify their long-term effects and inform mitigation strategies that balance human well-being with ecological integrity.

3.6.1 Ecosystem, climate, and human health implications of micro- and nanoplastics

Microplastics (MPs) and nano plastics (NPs) are increasingly recognized as pervasive pollutants that disrupt ecosystem services and contribute to climate change. In terrestrial ecosystems, these particles alter soil temperature dynamics, particularly in agricultural regions facing extreme climatic conditions. Fluctuations in topsoil and subsoil temperatures affect decomposition rates, nutrient cycling, and the survival of soil-dwelling organisms. Such changes can lead to severe consequences including soil erosion, forest fires, and desertification, resulting in biodiversity loss and degradation of ecosystem services. MPs and NPs also interfere with the reproductive and feeding behaviors of soil fauna, such as arthropods and reptiles, by influencing soil temperature and physicochemical conditions. For instance, soil temperature plays a critical role in egg hatching and sex determination in reptiles, meaning that plastic pollution can directly impact population dynamics and species survival.

In aquatic ecosystems, the intrusion of MPs has reached alarming levels, with annual estimates ranging from 4.8 to 12.7 million metric tons. These particles act as stressors, altering ocean temperature, acidification, and circulation patterns. Their presence in sea ice and surface waters affects marine species differently, with pelagic organisms ingesting more MPs than benthic species. This ingestion leads to biomagnification, genetic and phenotypic changes in aquatic invertebrates, and compromised resilience to climate-induced stressors. MPs also exacerbate eutrophication by promoting microbial processes that release greenhouse gases such as CO_2, CH_4, and N_2O. These emissions contribute to global warming and disrupt the marine carbon pump, which is essential for regulating atmospheric CO_2. Additionally, MPs influence solar radiation penetration in ocean layers, triggering climate feedback mechanisms and altering biogeochemical cycles.

The economic impact of MPs pollution is substantial, with estimated losses in ecosystem services ranging from 1%–5%, equivalent to approximately US$2.5 trillion. MPs and NPs act as vectors for toxic chemicals and invasive microbes, leading to shifts in community structure, biodiversity loss, and potential disease outbreaks. Their ability to breach biological systems and transfer through food chains raises concerns about future pandemics linked to plastic pollution. Atmospheric MPs and NPs also pose threats to pollination. Their size similarity to pollen grains allows them to obstruct pollination processes, particularly in tropical regions rich in endemic plant species. This interference can reduce seed banks and increase the risk of local extinctions. Inhalation of airborne MPs has been linked to respiratory issues, cancer, and other health complications, especially in regions with limited healthcare infrastructure. Overall, MPs and NPs represent a triad of environmental stressors-alongside climate change and overfishing-that collectively undermine marine ecosystems and human well-being. Their widespread presence, ecological toxicity, and potential to disrupt public health systems highlight the urgent need for global action and further research to mitigate their impacts and safeguard sustainable development goals.

3.6.2 Plastic waste management: Socio-economic impacts and community engagement

Plastic pollution poses significant socio-economic challenges, particularly in coastal and marine environments. Marine organisms often ingest or become entangled in plastic debris, which reduces the efficiency of commercial fisheries and aquaculture operations, ultimately threatening fish stocks and disrupting the broader food web. This is particularly concerning given that seafood provides over 20% of the animal protein intake for approximately 1.4 billion people globally-nearly 19% of the world's population (Mouat et al., 2010). Beyond ecological impacts, plastic waste also affects human livelihoods and well-being. Coastal tourism, a vital economic sector for many countries, suffers when beaches are littered with marine debris. Tourists tend to avoid polluted shorelines, leading to reduced tourism revenue and broader socio-economic instability.

Urbanization and increased plastic consumption have overwhelmed waste management systems, especially in developing countries. Cultural, environmental, and institutional factors influence how waste is generated and managed. In wealthier nations, waste is typically managed at the municipal or regional level, with higher per capita waste generation. In contrast, developing countries often rely on biogenic materials and generate less waste, but face challenges due to limited infrastructure and informal waste sectors. Integrating informal waste workers, improving data collection, and promoting sustainable consumption are essential for effective waste governance (UNEP, 2018).

Country-specific examples highlight the diversity in waste management practices. South Africa successfully disposes of 85% of its waste in regulated sites, while Thailand, a major plastic producer, generates 29 kg of plastic per capita annually-double the global average-yet struggles with waste management. Cyprus demonstrates strong performance, with only 7% of waste mismanaged. Mozambique, however, faces a critical situation: 99% of its plastic waste is untreated, and only 1% is recycled. With no domestic plastic production, the country imports all plastic products and releases an estimated 17 kilotons of plastic into waterways annually. Menorca stands out for its relatively effective waste management, collecting 90% of its plastic waste despite high per capita generation (111 kg/year), largely driven by tourism. In contrast, Vietnam's rapid industrial growth has led to excessive plastic stockpiling and open burning, which releases harmful pollutants like dioxins and contributes to climate change. Kenya, which imports 98% of its plastic, also lacks proper waste collection and disposal infrastructure, exacerbating environmental and public health risks. These examples underscore the urgent need for tailored, community-driven strategies to manage plastic waste. Solutions must address local socio-economic conditions, improve infrastructure, and promote public awareness to mitigate the environmental and health impacts of plastic pollution.

3.6.3 Role of circular economy and life cycle assessment in managing plastic products

The circular economy (CE) offers a transformative approach to managing plastic products by shifting away from the traditional linear model of "take, make, dispose" toward a system that emphasizes reducing, reusing, recycling, and recovering materials throughout production, distribution, and consumption (Fig. 3.4). This model seeks to eliminate the concept of "end-of-life" by continuously cycling resources back into the economy, thereby minimizing waste and environmental impact. CE operates across multiple scales: at the micro-level (individual products, companies, and consumers), meso-level (eco-industrial parks), and macro-level (cities, regions, nations, and globally), promoting sustainable development through environmental stewardship, economic resilience, and social equity.

Complementing CE, Life Cycle Assessment (LCA) is a scientific tool used to evaluate the environmental impacts and resource use associated with a product throughout its entire life cycle-from raw material extraction to manufacturing, usage, reuse, and final disposal. For plastic products, LCA identifies five key stages: sourcing raw materials for plastic pellets, manufacturing, usage/reuse,

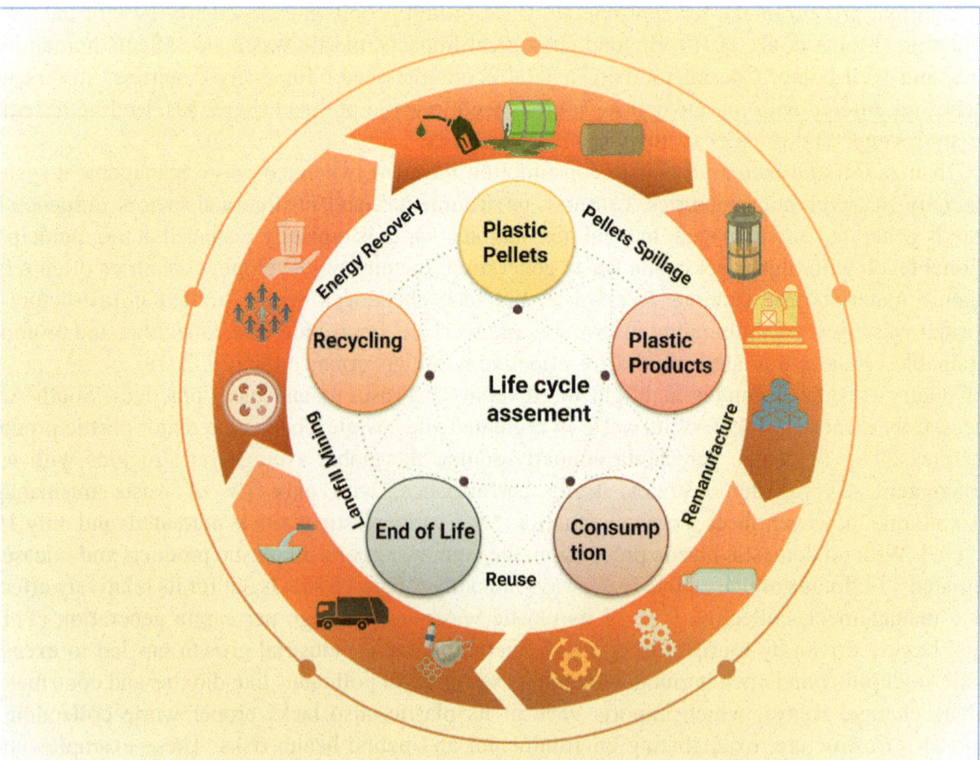

FIGURE 3.4

Conceptual framework highlighting life cycle assessment and circular economy principles in plastic product management.

end-of-life processing, and environmental discharge. Each stage is analyzed for its contribution to macro- and microplastic waste generation. The primary goal of LCA is to assess the "cradle-to-grave" footprint of plastic products, enabling informed decisions that reduce environmental harm. Together, CE and LCA provide a comprehensive framework for understanding and mitigating the environmental consequences of plastic production and consumption. By recognizing the value lost at the end-of-life stage-whether through wasted resources, underutilized capacity, or embedded energy-these approaches help redirect materials back into productive use, fostering a more sustainable and circular plastic economy.

3.6.4 Circular economy and life cycle assessment in plastic end-of-life management

At the end of their lifecycle, plastic products are typically landfilled, incinerated, or recycled. Globally, between 22% and 43% of plastic waste ends up in landfills-a reflection of the prevailing linear economy model. In India, approximately 65% of plastic waste is landfilled, while the European Union reports a lower figure of 24.9% (Gourmelon, 2015). In the United States, plastic waste in landfills increased from 20 million tons in 2000 to 27 million tons by 2018. Landfilling, although common, is the least preferred option due to its environmental drawbacks, including land use, groundwater contamination, and the long-term persistence of plastics, which can take thousands of years to degrade. Landfill mining has emerged as a strategy to recover resources, but most plastics retrieved are too degraded for recycling and pose risks such as soil infertility and air pollution from accidental burning (Sarker, 2011). Incineration is another method used to manage plastic waste, particularly in cases where land availability is limited. Around 25% of plastic waste in India is incinerated, often to reduce landfill volume. While incineration can generate energy, it is costly and environmentally contentious due to emissions. However, advanced incineration facilities can convert waste into energy and repurpose the resulting ash for construction materials. Several European countries, including Switzerland, Austria, and Denmark, have significantly reduced landfilling by diverting over 95% of plastic waste to recycling or energy recovery (Steensgaard et al., 2017).

Recycling remains a critical yet underutilized solution. Globally, only about 10% of plastic waste is recycled. In the U.S., recycling rates have stagnated around 8.7%, while India reports a higher rate of 60% for recyclable plastics. The recyclability of plastics depends on factors such as the presence of additives and contamination. PET and HDPE bottles are among the most recycled plastics due to their high demand and ease of processing. In Europe, 75.1% of collected plastic waste is recovered 42.6% through recycling and 32.5% through energy recovery (Plastics Europe, 2020). However, complex materials like tires and multilayer packaging remain difficult to recycle. Circular economy (CE) strategies offer alternatives, such as reusing tires in construction or as fuel due to their high calorific value. Bio-based plastics, derived from renewable sources like corn and sugarcane, present a promising alternative. These materials can be composted or processed through anaerobic digestion, chemical recycling, or waste-to-energy conversion. Although bio-based plastics have a slightly higher lifecycle cost-about 11% more than conventional plastics-they offer significant environmental benefits, including reduced greenhouse gas emissions and lower non-renewable energy use (Blanc et al., 2019).

Life Cycle Assessment (LCA) studies confirm that bioplastics have a 12% lower global warming potential compared to traditional plastics. Innovative approaches such as microbial degradation and mechano-biocatalytic process are being explored to enhance plastic breakdown and valorization.

These methods use bacteria and enzymes to depolymerize plastics breakdown and valuable raw materials, supporting the CE model. Accidental discharge of plastic waste into the environment remains a concern, with studies reporting 23,150 tons of plastic debris in the Mediterranean Sea and 8% of plastic bags in the EU ending up in oceans-79% of which are polyethylene (Sundt et al., 2014).

The transition from a linear to a circular economy is essential to address the environmental and economic challenges posed by plastic waste. CE promotes the 5R principles-reduce, reuse, recycle, recover, and redesign-aiming to minimize waste and extend the lifecycle of materials. Countries like China, Japan, the UK, and Sweden have adopted CE frameworks to reduce emissions and promote sustainable development. Green chemistry complements CE by eliminating toxic substance from plastic production and encouraging eco-friendly alternatives to single-use plastics. Together, CE and LCA provide a roadmap for reducing micro and nano plastics pollution, conserving resources, and mitigating climate change.

3.7 Summary and conclusions

Plastic pollution presents a multifunctional global challenge with far-reaching ecological, environmental, and socio-economic consequences. Its impacts are deeply intertwined with climate change and the achievement of the United Nations Sustainable Development Goals (SDGs). Despite the existence of plastic waste management policies in many countries, enforcement and implementation remain inconsistent and insufficient. This underscores the urgent need for a paradigm shift from the traditional linear economy-where products are used and discarded-to a circular economy (CE) model that emphasizes reduction, reuse, recycling, and recovery of materials. Life cycle assessment (LCA) plays a critical role in evaluating the environmental impacts of plastic products from production of disposal, offering insights that can inform more sustainable practices. Integrating CE and LCA approaches can help close the loop on plastic use, reduce waste generation, and minimize the release of microplastic and nano plastics into the environment.

Given the pervasive nature of plastic pollution, it must be addressed with the same urgency as other global crises such as climate change and pandemics. The SDGs offer a comprehensive framework to tackle plastic-related issues, particularly in areas like food security, public health, and sustainable economic development. To effective combat monitoring systems, technological innovation, and international collaboration-including the active participation of citizen science.

Promoting research into biodegradable alternatives, advancing recycling technologies, and fostering innovation in material science are essential steps forward. Equally important is the empowerment and education of communities to adopt sustainable behaviors and reduce reliance on single-use plastics. A collective, informed, and proactive approach is vital to mitigate the impacts of plastics pollution and ensure a healthier, more sustainable future for both people and the planet.

References

Akhbarizadeh, R., Moore, F., Keshavarzi, B., 2019. Investigating microplastics bioaccumulation and biomagnification in seafood from the Persian Gulf: a threat to human health? Food Addit. Contam. 36 (11), 1696–1708. https://doi.org/10.1080/19440049.2019.1649473.

Allen, S., Allen, D., Moss, K., Roux, G.L., Phoenix, V.R., Sonke, J.E., 2020. Examination of the ocean as a source for atmospheric microplastics. PLoS One 15 (5), e0232746. https://doi.org/10.1371/journal.pone.0232746.

Arat, S.A., 2024. An overview of microplastic in marine waters: sources, abundance, characteristics and negative effects on various marine organisms. Desalin. Water Treat. 317, 100138. https://doi.org/10.1016/j.dwt.2024.100138.

Barboza, L.G.A., Vethaak, A.D., Lavorante, B.R., Lundebye, A., Guilhermino, L., 2018. Marine microplastic debris: an emerging issue for food security, food safety and human health. Mar. Pollut. Bull. 133, 336–348. https://doi.org/10.1016/j. marpolbul.2018.05.047.

Besseling, E., Foekema, E.M., Van Franeker, J.A., Leopold, M.F., Kühn, S., Rebolledo, E.L.B., Heße, E., Mielke, L., Ijzer, J., Kamminga, P., Koelmans, A.A., 2015. Microplastic in a macro filter feeder: humpback whale Megaptera novaeangliae. Mar. Pollut. Bull. 95 (1), 248–252. https://doi.org/10.1016/j.marpolbul.2015.04.007.

Bhatt, V., Chauhan, J.S., 2023. Microplastic in freshwater ecosystem: bioaccumulation, trophic transfer, and biomagnifcation. Environ. Sci. Pollut. Res. 30 (4), 9389–9400. https://doi.org/10.1007/s11356-022-24529-w.

Blanc, S., Massaglia, S., Brun, F., Peano, C., Mosso, A., Giuggioli, N.R., 2019. Use of bio-based plastics in the fruit supply chain: an integrated approach to assess environmental, economic, and social sustainability. Sustainability 11, 2475.

Catarino, A.I., Macchia, V., Sanderson, W.G., Thompson, R.C., Henry, T.B., 2018. Low levels of microplastics (MP) in wild mussels indicate that MP ingestion by humans is minimal compared to exposure via household fibres fallout during a meal. Environ. Pollut. 237, 675–684. https://doi.org/10.1016/j.envpol.2018.02.069.

Chen, E.Y., Lin, K.T., Jung, C.C., Chang, C.L., Chen, C.Y., 2022. Characteristics and influencing factors of airborne microplastics in nail salons. Sci. Total Environ. 806, 151472. https://doi.org/10.1016/j.scitotenv.2021.151472.

Cole, M., Lindeque, P.K., Fileman, E., Clark, J., Lewis, C., Halsband, C., Galloway, T.S., 2016. Microplastics Alter the properties and sinking rates of zooplankton Faecal pellets. Environ. Sci. Technol. 50 (6), 3239–3246. https://doi.org/10.1021/acs. est.5b05905.

Curren, E., Kuwahara, V.S., Yoshida, T., Leong, S.C.Y., 2021. Marine microplastics in the ASEAN region: a review of the current state of knowledge. Environ. Pollut. 288, 117776. https://doi.org/10.1016/j.envpol.2021.117776.

De a Miranda, D., De Carvalho-Souza, G.F., 2016. Are we eating plastic-ingesting fish? Mar. Pollut. Bull. 103 (1–2), 109–114. https://doi.org/10.1016/j. marpolbul.2015.12.035.

Di Beneditto, A.P.M., Ramos, R.M.A., 2014. Marine debris ingestion by coastal dolphins: what drives differences between sympatric species? Mar. Pollut. Bull. 83 (1), 298–301. https://doi.org/10.1016/j.marpolbul.2014.03.057.

Ding, R., Li, Q., Wang, K., Tian, J., Lu, L., Li, W., Xu, L., 2024. Occurrence and distribution of microplastics in the adjacent environment of Yellow River Delta, China. Mar. Pollut. Bull. 199, 116019. https://doi.org/10.1016/j.marpolbul.2023.116019.

Fang, M.M., 2023. China's Battle against marine plastic pollution at the local level: a case study of Sanya city, Hainan Province. Ocean Yearb. Online 37 (1), 249–275. https://doi.org/10.1163/22116001-03701013.

Farrell, P., Nelson, K., 2013. Trophic level transfer of microplastic: *Mytilus edulis* (L.) to *Carcinus maenas* (L.). Environ. Pollut. 177, 1–3. https://doi.org/10.1016/j.envpol.2013.01.046.

Frohlich, E., Meindl, C., Wagner, K., Leitinger, G., Roblegg, E., 2014. Use of whole genome expression analysis in the toxicity screening of nanoparticles. Toxicol. Appl. Pharmacol. 280 (2), 272–284. https://doi.org/10.1016/j.taap.2014.07.017.

Ghosh, S., Sinha, J.K., Ghosh, S., Vashisth, K., Han, S., Bhaskar, R., 2023. Microplastics as an emerging threat to the global environment and human health. Sustainability 15 (14), 10821. https://doi.org/10.3390/su151410821.

Gourmelon, G., 2015. Global plastic production rises, recycling lags. In: Vital Signs: The Trends that Are Shaping Our Future. Worldwatch Institute, Washington, DC, USA, pp. 91–95. https://www.plastic-resource-center.com/wp-content/uploads/2018/11/Global-Plastic-Production-RisesRecycling-Lags.pdf. (Accessed 25 June 2021).

Guo, X., Wang, J., 2019. The chemical behaviours of microplastics in marine environment: a review. Mar. Pollut. Bull. 142, 1–14.

Guven, O., Gokdag, K., Jovanovic, B., Kıdeys, A.E., 2017. Microplastic litter composition of the Turkish territorial waters of the Mediterranean Sea, and its occurrence in the gastrointestinal tract of fish. Environ. Pollut. 223, 286–294. https://doi.org/10.1016/j.envpol.2017.01.025.

Hawke, A.M., Trujillo, J.E., Oey, I., Giteru, S.G., Allan, B.J., 2024. Exposure to petroleum-derived and biopolymer microplastics affect fast start escape performance and aerobic metabolism in a marine fish. Sci. Total Environ. 906, 167423. https://doi.org/10.1016/j.scitotenv.2023.167423.

Hossain, M.S., Rahman, M.S., Uddin, M.N., Sharifuzzaman, S., Chowdhury, S.R., Sarker, S., Chowdhury, M.S.N., 2020. Microplastic contamination in Penaeid shrimp from the northern bay of Bengal. Chemosphere 238, 124688. https://doi.org/10.1016/j.chemosphere.2019.124688.

Hu, C.J., Garcia, M.A., Nihart, A., Liu, R., Yin, L., Adolphi, N., Gallego, D.F., Kang, H., Campen, M.J., Yu, X., 2024. Microplastic presence in dog and human testis and its potential association with sperm count and weights of testis and epididymis. Toxicol. Sci. 200 (2), 235–240. https://doi.org/10.1093/toxsci/kfae060.

Huang, X., Chen, Y., Meng, Y., Liu, G., Yang, M., 2022. Are we ignoring the role of urban forests in intercepting atmospheric microplastics? J. Hazard. Mater. 436, 129096. https://doi.org/10.1016/j.jhazmat.2022.129096.

Jaganmohan, 2024. Annual Production of Plastics Worldwide from 1950 to 2022. https://www.statista.com/statistics/282732/global-production-of-plastics-since-1950/. (Accessed 2 February 2024).

Kahane-Rapport, S.R., Czapanskiy, M.F., Fahlbusch, J.A., Friedlaender, A.S., Calambokidis, J., Hazen, E.L., Goldbogen, J.A., Savoca, M.S., 2022. Field measurements reveal exposure risk to microplastic ingestion by filter-feeding megafauna. Nat. Commun. 13 (1). https://doi.org/10.1038/s41467-022-33334-5.

Kanhai, L.D.K., Gardfeldt, K., Krumpen, T., 2020. Microplastics in sea ice and seawater beneath ice floes from the Arctic Ocean. Sci. Rep. 10, 5004. https://doi.org/10.1038/s41598-020-61948-6.

Li, C., Wang, X., Liu, K., Zhu, L., Wei, N., Zong, C., Li, D., 2021. Pelagic microplastics in surface water of the eastern Indian Ocean during monsoon transition period: abundance, distribution, and characteristics. Sci. Total Environ. 755, 142629. https://doi.org/10.1016/j.scitotenv.2020.142629.

Masry, M., Rossignol, S., Temime Roussel, B., Bourgogne, D., Bussiere, P.O., Rmili, B., Wong-Wah-Chung, P., 2021. Experimental evidence of plastic particles transfer at the water-air interface through bubble bursting. Environ. Pollut. 280, 116949. https://doi.org/10.1016/j.envpol.2021.116949.

Miller, M.E., Hamann, M., Kroon, F.J., 2020. Bioaccumulation and biomagnification of microplastics in marine organisms: a review and meta-analysis of current data. PLoS One 15 (10), e0240792. https://doi.org/10.1371/journal.pone.0240792.

Miller, M.E., Motti, C.A., Hamann, M., Kroon, F.J., 2023. Assessment of microplastic bioconcentration, bioaccumulation and biomagnification in a simple coral reef food web. Sci. Total Environ. 858, 159615. https://doi.org/10.1016/j. scitotenv.2022.159615.

Mouat, J., Lozano, R.L., Bateson, H., 2010. Economic Impacts of Marine Litter; Kommunenes Internasjonale Miljøorganisasjon: Esbjerg, Denmark.

Naidoo, T., Glassom, D., 2019. Decreased growth and survival in small juvenile fish, after chronic exposure to environmentally relevant concentrations of microplastic. Mar. Pollut. Bull. 145, 254–259. https://doi.org/10.1016/j.marpolbul.2019.02.037.

Nakano, H., Alfonso, M.B., Jandang, S., Imai, K., Arakawa, H., 2024. Microplastic pollution indexes in the coastal and open ocean areas around Japan. Reg. Stud. Mar. Sci. 69, 103287. https://doi.org/10.1016/j.rsma.2023.103287.

Nelms, S.E., Barnett, J., Brownlow, A., Davison, N.J., Deaville, R., Galloway, T.S., Lindeque, P.K., Santillo, D., Godley, B.J., 2019. Microplastics in marine mammals stranded around the British coast: ubiquitous but transitory? Sci. Rep. 9 (1). https://doi.org/10.1038/s41598-018-37428-3.

Nizzetto, L., Langaas, S., Futter, M., 2016. Pollution: do microplastics spill on to farm soils? Nature 537 (7621), 488. https://doi.org/10.1038/537488b.

Nolte, T.M., Hartmann, N.B., Kleijn, J.M., Garnæs, J., van de Meent, D., Jan Hendriks, A., et al., 2017. The toxicity of plastic nanoparticles to green algae as infuenced by surface modifcation, medium hardness and cellular adsorption. Aquat. Toxicol. 2017 (183), 11–20. https://doi.org/10.1016/j.aquatox.2016.12.005.

OECD, 2024. Policies to Reduce Microplastics Pollution in Water. www.oecd.org/en/publications/policies-to-reduce-microplastics-pollution-in-water_7ec7e5ef-e n.html.

Pastorino, P., Anselmi, S., Zanoli, A., Esposito, G., Bondavalli, F., Dondo, A., Pucci, A., Pizzul, E., Faggio, C., Barcelo, D., Renzi, M., Prearo, M., 2023. The invasive red swamp crayfish (*Procambarus clarkii*) as a bio-indicator of microplastic pollution: insights from Lake Candia (northwestern Italy). Ecol. Indic. 150, 110200. https://doi.org/10.1016/j.ecolind.2023.110200.

Peng, X., Jiang, Y., Chen, Z., Osman, A.I., Farghali, M., Rooney, D.W., Yap, P.S., 2023. Recycling municipal, agricultural and industrial waste into energy, fertilizers, foodand construction materials, and economic feasibility: a review. Environ. Chem. Lett. 21 (2), 765–801. https://doi.org/10.1007/s10311-022-01551-5.

Pereira, L.G., Ferreira, G.V., Justino, A.K., De Oliveira, K.M.T., De Queiroz, M.T., Schmidt, N., Fauvelle, V., Carvalho, V.L., Lucena-Fredou, F., 2023. Exploring microplastic contamination in Guiana dolphins (Sotalia guianensis): insights into plastic pollution in the southwestern tropical Atlantic. Mar. Pollut. Bull. 194, 115407. https://doi.org/10.1016/j.marpolbul.2023.115407.

Plastic Europe, 2020. Plastics—The Fact of 2019: An Analysis of European Plastics Production, Demand and Waste Data. Available online: https://www.plasticseurope.org/application/files/9715/7129/9584/FINAL_web_version_Plastics_the_facts2019_14102019.pdf. (Accessed 24 May 2021).

Ratnasari, A., Zainiyah, I.F., Hadibarata, T., Yan, L.Y., Sharma, S., Thakur, S.S., 2024. The crucial Nexus of microplastics on ecosystem and climate change: types, source, impacts, and transport. Water Air Soil Pollut. 235 (5). https://doi.org/10.1007/s11270-024-07103-7.

Ryan, A.C., Allen, D., Allen, S., Maselli, V., LeBlanc, A., Kelleher, L., Krause, S., Walker, T.R., Cohen, M., 2023. Transport and deposition of ocean-sourced microplastic particles by a North Atlantic hurricane. Commun. Earth Environ. 4 (1). https://doi.org/10.1038/s43247-023-01115-7.

Saha, S., Chandra Saha, N., Chatterjee, A., Banerjee, P., Garai, P., Sharma, P., Patnaik, L., Nayak, S., Dhara, K., Chukwuka, A.V., Faggio, C., 2023. Integrated multi-biomarker responses in Mozambique tilapia, *Oreochromis mossambicus* under acute and chronic Diazinon® exposures. Chem. Ecol. 39 (3), 235–255. https://doi.org/10.1080/02757540.2023.2178649.

Salomon, J.J., Ehrhardt, C., 2011. Nanoparticles attenuate P-glycoprotein/MDR1 function in A549 human alveolar epithelial cells. Eur. J. Pharm. Biopharm. 77 (3), 392–397. https://doi.org/10.1016/j.ejpb.2010.11.009.

Santos, M.R., Martins, H.R., Bolten, A.B., Bjorndal, K.A., 2017. Plastic ingestion in oceanic-stage loggerhead sea turtles (*Caretta caretta*) off the North Atlantic subtropical gyre. Mar. Pollut. Bull. 121 (1–2), 222–229. https://doi.org/10.1016/j.marpolbul.2017.06.008.

Sarker, M., 2011. Converting waste plastic to hydrocarbon fuel materials. Energy Eng 108, 35–43.

Savoca, S., Bottari, T., Fazio, E., Bonsignore, M., Mancuso, M., Luna, G.M., Romeo, T., D'Urso, L., Capillo, G., Panarello, G., Greco, S., Compagnini, G., Lanteri, G., Crupi, R., Neri, F., Spano, ' N., 2020. Plastics occurrence in juveniles of Engraulis encrasicolus and Sardina pilchardus in the southern Tyrrhenian Sea. Sci. Total Environ. 718, 137457. https://doi.org/10.1016/j.scitotenv.2020.137457.

Savoca, S., Capillo, G., Mancuso, M., Bottari, T., Crupi, R., Branca, C., Romano, V., Faggio, C., D'Angelo, G., Spano, ' N., 2019. Microplastics occurrence in the Tyrrhenian waters and in the gastrointestinal tract of two congener species of seabreams. Environ. Toxicol. Pharmacol. 67, 35–41. https://doi.org/10.1016/j.etap.2019.01.011.

Shahul Hamid, F., Bhatti, M.S., Anuar, N., Anuar, N., Mohan, P., Periathamby, A., 2018. Worldwide distribution and abundance of microplastic: how dire is the situation? Waste Manag. Res. 36 (10), 873–897. https://doi.org/10.1177/0734242x18785730. The Journal for a Sustainable Circular Economy.

Smith, M., Love, D.C., Rochman, C.M., Nef, R.A., 2018. Microplastics in seafood and the implications for human health. Curr. Environ. Health Rep. 5 (3), 375–386.

Steensgaard, I.M., Syberg, K., Rist, S., Hartmann, N.B., Boldrin, A., Hansen, S.F., 2017. From macro-to microplastics-Analysis of EU regulation along the life cycle of plastic bags. Environ. Pollut. 224, 289–299.

Sundt, P., Schulze, P.-E., Syversen, F., December 4, 2014. Sources of Microplastic-Pollution to the Marine Environment. Available online: https://www.miljodirektoratet.no/globalassets/publikasjoner/M321/M321.pdf.

Tang, C.H., Lin, C.Y., Li, H.H., Kuo, F.W., 2024. Microplastics elicit an immune-agitative state in coral. Sci. Total Environ. 908, 168406. https://doi.org/10.1016/j. scitotenv.2023.168406.

UNEP, 2023. Picking up Litter: Pointless Exercise or Powerful Tool in the Battle to Beat Plastic Pollution? Retrieved August 15, 2024, from. https://www.unep.org. https://www.unep.org/news-and-stories/story/picking-litter-pointless-exercise-or-powerful-tool-battle-beat-plastic.

UNEP, 2018. Single-Use Plastics: A Roadmap for Sustainability. United Nations Environment Programme (UNEP), Nairobi, Kenya.

Woodall, L.C., Sanchez-Vidal, A., Canals, M., Thompson, R.C., 2014. The deep sea is a major sink for microplastic debris. R. Soc. Open Sci. 1 (4), 140317. https://doi.org/10.1098/rsos.140317.

Wright, S.L., Kelly, F.J., 2017. Plastic and human health: a Micro issue? Environ. Sci. Technol. 51 (12), 6634–6647. https://doi.org/10.1021/acs.est.7b00423.

Xiang, Y., Jiang, L., Zhou, Y., Luo, Z., Zhi, D., Yang, J., Lam, S.S., 2022. Microplastics and environmental pollutants: key interaction and toxicology in aquatic and soil environments. J. Hazard. Mater. 422, 126843. https://doi.org/10.1016/j.jhazmat.2021.126843.

Yu, Q., Hu, X., Yang, B., Zhang, G., Wang, J., Ling, W., 2020. Distribution, abundance and risks of microplastics in the environment. Chemosphere 249, 126059. https://doi.org/10.1016/j.chemosphere.2020.126059.

Zhang, N., Zhang, C., Qin, Y., Wang, J., Ge, X., Li, H., Dai, Y., Aruffo, E., 2024. A review of atmospheric microplastics: sources, characteristics, and detection method. Curr. Pollut. Rep. https://doi.org/10.1007/s40726-024-00316-8.

Zhou, C., Jiaqi, B., Zhong, S., 2024. Microplastics in water: a review of characterization and removal methods. Sustainability 16 (10), 4033.

CHAPTER

4

Impact of microplastics on biofilm formation, pollutant complexation, and greenhouse gas emissions

Melvin S. Samuel

Department of Civil, Construction, and Environmental Engineering, Marquette University, Milwaukee, WI, United States

Highlights

- Microplastics act as persistent substrates that promote dense microbial biofilm formation
- Plastisphere biofilms enhance the adsorption and transport of organic and inorganic pollutants
- Pollutant-Microplastic complexes increase contaminant bioavailability and ecological risk
- Microbial degradation of Microplastics alters carbon and nitrogen cycling pathways

4.1 Introduction

Microplastics (MPs) are introduced into aquatic environments through multiple pathways, including municipal and industrial effluents, surface runoff, atmospheric deposition, particularly in remote regions and effluent from wastewater treatment facilities (Alvim et al., 2020). Wetlands, due to their high particulate retention capacity, serve as natural filters and are ideal for studying the fate and behavior of MPs. The mobility of MPs in porous media is influenced by environmental conditions; for example, Dong et al. (2021) observed that higher pH, lower ionic strength and elevated humic acid concentrations enhance MP transport. Particle morphology also plays a role; irregularly shaped MPs with jagged edges tend to settle in aquatic substrates, whereas spherical particles remain suspended near the surface (Horton and Dixon, 2018). Hydrodynamic factors such as eddy diffusivity influence the dispersion and spatial distribution of MPs, affecting their aggregation and disintegration patterns. MPs undergo various transport processes, including advection, suspension, dispersion and sedimentation. Their vertical migration is more pronounced in sediments than in soil (Yuan et al., 2022). High flow velocities facilitate long distance transport, while denser polymers like PET exhibit limited mobility and are more prone to sedimentation compared to lighter polymers like PET exhibit limited mobility and are more to sedimentation compared to lighter polymers like PP (Phuong et al., 2022). Integrating hydrodynamic models with statistical simulation can help elucidate the influence of MP's physical properties on their environmental fate.

In remote regions such as the Tibetan Plateau, Dong et al. (2021) estimated atmospheric deposition of MPs to be approximately 3.3 tons during the monsoon season, with glacial and nonglacial runoff contributing 522 and 41 kg, respectively. Annual atmospheric deposition in glacial zones was around 500 kg, with 8% released during melt periods. In tropical regions, deposition rates were significantly

Microplastics in Aquatic Environments. https://doi.org/10.1016/B978-0-443-45619-0.00013-3

higher, particularly in urbanized Southeast Asia, and were not strongly influenced by rainfall or wind patterns (Truong et al., 2021). Once in aquatic systems, MPs serve as substrates for microbial colonization, promoting biofilm formation and facilitating horizontal gene transfer, which alters their surface characteristics. Gong et al. (2019) identified 20 dominant bacterial genera associated with polyethlene (PE) surfaces that act as vectors for human pathogens. These biofilms enhance microbial nutrient uptake and proliferation (Ma et al., 2022). The interaction between MPs and environmental contaminants is governed by surface properties such as crystallinity, functional groups, and specific surface area (You et al., 2021). Aged PE particles, due to increased surface polarity from oxidative weathering, exhibit stronger affinities for heavy metals and persistent organic pollutants like organochlorines and PCBs (Enfrin et al., 2019).

Environmental parameters such as neutral pH, elevated salinity (>3 mg/L), and high dissolved organic matter can alter the surface charge and microbial composition of MPs biofilms, reducing their capacity to adsorb organic pollutants. However, microbial communities within these biofilms can degrade MPs and associated contaminants into greenhouse gases (GHSs) such as CH_4, CO_2, ethylene, N_2O, contributing to climate change (Seeley et al., 2020). The adsorption of PCBs onto MPs biofilm has been shown to enhance GHG emissions by disrupting microbial carbon and nitrogen metabolism in both water columns and sediments (Zhang et al., 2023). Despite growing awareness, comprehensive studies on the ecological impacts of MPs biofilm complex remain limited. While several investigations have highlighted the role of MPs in transporting toxic pollutants (Yu et al., 2024), few have thoroughly examined the synergistic effects of MPs biofilms with cocontaminants. There is a notable gap in understanding the mechanisms by which aging or biodegradation alters MPs chemical transport properties. Furthermore, the absence of standardized methodologies for qualitative and quantitative assessment of MPs biofilms hinders ecological risk evaluations.

This section aims to bridge these gaps by critically analyzing: (a) the mechanisms and environmental drivers of MPs formation, (b) the adsorption behavior of organic and inorganic pollutants on MPs biofilms and their ecological consequences, (c) the role of MPs biofilms in GHG emissions, and (d) their ecotoxicological effects. The insights provided can inform the development of targeted cleanup strategies and policy frameworks for monitoring and mitigating MPs pollution in aquatic environments.

4.2 Microbial colonization on microplastic surfaces

The intricate relationship between MPs and microbial communities has led to the emergence of the term "*Plastisphere*", describing the unique ecological niche formed when microbes colonize MPs surfaces. These biofilms exhibit higher structural uniformity but lower microbial abundance compared to the surrounding environments. The plastisphere is known to harbor antibiotic-resistant genes and mobile genetic elements, largely due to the selective adsorption of antibiotics onto the MPs surface within biofilm (Parveen et al., 2022). This microbial colonization not only alters the physicochemical properties of MPs but also facilitates their biodegradation (Dong et al., 2021). Biofilm formation on MPs occurs through a mutistage process. Initially, microbial attachment is mediated by physicochemical interaction. Lifshitz van der Waals interactions initiate contact at distances greater than 20 nm, followed by electrostatic interactions at $\leq$ 20 nm, and finally, strong adhesion at 5 nm via specific microbial surface receptors. This adhesion progresses from a reversible phase where microbes reach MPs via Brownian motion and hydrophobic or electrostatic forces to an irreversible phase where microbes reach MPs via Brownian motion and hydrophobic or electrostatic forces, to an irreversible phase involving ionic, covalent, and hydrogen bonding. Bacterial flagella and extracellular polymeric substances (EPS) help overcome energy barriers to establish stable adhesion.

Once attached, microbes secrete EPS composed of carbohydrates, lipids, proteins, nucleic acids, and polysaccharides. The composition of EPS varies with bacterial species, MPs characteristics, and environmental conditions. EPS can be categorized into soluble forms comprising macromolecules and colloids, and bound forms, which include capsular polymers that shield microbes from stressors like shear forces, abrasion, and UV degradation. EPS-coated MPs become sticky, promoting hetero-aggregate formation and rapid adsorption of chemicals and microbes (Michels et al., 2018). As biofilms mature, they transition from two-dimensional to three-dimensional structures, incorporating organic and inorganic debris. This morphological evolution enhances microbial resilience, reduces cell detachment, and facilitates quorum sensing for intercellular communication. Notably, the microbial composition of MPs-biofilms differs significantly from natural aquatic matrices, showing distinct diversity, richness, and functional traits. Biofilm development is accelerated under high nutrient availability and optimized flow conditions that enhance mass transfer.

Cyanobacteria dominate the microbial communities of MPs, while Actinobacteria are more prevalent in surrounding waters. Early colonizers such as Bacteriodetes and Proteobacteria play key roles in EPS production. Gong et al. (2019) identified that half of the top 20 bacterial genera on low-density polyethene MPs were pathogenic, including Chryseobacterium (nosocomial), Agrobacterium (plant), and Flavobacterium (Fish pathogens)(Tu et al., 2020). observed diverse bacterial morphologies, rod, coccus, disc, and filamentous, on MPs surfaces, with biofilm accumulation increasing over time but decreasing with depth. Biofilm formation modifies MPs by reducing hydrophobicity and increasing hydrophilic functional groups like C=O and C–O, indicating the involvement of aromatic and aliphatic compounds in degradation pathways. Structural changes such as cracks, pores, and grooves induced by biofilms further increase MPs' surface area and density (Guan et al., 2020). This complex interplay between MPs and microbial biofilms underscores the need for deeper investigation into their ecological roles, especially in pollutant adsorption, biodegradation, and potential impacts on aquatic ecosystems.

4.2.1 Drivers of MP biofilm formation and their ecological implications

The formation of biofilms on MPs is governed by a complex interplay of factors related to the physical characteristics of the MPs, environmental conditions, and microbial traits. Environmental parameters contribute approximately 31% to biofilm development, while MPs' intrinsic properties account for around 16%, with microbial characteristics comprising the remainder of the influence. Among MP features, shape, color, polymer type, surface chemistry, hydrophilicity, texture, and particle size are critical in determining microbial adhesion and biofilm structure. Hydrodynamic conditions significantly affect biofilm formation. High water flux may inhibit microbial attachment, whereas elevated flow rates promote the development of thinner, denser biofilms. Seasonal temperature fluctuations also play a role; summer conditions enhance microbial diversity and biomass on MPs compared to ambient water. Biofilm formation by *Staphylococcus aureus* and *Pseudomonas aeruginosa* is modulated by oxidative stress enzymes under temperature induced MPs deformation. Jin et al. (2020) reported increased bacterial activity at 31.5 ± 1.07 in summer, which declined to 13.3 ± 2.49 °C in winter, highlighting seasonal variability in biofilm composition.

Environmental factors such as dissolved oxygen and light exposure further influence microbial metabolism and enzyme activity. Nutrient availability, particularly carbon, nitrogen, and phosphorous essential for biofilm growth. While unstable carbon sources lead to sparse biofilm formation, nitrogen and phosphorus concentrations are positively correlated with biofilm development. Salinity generally inhibits biofilm growth but can enhance colonization by halophilic bacteria like *Vibrio* spp., while proliferating 2–10 times more in saline conditions (>10 mg/L) (Li et al., 2019). pH also modulates microbial interactions, with different bacteria exhibiting preferences for acidic, neutral, or alkaline environments (Ramsperger et al., 2020). MPs particle size and density are additional determinants. Smaller particles (75 μm) support more extensive biofilm coverage and heteroaggregate formation than larger particles (4 mm). MPs with densities greater than water, such as PET and PVC, tend to sink more rapidly after biofilm formation due to altered buoyancy (Ayush et al., 2022). Surface chemistry also matters; aromatic functional groups can enhance bacterial adhesion but may hinder biofilm maturation. Color influences microbial colonization, blue-colored MPs exhibit higher functional group diversity and lower *Chloroflexi* abundance compared to yellow or transparent MPs (Zhang et al., 2021a,b). *Pseudomonas aeruginosa* shows a strong affinity for fluorescent blue MPs, producing 1.6–2.4 times more protein-rich biomass on blue or white MPs (Fajardo et al., 2023). Microbial community composition within biofilms affects MPs vertical transport in aquatic systems. MPs colonized by *Alcanivorax* sink due to biosurfactant production, whereas those associated with *Synechococcus* and *Anabaena* remain suspended (Pete et al., 2023). Overall, biofilm formation alters MPs physicochemical properties, influencing their interactions with organic and inorganic pollutants and affecting their bioavailability and ecological impact in aquatic environments.

4.2.2 Microplastics as emerging reservoirs for antibiotic resistance genes

MPs are increasingly acknowledged as efficient vectors for pathogenic microorganisms in aquatic environments. When these particles are ingested by aquatic organisms, the microbes adhere to their subsequent consequences for human health through bioaccumulation and biomagnification along trophic chains (Junaid et al., 2022). Concurrently, MPs have gained attention as important reservoirs and dissemination platforms for antibiotic resistance genes (ARGs) (Jia et al., 2024). For instance, Silva et al. demonstrated that antibiotic-resistant bacteria (ARB) are present at markedly higher abundance on MPs compared with many natural substrates cooccurring in the same aquatic system (Silva et al., 2023). In addition to their microbial association, MPs and their degradation products have been linked to carcinogenic, mutagenic, and reproductive toxicity outcomes, and MP particles have been identified in multiple human tissues, including liver, kidney, and testes (La Porta et al., 2023). In aquatic organisms, ingestion of MPs has likewise been associated with gastrointestinal damage and impaired gut functions. Moreover, MPs can serve as a cotransport agent for chemical pollutants by interacting with hydrophobic contaminants (PAHs), and organochlorine pesticides (e.g., DDT), thereby intensifying ecological risk in marine systems. Notably, MPs also create a novel ecological niche for microorganisms (Yu et al., 2022). In freshwater, estuarine, and marine environments, a wide range of microbial assemblages has consistently been observed colonizing MP surfaces. Microorganisms often exhibit targeted movement toward nutrient-rich substrates and rapidly adhere to MPs, initiating the development of biofilms. Once established, biofilms enhance microbial access to nutrients and provide protection against environmental stressors, leading to microbial enrichment on MP surfaces (Chen et al., 2021). Biofilm formation can further modify the buoyancy and density of particles, thereby altering their transport and fate within the water column. Additionally, extracellular enzymes produced by plastisphere-associated microbes may fragment MPs into smaller oligomeric units and alter surface chemical characteristics. Given that nutrients are frequently limited in aquatic ecosystems, increasing attention has been directed toward the hypothesis that microorganisms may exploit plastics as potential carbon and energy sources. In theory, complete mineralization would result in the production of CO_2, H_2O, and microbial biomass, offering an appealing long-term mitigation strategy (Chen et al., 2021). Nevertheless, despite encouraging laboratory-scale findings, the underlying mechanisms and environmentally relevant rate of MP biodegradation remain poorly constrained.

Considering the magnitude of global plastic contamination, there is an urgent demand to assess both engineering-based and biologically driven remediation strategies. Emerging technologies-including membrane bioreactors, synthetic biology approaches, organosilane-based treatments, biofilm-assisted remediation, and nanomaterial-enhanced capture methods-have shown promise in improving MP removal efficiencies (Ali et al., 2024). At the same time, the microbial aspect of MP pollution in aquatic systems remains far less explored than that in terrestrial environments. Accordingly, this review emphasizes:

- The effects of MPs on aquatic microorganisms at the cellular and community scale.
- The microbial potential for MP biodegradation and biomineralization; and
- Ecosystem-specific interactions between MPs and microbial communities.

4.3 Interactions between microplastics and microorganism

MPs and microorganisms exhibit bidirectional and closely interconnected interactions (Fig. 4.1). MPs can influence microbial physiology, alter community composition, and affect ecosystem-level processes. On the other hand, microorganisms can modify MP characteristics through biofilm formation, the release of metabolic products, and enzymatic transformations. Exposure to MPs has been associated with both inhibitory and stimulatory effects across bacteria, fungi, and algae (Sourkova et al., 2021). MPs can provide stable substrates that facilitate biofilm development and selectively enrich specific microbial taxa, while certain microorganisms are able to fragment MPs into smaller particles, including nanoplastics (NPs) and, in some cases metabolize plastic-derived carbon via specialized extracellular and intracellular enzymatic pathways. Furthermore, microbial metabolites such as proteins, lipids, and polysaccharides may absorb onto the plastic surface, forming an "eco-corona" that alters particle surface properties, biocompatibility, and toxicity.

The following sections synthesize recent findings regarding (i) the impacts of MPs on microbial activity and (ii) microbial-driven modification of MP physicochemical behavior.

4.3.1 Influence of microplastics on microorganisms

4.3.1.1 Effects on bacterial and fungal cells

MPs can affect microbial cells through both beneficial and detrimental mechanisms (Fig. 4.2). One of the most common initial responses is enhanced secretion of extracellular polymeric substance (EPS).

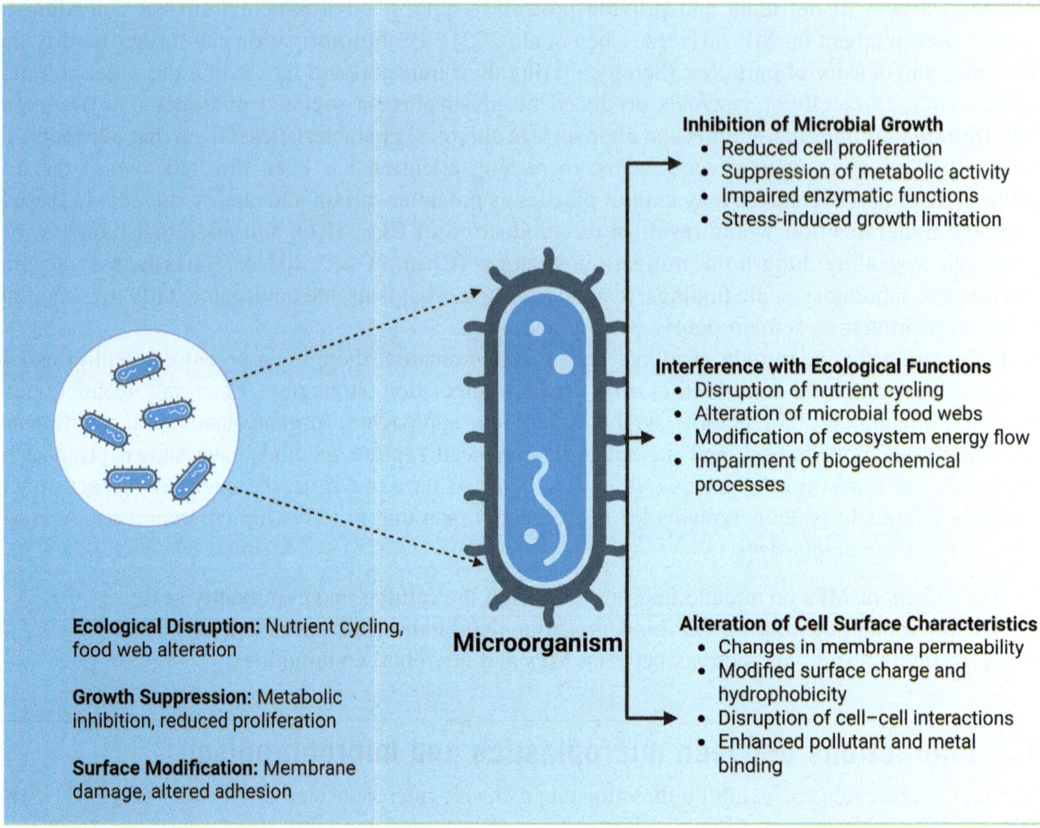

FIGURE 4.1

Cellular impact of Microplastics on microorganisms.

Upon attachment to MP surfaces, microorganisms increase EPS production, which reinforces adhesion and facilitates biofilm maturation. Well-developed biofilms can confer advantages by improving colonization efficiency, increasing nutrient acquisition, and offering partial protection from external stressors, including antibiotic exposure (Rather et al., 2021). In many environments, extracellular proteins constitute a substantial component of EPS and appear to be closely linked to microbial responses following MP exposure (Wu et al., 2023). Conversely, MPs may compromise microbial structural integrity. Numerous studies suggest that MPs-particularly smaller-sized particles, can disrupt cell envelopes and damage membrane architecture, potentially resulting in functional impairment and cell mortality. Elevated membrane permeability and other indicators of cellular damage have been documented across a broad range of microbial taxa (Liu et al., 2021). Additionally, because MPs are capable of accumulating environmental contaminants, their internalization by microbial cells may promote the simultaneous delivery of sorbed pollutants, leading to additive or synergistic toxic effects (Zong et al., 2024).

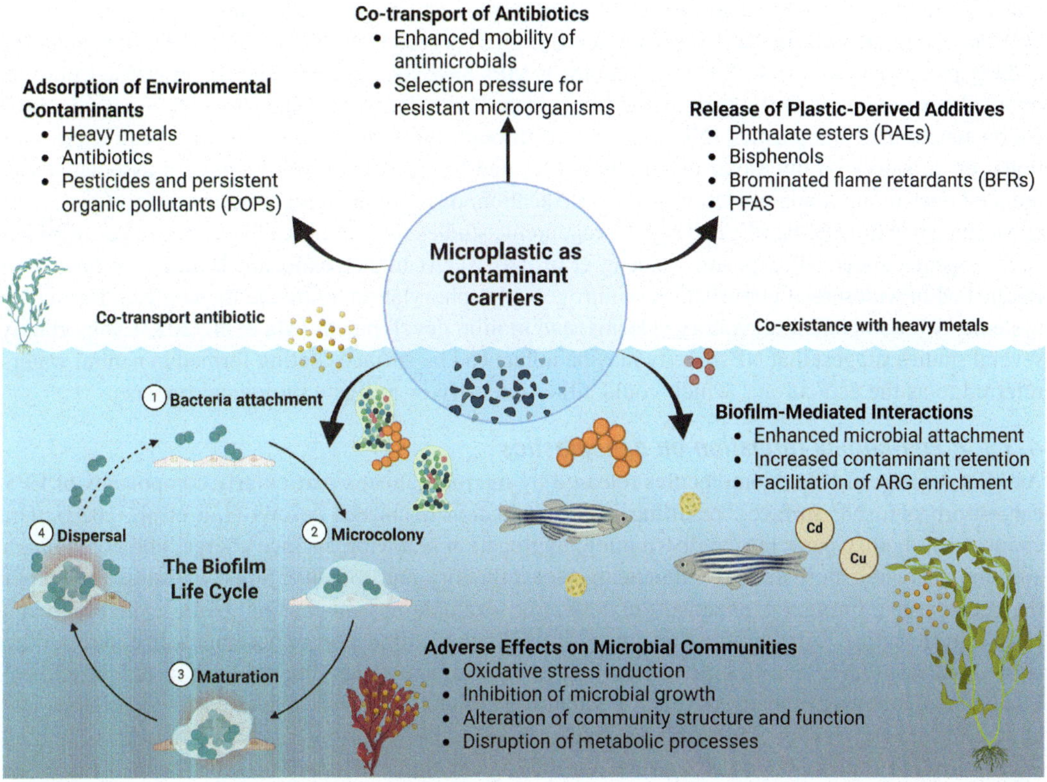

FIGURE 4.2

Schematic diagram of the impact mechanisms of microplastics as contaminant carriers on microbial communities.

Oxidative stress represents a key toxicity mechanism frequently associated with MP exposure, involving the generation of reactive oxygen species (ROS) both outside and within microbial cells. Environmental weathering of MPs, including ultraviolet irradiation and mechanical abrasion, can induce surface oxidation and the formation of reactive species in surrounding waters (Zhu et al., 2020). Many investigations also report increased intracellular ROS levels in microbes exposed to MPs (Sun et al., 2018). However, the specific pathways responsible remain insufficiently understood and may involve cooccurring stressors such as heavy metals and persistent organic pollutants (POPs) associated with MP surfaces (Zhou et al., 2024). Excessive ROS can overwhelm antioxidant defense systems, disrupt redox homeostasis, damage cellular macromolecules, and trigger apoptosis like response (Wei et al., 2019). In eukaryotic microorganisms, such as fungi and algae, MP exposure may additionally impair mitochondrial function and oxidative phosphorylation's, thereby exacerbating oxidative stress.

Water sludge and engineered bioprocess systems offer further insight into MP-microbes' interactions. Polyethylene terephthalate (PET) MPs have been shown to stimulate sulfur-related microbial activity in sludge reactors, enhancing chemical oxygen demand removal and sulfate transformation, potentially due to increased EPS production and enrichment of sulfate-reducing bacteria. Comparable stimulatory effects have been observed for polyethylene (PE) MPs in sulfur cycling processes (Xu et al., 2024). In contrast, MPs have also been linked to suppressed nitrogen removal performance, partly attributable to decreased abundance of genes associated with denitrification functions. Nanoplastics (NPs), generated through the fragmentation of MPs, can exert additional disturbances on microbial metabolism. Elevated NP concentrations have been associated with reduced rates of ammonia oxidation and denitrification, disruptions in membrane transport processes, and diminished lipid metabolic activity. Metagenomic studies further indicate declines in key nitrogen-cycling genes under NP exposure (Huang et al., 2023). At higher exposure levels, NPs have been implicated in widespread perturbation of nitrogen, phosphorylation, pyruvate metabolism, transporter systems, two-component regulatory systems, and biofilm development (Ma et al., 2022). Importantly, several studies suggest that NP toxicity may be influenced by surface coating formed in natural waters, referred to as the eco-corona, which could either intensify or mitigate biological impacts.

4.3.1.2 Eco-corona formation on nanoplastics

As MPs degrade in NPs, biomolecules released by microorganisms-particularly components of EPS-can absorb onto NP surfaces, resulting in the formation of an eco-corona (Liu et al., 2022). This coating can modify particle size distribution, aggregation behavior, surface charge, and contaminant sorption characteristics, thereby influencing bioavailability and toxicity. Some investigations report that eco-corona formation enhances toxicity, for example, by increasing particle ingestion and retention, elevating oxidative stress response, or facilitating cotransport of pollutants. In contrast, other studies indicate that EPS- or humic like coatings may attenuate toxicity by reducing direct particle-cell interactions, lowering oxidative stress, or enhancing cellular tolerance under specific conditions (Wu et al., 2019a). These divergent effects are likely governed by variations in EPS composition, ambient water chemistry, particle properties, and the biological traits of exposed organisms. As a result, accurately predicting eco-corona-mediated effects in natural aquatic systems remains a significant challenge.

4.3.1.3 Effects on algal cells

MPs and nanoplastics (NPs) can disrupt algal growth, metabolic activity, and photosynthetic processes, though the responses are influenced by polymer type, particle size, environmental aging, and exposure conditions. For instance, polyvinyl chloride (PVC) MPs have been observed to reduce algal biomass and modify both the production and composition of dissolved organic matter (DOM), particularly when the plastics are environmentally aged (Xiao et al., 2023). Experiments with marine microalgae have reported inhibited growth accompanied by changes in photosynthetic performance, fatty acid composition, and lipid profiles, suggesting that MPs can interfere with cellular energy balance and membrane integrity. Photosynthetic efficiency, including chlorophyll content and photosystem II (PSII) activity, has been reported to decline following MP exposure in some algal species (Wu et al., 2019b). Elevated temperature may exacerbate MP-induced stress in certain taxa by enhancing oxidative burdens and destabilizing electron transport mechanism (Hou et al., 2023). In cyanobacteria, exposure to NPs has been linked to alterations in secondary metabolite production (e.g., microcystins) along with changes in photosynthetic efficiency and oxidative stress signaling. However, inhibitory effects are not universal. In some cases, MPs have been observed to stimulate algal growth or photosynthesis, potentially due to the leaching of additives or low-dose effects that induce hormesislike response (Chae et al., 2020). These variable outcomes underscore the importance of employing environmentally relevant exposure scenarios and considering species-specific and polymer-specific responses.

4.3.1.4 Selective enrichment of specific microorganisms (plastisphere effects)

Microbial biofilms that develop on MPs, commonly referred to as the plastispher-frequently exhibit composition distinct from surrounding water or sediment microbial communities (Fig. 4.3). MPs can absorb organic matter and trace nutrients, creating localized zones of enrichment that attract microorganism's vis chemotaxis and facilitate colonization (Berg, 1975). Autotrophic organisms, including diatoms and cyanobacteria, as well as photoheterotrophic organisms such as *Erythrobacte*r, and *Roseobacter,* can serve as foundational contributors of carbon and nutrients within these biofilms. Heterotopic taxa commonly identified in plastisphere communities include genera such as *Pseudomonas, Bacillus,* and *Rhodococcus,* though the precise composition varies depending on environmental conditions and polymer type. Evidence also indicates selective colonization patterns, where

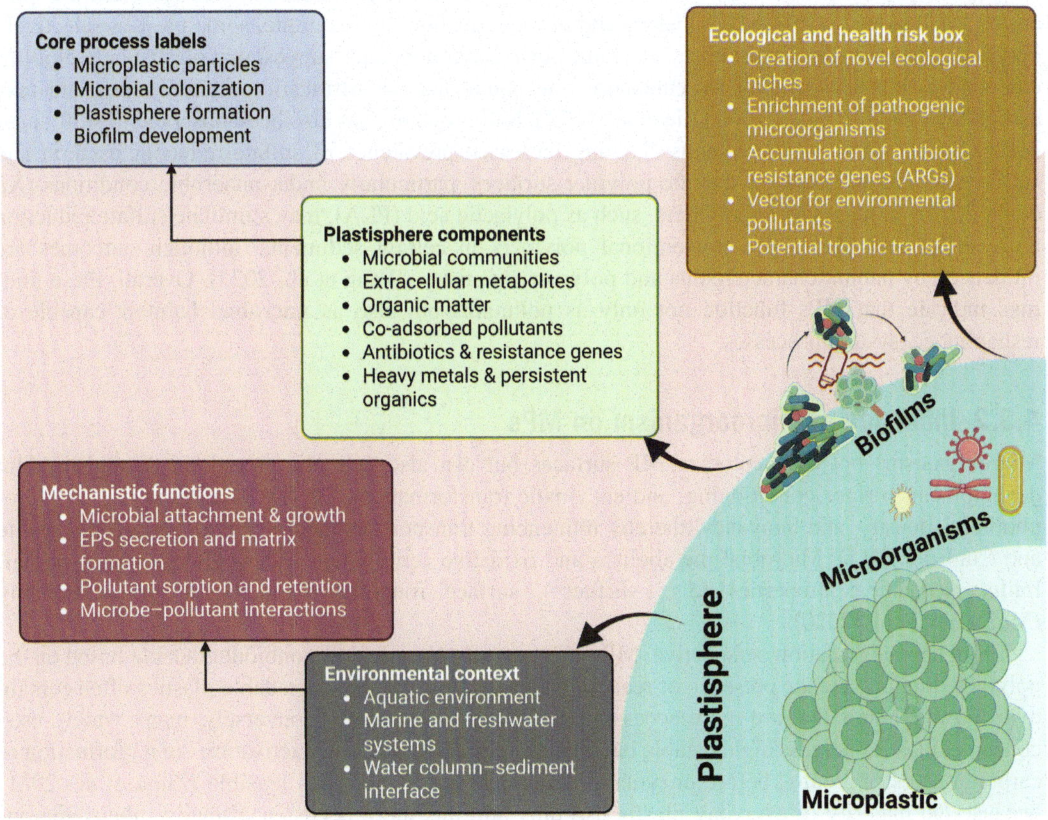

FIGURE 4.3

Microplastic-associated plastisphere and its ecological and health risks.

bacterial groups preferentially adhere to specific polymers. The development of biofilms can enhance secondary colonization and drive community succession (Amaral-Zettler et al., 2020). Concurrently, MP-associated stressors-including polymer additives, coabsorbed contaminants, interspecies competition, and environmental pressure-promote selective survival, resulting in plastisphere communities that differ from those on natural substrate.

A major concern is the enrichment of pathogenic bacteria and resistance genes on MPs. Certain pathogens are more tolerant of contaminants and environmental stress than many sensitive microbes, so MP-bound pollutants may indirectly favor their persistence and dominance (Rodrigues et al., 2019). Dense biofilm architecture further facilitates horizontal gene transfer, promoting the exchange of antibiotic resistance genes (ARGs) through plasmids and other mobile genetic elements, thereby increasing ARG prevalence in plastisphere communities (Syranidou and Kalogerakis, 2022). Elevated relative abundance of genera encompassing pathogenic species (e.g., *Vibrio, Arcobacter, Pseudomonas, Shewanella)* and enriched resistomes shifts may disrupt food webs and negatively affect aquatic organisms, contributing to oxidative stress, tissue damage, dysbiosis and metabolic disturbances (Rakib et al., 2023).

Beyond pathogens, MPs can also promote microbes involved in essential biogeochemical processes, such as hydrocarbon degradation and nutrient cycling. For example, some plastics selectively enrich hydrocarbon-degrading taxa, and studies in aquaculture ponds suggest that plastisphere biofilms can modify nitrogen cycling by enhancing the abundance of nitrification-related genes and taxa associated with nitrite oxidation (Xie et al., 2021). Sulfur cycling has also been observed to shift under MP exposure, with certain systems showing increased prevalence of sulfate-reducing bacteria and sulfur assimilation genes on specific polymer surfaces, particularly under anaerobic conditions (Xu et al., 2024). Biodegradable polymers, such as polylactic acid (PLA), may stimulate sulfate reduction more effectively than some conventional polymers in certain sediments, although outcomes are influenced by habitat characteristics and polymer chemistry (Wang et al., 2023). Overall, these findings indicate that MPs function not only as pollutants but also as microbial habitats capable of reshaping ecosystem processes.

4.3.2 Influence of microorganism on MPs

Microorganisms not only colonize MP surfaces but can also actively alter them through biofilm development, surface conditioning, and enzymatic transformations. Biofilms can modify MP hydrophobicity, density, and buoyancy, thereby influencing transport and sedimentation patterns (Lobelle and Cunliffe, 2011). Microbial metabolism and oxidative activity can further alter polymer crystallinity, mechanical properties (e.g., stiffness), surface roughness, and chemical functionality (McGivney et al., 2020).

From a biodegradation perspective, MPs are generally classified as nonbiodegradable based on the polymer structure and the presence of reactive functional groups. Biodegradable plastics often contain hydrolyzable bonds that are more accessible to microbial enzymes. Conversely, many widely used consumer plastics possess highly stable backbones, requiring oxidative "activation" (e.g., formation of carbonyl or ester groups) before enzymatic depolymerization becomes feasible (Zhou et al., 2021) because the majority of everyday plastic use falls into this more persistent category; their environmental durability poses a significant challenge (Zhang et al., 2021b). Degradation outcomes and polymer chemical properties (Miri et al., 2022).

4.3.2.1 How microbial MP degradation is measured

Laboratory investigation employs multiple approaches to evaluate microbial MP degradation (Paco et al., 2017). Mass loss is frequently measured, through results may be confounded by leaching of additives or particle fragmentation (Park and Kim, 2019). Consequently, studies often combine weight measurements with analytical techniques that detect polymer transformation. Fourier-transform infrared (FTIR) spectroscopy is used to monitor alterations in surface functional groups indicative of oxidation or bond cleavage (Hou et al., 2022). Changes in crystallinity, often reflecting preferential loss of amorphous regions, can be accessed via differential scanning calorimetry (DSC). Gel permeation chromatography coupled with advanced mass spectrometry can track shifts in molecular weight distributions as polymers degrade into smaller fragments (Xiang et al., 2023). Stable isotope techniques, such as $_{13}C$ labeled polymers, provide robust evidence of mineralization by tracing $_{13}C$ incorporation into CO_2/dissolved in organic carbon pools and microbial biomass. Microscopy, particularly scanning electron microscopy (SEM), offers visual confirmation of surface pitting, cracking, and erosion consistent with biodegradation (Jebashalomi et al., 2024).

4.3.2.2 Fungal biodegradation of MPs

Fungi are frequently recognized as promising agents for plastic transformation due to their extensive extracellular enzyme systems, which evolved to degrade complex natural polymers like lignin. Many fungal species tolerate harsh environments and can colonize contaminated habitats, making them relevant to ecosystems impacted by plastic pollution (Sangale et al., 2019). Recent studies have demonstrated fungal activity against polymers such as polystyrene (PS), polyethylene (PE), polypropylene (PP), polybutylene adipate terephthalates (PBAT), and polyethylene terephthalate (PET). Screening efforts often focus on isolates from landfills, coastal areas, and plastic-rich soils, where prior exposure may select for [plastic-tolerant microbial communities (Kim et al., 2022). Genera including *Aspergillus, Penicillium,* and *Cladosporium* are commonly reported and often exhibit measurable impacts on MP structure and mass over time. Mechanistically, fungi attach to MPs via hyphal networks and secrete surface-active compounds, subsequently releasing extracellular enzymes (e.g., oxidoreductase and hydrolases) that cleave polymer bonds, producing smaller oligomers and monomers that can be assimilated (Sanchez, 2020).

Reported degradation efficiencies vary widely, depending on polymer chemistry, pretreatment methods (e.g., UV oxidation), incubation duration, and experimental design. Polyolefins (PE, PP) are particularly recalcitrant due to their strong carbon-carbon backbones and low functionalization, often requiring oxidative pretreatments to enhance surface reactivity prior to enzymatic attack (Lv et al., 2024). Accordingly, many studies incorporate pretreatment strategies such as UV radiation, heat, or chemical oxidants to generate functional groups that increase microbial accessibility and accelerate depolymerization.

4.3.2.3 Bacterial biodegradation of MPs

Bacteria are generally considered less efficient than fungi at degrading polymers, but their high abundance, metabolic versatility, and rapid adaptability make them important contributors to MP transformation, particularly within biofilm communities. Initial bacterial colonization involves surface attachment mediated by cell surface hydrophobicity, followed by extracellular polymeric substance (EPS) production and biofilm formation. Extracellular enzymes can cleave polymers into smaller fragments that may cross cell membranes, allowing intracellular metabolism and partial mineralization

(Dai et al., 2022). Bacterial degradation pathways differ depending on the polymer type. For polyethylene terephthalates (PET), a widely studied route involves extracellular enzymes such as PETase and MHETase, as demonstrated in *Ideonella sakaiensis,* which hydrolyze PET to MHET and subsequently to terephthalate and ethylene glycol (Liu et al., 2019). In contrast, the breakdown of more inert polymers is often limited by the initial oxidative modification of the surface.

Although many bacterial strains exhibit modest polymer mass loss, some studies report comparatively high degradation under favorable conditions, such as UV-pretreated plastics, optimized incubation parameters, or enriched microbial consortia. Differences between fungal and bacterial degradation largely reflect cellular physiology: fungi can engage small particles via complex intracellular trafficking, whereas bacteria rely predominantly on surface-bound processes and transport mechanisms rather than true endocytosis. Notably, intracellular mineralization pathways for many plastics in bacteria remain poorly understood, highlighting a critical area for future mechanistic research.

4.3.2.4 Bioflocculants and microbial aggregation as a removal pathway

Beyond biodegradation, flocculation-based strategies provide a practical means for MP removal in wastewater and aquaculture systems (Xue et al., 2021). While chemical coagulants can significantly reduce MP concentrations, concerns over secondary pollution and toxicity have spurred interest in bioflocculants—microbially produced natural polymers (Rajala et al., 2020). Bioflocculants, often composed of EPS like polysaccharides or mixed biopolymers, can aggregate MPs and NPs into flocs that are most readily separated via sedimentation or filtration. Research has demonstrated that EPS derived from bacteria and microalgae can cluster MPs/NPs even at low-to-moderate concentrations, including scenarios where the particles themselves inhibit microbial growth. Certain approaches envision circular systems in which precipitated flocs are processed to regenerate flocculants to recover valuable materials (Zhao et al., 2024). Although biodegradation remains the most widely discussed biological removal mechanism, bioflocculation represents a complementary, near-term solution for treatment facilities — especially in cases where mineralization proceeds slowly. Future work should focus on identifying robust bioflocculation producers, optimizing floc performance under realistic wastewater conditions, and assessing the safety and sustainability of resultant flocs.

4.3.3 Influencing factors governing microbial colonization on MPs

Microbial colonization of MPs surfaces is a dynamic process shaped by multiple environmental and material-specific factors. In aquatic ecosystems, the development, architecture, and persistence of MP-associated biofilms arise from complex interactions among microbial communities and abiotic conditions. Variations in local environment, nutrient availability, salinity, temperature, polymer chemistry, and surface characteristics collectively influence biofilm formation, microbial diversity, and functional potential.

4.3.3.1 Environmental factors and spatial heterogeneity

Spatial heterogeneity in aquatic habitats strongly affects biofilm formation on MPs. The ecological functions and microbial composition associated with MPs are closely tied to regional environmental conditions, leading to substantial variability in biofilm structure across different water bodies. Observable differences include biofilm structure across different water bodies. Observable differences include biofilm thickness and biomass, which may range from tens to hundreds of nanometers depending on the system. Freshwater and wastewater settings often support distinct biofilm dry masses, reflecting differences in organic matter load, hydrodynamics, and microbial source communities. Microbial composition on MPs also varies significantly across geographic regions. Comparative studies of coastal ports, estuaries, and bays show that dominant bacterial families differ among sites, emphasizing the influence of local environmental parameters. Likewise, MPs of the same polymer type can host distinct microbial communities depending on the aquatic system in which they are deployed. For instance, polystyrene (PS) biofilms in wastewater environments frequently exhibit a higher abundance of specific Proteobacteria classes compared to those habitats.

4.3.4 Proteobacteria classes compared to those in riverine habitats

Hydrodynamic conditions further regulate biofilm development by affecting the transport of solutes from the surrounding water into the biofilm matrix. Flow regimes and channel morphology control the diffusion of dissolved organic matter (DOM) and nutrients. When solute transport becomes limiting, microbial metabolism may be constrained by diffusion rather than substrate-enhanced DOM uptake broadening the chemical diversity of organic compounds assimilated by millions, thereby increasing microbial functional diversity.

4.3.4.1 Nutrient availability

Nutrient levels, particularly total nitrogen (TN) and total phosphorus (TP), are key drivers of MP biofilm formation and microbial community composition. Elevated nutrient concentrations generally accelerated biofilm establishment, enhanced growth rates, and increased biomass accumulation on MPs. In nutrient-rich freshwater systems, MPs tend to support denser biofilms compared to oligotrophic environments. Nutrient status also influences biofilm composition and metabolic pathways. In mesotrophic and dystrophic lakes, MP biofilms are often dominated by autotrophic taxa such as diatoms and green algae, whereas eutrophic systems favor heterotrophic bacteria and ciliates. These compositional shifts affect biofilm productivity and carbon metabolism. The availability of inorganic nitrogen and phosphorus further regulates the balance between autotrophic carbon fixation and heterotrophic assimilation, determining how newly fixed organic carbon is retained within microbial biomass.

Importantly, MP biofilms can also influence nutrient dynamics in the surrounding water. Biofilm-coated MPs can transport and release nitrogen and phosphorus, potentially affecting nutrient cycling in the adjacent ecosystem. Co-occurring pollutants, such as heavy metals, may disrupt these processes by altering microbial communities and reducing nutrient removal efficiency. These observations indicate that nutrient-biofilm interactions on MPs are closely interlinked with contaminant exposure and microbial composition.

4.3.4.2 Salinity effects

Salinity is another critical factor influencing microbial colonization on MPs. High salinity levels generally suppress microbial growth and biofilm density due to the osmotic stress, yet they may enhance microbial diversity by selecting for salt-tolerant taxa. Studies consistently report an inverse relationship between salinity and biofilm growth rate, alongside a positive relationship between salinity and biofilm growth rate, alongside a positive relationship between salinity and bacterial diversity. Variations in salinity, particularly in estuarine systems influenced by freshwater inflow, can substantially alter MP-associated microbial communities. Dominant bacterial groups shift following a reduction in salinity caused by river discharge, reflecting the sensitivity of plastisphere communities to salinity gradients. Salinity also affects the sorption behavior of organic contaminants on MPs by altering ionic strength and competitive interactions, indirectly influencing microbial exposure to coadsorb pollutants. Despite these findings, mechanistic links between salinity, pollutant sorption, and microbial responses remain incompletely resolved.

4.3.4.3 Temperature influences

Temperature significantly regulates microbial physiology, enzymatic activity, and seasonal biofilm dynamics on MPs. As MPs persist in the environment over a long period, seasonal temperature fluctuation shapes microbial diversity and biofilm succession. Warmer conditions generally enhance metabolic rates, resulting in increased biofilm biomass and growth during summer months, which are characterized by higher temperatures, longer daylight, and lower dissolved oxygen levels. Interestingly, temperature effects on microbial density and diversity are not always aligned. Elevated temperature may reduce overall bacterial density while simultaneously increasing community diversity. Temperature also modulates interactions between MPs and environmental pollutants. Some studies reported reduced adoption of certain organic forces, whereas others suggest that the adsorption of persistent organic pollutants may be relatively temperature independent. These contrasting results highlight the need for further investigation into temperature-driven MP-pollutant interactions.

4.3.4.4 Influence of MP matrix type

The type of polymer comprising MPs significantly affects patterns of microbial colonization. Distinct plastic polymers selectively favor certain microbial taxa due to differences in surface chemistry, hydrophobicity, and additive content. Evidence indicates that specific bacterial phyla preferentially attach to polymers, demonstrating substrate selectivity within plastisphere communities. Among frequently studied polymers, polyvinyl chloride (PVC) generally supports greater biofilm biomass and enhanced extracellular polymeric substance (EPS) production relative to polyethylene, polypropylene, polystyrene, and polyethylene terephthalate, especially under contaminated conditions. This increased colonization may be associated with chlorine-containing compound release and the elevated additive content characteristic of PVC. Plastic additives further influence microbial proliferation and community composition. Many plastics include plasticizers, stabilizers, and other low-molecular-weight compounds intended to improve material performance. PVC, in particular, often contains substantial amounts of phthalates and ester-based plasticizers, which can serve as supplemental carbon sources for microbes. These compounds can stimulate biofilm formation and modify microbial metabolic functions, contributing to polymer-specific differences in plastisphere structure.

4.3.4.5 Surface characteristics of MPs

The surface traits of MPs are crucial during initial microbial attachment. Features such as surface charge, hydrophobicity, roughness, and hardness influence bacterial adhesion and biofilm stability. Electrostatic forces are particularly important, as surface charge dictates the strength of attraction between microbial cells and plastic substrates. Negatively charged polymers, such as polyethylene and polystyrene, generally discourage bacterial adhesion compared to positively charged surfaces, which can disrupt microbial cell membranes. Hydrophobicity also affects colonization dynamics. Early biofilm formation is often facilitated on less hydrophobic surfaces due to higher availability of polar functional groups, whereas mature biofilms typically exhibit stronger adhesion on hydrophobic surfaces. Nevertheless, hydrophobicity alone appears to have a relatively minor impact compared to other surface features. Surface roughness enhances microbial attachment by increasing available surface area and creating microenvironments that trap nutrients and reduce shear stress. Rough surfaces consistently support higher microbial density and accelerate biofilm formation. Additionally, softer particles generally promote bacterial adhesion more than harder substrates, possibly due to differences in quorum-sensing regulation and surface-associated gene expression. However, the role of surface hardness in MP colonization remains underexplored and requires further study.

4.4 Formation of pollutant-MP complexes

4.4.1 Adsorption dynamic of inorganic pollutants on MPs associate biofilms

The widespread formation of biofilm on MP surfaces significantly alters their adsorption behavior, particularly in relation to inorganic pollutants. These biofilms, composed of extracellular polymeric substances (EPS), introduce functional groups such as carboxyl, hydroxy, acetyl, phosphate and sulfide, which facilitate complexation and coordination with inorganic compounds. The overall adsorption mechanism includes physical adsorption, biosorption, and chemisorption, each governed by intermolecular interactions and influenced by MP characteristics, biofilm maturity, environmental conditions, and the nature of the adsorbate (Pan et al., 2023).

Biofilms enhance the adsorption of heavy metals through multiple concurrent mechanisms, including electrostatic attraction, ion exchange, surface precipitation, and coordination bonding. Ultraviolet exposure increases surface porosity on MPs, promoting metal uptake and aggregation (Zhou et al., 2022; Johansen et al., 2019) reported that high-density polyethene (HDPE) MPs exhibited greater radionuclide adsorption than their biofilm-coated counterparts, which contained aluminum, silicon, and oxygen elements that facilitated clay particle capture. In natural environments, MPs-biofilms adsorbed cesium (Cs) at 80 mL/g in freshwater and strontium (Sr) at 5 mL/g in estuarine conditions, values significantly lower than sediment-based MPs by two to three orders of magnitude.

The adsorption of cadmium (Cd) on MPs-biofilm was found to increase with pH from two to 6, then decline between pH 6.0 and 9.0. Multivariate analyses revealed elevated concentrations of metals such as Cu, U, Co, Pb, Al, Mg, K and Mn on MPs-biofilms, with levels rising as biofilms aged (Richard et al., 2019). Particle size also plays a crucial role; Wu et al. (2022) demonstrated that the MPs around 75 μm had significantly higher adsorption capacities for Cr(VI) and Cu compared to larger MPs (4 mm). Postbiofilm formation, adsorption increased by 205.8% for Zn and 701.8% for Cu, attributed to enhanced functional group availability and positive zeta potential facilitating electrostatic interactions. Li et al. (2022) found that lead Pb(II) adsorption on biofilm-coated polybutylene succinate (PBS) MPs was nearly 10 times higher than on virgin PBS (647.09 μg/g vs. 64.13 μg/g), with adsorption behavior best described by the Frendlich and Temkin isotherm models. Increased ionic strength and acidic conditions ($pH < 7.0$) were shown to reduce Pb(II) adsorption, while higher pH promoted the formation of crystalline oxide precipitates. Remarkably, Pb(II) remained bound to MPs even after biofilm removal, suggesting strong EPS-mediated complexation. Langmuir model analysis ranked Pb(II) adsorption capacity as follows: suspended solids > sediments > MPs biofilm > virgin MPs (Qi et al., 2021). Additionally, biofilm facilitates the formation of hydrous oxides of manganese and iron, which coprecipitation with heavy metals, enhancing their retention on MPs surfaces (Richard et al., 2019). Overall, the adsorption process on MPs-biofilms is multilayered and heterogenous, with biofilm complexity playing a pivotal role in pollutant binding and environmental fate. These findings underscore the importance of understanding MPs-biofilm interactions to assess their ecological risks and inform mitigation strategies.

4.4.2 Competitive adsorption and organic pollutant association on MP associated biofilms

The adsorption behavior of MPs coated with biofilms is significantly influenced by competitive interactions between coexisting pollutants. The preadsorbed copper ions (Cu^{2+}) on MPs-biofilm surfaces were desorbed upon the introduction of tetracycline, indicating competitive binding and pH-dependent complexation. Biofilm-coated poly(lactic acid) MPs exhibited a higher monolayer adsorption capacity for Cu^{2+} (151.08 μg/g) compared to virgin MPs (1045.67 μg/g), although fulvic acid was found to inhibit Cu^{2+} adsorption due to its strong complexation potential (Sun et al., 2022). Interestingly, the removal of biofilms exposed aged surface areas that enhanced Cu^{2+} adsorption by 32.26%, yet adsorption efficiency was lowest within the pH range of 4.5–6.0.

Persistent organic pollutants (POPs), characterized by long environmental half-lives and high bioaccumulation potential, also interact strongly with MPs biofilms. The growth of biofilms increases the Brunauer-Emmett-Teller (BET) surface area and water contact angle of MPs, enhancing their hydrophobicity and facilitating the adsorption of organic compounds (Jose and Jordao, 2022). The fluorescent protein-like substances in biofilm EPS reduced MPs zeta potential, thereby improving the electrostatic adsorption of antibiotics and emerging contaminants (ECs).

Norfloxacin adsorption across different biofilm-coated MPs PVC, HDPE, and PA, reporting adsorption efficiencies of 50.6%, 46%, and 24.1%, respectively. These interactions were governed by electrostatic forces, hydrophobic interactions, and hydrogen bonding. Aging-induced reductions in surface crystallinity further enhanced MPs adsorption capacity (Huffer and Hofmann, 2016). Rapid adsorption kinetics for ECs such as diclofenac and miconazole, reaching pseudoequilibrium within 72 h. However, MPs biofilms reduced EC adsorption by a factor of 3.8 compared to virgin MPs. Lower adsorption rates for polycyclic aromatic hydrocarbons (PAHs) such as phenanthrene, fluoranthene, benzanthracene, and chrysene during summer, with increased sorption observed in winter. Metal organic interactions further complicate MPs biofilm adsorption behavior. The oxytetracyline adsorption on virgin MPs and MPs biofilms reached 692.05 μg/g and 1396.2 μg/g, respectively, with Cu^{2+} enhancing adsorption by 431.1% and 165.9%. The complexation between Cu^{2+} and MPs biofilm inhibited oxytetracycline desorption, demonstrating strong retention (Sun et al., 2023).

Biofilms formed on MP surfaces have emerged as significant ecological vectors, particularly in the propagation of antibiotic resistance genes (ARGs) within aquatic ecosystems. These biofilms provide a stable and protective microenvironment that facilitates microbial colonization and genetic exchange. MP biofilms harbor a higher abundance of ARGs compared to quartzite biofilms, with ARG concentrations increasing proportionally to ambient antibiotic levels. This phenomenon is attributed to horizontal gene transfer mechanisms, conjugation, transformation, and transduction, where mutations in noncoding DNA sequences enable replication and differentiation of ARGs across diverse microbial taxa (Arnold et al., 2022). The presence of ARGs on MPs' biofilms poses a substantial threat to aquatic biodiversity and ecosystem health, necessitating targeted mitigation strategies to preserve ecological integrity. Beyond genetic risks, MPs biofilms also influence biogeochemical processes. Certain bacteria within these biofilms possess metabolic capabilities to degrade MPs and associated organic pollutants into dissolved organic carbon (DOC), methane (CH_4), carbon dioxide (CO_2), and ethylene (C_2H_4), thereby contributing to atmospheric carbon emissions and altering the natural carbon cycle (Galgani et al., 2019). These dual roles, facilitating ARG dissemination and disrupting carbon cycling, underscore the urgent need for comprehensive research and policy interventions to monitor, manage, and remediate MPs biofilm contamination in aquatic environments. These findings underscore the multifaceted role of MPs biofilms in pollutant dynamics, highlighting their influence on competitive adsorption, retention, and environmental mobility of both organic and inorganic contaminants. Understanding these interactions is crucial for assessing ecological risks and developing effective mitigation strategies.

4.4.3 Biofilm-mediated enrichment of organic pollutants on MPs

Biofilms associated with MPs play a central role in the enrichment, transformation, and fate of organic pollutants in aquatic and terrestrial environments. Organic contaminants are typically hydrophobic and lipophilic, favoring partitioning onto solid surfaces rather than remaining dissolved in water. As these compounds circulate through freshwater and marine systems, they readily associate with organic matter, sediments, and synthetic polymers like MPs. Many organic pollutants are persistent, mobile and bioaccumulative, posing significant ecological and human health risks. Frequently detected pollutant classes on plastic include polychlorinated biphenyls (PCBs), polycyclic aromatic hydrocarbons (PAHs), organochlorine pesticides such as DDTs, polybrominated diphenyl ethers (PBDEs), alkylphenols, and bisphenol A (BPA), with concentrations ranging from trace amounts to several milligrams per kilogram. The accumulation of persistent organic pollutants (POPs) on MPs is influenced not only by the physicochemical properties of the plastic surface but also by microbial biofilms. Biofilms modify MP surfaces by increasing roughness, surface area, and hydrophobicity, enhancing their capacity to adsorb organic pollutants. Microorganisms within these biofilms can actively transform, metabolize, or redistribute adsorbed contaminants. Since environmental monitoring typically captures cooccurrence rather than mechanistic interactions, laboratory and field studies are often used to elucidate adsorption processes and microbial influences on pollutant behavior.

4.4.3.1 Polychlorinated biphenyls

PCBs are among the most frequently reported POPs on MPs. Biofilms influence PCB behavior primarily through microbial metabolism at the plastic interference. Reductive dichlorination by specialized organohalide-respiring microbes can alter PCB composition, toxicity, and bioavailability. Studies indicate that PCBs adsorbed onto sterile MPs undergo little to no dichlorination, whereas significant transformation occurs on MPs colonized by microbial communities, highlighting the role of biofilm-associated microbes in accelerating PCB metabolism. Several bacterial taxa within MP biofilms possess enzymes capable of reductively dehalogenating chlorinated aromatic compounds. These microbial processes can convert highly chlorinated PCBs into less chlorinated forms, modifying environmental behavior and biological effects. Biofilm-coated MPs often exhibit increased PCB adsorption due to greater surface heterogeneity and hydrophobic domains formed by EPS, further by EPS, further concentrating contaminants at the biofilm-plastic interface.

Ecological outcomes of PCB enrichment on MPs are complex and context dependent. In some freshwater sediments, MPs contaminated with PCBs have been shown to suppress key microbial genes involved in methane production, reducing methane emissions. Conversely, other studies indicate that MPs can stimulate methane release and inhibit nitrate reduction in anoxic PCB-contaminated soils by altering microbial community composition. These biogeochemical cycles depend on environmental conditions, microbial assemblages, and redox status. Some evidence indicates that while MPs may influence PCB bioavailability, they do not always alter dichlorination extent, emphasizing the need for further investigation of how plastic surfaces modulate microbial PCB metabolism.

4.4.3.2 Polycyclic aromatic hydrocarbons

PAHs are another important class of hydrophobic pollutants commonly enriched on MPs with strong mutagenic and carcinogenic properties. Sorption of PAHs onto MPs is primarily governed by hydrophobic interactions, though molecular structure and polarity strongly affect adsorption efficiency. High-molecular-weight PAHs generally show greater affinity for plastics than lower-molecular-weight compounds due to increased hydrophobicity. Polymer type strongly affects PAH adsorption. For instance, polystyrene often exhibits higher adsorption capacity for specific PAHs compared with PVC, partly due to π-π interactions between polymer aromatic rings and PAH molecules. Noncovalent interactions, including hydrogen and halogen bonding, can further enhance adsorption, especially on plastics with polar functional groups. Biodegradable plastics with oxygenated surface functionalities may strongly interact with polar organic compounds via hydrogen bonding, leading to higher adsorption than conventional polymers.

Microbial communities within MP biofilms can both accumulate and degrade PAHs. Numerous bacteria and fungi on plastic surface possess enzymatic pathways for metabolizing PAHs, utilizing them as carbon and energy sources. Certain microbial species can persist on PAH-contaminated MPs for extended periods, indicating roles in both pollutant degradation and microbial survival under contaminated conditions. Biodegradation efficiency varies with PAH molecular weight; microorganisms preferentially degrade lower-molecular-weight PAHs, while higher-molecular-weight compounds often persist. PAH enrichment on MPs also promotes selective colonization of PAH-degrading microbes. Bacterial families frequently detected in PAH-rich plastisphere communities include Erythrobacteraceae, Rhodobacteraceae, Sphingomonadaceae, Bacillaceae, Moraxellaceae, and Pseudomonadaceae. Their presence significantly influences PAH transformation rates, underscoring MPs' roles as both pollutant carriers and bioreactive substrates.

4.4.3.3 Organochlorine pesticides

Organochlorine pesticides (OCPs) readily associate with MPs due to their small size, hydrophobicity and large surface area. When MPs cooccur with OCPs, pesticides can accumulate on plastic surfaces, acting as mobile reservoirs that redistribute contaminants across water, sediments, and biota. Laboratory studies suggest that adsorption is dominated by surface sorption rather than penetration into the polymer matrix for some pesticide-polymer contaminations, indicating that interfacial processes primarily govern uptake under environmentally relevant conditions. This cotransport phenomenon can exacerbate "mixed pollution", as plastics not only concentrate OCPs but also prolong their residence time and broaden their exposure range. Biofilms further influence OCP behavior by modifying the plastic surface and boundary-layer chemistry. Microbial colonization can drive interfacial reactions and biodegradation, turning MPs into bioactive microreactors rather than passive carriers. For instance, exposure to certain antimicrobials can reshape plastisphere communities by enriching taxa capable of degrading these compounds. Legacy pesticides, such as DDTs, can undergo biofilm-associated microfilm transformations, particularly under sulfate-reducing conditions, facilitating dichlorination and other pathways that are otherwise slow in nutrient-limited settings. Biofilms may also increase apparent dissolution and desorption flux of hydrophobic pesticides from polyethylene, raising their bioavailability in surrounding water and affecting contaminant partitioning among plastics, porewater, and sediments. Despite these insights, mechanistic understanding remains incomplete, including how biofilm composition regulates OCP sorption kinetics, whether microbial transformation reduces or enhances toxicity and how combined processes – adsorption, desorption, diffusion, and biodegradation – determine net transport and ecological risk.

4.4.3.4 Antibiotics

Antibiotics, extensively used in medicine, livestock, and agriculture, can enter aquatic ecosystems and disrupt microbial community composition and ecosystem processes. Many antibiotics adsorb onto MP surfaces through various noncovalent forces, including hydrogen bonding, van der Waals interactions, electrostatic attraction or repulsion, and hydrophobic partitioning. The dominant adsorption mechanism depends on antibiotic speciation, polymer type, and water chemistry, including pH, ionic strength, and dissolved organic matter. MPs thus act not only as passive carriers but also as selective substrates capable of concentrating antibiotic mixtures under specific conditions. Biofilm formation on MPs often enhances antibiotic adsorption by introducing additional functional groups and binding domains. As biofilms mature, oxygen-containing groups (e.g., carbonyls) and extracellular polymeric substances (EPS) accumulate, increasing hydrophilicity and creating ligand-rich sites. EPS components, such as polysaccharides and proteins, containing carboxyl, hydroxyl, and amide groups, form extensive hydrogen-bond networks with antibiotics. Microbial weathering can also increase surface roughness and effective surface area, further promoting adsorption. For tetracyclines, adsorption is influenced by electrostatic interactions, hydrogen bonding, and π-π interactions, with biofilms enhancing these effects. For fluoroquinolones such as norfloxacin, hydrogen bonding and electrostatic forces dominate, while hydrophobic contributions vary with polymer type and water chemistry.

Antibiotic adsorption and microbial colonization, creating feedback that affects pollutant fate and microbial ecology. High antibiotic loads on MPs can select for tolerant taxa and reshape plastisphere composition. Under combined stressors (MPs plus antibiotics), microbial cocultures may exhibit enhanced survival, partly due to the protective microenvironment of biofilms that reduces effective antibiotic exposure. Subinhibitory antibiotic concentrations can also "train" microbial communities by altering metabolic allocation and increasing biofilm-related gene expression, promoting persistence and colonization on MP surfaces. MP-associated biofilms can become hotspots for antibiotic resistance genes (ARGs), with enhanced adsorption supporting enrichment of tetracycline-resistance genes and increasing horizontal gene transfer potential. ARG profiles vary by polymer type, and the long-term role of antibiotic-loaded MPs as net sources or sinks of resistance determinants remains uncertain.

4.4.3.5 Heavy metals

Heavy metals from industrial and urban activities persist in aquatic systems and are toxic even at low concentrations. MPs can accumulate metals at levels far above surrounding waters, and biofilms further enhance this capacity. Biofilms after surface charge introduce functional groups and provide sorption and precipitation microenvironments, making biofilm-coasted MPs effective "metal carriers" that influence exposure pathways and metal transport between water and sediments.

4.4.3.5.1 Anionic metals: Chromium and Arsenic

Chromium adsorption is strongly influenced by polymer type and biofilm chemistry. Different plastics (e.g., PE, PET, and PP) exhibit varied Cr uptake due to differences in surface functional groups and texture rather than particle size alone. Biofilms generally enhance chromium adsorption by introducing reactive ligands (hydroxyl, carboxyl, sulfhydryl) and modifying surface electrostatics. In such cases, Cr(VI) is partially reduced to Cr(III), mediated by biofilm functional groups, which then bind more strongly to negatively charged sites. pH affects these processes, with low pH suppressing binding due to competition with protons, while higher pH increases negative charge density and enhances adsorption.

Arsenic behavior depends on speciation: As(III) is typically neutral, while As(V) is negatively charged, making adsorption highly pH- and electrostatics-dependent. Microbial oxidation can increase As(V) availability for adsorption. However, precise mechanistic understanding is limited because many studies do not fully characterize biofilm composition, redox gradients, or microbial taxa and enzymes involved. The toxicological consequences of metal-loaded MPs remain incompletely resolved, with combined exposures often increasing organismal stress, through the mechanisms-altered bioavailability, ingestion, oxidative stress or membrane effects-are not fully clarified.

4.4.3.5.2 Cationic metals: Lead and copper

Lead (Pb^{2+}) is highly enriched on biofilm-coated MPs across polymers. Binding occurs through electrostatic attraction enhanced by negative surface charge and through complexation with polar functional groups (-OH, –COOH, $-NH_2$) in EPS. Biofilm mineral phases such as sulfides, phosphates, and sulfates can further immobilize Pb through microprecipitation. Spectroscopic analyses show Pb colocalizes with biofilm-associated elements, indicating EPS and biofilm chemistry drive adsorption rather than pristine polymer properties alone.

Copper (Cu^{2+}) adsorption is also enhanced by biofilms, though typically less than Pb due to lower electrostatic affinity. Biofilms increase negative charge and provide additional complex sites, while environmental ionic strength modulates uptake through competition with cations (e.g., Na^{+}). EPS containing amino, carbonyl, and oxygenated groups further facilitate Cu binding. Polymer type and surface characteristics remain influential, with polar or high-surface area plastics showing stronger adsorption. Biological feedback occurs as low Cu levels can support microbial diversity, whereas Pb toxicity may reduce diversity and alter EPS composition over time.

Across metal classes, three biofilm effects dominate enhanced adsorption:

1. abundant ionizable functional group that complexes metals.
2. increased surface heterogeneity and microdomains creating binding hotspots; and
3. microbial weathering and enzymatic erosion, exposing additional functional groups.

Metal uptake on biofilm-coated MPs reflects combined contributions of electrostatics, ligand complexation, redox transformations (for redox-active metals), microprecipitation, and environmental conditions (pH, ionic strength).

4.4.3.6 Radionuclides

Radionuclides in aquatic systems can also associate with MPs. Their adsorption is sensitive to pH and ion strength and involves electrostatic interactions, surface charge neutralization, pore filling, ion-dipole forces, van der Waals interactions, and functional group binding. Biofilms can promote attachment of mineral–or clay-rich aggregates, increasing sorption capacity for radionuclides such as radiocesium and radiostrontium. Biofilm-coated MPs thus act as multi-phase carriers, combining polymer surfaces, microbial matrices, and mineral coating. Mechanistic understanding linking polymer surface properties, biofilm composition, and radionuclide sorption remains limited.

4.4.3.7 Perspective and remaining knowledge gaps

Although MPs are widely studied, the role of biofilms remains comparatively underexplored. Biofilm growth can significantly modify MP properties (surface chemistry, roughness, density/buoyancy) and enhance pollutant capture, transformation, and transport. Yet mechanistic and ecological uncertainties persist. Future research should move beyond describing plastisphere communities to integrate how biofilm-MP interactions influence contaminant fate, exposure pathways, and biological effects in real-world aquatic systems. Expanding biological scope is critical. Current knowledge mainly reflects bacterial colonization, while fungi, algae, protozoa, and mixed consortia likely influence early attachment, EPS production, enzymatic potential, and eco-corona formation, all affecting MP-pollutant interactions. Cross-kingdom studies are needed to understand biofilm differences across polymer types, habitats, and seasons.

Research should also cover diverse environmental compartments, including rivers, lakes, reservoirs, estuaries, and urban inland waters where MPs cooccur with microbes, pharmaceuticals, metals, and industrial organic. Multi-site, comparative studies linking microbial composition to hydrodynamics, nutrient regimes, and DOM gradients are necessary.

Another gap is the lack of standardized evaluation of pollutant adsorption and release under realistic conditions. Many studies examine single pollutants in simplified media, but natural environments present complex mixtures of POPs, antibiotics, metals, PFAS, and organic matter. Experiments should integrate temperature, salinity, pH, ionic strength, and polymer aging, paired with in vivo or mesocosm studies to assess whether biofilm-enhanced sorption affects uptake, toxicity, and trophic transfer.

Finally, the combined toxicity of MPs and biofilms is not fully understood. Biofilms may sequester pollutants, reducing dissolved concentrations, but can also increase ingestion, promote pathogens' persistence, and enhance ARG transfer. Mechanistic studies linking biofilm composition, pollutant load, and host responses (oxidative stress, immunomodulation, gut microbiome shifts, inflammation, reproduction) are needed, alongside harmonized risk metrics, a monitoring framework, and practical mitigation strategies.

Addressing these gaps-beyond bacteria, across habitats, using realistic mixtures, and integrating toxicity studies-will support predictive understanding of biofilm-mediated MP behavior and impacts and guide effective mitigation approaches.

4.5 Characterization of microplastic-associated biofilms

Understanding the complex interactions between MPs and biofilms is essential for assessing their environmental behavior and pollutant dynamics. These interactions influence pollutant degradation, migration, and bioavailability, thereby affecting treatment outcomes and ecological impacts. MPs physical attributes, such as size, shape, and surface chemistry, play a critical role in determining their biological interactions and pollutant transport pathways. To comprehensively evaluate these effects, a suite of characterization techniques is employed, broadly categorized into morphological analysis, chemical composition profiling, and bioinformatics assessment.

4.5.1 Morphological characterization

Scanning Electron Microscopy (SEM): Reveals surface topography and microbial colonization patterns in MPs. Biofilms on MPs often display distinct morphologies compared to those on natural substrates.

Atomic Force Microscopy (AFM): Measures cell surface interactions via force-distance curves, providing insights into biofilm adhesion mechanisms and structural evolution.

Confocal Laser Scanning Microscopy (CLSM): Enables spatial and temporal visualization of biofilm architecture.

4.5.2 Chemical composition analysis

Fourier Transform Infrared Spectroscopy (FTIR): Identifies functional groups associated with biofilm biomass on MPs. Characteristic absorption bands appear at 3600–3000 cm^{-1}, 1550 cm^{-1}, 1250–950 cm^{-1}, and 500 cm^{-1} (Battulga et al., 2022).

Raman Spectroscopy: Offers high-resolution spectral data with improved signal-to-noise ratios. Spectra from biofilm-covered and bare regions are compared to detect compositional changes.

Laser Induced Breakdown Spectroscopy (LIBS): Detects elemental composition of MPs and biofilms, including biogenic elements like calcium and magnesium. Variations in CN and C2 band intensities indicate biofilm presence (Porizka et al., 2023).

4.5.3 Bio-information analysis

Crystal Violet Staining: Quantifies surface biofilm biomass by measuring absorbance at 595 nm using UV-Vis spectrophotometry.

Microbial Community Sequencing: 16S rRNA and Long-Read Sequencing: Provides species and strain level resolution of biofilm communities. Dominant taxa such as *Alphaproteobacteria*, *Gammaproteobacteria*, and *Bacteroidia* are commonly found on polyethylene (PE) biofilms.

MiniON Sequencing: Used to profile microbial diversity in HDPE, PET, and PS surfaces. *Limnobacter thiooxidans* was identified as a key colonizer, with strain selection varying by polymer type.

4.5.4 Integration and application

To ensure accurate and meaningful characterization, technique selection must align with the MPs particle size, sample complexity, and analytical goals. Combining morphological, chemical, and biological data enables a holistic understanding of MPs' biofilm systems, facilitating risk assessment, pollutant tracking, and the development of mitigation strategies. This integrated approach is vital for advancing our knowledge of MPs biofilm interactions and their implications for environmental health and sustainability.

4.6 Greenhouse gas emissions from microplastic degradation

4.6.1 Role of microplastic associated biofilms in greenhouse gas emission and climate impacts

Biofilm formed on MP surfaces significantly alters their physical behavior and environmental interactions, particularly in relation to greenhouse gas (GHG) emissions. The presence of biofilms changes MPs viscosity and buoyancy, reducing their hydrophobicity and floatation capacity, which causes them to sink deeper into aquatic systems (Sharma et al., 2023). This sinking disrupts vertical carbon distribution and impairs gas exchange and carbon sequestration, especially in marine environments. Additionally, floating MPs reflect and block sunlight, reducing phytoplankton photosynthesis by approximately 45%, further affecting carbon fixation (Sjollema et al., 2016).

Microbial communities within MPs biofilms promote the mineralization and degradation of MPs, releasing potent GHGs such as methane (CH_4), ethylene (C_2H_4), nitrogen (No_x) and carbon dioxide (CO_2). These emissions are driven by microbial processes including ammonification, nitrification, and denitrification. The presence of nitrite reductase genes (e.g., nirS) correlates with elevated nitrous oxide (N_2O) emissions. N_2 fluxes ranging from 29.85 to 276.6 μmol/m^2/day in inflow rivers of Taihu Lake, with dissolved N_2O concentrations between 45.71 and 132.42 nmol/L. The nitrifying genus *Nitrospira* and denitrifying genera *Flavobacterium* and *Dechloromonas* were identified as key contributors to these emissions.

Sediment-bound MPs biofilms further exacerbate GHG emissions. CO_2 and CH_4 concentrations ranging from 1522 to 2253 mol/g/L and 134–1276 μmol/g/L, respectively (Chen et al., 2022). In freshwater and saltwater sediments in China, freshwater systems show greater sensitivity. Smaller MPs, due to increased surface area from photodegradation and structural damage (e.g., cracks and pitsO, emit more hydrocarbons per unit mass (Zhang et al., 2020)). The solar exposure of virgin and aged PE led to CH_4 emissions of 5.8 nmol/g/day and C_2H_4 emissions of 14.5 nmol/g/day, with rates increasing over time (Royer et al., 2018). Given the significant role of MPs biofilm in GHG emissions and their potential contribution to climate change, urgent attention is needed to understand and mitigate these impacts. Strategies should include hierarchical management of MPs using the 5 R's: reduce, reuse, recycle, repair, and recover (Prata et al., 2019). Reducing MP production through design modifications for durability and promoting alternative materials can help curb environmental release (Kumar et al., 2023). These measures are essential for safeguarding ecological sustainability and addressing the climate implications of MP pollution.

4.6.2 Ecotoxicological impacts of microplastics associate biofilms in aquatic ecosystems

MPs associated biofilms serve as pervasive ecological niches that facilitate the colonization of pathogenic microorganisms, thereby disrupting ecosystem services and food web dynamics. Their presence in aquatic environments has been linked to bioaccumulation in various taxa, including phytoplankton, bivalve mollusks, microcrustaceans, amphibians, and aquatic mammals, with documented implications for human health (Jeyavani et al., 2022). MP biofilms promote the selective enrichment of antibiotic-resistant bacteria and facilitate horizontal gene transfer, intensifying their toxicological impact. The adsorption of pharmaceuticals such as ciprofloxacin onto MPs surfaces has been shown to exert antagonistic toxicity effects on cyanobacteria like *Synechocystis* sp. In zebrafish, MPs biofilm exposure induces oxidative stress in intestinal tissues, with altered responses observed under coexposure to oxytetracycline (Yu et al., 2024). Moreover, MPs biofilms modulate the bioavailability of cocontaminants such as cadmium and benzopyrene, influencing gene expression related to toxicity biomarkers in larval zebrafish (Natale et al., 2023). Biofilm exposure also increases goblet cell density in zebrafish intestines, leading to microbiome dysbiosis and colonization by pathogenic genera. In fish, MPs can penetrate the gill vasculature, circulate to the liver, cross the blood-brain barrier, and accumulate in gonadal tissues, potentially transferring to offspring. Pathogens such as Aeromonas and Salmonella have been identified within MPs' biofilms, contributing to fish mortality and disrupting population dynamics. In mussels, MPs tend to adhere strongly to tissues rather than being ingested, impairing immune responses (Song et al., 2022).

The physical attributes of MPs size, shape and color significantly influence ingestion rates. Smaller particles resembling food are readily consumed by species such as *Chironomus tepperi*, *Daphnia magna*, and *Chironomus riparius*, leading to suppressed growth and survival (Silva et al., 2019). Juvenile common goby preferentially ingests white MPs, likely due to their resemblance to brine shrimp, a primary food source. MPs biofilms also affect aquatic plants by obstructing root pores and inhibiting growth. MPs can alter the toxicity of cooccurring pollutants. Polyethylene (PE) exposure impairs photosynthesis in wheat seedlings, with phenanthrene coexposure exacerbating toxicity (Liu et al., 2021b). Conversely, polystyrene (PS) particles adsorbing glyphosate were found to mitigate its inhibitory effects on Microcystis aeruginosa growth (Zhang et al., 2018).

4.6.3 Prospective and future research direction for microplastics-associated biofilms

MP-associated biofilms represent both a potential solution and a growing concern in the context of plastic pollution and ecosystem health. These biofilms provide a conductive environment for microbial communities capable of secreting catabolic enzymes such as esterase, lipase, and hydroxylase that facilitate the biodegradation of MPs. Notably, microbial strains including *Pseudomonas aeruginosa*, Rhodococcus spp., Bacillus cereus, *Ideonella sakaiensis*, *Bacillus gottheilii*, and *Stenotrophomonas maltophilia* have demonstrated the ability to degrade polymers like low-density polyethylene (LDPE), polyethylene terephthalate (PET), and polystyrene (PS) (Yuan et al., 2020). These microbial processes offer a promising avenue for converting MPs into polyhydroxyalkanoates (PHAs), a class of biodegradable bioplastics, thereby contributing to a circular economy. This "waste to wealth" strategy aligns with global sustainability goals by addressing climate change, biodiversity loss, and pollution. However, the same biofilms can also harbor pathogenic bacteria and antibiotic resistance genes, posing ecological risks by altering microbial diversity, disrupting nutrient cycles, and potentially exacerbating climate change.

Aging of MPs further complicates their environmental behavior by increasing surface charge, polarity, and hydrophilicity, which enhances microbial colonization and the leaching of toxic additives (Luo et al., 2024). To mitigate these risks, further strategies should prioritize preventing MPs from entering natural ecosystems, promoting the development of biodegradable alternatives derived from renewable biomaterials like cellulose and lignin, and reducing greenhouse gas (GHG) emissions through sustainable material design. Innovative approaches, such as combined chemo-biological treatments including thermal processing and microbial fermentation, could be employed to remove MPs from wastewater before they reach open water bodies. Although MPs biofilms may enhance nitrification and denitrification in freshwater systems, these processes may be suppressed in the presence of aquatic vegetation (Chen et al., 2020). Optimizing MPs removal technologies could also reduce microbial loads and complex pollutant mixtures in aquatic environments.

4.6.4 Research priorities

Mechanistic Studies on Co-Transport: Investigate the mechanisms governing the cotransport of MPs biofilms with other pollutants, particularly biological colloids, in aquatic systems. Development of predictive mathematical models is essential for understanding MPs fate and deposition. *Realistic Aging Conditions*: Examine the ecological impacts of aged MPs, which better represent environmental conditions. Aging alters surface morphology and increases specific surface area, enhancing microbial colonization and adsorption of hydrophobic organic pollutants. *Temporal Microbial Dynamics*: Explore how microbial responses to MPs evolve over time, affecting their ecological roles. Long-term studies are needed to assess the biological and toxicological effects of MPs under realistic environmental conditions.

Functional Ecology of the Plastisphere: Assess whether MPs-associated microbial communities perform similar ecological functions such as nutrient cycling and primary production as those on natural substrates, especially under nutrient-limited conditions.

GHG Emissions from Resuspended MPs: Investigate the role of MPs resuspension in shallow lakes in promoting mineralization, nitrification, and denitrification, which contribute to CH_4, CO_2 and N_2O emissions. Quantitative assessments of GHG fluxes from MPs surfaces are crucial for understanding their contribution to climate change.

In addition to scientific research, effective mitigation will require coordinated efforts involving public-private partnerships. Strategies should align with international frameworks such as the Geneva Convention and focus on four key areas: (1) enhancing the efficiency and economics of plastic recycling, (2) minimizing plastic litter, (3) investing in technologies that convert MPs into energy or valuable materials, and (4) fostering global collaboration to eliminate plastic pollution.

4.7 Conclusion

Collectively, emerging evidence highlights MPs as active ecological niches that promote the enrichment and persistence of antibiotic resistance genes and antibiotic-resistant bacteria in aquatic environments. The physicochemical properties of MP surfaces facilitate biofilm formation, enhance microbial interactions, and create favorable conditions for horizontal gene transfer, thereby accelerating resistance dissemination beyond natural particulate substrates. As MPs are highly mobile and persistent, they represent an underrecognized vector linking pollution, microbial ecology, and public health risk. Addressing antibiotic resistance in aquatic systems, therefore, requires integrated mitigation strategies that consider MP pollution control alongside conventional antimicrobial management and wastewater treatment approaches.

References

Ali, S.S., Elsamahy, T., Al-Tohamy, R., Sun, J., 2024. A critical review of microplastics in aquatic ecosystems: degradation mechanisms and removal strategies. Environ. Sci. Ecotechnol. 21, 100427.

Alvim, C.B., Mendoza–Roca, J.A., Bes–Pia, A., 2020. Wastewater treatment plant as microplastics release source–quantification and identification techniques. J. Environ. Manag. 255, 109739.

Amaral-Zettler, L.A., Zettler, E.R., Mincer, T.J., 2020. Ecology of the plastisphere. Nat. Rev. Microbiol. 18, 139–151.

Arnold, B.J., Huang, I.T., Hanage, W.P., 2022. Horizontal gene transfer and adaptive evolution in bacteria. Nat. Rev. Microbiol. 20 (4), 206–218. https://doi.org/10.1038/s41579-021-00650-4.

Ayush, P.T., Ko, J.H., Oh, H.S., 2022. Characteristics of initial attachment and Biofilm Formation of *Pseudomonas aeruginosa* on microplastic surfaces. Appl. Sci. 12 (10). https://doi.org/10.3390/app12105245.

Battulga, B., Kawahigashi, M., Oyuntsetseg, B., 2022. Characterization of biofilms formed on polystyrene microplastics (PS-MPs) on the shore of the Tuul River, Mongolia. Environ. Res. 212 (Pt B), 113329. https://doi.org/10.1016/j.envres.2022.113329.

Berg, H.C., 1975. Chemotaxis in bacteria. Annu. Rev. Biophys. Bioeng. 4, 119–136.

Chae, Y., Hong, S.H., An, Y.-J., 2020. Photosynthesis enhancement in four marine microalgal species exposed to expanded polystyrene leachate. Ecotoxicol. Environ. Saf. 189, 109936.

Chen, X., Chen, X., Zhao, Y., Zhou, H., Xiong, X., Wu, C., 2020. Effects of microplastic biofilms on nutrient cycling in simulated freshwater systems. Scie. Total Environ. 719. https://doi.org/10.1016/j.scitotenv.2020.137276.

Chen, X., Wang, Y., Chen, S., Sun, Y., Tan, Q., Ding, Z., Lu, Y., Yu, Y., 2021. Microplastics as carbon–nutrient sources and shapers of microbial communities in stagnant water. J. Hazard Mater. 420, 126662.

Chen, Y., Gao, B., Yang, Y., Pan, Z., Liu, J., Sun, K., Xing, B., 2022. Tracking microplastics biodegradation through CO2 emission: role of photoaging and mineral addition. J. Hazard. Mater. 439, 129615.

Dai, S., Ye, R., Huang, J., Wang, B., Xie, Z., Ou, X., Yu, N., Huang, C., Hua, Y., Zhou, R., Tian, B., 2022. Lipid membrane interactions and cellular uptake of differentially charged nanoplastics in bacteria. J. Nanobiotechnol. 20, 191.

Dong, X., Zhu, L., Jiang, P., Wang, X., Liu, K., Li, C., Li, D., 2021. Seasonal biofilm formation on floating microplastics in coastal waters of intensified marinculture area. Marin. Poll. Bull. 171, 112914.

Enfrin, M., Dumée, L.F., Lee, J., 2019. Nano/microplastics in water and wastewater treatment processes – origin, impact and potential solutions. Water Res. 161, 621–638. https://doi.org/10.1016/j.watres.2019.06.049.

Fajardo, C., Sánchez-Fortún, S., Videira-Quintela, D., Martin, C., Nande, M., D́ors, A., Costa, G., Guillen, F., Montalvo, G., Martin, M., 2023. Biofilm formation on polyethylene microplastics and their role as transfer vector of emerging organic pollutants. Environ. Sci. Pollut. Control Ser. 30 (35), 84462–84473. https://doi.org/10.1007/s11356-023-28278-2.

Galgani, L., Tsapakis, M., Pitta, P., Tsiola, A., Tzempelikou, E., Kalantzi, I., Esposito, C., Loiselle, A., Tsotskou, A., Zivanovic, S., Dafnomili, E., Diliberto, S., Mylona, K., Magiopoulos, I., Zeri, C., Pitta, E., Loiselle, S.A., 2019. Microplastics increase the marine production of particulate forms of organic matter. Environ. Res. Lett. 14 (12), 124085. https://doi.org/10.1088/1748-9326/ab59ca.

Gong, M., Yang, G., Zhuang, L., Zeng, E.Y., 2019. Microbial biofilm formation and community structure on low–density polyethylene microparticles in lake water microcosms. Environ. Poll. 252, 94–102.

Guan, J., Qi, K., Wang, J., Wang, W., Wang, Z., Lu, N., Qu, J., 2020. Microplastics as an emerging anthropogenic vector of trace metals in freshwater: significance of biofilms and comparison with natural substrates. Water Res. 184, 116205. https://doi.org/10.1016/j.watres.2020.116205.

Horton, A.A., Dixon, S.J., 2018. Microplastics: an introduction to environmental transport processes. Wiley Interdiscip. Rev. Water 5 (2), e1268.

Hou, L., Xi, J., Liu, J., Wang, P., Xu, T., Liu, T., Qu, W., Lin, Y.B., 2022. Biodegradability of polyethylene mulching film by two Pseudomonas strains and their potential degradation mechanisms. Chemosphere 286, 131758.

Hou, X., Mu, L., Hu, X., Guo, S., 2023. Combined effects of warming and microplastic pollution on algal carbon and nitrogen cycling. J. Hazard Mater. 447, 130775.

Huang, S., Zhang, B., Zhao, Z., Yang, C., Zhang, B., Cui, F., Lens, P.N.L., Shi, W., 2023. Metagenomic insights into microbial community responses and nitrogen metabolic pathways under exposure to polystyrene micro- and nanoplastics in activated sludge systems. Water Res. 241, 120161.

Huffer, T., Hofmann, T., 2016. Sorption of non–polar organic compounds by micro–sized plastic particles in aqueous solution. Environ. Pollut. 214, 194–201.

Jebashalomi, V., Emmanuel Charles, P., Rajaram, R., 2024. Microbial degradation of low-density polyethylene and polystyrene by Bacillus cereus isolated from a plastic-polluted tropical coastal environment. Sci. Total Environ. 924, 171580.

Jeyavani, J., Sibiya, A., Bhavaniramya, S., Mahboob, S., Al-Ghanim, K.A., Nisa, Z.u., Riaz, M.N., Nicoletti, M., Govindarajan, M., Vaseeharan, B., 2022. Toxicity evaluation of polypropylene microplastic on marine microcrustacean *Artemia salina*: an analysis of implications and vulnerability. Chemosphere 296. https://doi.org/10.1016/j.chemosphere.2022.133990.

Jia, J., Liu, Q., Zhao, E., Li, X., Xiong, X., Wu, C., 2024. Biofilm formation on microplastics and interactions with antibiotics, antibiotic resistance genes and pathogens in aquatic environment. Eco-Environ. Health 3 (4), 516–528.

Jin, M., Yu, X., Yao, Z., Tao, P., Li, G., Yu, X., Zhao, J.-L., Peng, J., 2020. How biofilms affect the uptake and fate of hydrophobic organic compounds (HOCs) in microplastic: insights from an in situ study of Xiangshan Bay, China. Water Res. 184, 116118. https://doi.org/10.1016/j.watres.2020.116118.

Johansen, M.P., Cresswell, T., Davis, J., Howard, D.L., Howell, N.R., Prentice, E., 2019. Biofilm-enhanced adsorption of strong and weak cations onto different microplastic sample types: use of spectroscopy, microscopy and radiotracer methods. Water Res. 158, 392–400. https://doi.org/10.1016/j.watres.2019.04.029.

Jose, S., Jordao, L., 2022. Exploring the interaction between microplastics, polycyclic aromatic hydrocarbons and biofilms in freshwater. Polycyc. Aroma. Comp. 42, 2210–2221.

Junaid, M., Siddiqui, J.A., Sadaf, M., Liu, S., Wang, J., 2022. Enrichment and dissemination of bacterial pathogens by microplastics in the aquatic environment. Sci. Total Environ. 830, 154720.

Kim, S.H., Lee, J.W., Kim, J.S., Lee, W., Park, M.S., Lim, Y.W., 2022. Plastic-inhabiting fungi in marine environments and polycaprolactone degradation activity. Antonie Leeuwenhoek 115, 1379–1392.

Kumar, A., Mishra, S., Pandey, R., Yu, Z.G., Kumar, M., Khoo, K.S., Thakur, T.K., Show, P.L., 2023. Microplastics in terrestrial ecosystems: un-ignorable impacts on soil characterises, nutrient storage and its cycling. Trends Anal. Chem. 158. https://doi.org/10.1016/j.trac.2022.116869.

La Porta, E., Exacoustos, O., Lugani, F., Angeletti, A., Chiarenza, D.S., Bigatti, C., Spinelli, S., Kajana, X., Garbarino, A., Bruschi, M., 2023. Microplastics and kidneys: an update on the evidence for deposition of plastic microparticles in human organs, tissues and fluids and renal toxicity concern. Int. J. Mol. Sci. 24, 14391.

Li, W., Zhang, Y., Wu, N., Zhao, Z., Xu, W., Ma, Y., Niu, Z., 2019. Colonization characteristics of bacterial communities on plastic debris influenced by environmental factors and polymer types in the haihe estuary of bohai bay, China. Environ. Sci. Technol. 53 (18), 10763–10773. https://doi.org/10.1021/acs.est.9b03659.

Li, L., Li, F., Deng, M., Wu, C., Zhao, X., Song, K., Wu, F., 2022. Microplastics distribution characteristics in typical inflow rivers of Taihu lake: linking to nitrous oxide emission and microbial analysis. Water Res. 225, 119117.

Liu, C., Shi, C., Zhu, S., Wei, R., Yin, C.-C., 2019. Structural and functional characterization of polyethylene terephthalate hydrolase from Ideonella sakaiensis. Biochem. Biophys. Res. Commun. 508, 289–294.

Liu, X., Ma, J., Yang, C., Wang, L., Tang, J., 2021. Toxic effects of nano- and microplastics on an antibiotic-producing strain Streptomyces coelicolor M145. Sci. Total Environ. 764, 142804.

Liu, S., Junaid, M., Liao, H., Liu, X., Wu, Y., Wang, J., 2022. Eco-corona formation and associated ecotoxicological impacts of nanoplastics in the environment. Sci. Total Environ. 836, 155703.

Lobelle, D., Cunliffe, M., 2011. Early microbial biofilm formation on marine plastic debris. Mar. Pollut. Bull. 62 (1), 197–200. https://doi.org/10.1016/j.marpolbul.2010.10.013.

Luo, D., Wang, Z., Liao, Z., Chen, G., Ji, X., Sang, Y., Qu, L., Chen, Z., Wang, Z., Dahlgren, R.A., Zhang, M., Shang, X., 2024. Airborne microplastics in urban, rural and wildland environments on the Tibetan Plateau. J. Hazard Mater. 465, 133177. https://doi.org/10.1016/j.jhazmat.2023.133177.

Lv, S., Li, Y., Zhao, S., Shao, Z., 2024. Biodegradation of typical plastics: from microbial diversity to metabolic mechanisms. Int. J. Mol. Sci. 25, 593.

Ma, Y., Huang, J., Han, T., Li, R., Cao, M., Li, X., 2022. Comprehensive assessment of nanoplastics exposure in constructed wetland ecosystems: macrophyte physiology, microbial enzymology, community composition and metabolic functions. Chem. Eng. J. 434, 134592.

McGivney, E., Cederholm, L., Barth, A., Hakkarainen, M., Hamacher-Barth, E., Ogonowski, M., Gorokhova, E., 2020. Rapid physicochemical changes in microplastics induced by biofilm formation. Front. Bioeng. Biotechnol. 8, 205.

Michels, J., Stippkugel, A., Lenz, M., Wirtz, K., Engel, A., 2018. Rapid aggregation of biofilm–covered microplastics with marine biogenic particles. Proc. R. Soc. A B 285, 20181203.

Miri, S., Saini, R., Davoodi, S.M., Pulicharla, R., Brar, S.K., Magdouli, S., 2022. Biodegradation of microplastics: better late than never. Chemosphere 286, 131670.

Natale, M.D., Catarino, A.I., Summers, S., Boyle, D., Torri, M., Nicosia, A., Musco, M., Masullo, T., Russo, S., Bennici, C.D., Mazzola, A., Cuttitta, A., Henry, T.B., 2023. Influence of microplastic–associated biofilms on the bioavailability of a mixture of cadmium and benzo[a]pyrene by the analysis of biomarker gene expression in larval zebrafish. Ecol. Ind. 152, 110369.

Paco, A., Duarte, K., da Costa, J.P., Santos, P.S., Pereira, R., Pereira, M.E., Freitas, A.C., Duarte, A.C., Rocha-Santos, T., 2017. Biodegradation of polyethylene microplastics by the marine fungus Zalerion maritimum. Sci. Total Environ. 586, 10–15.

Pan, H., Zhao, X., Zhou, X., Yan, H., Han, X., Wu, M., Chen, F., 2023. Research progress on the role of biofilm in heavy metals adsorption-desorption characteristics of microplastics: a review. Environ. Pollut. 336, 122448. https://doi.org/10.1016/j.envpol.2023.122448.

Park, S.Y., Kim, C.G., 2019. Biodegradation of micro-polyethylene particles by bacterial colonization of a mixed microbial consortium isolated from a landfill site. Chemosphere 222, 527–533. https://doi.org/10.1016/j.chemosphere.2019.01.159.

Parveen, N., Ranjan, V.P., Chowdhury, S., Goel, S., 2022. Occurrence and potential health risks due to trihalomethanes and microplastics in bottled water. Environ. Eng. Sci. 39 (6), 523–534. https://doi.org/10.1089/ees.2021.0295.

Pete, A.J., Brahana, P.J., Bello, M., Benton, M.G., Bharti, B., 2023. Biofilm Formation influences the wettability and settling of microplastics. Environ. Sci. Technol. Lett. 10 (2), 159–164. https://doi.org/10.1021/acs.estlett.2c00728.

Phuong, N.N., Duong, T.T., Le, T.P.Q., Hoang, T.K., Ngo, H.M., Phuong, N.A., Pham, Q.T., Doan, T.O., Ho, T.C., Da Le, N., Nguyen, T.A.H., Strady, E., Fauvelle, V., Ourgaud, M., Schmidt, N., Sempere, R., 2022. Microplastics in Asian freshwater ecosystems: current knowledge and perspectives. Sci. Total Environ. 808. https://doi.org/10.1016/j.scitotenv.2021.151989.

Porizka, P., Brunnbauer, L., Porkert, M., Rozman, U., Marolt, G., Holub, D., Kizovský, M., Benesova, M., Samek, O., Limbeck, A., Kaiser, J., Kalcíkova, G., 2023. Laser-based techniques: novel tools for the identification and characterization of aged microplastics with developed biofilm. Chemosphere 313, 137373. https://doi.org/10.1016/j.chemosphere.2022.137373.

Prata, J.C., Patrício Silva, A.L., da Costa, J.P., Mouneyrac, C., Walker, T.R., Duarte, A.C., Rocha-Santos, T., 2019. Solutions and integrated strategies for the control and mitigation of plastic and microplastic pollution. Int. J. Environ. Res. Publ. Health 16 (13). https://doi.org/10.3390/ijerph16132411.

Qi, K., Lu, N., Zhang, S., Wang, W., Wang, Z., Guan, J., 2021. Uptake of Pb(II) onto microplastic-associated biofilms in freshwater: adsorption and combined toxicity in comparison to natural solid substrates. J. Hazard Mater. 411, 125115. https://doi.org/10.1016/j.jhazmat.2021.125115.

Rajala, K., Grönfors, O., Hesampour, M., Mikola, A., 2020. Removal of microplastics from secondary wastewater treatment plant effluent by coagulation–flocculation using iron-, aluminum- and polyamine-based chemicals. Water Res. 183, 116045.

Rakib, M.R.J., Sarker, A., Ram, K., Uddin, M.G., Walker, T.R., Chowdhury, T., Uddin, J., Khandaker, M.U., Rahman, M.M., Idris, A.M., 2023. Microplastic toxicity in aquatic organisms and aquatic ecosystems: a review. Water Air Soil Pollut. 234, 52.

Ramsperger, A.F.R.M., Stellwag, A.C., Caspari, A., Fery, A., Lueders, T., Kress, H., Loder, M.G.J., Laforsch, C., 2020. Structural diversity in early–stage biofilm formation on microplastics depends on environmental medium and polymer properties. Water 12, 3216.

Rather, M.A., Gupta, K., Mandal, M., 2021. Microbial biofilm: formation, architecture, antibiotic resistance, and control strategies. Braz. J. Microbiol. 52, 1701–1718.

Richard, H., Carpenter, E.J., Komada, T., Palmer, P.T., Rochman, C.M., 2019. Biofilm facilitates metal accumulation onto microplastics in estuarine waters. Sci. Total Environ. 683, 600–608. https://doi.org/10.1016/j.scitotenv.2019.04.331.

Rodrigues, J.P., Duarte, A.C., Santos-Echeandía, J., Rocha-Santos, T., 2019. Significance of interactions between microplastics and persistent organic pollutants in the marine environment: a critical overview. TrAC, Trends Anal. Chem. 111, 252–260.

Royer, S.-J., Ferrón, S., Wilson, S.T., Karl, D.M., Pardha-Saradhi, P., 2018. Production of methane and ethylene from plastic in the environment. PLoS One 13 (8), e0200574. https://doi.org/10.1371/journal.pone.0200574.

Sanchez, C., 2020. Fungal potential for the degradation of petroleum-based polymers: an overview of macro- and microplastic biodegradation. Biotechnol. Adv. 40, 107501.

Sangale, M.K., Shahnawaz, M., Ade, A.B., 2019. Potential of fungi isolated from mangrove rhizosphere soils at dumping sites to degrade polythene. Sci. Rep. 9, 5390.

Seeley, M.E., Song, B., Passie, R., Hale, R.C., 2020. Microplastics affect sedimentary microbial communities and nitrogen cycling. Nat. Commun. 11 (1). https://doi.org/10.1038/s41467-020-16235-3.

Sharma, S., Mohler, J., Mahajan, S.D., Schwartz, S.A., Bruggemann, L., Aalinkeel, R., 2023. Microbial biofilm: a review on formation, infection, antibiotic resistance, control measures, and innovative treatment. Microorganisms 11 (6). https://doi.org/10.3390/microorganisms11061614.

Silva, C.J.M., Silva, A.L.P., Gravato, C., Pestana, J.L.T., 2019. Ingestion of small-sized and irregularly shaped polyethylene microplastics affect Chironomus riparius life-history traits. Sci. Total Environ. 672, 862–868. https://doi.org/10.1016/j.scitotenv.2019.04.017.

Silva, I., Rodrigues, E.T., Tacao, M., Henriques, I., 2023. Microplastics accumulate priority antibiotic-resistant pathogens: evidence from the riverine plastisphere. Environ. Pollut. 332, 121995.

Sjollema, S.B., Redondo-Hasselerharm, P., Leslie, H.A., Kraak, M.H.S., Vethaak, A.D., 2016. Do plastic particles affect microalgal photosynthesis and growth? Aquat. Toxicol. 170, 259–261. https://doi.org/10.1016/j.aquatox.2015.12.002.

Song, X., Wen, Z., Cui, H., Liu, M., Gao, T., Li, A., Gao, Z., 2022. Interactions of microplastics with organic, inorganic and bio-pollutants and the ecotoxicological effects on terrestrial and aquatic organisms. Sci. Total Environ. 838 (Pt 2), 156068. https://doi.org/10.1016/j.scitotenv.2022.156068.

Sourkova, M., Adamcova, D., Vaverkova, M.D., 2021. Influence of microplastics derived from ground tyres on acute and subchronic toxicity and microbial respiration of soil. Environments 8, 128.

Sun, X., Chen, B., Li, Q., Liu, N., Xia, B., Zhu, L., Qu, K., 2018. Toxicities of polystyrene nano- and microplastics toward the marine bacterium Halomonas alkaliphila. Sci. Total Environ. 642, 1378–1385.

Sun, Y., Wang, X., Xia, S., Zhao, J., 2022. Cu(II) adsorption on poly(lactic acid) microplastics: significance of microbial colonization and degradation. Chem. Eng. J. 429, 132306.

Sun, Y., Peng, B., Wang, X., Li, Y., Wang, Y., Zhang, Y., Xia, S., Zhao, J., 2023. Adsorption and desorption mechanisms of oxytetracycline on poly(butylenes adipate–co–terephthalate) microplastics after degradation: the effects of biofilms, Cu (II), water pH, and dissolved organic matter. Sci. Total Environ. 863, 160866.

Syranidou, E., Kalogerakis, N., 2022. Interactions of microplastics, antibiotics and antibiotic resistant genes within WWTPs. Sci. Total Environ. 804, 150141. https://doi.org/10.1016/j.scitotenv.2021.150141.

Truong, T.-N.-S., Strady, E., Kieu-Le, T.-C., Tran, Q.-V., Thi-Minh-Tam, Le, Thuong, Q.-T., 2021. Microplastic in atmospheric fallouts of a developing Southeast Asian megacity under tropical climate. Chemosphere 272, 129874. https://doi.org/10.1016/j.chemosphere.2021.129874.

Tu, C., Chen, T., Zhou, Q., Liu, Y., Jing, W., Waniek, J.J., Luo, Y., 2020. Biofilm formation and its influences on the properties of microplastics as affected by exposure time and depth in the seawater. Sci. Total Environ. 734, 139237. https://doi.org/10.1016/j.scitotenv.2020.139237.

Wang, H., Yang, Q., Li, D., Wu, J., Yang, S., Deng, Y., Luo, C., Jia, W., Zhong, Y., Peng, P., 2023. Stable isotope and metagenomic evidence for microbial-mediated effects of microplastics on sulfur cycling in coastal sediments. Environ. Sci. Technol. 57, 1167–1176.

Wei, W., Zhang, Y.-T., Huang, Q.-S., Ni, B.-J., 2019. Polyethylene terephthalate microplastics affect hydrogen production from alkaline anaerobic fermentation of waste activated sludge by altering the viability and activity of anaerobic microorganisms. Water Res. 163, 114881.

Wu, J., Jiang, R., Lin, W., Ouyang, G., 2019a. Effects of salinity and humic acid on aggregation behavior and toxicity of polystyrene nanoplastics with different functional groups and surface charges. Environ. Pollut. 245, 836–843.

Wu, Y., Guo, P., Zhang, X., Zhang, Y., Xie, S., Deng, J., 2019b. Effects of microplastic exposure on the photosynthetic system of freshwater algae. J. Hazard Mater. 374, 219–227.

Wu, C., Tanaka, K., Tani, Y., Bi, X., Liu, J., Yu, Q., 2022. Effect of particle size on the colonization of biofilms and the potential of biofilm–covered microplastics as metal carriers. Sci. Total Environ. 821, 153265.

Wu, T., Zhong, L., Ding, J., Pang, J.-W., Sun, H.-J., Ding, M.-Q., Ren, N.-Q., Yang, S.-S., 2023. Microplastics perturb nitrogen removal, microbial community structure and metabolic mechanisms in biofilm systems. J. Hazard Mater. 458, 131971.

Xiang, P., Zhang, Y., Zhang, T., Wu, Q., Zhao, C., Li, Q., 2023. A novel bacterial consortium for efficient degradation of polystyrene microplastics. J. Hazard Mater. 458, 131856.

Xiao, Y., Wang, Q., Li, P., Xu, M., Zhou, Y., Li, H., Yan, W., Liu, C., Vähätalo, A.V., 2023. Impact of light-aged microplastics on microalgal production of dissolved organic matter. Sci. Total Environ. 889, 164312.

Xie, H., Chen, J., Feng, L., He, L., Zhou, C., Hong, P., Sun, S., Zhao, H., Liang, Y., Ren, L., Zhang, Y., Li, C., 2021. Chemotaxis-selective colonization of mangrove rhizosphere microbes on different microplastics. Sci. Total Environ. 752, 142223.

Xu, Y., Fu, Q., He, D., Yang, F., Ma, X., Wang, Y., Liu, Z., Liu, X., Wang, D., 2024. Exposure to polyethylene microplastics affects sulfur migration and transformation in anaerobic systems. J. Hazard Mater. 134520.

Xue, J., Peldszus, S., Van Dyke, M.I., Huck, P.M., 2021. Removal of polystyrene microplastic spheres from surface waters using alum-based coagulation–flocculation–sedimentation. Chem. Eng. J. 422, 130023.

You, X., Cao, X., Zhang, X., Guo, J., Sun, W., 2021. Unraveling individual and combined toxicity of nano/microplastics and ciprofloxacin to Synechocystis sp. at the cellular and molecular levels. Environ. Int. 157, 106842. https://doi.org/10.1016/j.envint.2021.106842.

Yu, H., Liu, M., Gang, D., Peng, J., Hu, C., Qu, J., 2022. Polyethylene microplastics interfere with the nutrient cycle in water–plant–sediment systems. Water Res. 214, 118191.

Yu, Z., Qiu, D., Zhou, T., Zeng, L., Yan, C., 2024. Biofilm enhances the interactive effects of microplastics and oxytetracycline on zebrafish intestine. Aqua. Toxicol. 270, 106905.

Yuan, J., Ma, J., Sun, Y., Zhou, T., Zhao, Y., Yu, F., 2020. Microbial degradation and other environmental aspects of microplastics/plastics. Sci. Total Environ. 715, 136968. https://doi.org/10.1016/j.scitotenv.2020.136968.

Yuan, Y., Qin, Y., Wang, M., Xu, W., Chen, Y., Zheng, L., Chen, W., Luo, T., 2022. Microplastics from agricultural plastic mulch films: a mini-review of their impacts on the animal reproductive system. Ecotoxicol. Environ. Safety 244. https://doi.org/10.1016/j.ecoenv.2022.114030.

Zhang, Q., Qu, Q., Lu, T., Ke, M., Zhu, Y., Zhang, M., Zhang, Z., Du, B., Pan, X., Sun, L., Qian, H., 2018. The combined toxicity effect of nanoplastics and glyphosate on Microcystis aeruginosa growth. Environ. Pollut. 243, 1106–1112. https://doi.org/10.1016/j.envpol.2018.09.073.

Zhang, Y., Liang, J., Zeng, G., Tang, W., Lu, Y., Luo, Y., Xing, W., Tang, N., Ye, S., Li, X., Huang, W., 2020. How climate change and eutrophication interact with microplastic pollution and sediment resuspension in shallow lakes: a review. Sci. Total Environ. 705, 135979.

Zhang, B., Yang, X., Liu, L., Chen, L., Teng, J., Zhu, X., Zhao, J., Wang, Q., 2021a. Spatial and seasonal variations in biofilm formation on microplastics in coastal waters. Sci. Total Environ. 770, 145303.

Zhang, S., Wang, J., Yan, P., Hao, X., Xu, B., Wang, W., Aurangzeib, M., 2021b. Non-biodegradable microplastics in soils: a brief review and associated challenges. J. Hazard Mater. 409, 124525.

Zhang, G., Liu, D., Lin, J., Kumar, A., Jia, K., Tian, X., Yu, Z., Zhu, B., 2023. Priming effects induced by degradable microplastics in agricultural soils. Soil Biol. Biochem. 180, 109006. https://doi.org/10.1016/j.soilbio.2023.109006.

Zhao, H., Sun, S., Cui, Y., Ullah, M.W., Alabbosh, K.F., Elboughdiri, N., Zhou, J., 2024. Sustainable production of bacterial flocculants from nylon-6,6 microplastic hydrolysate using Brucella intermedia ZL-06. J. Hazard Mater. 465, 133435.

Zhou, D., Chen, J., Wu, J., Yang, J., Wang, H., 2021. Biodegradation and catalytic–chemical degradation strategies for mitigating microplastic pollution. Sustain. Mater. Technol. 28, e00251.

Zhou, D., Cai, Y., Yang, Z., 2022. Key factors controlling transport of micro– and nanoplastic in porous media and its effect on coexisting pollutants. Environ. Poll. 293, 118503.

Zhou, S., Wang, L., Liu, J., Zhang, C., Liu, X., 2024. Toxic effects of microplastics and influencing factors on microorganisms in biological wastewater treatment units. Water Sci. Technol. 89, 1539–1553.

Zhu, K., Jia, H., Sun, Y., Dai, Y., Zhang, C., Guo, X., Wang, T., Zhu, L., 2020. Long-term phototransformation of microplastics under simulated sunlight irradiation in aquatic environments: roles of reactive oxygen species. Water Res. 173, 115564.

Zong, L., Wang, X., Huo, M., Yi, F., Huang, S., Ling, T., Fang, Y., Ma, F., Zhang, X., Guan, M., 2024. Synergistic toxicity mechanisms of nano- and microplastics with triclosan revealed by a dose-dependent functional genomics approach in *Saccharomyces cerevisiae*. Chemosphere 362, 142629.

CHAPTER

Detection, monitoring, and analytical techniques for microplastic assessment

5

Melvin S. Samuel

Department of Civil, Construction, and Environmental Engineering, Marquette University, Milwaukee, WI, United States

Highlights

- Reviews state-of-the-art spectroscopy and microscopy techniques for Microplastic detection and quantification.
- Compare advances in field sampling, laboratory extraction, and analytical workflows across environmental matrices.
- Evaluate emerging tools such as automated imaging, chemometrics, and bioinformatics assisted analysis.
- Identifies key methodologies gaps and inconsistencies limiting global comparability of Microplastic data
- Process pathways toward standardized, reproducible framework to support monitoring and policy development.

5.1 Introduction

The reliability of MP data in aquatic environments is intrinsically linked to the sampling methodology employed. While commonly adopted sampling techniques offer distinct benefits, they also present limitations that influence data quality and interpretation. Variability in MP abundance estimates across studies is often attributed to inconsistencies in sampling gear, mesh sizes, and fluid collection systems, thereby complicating cross-comparison and standardization efforts (Priya et al., 2022). Post-collection, samples typically undergo chemical digestion to eliminate organic matter, followed by separation using density gradient centrifugation and membrane filtration. These preparatory steps are critical for isolating MP particles from complex environmental matrices. Subsequent identification of polymer types relies on a suite of analytical techniques, including spectroscopic, thermal, and chemical methods. Notably, Fourier-transform infrared spectroscopy (FTIR), Raman spectroscopy, optical photothermal infrared spectroscopy (O-PTIR), and pyrolysis-gas chromatography/mass spectrometry (Py-GC/MS) are widely utilized for polymer characterization and surface analysis.

Despite their analytical precision, these methods are often constrained by high operational costs, labor-intensive protocols, and limited applicability for in situ detection. Such challenges underscore the need for more accessible and scalable approaches to MP analysis. Moreover, the absence of harmonized protocols across research initiatives hampers the comparability of datasets, necessitating the development of standardized, cost-effective methodologies that facilitate consistent monitoring

Microplastics in Aquatic Environments. https://doi.org/10.1016/B978-0-443-45619-0.00007-8

across spatial and temporal scales. To address discrepancies arising from diverse sampling and reporting practices, rescaling frameworks have been proposed. These approaches aim to normalize data across varying particle size ranges and measurement units, thereby enhancing the interoperability and utility of existing datasets for meta-analyses and policy formulation.

5.2 Spectroscopy and microscopy-based techniques

Spectroscopy and microscopy-based techniques form the backbone of MP detection, identification, and risk interpretation across aquatic environments (Table 5.1). Microscopic approaches, including optical and fluorescence microscopy, enable the initial visualization, sizing, and morphological classification of MPs, while advanced tools such as scanning electron microscopy (SEM) coupled with energy-dispersive X-ray spectroscopy (EDX) provide high-resolution surface characterization and elemental composition. Spectroscopic techniques, particularly Fourier transform infrared spectroscopy (FTIR), micro-FTIR, Raman spectroscopy, and pyrolysis gas chromatography/mass spectrometry (Py-GC/MS), allow reliable polymer identification and chemical fingerprinting of MPs and their associated additives or sorbed contaminants. These analytical methods are essential for inking MP characteristics to environmental behavior, as illustrated in the accompanying schematic overview, which depicts how meteorological events, hydrodynamics, wastewater inputs, and human activities govern the spatial and temporal distribution of MPs in aquatic systems (Fig. 5.1).

5.2.1 Optical microscopy

Optical Microscopy remains a widely utilized technique for the preliminary identification of MPs, relying on visual assessment of particle morphology, including size, color, and shape. It is particularly effective for particles within 1–5 mm range; however, its resolution is fundamentally constrained by the diffraction limit of visible light, typically around 200 nm, rendering it inadequate for detecting submicron particles. The method is also labor-intensive, requiring manual inspection and classification, which limits throughput and reproducibility.

In contrast, Laser Direct Infrared Imaging (LDIR) integrates high-resolution imaging with infrared spectroscopy, enabling automated and rapid detection of MPs. LDIR circumvents the limitations of manual microscopy by providing chemical specificity and spatial resolution simultaneously, thereby enhancing analytical efficiency and reliability.

5.2.2 Fluorescence microscopy

Fluorescence microscopy employs fluorescent dyes to visualize and identify MPs under specific excitation wavelength. This approach facilitates the detection of smaller particles and supports image-based quantification. MPs combined with the selective fluorescence dyes such as Nile Red, MPs is visualized and quantified in water samples, making this approach attractive for large scale monitoring studies. Several investigations have demonstrated the applicability of fluorescence-based methods for both detection and quantity in water samples, making this approach attractive for large scale monitoring studies. Several investigations have demonstrated the applicability of fluorescence-based method for both detection and quantification of MP particles. Despite these, fluorescence microscopy has certain limitations. While it is relatively inexpensive and faster than spectroscopic

Table 5.1 Comparison of various methods for detection of microplastics.

S. No	Techniques	Principle	Detection limit	Advantages	Disadvantages
1	Optical microscopy	Microplastics are counted based on shapes, colors, thickness, sheen and/or with a hot pin	>500 μM	Non-destructive and less expensive	Polymer type cannot be identified, lower automation
2	Fluorescence microscope	Microplastics are counted based on shapes, colors, thickness, sheen and/or with a hot pin	>500 μM	Non-destructive and less expensive	Polymer type cannot be identified, lower automation
3	μ-FTIR	Pre-selected microplastics or a proportion of pretreated samples are identified according to the changes in the dipole moment of functional groups in the MPs	>100 μM	High automation Faster analysis Non-destructive	Comparatively low identification of MP types A limited quantity of ambiguous particles can be analyzed under a single-point analysis mode
4	LC/MS	Samples are solvent phase separated by LC and then passed to the mass spectrometer for detection	>5 mg	Less sample preparation	Only a sample amount of sample can be analyzed at a time
5	LDIR	Rapid imaging and full-spectrum analysis via quantum cascade laser (QCL).	>20 μM	Non-destructive, higher automation, fully integrated workflow, no liquid nitrogen needed	Expensive

techniques, it is prone to errors arising from the co-staining of natural organic matter, which can lead to overestimation or misidentification of MPs. In addition, rubber-based polymers are generally not detectable using this approach, as they do not exhibit fluorescence under typical analytical conditions. Beyond simple quantification, fluorescence spectroscopy can also provide information on polymer characteristics. Emission spectra analysis has shown that different classes of MPs exhibit distinct fluorescence responses. Polar polymers such as polypropylene (PP) and polyethylene (PE) typically show emission maxima in the range of approximately 420–520 nm, whereas non-polar polymers

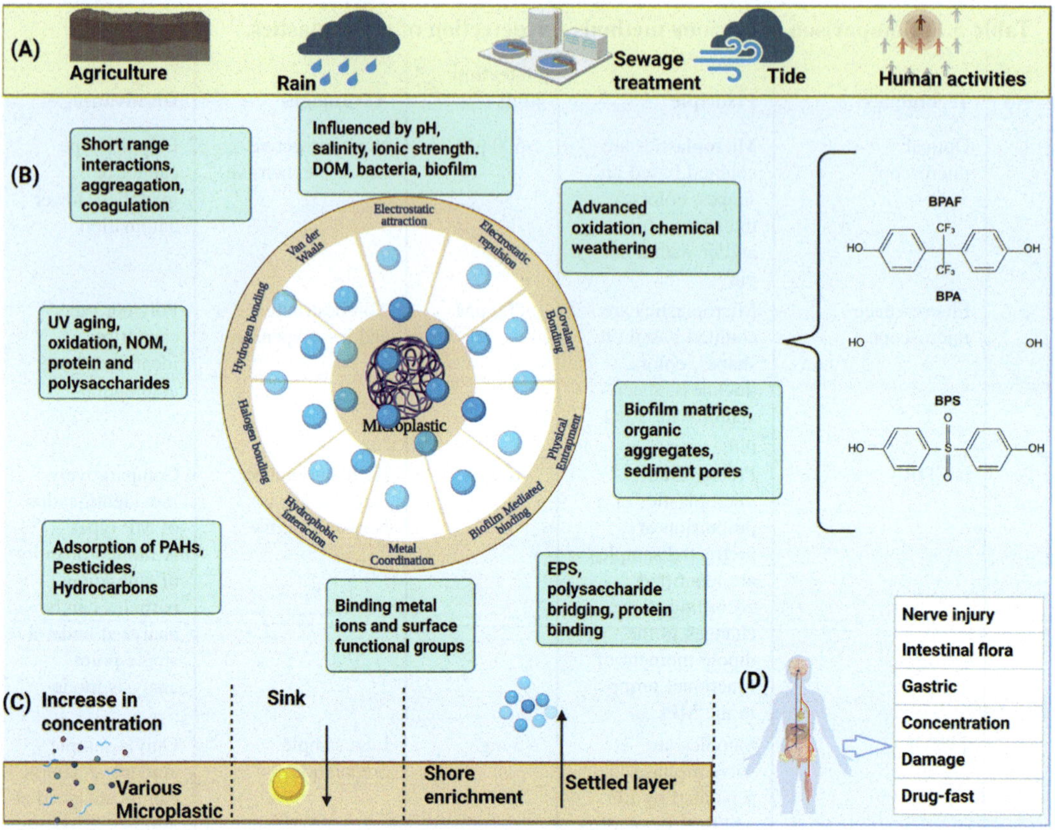

FIGURE 5.1

Overview of extraction, purification, and identification techniques for roadside microplastics. The schematic shows density separation usi nb ng salt solutions or oils, followed by organic matter removal via chemical or enzymatic digestion. Microplastics are subsequently identified and characterized using spectroscopic, microscopic, and chromatographic methods (e.g., FTIR, Raman, SEM/EDX, GC–MS).

display broader emission features at wavelengths above 520 nm. These spectral differences offer additional potential for polymer differentiation, although confirmatory techniques are often required for accurate identification. However, its accuracy is compromised by the non-specific binding of dyes to various polymer types, which may result in false positives or misidentification (Tamminga, 2017). LDIR offers a more robust alternative by eliminating the need for chemical staining. It directly interrogates the chemical structure of particles using infrared light, thereby reducing analytical variability and improving specificity.

5.2.3 Raman spectroscopy

Raman spectroscopy has emerged as a powerful tool for MP analysis due to its capacity to detect molecular vibrations via inelastic scattering of monochromatic light. This technique provides high-resolution spectral data and is particularly effective for identifying particles smaller than 1 μm, including nanoplastics (Fischer et al., 2016). Raman spectroscopy is a vibrational spectroscopic technique widely employed to characterize the molecular structure and chemical composition of materials. It is based on the Raman effect, in which a small fraction of incident light is scattered with a shift in wavelength due to interactions with molecular vibrations. These wavelength shifts generate a characteristic spectral fingerprint that reflects the molecular bonds and structural features of the analyzed material. In Raman analysis, a monochromatic laser is focused onto the sample surface, and the scattered light is collected and analyzed to produce a Raman spectrum. Each polymer exhibits a unique set of vibrational modes, enabling reliable identification of different plastic types based on their spectral signatures. Raman spectroscopy is applicable to solids, liquids, and gases, and is particularly effective for studying polymers, composites, and biological materials. The technique has become increasingly important in MP research due to its high chemical specificity and ability to analyze particles at very small sizes (Jorquera et al., 2022). Raman spectroscopy can detect and identify MPs at micrometer and even sub-micrometer scales, making it particularly valuable for studying fine particles that are difficult to characterize using FTIR alone. In addition to polymer identification, Raman spectra can provide insights into particle crystallinity, structural degradation, and the presence of additives or impurities.

One of the key advantages of Raman spectroscopy is its non-destructive nature. Because only low-intensity laser light is required, samples remain largely unaltered during analysis, allowing for repeated measurements or complementary testing. The technique is also relatively insensitive to water, which makes it suitable for analyzing wet samples or materials with high moisture content. Compared with FTIR, Raman spectroscopy generally requires less sample preparation and can be applied directly to individual particles. However, several challenges limit the routine application of Raman spectroscopy in environmental MP analysis. Accurate measurements typically require clean particle surfaces, which can be difficult to achieve when MPs are coated with organic matter, biofilms, or inorganic debris. Fluorescence interference from pigments, stabilizers, or biological residues can obscure Raman signals and reduce spectral quality. In addition, laser-induced heating may alter or damage sensitive polymer materials during prolonged exposure. Standard Raman spectroscopy may also struggle to detect extremely small particles due to weak signal intensity. To address this limitation, micro-Raman spectroscopy has been developed, enabling reliable identification of MPs at sizes well below 50 μm and, in some cases, down to a few micrometers. Combined analytical workflows that integrate micro-FTIR for larger particles and micro-Raman for finer fractions are increasingly adopted to achieve comprehensive size coverage. When compared with FTIR spectroscopy, Raman spectroscopy offers complementary strengths. While FTIR measures infrared absorption, Raman detects light scattering, allowing it to identify molecular groups that may be less prominent in infrared spectra. Raman analysis is also less affected by water interference. However, FTIR spectra are often easier to interpret, particularly for complex mixtures with overlapping vibrational bands. Raman spectroscopy typically requires longer acquisition times and is more susceptible to fluorescence effects, which can complicate data interpretation. Overall, Raman spectroscopy represents a powerful and versatile tool for MP identification and characterization. Although it is not without limitations, its high spatial resolution, chemical specificity, and compatibility with small particle sizes make it an essential component of modern MP analytical frameworks, particularly when used in combination with FTIR and microscopic techniques.

5.2.4 Thermal analysis

Thermal analysis techniques, including differential scanning calorimetry (DSC) and thermogravimetric analysis (TGA), are employed to assess the thermal behavior of MPs. These methods provide insights into polymer stability, melting points, and decomposition profiles, which are useful for material classification (Kosore et al., 2022). However, thermal analysis is inherently destructive and lacks the specificity required to distinguish chemically similar polymers. It also demands extensive sample preparation and cannot preserve samples for subsequent analysis. In contrast, LDIR offers a non-destructive alternative that retains sample integrity while delivering precise chemical characterization. Its rapid acquisition and automation make it a superior choice for routine environmental assessments.

5.2.5 Pyrolysis-gas chromatography/mass spectrometry

Pyrolysis-gas chromatography/mass spectrometry (Pyr-GC/MS) is a robust analytical technique employed for molecular-level characterization of MPs. It operates by thermally decomposing polymers into volatile fragments, which are subsequently separated via gas chromatography and identified through mass spectrometry. This method provides highly detailed compositional data, including polymer type and additive profiles, making it particularly valuable for forensic and source tracking applications (Lam et al., 2022). Despite its analytical depth, Pyr-GC/MS is constrained by several operational limitations. It requires stringent experimental conditions, extensive sample preparation, and specialized instrumentation. These factors contribute to low throughput and hinder its scalability for routine environmental monitoring. In contrast, Laser Direct Infrared Imaging (LDIR) offers a more efficient alternative, balancing chemical specificity with operational simplicity. LDIR enables rapid, non-destructive analysis of large sample volumes, making it well suited for high throughput screening in environmental matrices (Law et al., 2010).

5.2.5.1 Scanning electron microscopy with energy-dispersive X-ray spectroscopy (SEM-EDS)

SEM-EDS combines high-resolution imaging with elemental analysis, allowing for detailed morphological and compositional characterization of MPs. It is particularly effective for assessing surface and inorganic additives. However, SEM-EDS demands complex sample preparation including drying, sputter coating, and embedding which can introduce artifacts and alter particle morphology (Primpke et al., 2020). Moreover, SEM-EDS is time intensive and less amendable to large scale monitoring due to its manual workflow and limited chemical specificity for organic polymers. IN contrast, LDRI provides automated imaging and infrared based chemical identification without extensive sample manipulation. Its non-destructive nature and compatibility with diverse environmental matrices including sediments, water, air, and biota make it a practical tool for comprehensive MP assessment.

5.2.6 Liquid chromatography/mass spectrometry

Liquid chromatography/mass spectrometry (LC/MS) is employed for the target detection of specific plastic additives and polymer fragments, particularly in complex matrices such as wastewater, dust, and indoor air. It offers high sensitivity and quantitative accuracy but is inherently destructive and requires substantial sample volumes for reliable analysis. Additionally, LC/MS is limited to known compounds and may not capture the full spectrum of polymer diversity in environmental samples. LDIR surpasses LC/MS versatility by enabling simultaneous identification of multiple polymer types without prior knowledge of their chemical structure. Its rapid scanning and minimal sample preparation requirements facilitate large scale environmental monitoring. Furthermore, LDIR's performance metrices-precision (repeatability under consistent conditions) and accuracy (closeness to true value) align with the Guide to the Expression of Uncertainty in Measurement (GUM), ensuring methodological robustness and data reliability.

5.2.7 Laser direct Infrared Imaging

Laser Direct Infrared Imaging (LDIR) has emerged as a cutting-edge analytical platform for the detection and characterization of MP across diverse environmental matrices, including marine ecosystems, terrestrial substrates, groundwater, and biological tissues. Unlike conventional spectroscopic techniques, LDIR integrates a high-resolution visible imaging system with a quantum cascade laser (QCL), enabling the detection of particles as small as 10 μm with exceptional spectral fidelity and spatial precision (Muller et al., 2020). LDIR distinguishes itself from traditional methods such as micro-Fourier Transform Infrared (μ-FTIR) and Raman spectroscopy by offering rapid acquisition of full infrared spectra typically within one second per particle compared to the extend processing times required by legacy techniques. This high throughput capability is particularly advantageous for large scale environmental monitoring, where sample volumes are extensive, and particle concentrations may be low (Lopez-Rosales et al., 2024). A key operational advantage of LDIR lies in its automated scanning protocol, which surveys the entire sample surface to locate MP particles prior to spectral imaging. This contrasts with μ-FTIR and Raman systems that often rely on predefined regions of interest, potentially overlooking relevant particles or spending time on irrelevant areas. The comprehensive scanning approach of LDIR enhances analytical efficiency and minimizes redundancy, especially in samples with sparse MP distribution (Zhang et al., 2018).

Lopez-Rosales et al., (2024), reported the use of LDIR for the examination of MPs present in fish gastrointestinal tissues, where multiple digestion strategies were systematically evaluated. Their findings showed that a combined enzymatic-oxidative digestion approach, integrated with automated solvent evaporation and subsequent LDIR analysis, provided a robust and time efficient workflow. This protocol achieved nearly complete recovery for larger MPs (>500 μm) and approximately 75% recovery for fine fibers with diameters around 10 μm, highlighting the capability of LDIR to accurately quantify MPs spanning a broad size range. Beyond biotic samples, LDIR has been successfully applied to environmental monitoring studies. Measurements of airborne and waterborne MPs revealed concentrations ranging from 8–132 particles per cubic meter, with the majority of detected particles exceeding 100 μm in diameter. The technique proved effective in identifying diverse polymer types, including polyurethane (PU), polyethylene terephthalate (PET), and polypropylene (PP), confirming is suitability for comprehensive polymer profiling in environmental samples.

Further methodological advancements were reported by Hildebrandt et al. (2020); coupled microwave-assisted enzymatic-oxidative digestion with LDIR imaging to analyze MPs across multiple environmental compartments. This approach enabled the identification of 20 distinct polymer categories, with PET and PU being the most prevalent. Notably, over 90% of the detected MPs were smaller than 100 μm, underscoring the sensitivity of LDIR for fine scale particle detection. The study also revealed pronounced spatial and temporal variability in MPs distribution, emphasizing the methods utility for large scale environmental surveillance.

Collectively, these investigations underline the advantages of LDIR as a rapid, non-destructive, and highly precise analytical technique for MPs detection and quantifications. However, they also highlight the critical role of sample preparation in ensuring analytical accuracy. Efficient digestion and isolation of MPs from complex matrices such as biological tissues or sediments are essential to prevent polymer degradation, false negatives, or inaccurate particle counts. Optimized digestion protocols are therefore indispensable for fully exploiting the analytical capabilities of LDIR spectroscopy. The ecological relevance of LDIR based assessments has been demonstrated in sediment studies from major riverine and coastal systems. Rochman et al. (2019), used LDIR to evaluate MPs contamination and associated ecological risks in sediments from the Yellow River and the Yellow Sea. Their results showed substantially higher MPs abundance in marine sediments compared with riverine samples, with average concentrations exceeding 50,000 particles per kilogram of dry sediment in coastal areas. Silicone, fluororubber, and polypropylene were the dominant polymers, and more than 90% of particles were smaller than 100 μm, indicating a high potential for ecological exposure and risk.

LDIR has also been applied to assess contamination of MPs in food-related matrices. Pereira et al. (2023), investigated the influence of crushed and washed sea salt processing on MPs residues. Although processing significantly reduced MPs abundance, both crude and commercial salts still contain detectable levels of plastic particles. Polyamide and polyurethane were the most common polymers, while polypropylene contamination was largely attributed to post-processing and packaging stages, suggesting opportunities for further mitigation. Freshwater systems have similarly been evaluated using LDIR. Studies on reservoir sediments revealed that irregular fragments and fibrous MPs dominated, with particles larger than 50 μm accounting for most of the total abundance. Polyurethane and polyamide were frequently identified, and higher accumulation was observed near inflow zones, raising concerns about localized ecological impacts in freshwater environments. In agricultural soils, LDIR has enabled the simultaneous analysis of MPs and associated plastic additives. Elevated MPs concentrations were reported in greenhouse soils compared with open-field soils, while phthalate ester levels varied regionally. These findings indicate that MPs abundance is largely governed by external inputs, whereas additive concentrations are influenced by both input and degradation process. Additionally, sediment and urban freshwater studies have confirmed the high sensitivity of LDIR, with reliable detection of particles as small as 20 μm and recovery rates comparable to established FTIR and Raman techniques. Long-term agricultural practices, such as plastic film mulching were associated with exceptionally high MPs loads, with particles predominantly falling within the 10–500 μm size range and encompassing a wide variety of polymer types. LDIR has also been employed to assess MPs contamination in wastewater effluents and groundwater systems. Wastewater treatment plant effluents showed seasonal variability in MPs concentrations, with larger particles and polymers such as silicone, rubber, PTFE, and PU being predominant. Groundwater investigations consistently detected MPs across sampling sites, with PET, PU, PE, and PVC frequently observed and particle sizes largely below 100 μm. Despite its strengths, LDIR does present minor limitations. The closely situated particles may be interpreted as a single entity die to the imaging algorithm's resolution constraints. In contrast, μ-FTIR can resolve adjacent particles more effectively by analyzing their full surface contours. Additionally, LDIR operates within a narrow infrared spectral range (1800–900 cm^{-1}), which may exclude certain vibrational modes. However, these limitations are offset by LDIR's superior automation, speed, and scalability, making it a highly effective tool for environmental MP surveillance. Overall, the collective body of evidence confirms LDIR spectroscopy as a powerful and versatile tool for MPs research. Its high sensitivity, broad polymer identification capability, and non-destructive nature make it particularly well suited for integrated assessments of MPs contamination across terrestrial, aquatic, and atmospheric environments.

5.2.8 Fourier-transform infrared spectroscopy

FTIR spectroscopy is a cornerstone technique for polymer identification, leveraging the unique infrared absorption spectra of molecular bonds. When coupled with spectral libraries, FTIR enables accurate classification of common polymers such as polyethylene (PE) and polypropylene (PP). Micro-FTIR variants, including attenuated total reflectance (ATR) and transmission modes, enable the detection of particles as small as 2 μm. Focal Plane Array (FPA)-based FTIR imaging further enhances analytical resolution and throughput by allowing simultaneous mapping of particle distributions across sample surfaces. These capabilities render FTIR suitable for diverse environmental matrices such as sediments, water, and biota, with demonstrated reproducibility in inter-laboratory validations (Fischer et al., 2016). However, FTIR is constrained by its relatively slow processing speed and limited spatial resolution, particularly when analyzing large sample volumes or submicron particles. Sample preparation can be labor-intensive, and micro-FTIR imaging may require several hours per sample. In contrast, Laser Direct Infrared Imagining (LDIR) offers rapid, automated chemical imaging, acquiring detailed compositional data within seconds per particle. This significantly improves analytical throughput and reduces manual intervention, making LDIR a more scalable solution for environmental monitoring (Agilent, 2025).

5.3 Other analytical techniques for MP identification

In addition to Fourier Transform Infrared (FTIR) spectroscopy, several other analytical techniques are widely applied for the identification and characterization of MPs. These methods may be used independently or in combination with FTIR, depending on the objectives of the study, characteristics of the sample matrix, and the type of polymers being investigated. Like FTIR, each technique presents specific advantages and limitations, and no single method is universally applicable to all MP analyses. Consequently, the selection of an appropriate analytical approach should be guided by the research question, particle size range, polymer composition, and required level of chemical specificity. In many studies, complementary analytical techniques are employed to improve analytical confidence. For example, chromatographic methods such as chromatography mass spectrometry (GC-MS) are frequently used alongside FTIR to target specific additives or degradation products. In one such application, GC-MS was utilized to quantify phthalic acid esters, while FTIR was applied to identify polymer types in biological samples. This integrated strategy allows researchers to capture both polymer identity and associated chemical constituents. The following sections summarize some of the most used alternative techniques for MPs identification and quantification, with a particular focus on optical methods.

5.3.1 Optical techniques

Optical approaches are among the most frequently applied in MPs research due to their simplicity, rapid analysis, and non-destructive nature. These techniques enable direct observation of particles and are typically used during the initial stages of MPs analysis, regardless of the downstream analytical methods employed (Huang et al., 2022). Visual inspection is often considered an essential step, as it allows for preliminary screening, counting, and classification of suspected MP particles prior to chemical confirmation. Light-based microscopy, including stereomicroscopy and compound microscopy, is widely used to examine particle characteristics such as size, shape, color, and surface texture (Mu et al., 2022). MPs are commonly categorized into morphological classes, including fragments, fibers, pellets, films, and foams, and particle dimensions are usually measured along the longest axis. In many studies, stereomicroscopy is applied as a first screening step, followed by more advanced techniques such as micro-FTIR, Raman spectroscopy, or electron microscopy for polymer confirmation.

Optical techniques are valued for their operational simplicity and high throughput. Unlike spectroscopic methods, they do not require extensive operator training and can be applied efficiently to large numbers of samples. Additionally, optical observations provide useful contextual information that may help infer potential sources of MPs based on particle appearance and coloration. As these methods are non-destructive, analyzed particles can be retained for subsequent chemical or spectroscopic analysis, making optical microscopy a flexible component of multi-technique workflows. Despite these advantages, optical methods have notable limitations. Their applicability decreases significantly for particles smaller than approximately 100 μm, where accurate visual identification becomes challenging. Transparent or weakly colored MPs are particularly difficult to distinguish from natural materials using optical techniques alone. Furthermore, optical microscopy provides no direct information on polymer chemistry, which limits its ability to confirm the identity of MPs without complementary analytical tools. Another important constraint is the subjectivity associated with visual identification. Analytical outcomes may vary depending on the experience of the examiner, the complexity of the sample matrix, particle morphology, and the resolution of the optical equipment used. As a result, optical techniques are most reliable when combined with spectroscopic or electron-based methods that provide chemical or elemental confirmation, especially for small or ambiguous particles (Liu et al., 2023). Nevertheless, in certain applications, optical microscopy alone has been used for MP quantification, particularly in large-scale monitoring studies where rapid screening is prioritized over detailed chemical characterization. While such approaches may be appropriate for preliminary assessments, they should be interpreted cautiously and, where possible, validated using complementary analytical techniques.

5.3.2 Scanning electron microscopy

Scanning Electron Microscopy (SEM) is a well-established analytical technique that has been extensively applied across scientific disciplines since the development of electron microscopy in the early twentieth century. In MP research, SEM has emerged as a valuable tool for detailed morphological characterization and surface analysis, complementing spectroscopic techniques used for polymer identification. Its ability to generate high-resolution images makes SEM particularly useful for examining the surface features and physical condition of MP particles. SEM enables detailed visualization of MP surfaces at a spatial resolution far beyond that of conventional optical microscopy. Because the wavelength of the electron beam is significantly shorter than that of visible light, SEM can resolve particles at the nanometer scale, including fine surface textures and microstructural features that are often undetectable using light-based techniques (Sadri and Thompson, 2014). This capability is especially important for studying weathered, fragmented, or highly degraded MPs.

A major advantage of SEM is its frequent integration with energy-dispersive X-ray spectroscopy (EDS). When used together, SEM-EDS provides not only morphological information but also qualitative and semi-quantitative data on elemental composition. This combined approach has been widely adopted in MP studies to detect inorganic elements and trace metals associated with plastic particles. Such analyses have revealed the presence of elements including titanium, aluminum, copper, and zinc on MP surfaces, highlighting their potential role as carriers of inorganic contaminants in environmental and biological samples. The operational principle of SEM involves scanning a focused beam of electrons across the sample surface. Interactions between the incident electrons and the sample generate secondary electrons, backscattered electrons, and characteristic X-rays. These emitted signals are collected by dedicated detectors and converted into high-resolution images and elemental spectra. Secondary electrons primarily provide topographical information, while backscattered electrons and X-ray emissions are useful for compositional analysis (Primpke et al., 2020).

Sample preparation is a critical step in SEM analysis of MPs. Prior to imaging, MP particles must be isolated from the surrounding matrix, such as water, sediment, or biological tissue. This typically involves a combination of filtration, chemical digestion, and density separation. Once extracted, particles are mounted onto sample stubs using conductive carbon tape or adhesive substrates. Because most plastics are inherently non-conductive, samples are often coated with a thin layer of conductive material, such as gold, platinum, or carbon, to prevent surface charging during imaging.

SEM is frequently used alongside FTIR or Raman spectroscopy to provide a more comprehensive assessment of MP contamination. While spectroscopic techniques are essential for polymer identification, SEM offers detailed insight into particle size, shape, surface roughness, cracking, pitting, and other indicators of environmental weathering or chemical degradation. This combined approach enhances confidence in MP classification and interpretation of environmental fate. Compared to optical microscopy, SEM provides superior spatial resolution and allows precise examination of particles down to the nanometer scale. This makes it particularly valuable for analyzing MPs in complex matrices or in samples containing fine particles and impurities. Additionally, SEM-EDS and related techniques such as X-ray diffraction can provide information on crystalline phases and elemental associations that are not accessible through FTIR analysis alone.

Despite its strengths, SEM has several limitations. The requirement for conductive samples often necessitates additional preparation steps, increasing analysis time and cost. Advanced SEM variants, such as environmental SEM or field emission (FE-SEM), can reduce or eliminate the need for conductive coating, but these instruments are expensive and not universally available. Furthermore, SEM operation and data interpretation require specialized training and technical expertise. Another important limitation is that SEM, even when coupled with EDS, cannot reliably identify polymer types or organic additives withing MPs. Elemental analysis alone is insufficient to distinguish between different polymer matrices, making SEM unsuitable as a standalone technique for comprehensive MP identification. As a result, SEM is most effective when used as part of a multi-method analytical framework. Overall, SEM plays a crucial role in MP research by providing high-resolution morphological and surface information that complements chemical identification methods. While it cannot replace spectroscopic techniques for polymer characterization, its ability to reveal physical degradation, surface interactions, and elemental associations makes it an indispensable tool in advanced MP analysis.

5.3.3 X-ray fluorescence

X-ray fluorescence (XRF) is a rapid, non-destructive analytical technique widely used for determining the elemental composition of solid materials. In recent years, XRF has gained attention in MP research as a complementary tool for assessing the inorganic constituents associated with plastic particles in environmental samples. Its simplicity, speed, and minimal sample preparation requirements make it attractive for large-scale screening studies. The working principle of XRF is based on the interaction between high-energy X-rays and the atoms present on a sample surface. When an MP particle is irradiated with an incident X-ray beam, inner-shell electrons are displaced, leading to the emission of secondary X-rays with characteristic energies. This emission process, referred to as X-ray fluorescence, produces an elemental spectrum in which peak positions correspond to specific elements, while peak intensities provide semi-quantitative information on elemental abundance (XRF, 2017).

XRF analysis can be performed using either energy-dispersive (EDXRF) or wavelength-dispersive (WDXRF) instruments. In EDXRF systems, fluorescent X-rays are classified based on their energy, whereas WDXRF instruments separate emitted radiation according to wavelength using diffraction crystals. Although both approaches rely on the same physical principles, they differ in resolution, sensitivity, and analytical throughput. EDXRF instruments, including portable devices, are particularly well suited for field-based or rapid assessments of plastic materials. In MP studies, XRF has been used to examine elemental signatures associated with polymer matrices, pigments, stabilizers, and inorganic fillers. It is especially valuable for detecting elements such as chlorine, bromine, zinc, titanium, and lead, which may originate from additives or sorbed contaminants. Because the technique does not damage the sample, XRF can be applied prior to further analyses such as spectroscopy or microscopy (Xu et al., 2019). Despite these advantages, XRF has inherent limitations when applied to MP identification. The technique provides information only on elemental composition and does not directly resolve polymer structure or chemical bonding. As a result, XRF cannot reliably differentiate between polymer types with similar elemental profiles, nor can it unambiguously identify organic additives such as plasticizers or dyes (Yadav et al., 2023). Consequently, XRF is best used as a complementary method alongside spectroscopic techniques rather than as a standalone tool for polymer identification.

5.3.4 Dynamic light scattering

Dynamic Light Scattering (DLS) is an optical technique increasingly applied in MP research, particularly for assessing particle size distributions in liquid suspensions. While DLS has been used for MP characterization, it is regarded as especially valuable for the detection and analysis of nano plastics due to its high sensitivity to small particle size variations. The technique operates on the principle of Brownian motion. When particles suspended in a liquid are illuminated by a coherent laser beam, they scatter light in all directions. The continuous random movement of particles causes time dependent fluctuations in the intensity of the scattered light (Fu et al., 2020). These fluctuations are mathematically analyzed to calculate the hydrodynamic diameter of particles and their size distribution within the sample. DLS is most effective when analyzing relatively uniform or moderately polydisperse particle populations. In samples containing a wide range of particle sizes, larger particles tend to dominate the scattering signal, which can mask the presence of smaller particles and result in biased size distributions. For this reason, DLS is best suited for samples with relatively narrow size ranges or for preliminary screening purposes. One advantage of DLS is its ability to distinguish between spherical particles, such as plastic microbeads, and irregularly shaped fragments based on differences in hydrodynamic behavior. The technique is also capable of detecting MPs at very low concentrations, making it useful for environmental samples where particle abundance is limited. Because of its high sensitivity and non-invasive nature, DLS can detect subtle changes in particle aggregation, dispersion, or size evolution over time (Caputo et al., 2021). Compared to FTIR spectroscopy, DLS requires minimal sample preparation and does not involve chemical analysis, which significantly reduces analysis time. This makes DLS particularly useful in studies where particle size, concentration, or aggregation behavior is the primary focus rather than polymer identification. The method is also well suited for real-time monitoring of particle dynamics under varying environmental conditions, such as changes in salinity, pH, or temperature.

Despite these advantages, DLS has notable limitations. The technique does not provide information on polymer composition, chemical structure, or surface functional groups, all of which can be obtained using FTIR or Raman spectroscopy. DLS is also less effective for opaque, highly pigmented, or strongly absorbing particles and is generally unsuitable for analyzing individual particles (Chicea, 2020). Furthermore, highly heterogeneous samples with broad size distributions can yield inaccurate or misleading results. As a result, DLS is typically used as a complementary technique rather than a standalone method for comprehensive MP characterization.

5.3.5 Laser-induced breakdown spectroscopy

Laser-Induced Breakdown Spectroscopy (LIBS) is an emerging analytical technique that has recently attracted interest for MP detection and screening, although it is less commonly discussed in comparison to established methods such as FTIR and Raman spectroscopy. LIBS is based on rapid elemental analysis and is particularly valued for its speed, minimal sample preparation, and potential for in situ measurements. In LIBS analysis, a high-energy laser pulse is focused onto a small area of the sample surface. This intense energy input causes localized vapourization and ionization of the material, forming a short-lived plasma plume. As the plasma cools, excited atoms and ions emit light at characteristic wavelengths corresponding to their elemental composition (Chen et al., 2020b). The emitted spectrum is collected by a spectrometer and compared with reference spectra to identify the elements present in the sample.

For MP analysis, samples are typically collected, cleaned, and dried prior to LIBS measurement. The prepared sample is placed on a target stage, and laser pulses are applied sequentially to generate emission spectra. LIBS can detect a wide range of plastic materials, including polyethylene, polypropylene, and polystyrene, based on their elemental signatures and additive-related components. One of the main advantages of LIBS is its rapid analysis time. Unlike FTIR or Raman spectroscopy, which may require extensive sample preparation and longer acquisition times, LIBS can provide near-instantaneous results. This makes it particularly suitable for rapid screening, on-site analysis, and investigations conducted in remote or challenging environments where laboratory-based techniques are impractical (Chen et al., 2020a). LIBS is also capable of analyzing very small sample quantities, which is beneficial for trace-level MP detection.

Despite its potential, LIBS has important limitations that restrict its standalone application. The technique primarily provides elemental information and does not yield detailed molecular or polymer-specific structural data. Consequently, LIBS cannot reliably identify polymer types or distinguish between plastics with similar elemental compositions. In addition, information on organic additives, plasticizers, and stabilizers is often incomplete or ambiguous compared to FTIR-based methods. Due to these constraints, LIBS is most effectively used in combination with complementary analytical techniques. When integrated with spectroscopic or microscopic approaches, LIBS can enhance overall analytical efficiency by enabling rapid localization and preliminary identification of MP particles before more detailed chemical characterization (Brunnbauer et al., 2022). Continued methodological development and standardization may further improve its applicability in future MP monitoring frameworks.

5.3.6 Pyrolysis gas chromatography–mass spectrometry

Pyrolysis Gas Chromatography Mass Spectrometry (Py-GC-MS) is a powerful analytical technique increasingly used for the identification and quantification of MPs in complex environmental and biological samples. The method is particularly valuable for determining polymer composition and additive content, making it one of the most chemically informative tools available for MP analysis. Py-GC-MS is based on thermal degradation (pyrolysis), in which a sample is rapidly heated to high temperatures in an oxygen free environment. Under these conditions, polymer chains break down into smaller, characteristic molecular fragments. These pyrolysis products are subsequently separated by gas chromatography and identified using mass spectrometry (Pico and Barceló, 2020). Because different polymers generate unique thermal degradation profiles, the resulting chromatograms and mass spectra can be used accurately identify the type of plastic present.

One advantage of Py GC-MS is that it requires relatively minimal sample preparation compared to spectroscopic methods. Once isolated, MP particles can be directly analyzed without extensive pre-treatment. The technique can detect specific polymers as well as their monomers, oligomers, and characteristic chemical bonds, providing detailed insight into polymers composition and degradation pathways. This level of molecular specificity enables precise identification of MP sources and aging processes. Py GC-MS is also highly sensitive and can detect trace quantities of MPs, making it well suited for environmental monitoring where concentrations are often low. In addition to polymer identification, the technique can identify additives such as plasticizers, stabilizers, flame retardants, and dyes. Information on these additives is particularly important for assessing the potential environmental and toxicological impacts of MPs, as many additives are biologically active or hazardous (Primpke et al., 2020).

Compared to FTIR spectroscopy, Py-GC-MS is less affected by sample matrix interferences and can produce reliable results even in complex samples containing organic matter. However, in some cases, co-pyrolysis of organic residues may still interfere with signal interpretation, highlighting the importance of proper sample cleanup and data validation. Despite its strengths, Py-GC-MS has several limitations. Most notably, the technique is inherently destructive, meaning that samples cannot be reanalyzed or subjected to additional characterization after pyrolysis. This contrasts with non-destructive techniques such as FTIR and Raman spectroscopy. Py-GC-MS is also time-intensive, requires specialized instrumentation, and demands skilled operators for both instrument handling and data interpretation. As a result, it is not well suited for high-throughput analysis of large numbers of samples. Nevertheless, Py-GC-MS offers complementary advantages when used alongside spectroscopic and microscopic techniques. While FTIR and Raman spectroscopy are effective for particle-level screening and rapid identification, Py-GC-MS provides definitive chemical confirmation and quantitative polymer composition. Consequently, Py-GC-MS is often applied as a confirmatory or reference method in comprehensive MP studies, particularly those addressing complex environmental matrices or regulatory assessments.

5.3.7 Detection of microplastics by ICP-MS

Single-particle inductively coupled plasma mass spectrometry (SP-ICP-MS) has emerged as a promising analytical tool for the detection, sizing, and quantification of MPs (MPs) and nano plastics (NPTs) in environmental and consumer samples. Unlike spectroscopic techniques that rely on polymer-specific molecular vibrations, SP-ICP-MS enables particle-by-particle analysis based on elemental signals, allowing the enumeration of individual plastic particles even at very low concentrations. The feasibility of SP-ICP-MS for MP detection was first demonstrated by Bolea-Fernandez et al. (2020), who employed polystyrene (PS) microspheres doped with lanthanide nuclides (2.5 μm) dispersed in ultrapure water. In this study, MPs were detected by monitoring the minor carbon isotope 13C rather than 12C to minimize spectral interferences and background noise, alongside 165Ho from the lanthanide tracer. Instrumental parameters such as dwell time (100μs) and sample introduction flow rate were optimized to ensure discrete particle detection. A direct correlation between the number of 13C signal peaks and MP concentration was observed, and agreement with the 165Ho signal validated the approach, confirming the suitability of SP-ICP-MS for MP detection and quantification (Bolea-Fernandez et al., 2020).

Subsequent studies extended this methodology to real-world samples (Laborda et al., 2021). successfully applied 13C-based SP-ICP-MS to identify and quantify PS microparticles down to approximately 1.2 μm in personal care products and food-packaging-derived materials, highlighting the versatility of this technique beyond controlled laboratory systems. While 13C monitoring is effective in simple matrices, environmental samples such as seawater pose additional analytical challenges. High ionic strength in seawater leads to signal suppression and underestimation of particle size due to ion beam defocusing and salt deposition at the interface of the mass spectrometer. These matrix effects were shown to reduce apparent MP sizes by up to 28% when compared with measurements performed in ultrapure water (Bolea-Fernandez et al., 2020). Nevertheless, appropriate calibration using MP reference standards prepared in saline matrices has been demonstrated to correct these effects and improve size accuracy (Gonzalez-Herrera et al., 2026).

To further address spectral interferences and background noise, Gonzalez de Vega et al. (2026) introduced methodological improvements using tandem mass spectrometry (ICP-MS/MS) combined with collision–reaction gases. By comparing ^{12}C and ^{13}C detection under different gas conditions (H_2, O_2, and NH_3), the authors demonstrated that hydrogen was particularly effective in reducing interferences. The lowest size detection limits were achieved using H_2 as the collision gas, reaching approximately 0.62 μm in ultrapure water and 0.96 μm in seawater. Importantly, monitoring ^{12}C instead of ^{13}C improved ion transmission efficiency and reduced background noise, enhancing the applicability of SP-ICP-MS for MP analysis in complex environmental matrices (Gonzalez de Vega et al., 2021). To extend SP-ICP-MS toward nano plastics, innovative labeling strategies based on functionalized gold nanoparticles (AuNPs) have been developed. Bolea-Fernandez et al. (2020)

proposed a method in which nano plastics were modified to carry negatively charged surface groups, while AuNPs were functionalized to possess a positive surface charge. Electrostatic coupling between AuNPs and nano plastics generated nanocomposites that produced discrete gold signals during SP-ICP-MS analysis. This approach enabled selective detection and accurate quantification of NPTs in drinking, tap, and river water, with reported errors below 5%.

More comprehensive analytical workflows integrating pretreatment, isolation, labeling, and detection have since been reported. These typically involve filtration and acid digestion to remove coexisting inorganic nanoparticles, followed by cloud-point extraction to isolate NPTs from complex matrices. Subsequent in-situ growth of AuNPs on nano plastic surfaces significantly improved detection sensitivity, allowing quantification of particles across a wide size range from approximately 50–1200 nm (Bolea-Fernandez et al., 2020). Comparative evaluation of different metal-based labeling strategies including metal ions, hydrophobic organometallic compounds, and functionalized nanoparticles revealed that AuNP labeling provides the most reliable results. Metal ion labeling was found to be unstable due to desorption during dilution, while hydrophobic organometallic probes required additional purification steps that reduced analytical efficiency. In contrast, functionalized AuNPs enabled strong electrostatic interactions and stable binding, resulting in sensitive and accurate quantification of highly functionalized nano plastics (Bolea-Fernandez et al., 2020).

SP-ICP-MS has also been adapted for the analysis of biological samples (Lai et al., 2021). developed an improved method combining alkaline digestion, centrifugation, cloud-point extraction, and AuNP labeling to quantify PS nano plastics in organisms such as bacteria, algae, nematodes, and earthworms. Enhanced labeling efficiency led to significantly improved detection limits, enabling measurement of environmentally relevant nano plastic concentrations. More recently, SP-ICP-MS has been applied to MPs in the size range of 1–6 μm in consumer products such as teabags and face masks (Sakanupongkul et al., 2024). demonstrated that, after appropriate washing steps, SP-ICP-MS size estimates closely matched those obtained by scanning electron microscopy, confirming the reliability of the technique. While further refinement is required for routine nano plastic detection, these studies underscore the growing potential of SP-ICP-MS as a complementary tool for MP and nano plastic research. Overall, SP-ICP-MS provides a powerful alternative to conventional spectroscopic techniques, offering particle resolved quantification and improved sensitivity in complex matrices. When combined with appropriate calibration and labeling strategies, this technique represents a valuable addition to the analytical toolbox for environmental MP and nano plastic monitoring.

5.3.8 Determination of trace elements in microplastics by ICP-MS

Inductively coupled plasma mass spectrometry (ICP-MS) has been widely applied to quantify trace and heavy metals associated with MP particles, owing to its high sensitivity, multielement capability, and low detection limits. Recent studies have demonstrated that MPs can accumulate metals from surrounding environments and potentially act as secondary carriers, influencing the mobility and bioavailability of trace elements in soils, sediments, and aquatic organisms.

Zhou et al. (2019) investigated the occurrence of heavy metals in MPs and compared their concentrations with those present in surrounding soils. In their study, approximately 400 individual MP particles were isolated and subjected to ultrasonic cleaning to remove loosely attached impurities prior to metal extraction. The MPs were then treated with 20% aqua regia and agitated for 24 h at room temperature before ICP-MS analysis. Soil samples underwent microwave-assisted digestion using nitric and hydrofluoric acids. The concentrations of Cd, Cr, Pb, Ag, Cu, Sb, Hg, Fe, and Mn detected in MPs were substantially lower than those measured in soils, with Fe showing the highest abundance among all elements analyzed (Zhou et al., 2019). The authors attributed the higher metal content in soils to their greater surface area, higher porosity, and abundance of charged functional groups capable of binding metals. Nevertheless, a strong correlation between metal concentrations in MPs and soils was observed, suggesting that MPs may serve as indicators of environmental metal contamination and contribute to metal transport across soil matrices.

Similar findings were reported by Deng et al. (2020); examined trace metal accumulation in MPs and sediments using microwave-assisted acid digestion followed by ICP-MS determination. Their protocol involved digestion of MP samples with a mixture of hydrogen peroxide, nitric acid, and sulfuric acid, while sediment samples were treated with nitric acid, hydrofluoric acid, and perchloric acid. Chromium, nickel, copper, zinc, lead, arsenic, cadmium, and mercury were detected in both MPs and sediments. Although metal concentrations in MPs were generally lower than those in sediments, enrichment ratios varied widely depending on the element, with some metals such as Pb and Cd exhibiting comparable or even higher relative accumulation in MPs (Deng et al., 2020). These results indicate that while MPs may not represent the primary reservoir of metals in sediments, they can function as mobile vectors that redistribute metals within aquatic environments.

The interaction between MPs and trace elements has also been examined under biological exposure conditions. Zhu et al. evaluated metal accumulation in oysters to better understand in vivo interactions between MPs and trace elements. Freeze-dried oyster tissues were digested using nitric acid and hydrogen peroxide prior to ICP-MS analysis. Cadmium, chromium, lead, and copper were selected as target analytes, with copper showing the highest concentration and lead the lowest. Importantly, a significant positive correlation was observed between MP abundance in oyster tissues and trace metal concentrations, suggesting that MPs may facilitate the internal uptake of metals in marine organisms (Zhu et al., 2020). The authors proposed that MPs enter oysters primarily through gill filtration, allowing adsorbed metals to be transferred into soft tissues during ingestion and bioaccumulation.

Beyond environmental and biological matrices, the bio accessibility of metals associated with MPs has been evaluated using physiologically based extraction tests (PBETs). Smith and Turner investigated the mobilization of hazardous elements from MPs under simulated digestive conditions, employing both standard PBET and diet-adapted PBET (DA-PBET) protocols that included fish oil to mimic lipid-rich diets. The extracts were analyzed by ICP-MS for elements such as Br, Cd, Cr, Hg, Pb, and Sb. Their results demonstrated that acidic gastric conditions significantly enhanced metal release from MPs compared to neutral saline solutions. The inclusion of dietary oils further increased metal mobilization by altering plastic surface integrity and promoting the extraction of hydrophobic compounds (Fig. 5.3). These findings highlighted that metal bio accessibility from MPs depends strongly on digestive chemistry, exposure time, and dietary composition (Smith and Turner, 2020).

Holmes et al. further explored the bio accessibility of metals from MP pellets using a modified PBET designed to simulate avian digestion. Synthetic gastric fluid containing sodium chloride, pepsin, and hydrochloric acid was used to extract metals from polystyrene pellets under controlled temperature and agitation. Following extraction, residual MPs were digested using aqua regia to quantify remaining metal content. ICP-MS analysis revealed that iron, manganese, cobalt, and lead were the most abundant metals in the pellets, with bio accessibility values exceeding 60% for some elements. Although the measured concentrations were below toxicity thresholds for seabirds, the study demonstrated that a substantial fraction of metals can be released from MPs under realistic digestive conditions, potentially posing ecological risks depending on polymer composition and exposure scenarios (Holmes et al., 2020). A synthesis of recent ICP-MS-based studies indicates that most analytical methodologies focus on polypropylene (PP) and polyethylene (PE), which are among the most prevalent polymers in environmental MPs. Detection limits are typically below 0.18 μg/g, and digestion protocols generally aim to achieve complete metal recovery rather than selectively assessing bio accessible fractions. Due to the diverse origins and chemical compositions of MPs, interpretation of trace metal data remains challenging. Reported concentrations span several orders of magnitude, from rare earth elements present at <0.1 μg/g to elevated levels of Ni, As, Cd, and Pb in paint-derived MPs reaching up to 1800 μg/g (Zhou et al., 2019; Deng et al., 2020). Overall, these findings underscore the importance of ICP-MS as a robust analytical tool for elucidating the role of MPs in trace metal transport, accumulation, and potential ecological risk.

5.3.9 Determination of volatile organic compounds in microplastics

In addition to trace metals, volatile organic compounds (VOCs) represent an important class of pollutants associated with MPs (MPs). VOCs originate from both natural and anthropogenic sources. Natural emissions include vegetation, forest fires, and anaerobic processes in wetlands, whereas anthropogenic contributions arise predominantly from domestic and industrial activities, such as the manufacture and use of paints, coatings, waxes, disinfectants, cosmetics, and the combustion, storage, and distribution of hydrocarbon fuels (Atkinson and Arey, 2003). Common VOCs encountered in daily life include aromatic hydrocarbons, alkanes, aldehydes, terpenes, chlorinated hydrocarbons, glycol ethers, and ester compounds (Derwent et al., 2018).

Due to their high chemical reactivity, VOCs readily participate in radical-driven atmospheric reactions, contributing to air pollution and the formation of secondary organic aerosols (Atkinson and Arey, 2003). VOCs may also be generated during the degradation of organic matter in sludge through hydrothermal and thermal processes. Aged MPs, characterized by increased surface roughness and the presence of oxygen-containing functional groups, exhibit enhanced sorption capacity for VOCs, thereby facilitating their transport and redistribution in the environment. This interaction raises ecological and human health concerns, as VOC exposure has been linked to respiratory disorders, neurotoxicity, and endocrine disruption (Derwent et al., 2018).

A general analytical workflow for VOC determination in MPs is illustrated in Fig. 5.2. Comprises four key steps: sampling, pre-separation, extraction, and separation/detection. Sampling involves the careful collection of MPs from environmental matrices such as water, sediments, or biological tissues, with strict contamination control. Pre-separation aims to isolate MPs from mineral particles and organic matter, typically using density separation, flotation, and filtration techniques (Prata et al., 2020). VOC extraction is then performed using approaches such as headspace solid-phase microextraction (HS-SPME), purge and trap, or thermal desorption, which enable the release of VOCs adsorbed onto or embedded within polymer matrices (Pinto et al., 2019). Final separation and detection are generally achieved using chromatographic techniques coupled with mass spectrometry.

Gas chromatography mass spectrometry (GC-MS) remains the most widely employed technique for VOC analysis in MPs due to its sensitivity, selectivity, and robustness (Pinto et al., 2019). However, increasing interest in in situ and real-time monitoring has highlighted the need for portable analytical tools capable of detecting VOC emissions from plastic debris directly in environmental settings. The development of such tools would significantly improve the identification of macro- and MPs and

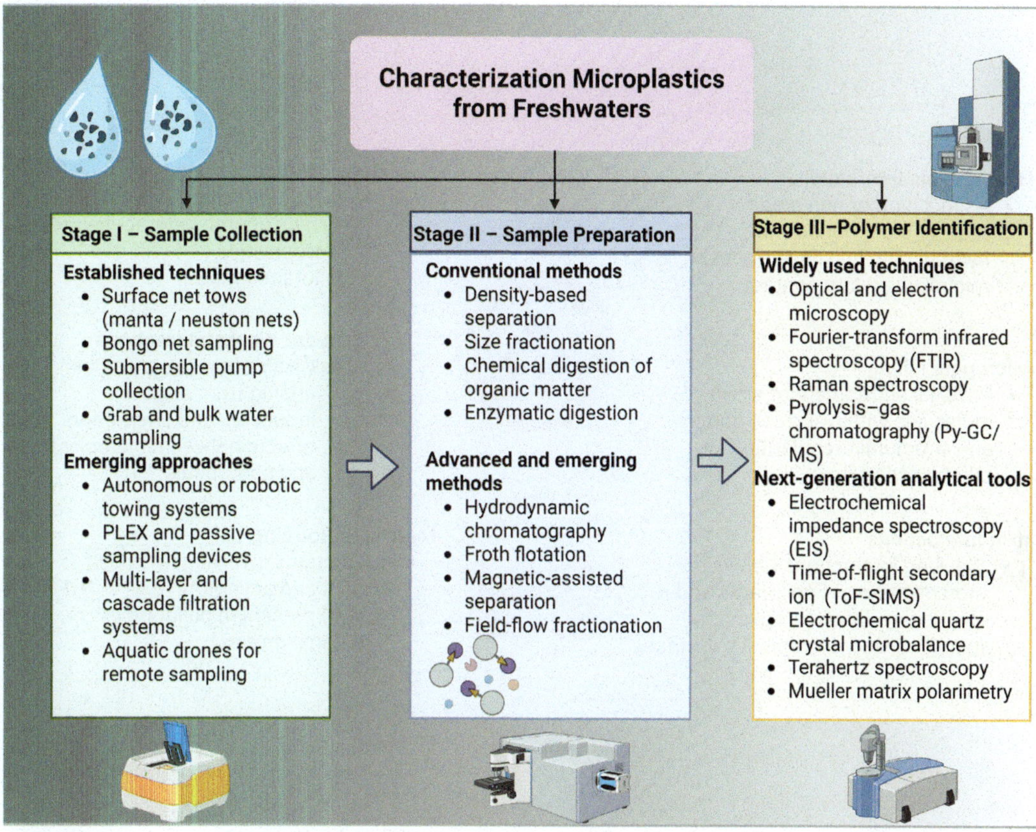

FIGURE 5.2

Characterization of Microplastics from Freshwater.

strengthen links between VOC emissions and the growing evidence of adverse biological interactions. Among portable technologies, Selected Ion Flow Tube–Mass Spectrometry (SIFT-MS) has emerged as a powerful technique for real-time qualitative and quantitative VOC analysis at trace levels. SIFT-MS employs ultra-soft chemical ionization using reagent ions such as H_3O^+, NO^+, and O_2^+, enabling compound-specific detection without the need for chromatographic separation (Smith and Spanel, 2011). This high selectivity makes SIFT-MS particularly attractive for monitoring VOCs released from plastics under environmental conditions.

Current research on VOC determination in MPs generally focuses on three major areas:

(i) optimization of analytical methodologies
(ii) assessment of VOC occurrence and distribution in environmental and biological matrices.
(iii) evaluation of toxicological and biological impacts. Penalver et al. developed a GC-MS method coupled with static headspace sampling (SHS-GC-MS) to investigate VOCs associated with MPs in animal tissues. This approach enabled both targeted and non-targeted analysis, allowing

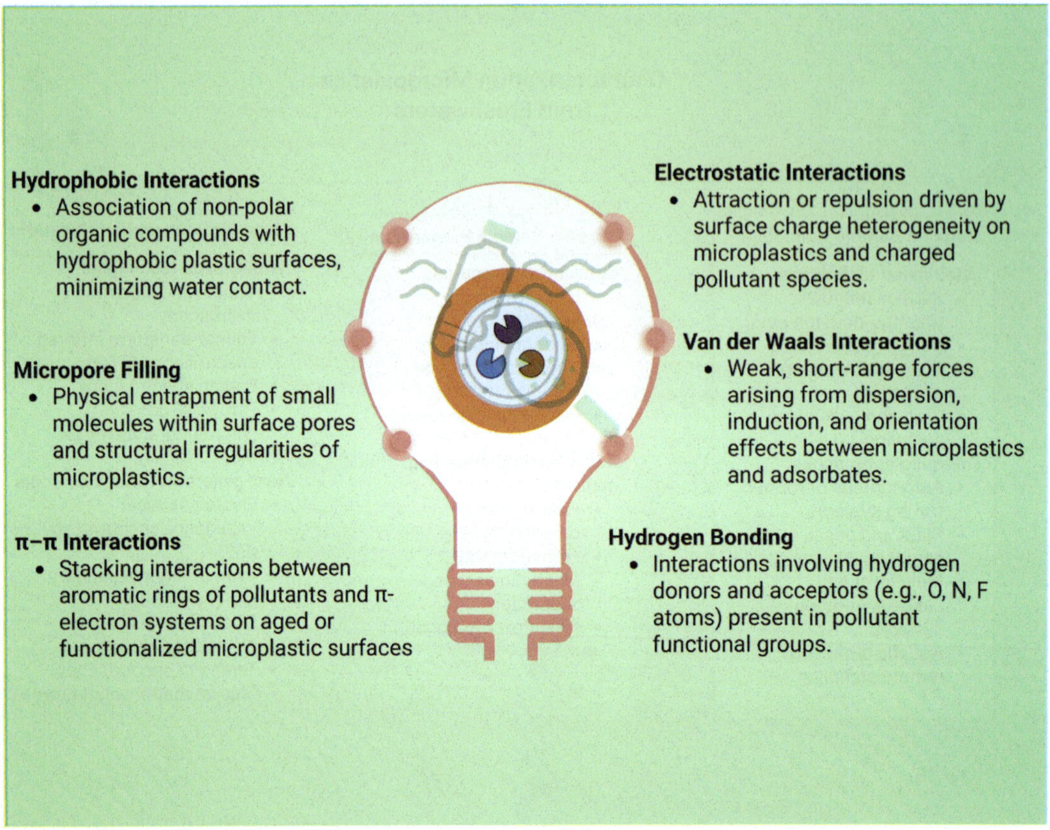

FIGURE 5.3

Types of interactions with microplastics.

the tentative identification of approximately 60 VOCs across multiple organs from different animal species. Targeted quantification of 25 compounds revealed concentrations ranging from 1.12 to 920 ng g^{-1}. The integration of chemometric tools and thermogravimetric analysis mass spectrometry (TGA-MS) further facilitated source attribution and assessment of bioaccumulation processes (Penalver et al., 2022). Thermogravimetric studies confirmed the plastic origin of detected VOCs, as samples lacking VOC signals showed no characteristic polystyrene (PS) fragments (m/z 78 and 104), whereas samples with elevated VOC concentrations exhibited increased release of PS-derived monomers. These findings demonstrated a strong correlation between VOC emissions and MP presence (Penalver et al., 2022).

Isotopic approaches have also been introduced to enhance source tracing and transformation studies of MPs. Le Juge et al. developed a purge-and-trap GC-MS method combined with combustion isotope ratio mass spectrometry (PT-GC-MS-C-IRMS) to determine the carbon isotopic composition

($\delta^{13}C$) of VOCs embedded in MPs. The method enabled simultaneous quantification and isotopic characterization of VOCs released from polymers such as PS, PET, and PLA. Distinct $\delta^{13}C$ signatures were observed among specific VOCs, highlighting the potential of compound-specific isotope analysis (CSIA) for plastic fingerprinting and tracking environmental degradation pathways (Le Juge et al., 2021). A notable outcome of this work was the significant isotopic offset between styrene monomers and bulk PS polymer, suggesting isotopic fractionation during polymer synthesis or diffusion processes. Optimization of sample mass and temperature improved sensitivity without affecting isotopic integrity, confirming the robustness of the method at ultra-trace levels. Overall, VOC $\delta^{13}C$ CSIA represents a promising complementary tool for elucidating the origin, transformation, and environmental fate of MPs (Le Juge et al., 2021).

The most frequently investigated polymers include PS, PP, PET, LDPE, and HDPE, while detected analytes range from aromatic hydrocarbons and aldehydes to ketones and halogenated esters. Reported concentrations span several orders of magnitude (approximately 0.0002–30 µg/g), with detection limits typically ranging from 2 to 1000 ng/L (Penalver et al., 2022). Investigated matrices include marine debris, sediments, food packaging materials, consumer products, biological tissues, fecal matter, and air samples. Collectively, these studies highlight the diversity of analytical strategies available and underscore the importance of tailoring methods to sample type, polymer composition, and target VOC's to better understand the environmental and health implications of VOC MP interactions. Fig. 5.4. Further highlights physicochemical interactions between MPs and co-contaminants, their transport pathways, accumulation and shoreline enrichment, and the resulting ecological and human health risks, emphasizing the importance of threshold concentrations and regulatory frameworks for effective pollution management.

5.3.10 Microplastic variability and challenges in multi-polymer detection

The accurate identification and classification of MPs in soil are strongly influenced by their wide variability in physical and chemical properties. Differences in polymer type, degree of weathering, size, color, and surface condition introduce significant complexity into detection workflows, particularly when spectroscopic and hyperspectral techniques are applied. Addressing this inherent heterogeneity is essential for developing robust analytical models capable of handling realistic environmental samples.

Several studies have highlighted the importance of incorporating diverse MP characteristics into model training datasets. It has been demonstrated that classification models trained exclusively on pristine polymers often perform poorly when applied to real-world samples, leading to substantial misclassification errors. To improve predictive performance, training datasets must include MPs representing a broad range of conditions, including pristine materials, weathered fragments, and consumer-derived plastics of varying sizes and colors. Such diversity enables models to better capture spectral variability associated with environmental aging and material degradation.

Shortwave infrared (SWIR) hyperspectral imaging (HSI) combined with partial least squares–discriminant analysis (PLS-DA) has shown promising performance for classifying agricultural plastics recovered from soil. Studies using both new and weathered agricultural plastics such as polyethylene (PE) irrigation hoses, ethylene-vinyl acetate (EVA) greenhouse films, and polypropylene (PP) mulching sheets have reported high classification accuracies for aged and unaged polymers. However, these investigations were largely limited to relatively large plastic fragments (>5 mm) and a

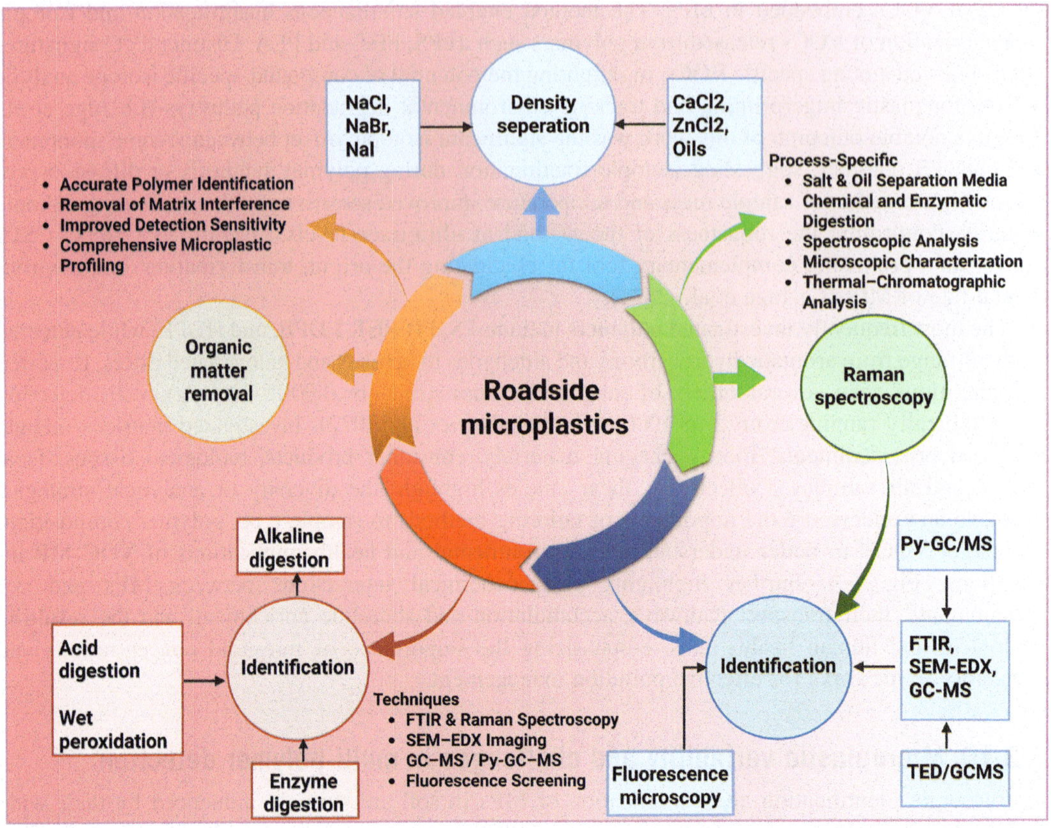

FIGURE 5.4

Schematic overview of the key factors controlling the spatial and temporal distribution of microplastics (MPs) in aquatic systems, including meteorological events, hydrodynamics, wastewater inputs, and human activities.

narrow range of colors, highlighting the need for further research involving smaller particles and more diverse color profiles.

Subsequent investigations extended these approaches to smaller particle sizes (0.5–5 mm) and more complex backgrounds containing foreign materials such as leaves, branches, rocks, and soil. These studies revealed several important detection limitations. Black and dark-colored MPs were consistently more difficult to identify due to low reflectance and spectral similarity to soil and shadowed regions. In contrast, brightly colored consumer plastics were more readily detected. Naturally weathered MPs extracted directly from soil exhibited lower detection accuracy compared with consumer-derived plastics, likely due to surface oxidation, biofilm coverage, and mineral adherence. Across studies, larger MPs were generally detected more reliably than smaller particles.

Polymer identification and particle quantification studies conducted in sand matrices using supervised classification approaches demonstrated high sensitivity and specificity when datasets incorporated sufficient variability in polymer type and color. However, gray and black polymers remained challenging to identify because of weak spectral features and reduced reflectance. These findings emphasize the importance of building comprehensive spectral libraries that capture both color and weathering effects to improve detection performance in natural soils.

Beyond MP variability, the presence of multiple polymer types within a single soil matrix introduces additional complexity. Quantitative studies comparing soils containing single versus multiple polymer types have shown that concentration prediction models generally perform better when only one polymer is present. Prediction accuracy often declines when multiple polymers coexist, as overlapping spectral features reduce model specificity. Notably, some polymers, such as polyethylene terephthalate (PET), may exhibit improved prediction performance in mixed-polymer systems, while others, including low-density polyethylene (LDPE) and polyvinyl chloride (PVC), tend to show reduced accuracy under the same conditions.

Qualitative classification studies using visible–near infrared hyperspectral imaging (Vis NIR HSI) have demonstrated that simultaneous detection of multiple polymer types in soil is feasible. Classification models built using decision trees, support vector machines, convolutional neural networks, and back-propagation neural networks have successfully differentiated soils containing single, dual, and triple polymer combinations. Although these results are encouraging, broader investigations involving additional polymer types, higher polymer diversity, and extended spectral ranges particularly in the SWIR region are required to improve model generalizability.

5.3.11 Influence of soil variability on microplastic detection

Soil is a highly heterogeneous matrix composed of mineral particles, organic matter, water, and varying chemical constituents, all of which can significantly influence spectroscopic detection of MPs. Soil moisture content, texture, and mineral composition affect light absorption, scattering, and reflectance, thereby altering spectral responses. Studies examining the influence of moisture content across different soil types including field soil, kaolinite, and Ottawa sand have shown that spectral responses often exhibit linear correlations with MP concentration under dry conditions. However, these relationships vary depending on polymer type. In clay-rich matrices such as kaolinite, contrasting trends have been observed among different polymers, with some showing positive correlations, others negative, and some exhibiting no clear relationship. Increasing soil moisture further disrupts these patterns, introducing nonlinear spectral behavior and reducing model reliability.

Comparative analyses using shortwave infrared (SWIR) and mid-infrared (MIR) spectroscopy have demonstrated that detection performance is strongly soil dependent. For SWIR spectroscopy, detection accuracy typically decreases from sandy to loamy to clay-rich soils. In contrast, MIR spectroscopy particularly when operated in attenuated total reflectance (ATR) mode often performs better in clay-rich soils. This difference is attributed to the strong binding capacity of clay minerals, which can coat MP surfaces and obscure spectral signals in reflectance-based measurements, while ATR-MIR remains sensitive to surface chemical features.

Investigations across different soil orders, including Alfisols, Inceptisols, Mollisols, and Vertisols, have shown that prediction accuracy for MP concentration varies slightly with soil type. While overall model performance remains high, subtle differences in calibration and validation metrics highlight the influence of soil composition on spectral responses. Emerging techniques such as terahertz spectroscopy have further underscored the importance of dataset diversity. Models trained on limited soil types show reduced accuracy when applied to unfamiliar soil matrices. However, incorporating soils from multiple regions into calibration datasets significantly improves predictive performance. These findings reinforce the necessity of developing comprehensive soil–MP spectral libraries that account for regional soil variability, mineralogy, and organic content.

5.3.12 Implications for future detection strategies

Collectively, current evidence demonstrates that both MP heterogeneity and soil variability exert strong influences on detection accuracy. Robust identification of MPs in soil therefore requires

1. diverse training datasets encompassing multiple polymer types, sizes, colors, and weathering states.
2. inclusion of soils representing a wide range of textures, mineral compositions, and moisture conditions.
3. integration of complementary spectroscopic techniques. The establishment of standardized, large-scale spectral libraries for both MPs and soils will be critical for advancing reliable, transferable detection models capable of addressing the complexity of real environmental systems.

5.3.13 Comparison of pointed spectroscopy and hyperspectral imaging

Both pointed spectroscopy and hyperspectral imaging (HSI) have been extensively employed for the detection of MPs in soil across a wide range of spectral regions, including the visible (Vis), near-infrared (NIR), shortwave infrared (SWIR), mid-infrared (MIR), and terahertz domains. Over the past decade, these techniques have evolved significantly, progressing from early Vis-NIR-SWIR spectroscopy studies to recent investigations utilizing miniaturized spectrometers and advanced imaging systems. The growing environmental concern associated with MP contamination has driven the demand for rapid, accurate, and ideally on-site detection methods. Recent advances in portable and miniaturized spectrometers have made in-field MP screening increasingly feasible. Studies evaluating multiple miniaturized SWIR spectrometers against benchtop FT-NIR instruments have demonstrated that several compact devices can achieve performance comparable to laboratory-based systems. Principal component analysis (PCA) of spectral datasets revealed well-structured and consistent clustering, highlighting the potential of miniaturized SWIR spectrometers for on-site screening and bulk quantification of soil MPs. Appropriate instrument selection and calibration are critical for maximizing detection accuracy under field conditions.

While miniaturized spectrometers typically operate at low to moderate spectral resolution and are well suited for rapid screening, ultra-high resolution pointed spectroscopy has shown superior sensitivity for detecting low MP concentrations. Studies using high-resolution Vis NIR SWIR pointed spectroscopy demonstrated the ability to identify and quantify soil MPs at concentrations below 1%, achieving strong predictive performance with coefficients of determination exceeding 0.9. These findings indicate that pointed spectroscopy is particularly effective for precise quantification when detailed spectral information is available. A key advantage of pointed spectroscopy is its ability to detect MPs directly within soil matrices without requiring prior extraction, enabling faster and less destructive analysis. However, this approach also presents notable limitations. The small measurement footprint, typically on the order of a few square centimeters, makes results sensitive to uneven mixing of soil and MPs. Consequently, limited spatial representativeness and reduced applicability to heterogeneous field conditions remain significant challenges. To address this, multiple spectral measurements at different positions are often recommended to improve representativeness and reduce sampling bias.

In contrast, hyperspectral imaging offers a substantially larger spatial coverage, allowing entire sample surfaces to be scanned within a short time. This broader coverage reduces sensitivity to uneven MP distribution and enables simultaneous analysis of multiple samples. Hyperspectral imaging systems have been shown to support more comprehensive soil MP assessments, particularly for mapping and spatial analysis. Comparative studies of hyperspectral systems operating in the VNIR and SWIR ranges indicate that SWIR-based systems generally outperform VNIR systems in terms of detection accuracy. VNIR hyperspectral imaging tends to be more susceptible to noise, color dependence, and limited penetration depth, which restricts its applicability to controlled conditions with uniform soil and MP colors. One of the unique strengths of hyperspectral imaging is its ability to generate pixel-level predictions, enabling the construction of classified images that reveal the spatial distribution of MPs across a sample surface. This spatial resolution allows not only MP detection but also preliminary assessments of particle morphology and size. Such capabilities are not achievable with pointed spectroscopy, which lacks spatial context. However, hyperspectral imaging primarily detects particles exposed at the surface and may struggle to identify MPs embedded within soil aggregates or covered by organic matter.

In addition to hyperspectral systems, multispectral imaging has emerged as a promising alternative for non-destructive soil MP detection. Multispectral cameras operate with a limited number of spectral bands, offering a balance between data acquisition speed and spectral resolution. Studies combining multispectral imaging in the Vis–NIR range with machine learning classifiers, such as support vector machines, have demonstrated effective quantification of MPs in soil. While multispectral approaches provide less detailed spectral information than hyperspectral systems, their faster processing and reduced data complexity make them attractive for large-scale or routine monitoring applications. Despite their advantages, hyperspectral techniques face several challenges. One persistent limitation is the difficulty in accurately identifying black MPs, which strongly absorb light and exhibit weak spectral signatures, particularly in shadowed or dark regions. Furthermore, both hyperspectral imaging and pointed spectroscopy rely heavily on large, diverse spectral libraries to achieve reliable performance. Comprehensive libraries incorporating multiple polymer types, sizes, colors, weathering states, and soil backgrounds are essential for improving model robustness and transferability.

Overall, pointed spectroscopy and hyperspectral imaging should be viewed as complementary rather than competing techniques. Pointed spectroscopy excels in rapid, sensitive quantification at low concentrations and is well suited for on-site screening, while hyperspectral imaging provides superior spatial coverage, mapping capability, and morphological insight. Integrating both approaches, along with continued development of standardized spectral libraries, represents a promising pathway toward reliable, scalable soil MP detection.

5.3.14 Limitations associated with instrumental detection

The application of pointed spectroscopy for MP detection is constrained primarily by its limited measurement footprint. Because spectra are collected over a relatively small surface area, uneven distribution of MPs within calibration samples can significantly influence detection accuracy. In natural environments, MPs are heterogeneously dispersed rather than uniformly mixed, further complicating quantitative analysis. This challenge can be partially mitigated by collecting multiple spectra at different locations within a sample or by increasing the overall sample size to improve representativeness. However, these approaches increase analysis time and may not fully resolve spatial heterogeneity issues.

Hyperspectral imaging (HSI), while offering broader spatial coverage, faces its own set of limitations. A well-documented challenge is the reliable identification of black MPs. Dark-colored polymers often contain soot or carbon black additives that strongly absorb incident radiation, resulting in low reflectance and featureless spectra. Consequently, hyperspectral algorithms may incorrectly classify shadows or other dark background materials as black plastics. The absence of distinct spectral features in these materials underscores the need for improved analytical strategies, such as advanced preprocessing techniques, hybrid spectral spatial models, or complementary sensing approaches, to enhance discrimination accuracy.

Particle size further limits the effectiveness of hyperspectral imaging. Environmental MPs span a wide size range, from micrometers to several millimeters. Most hyperspectral systems are optimized for detecting particles within the millimeter scale, typically between 0.5- and 5-mm smaller particles may remain undetected when their dimensions fall below the spatial resolution of the imaging system. Although modifying optical configurations can reduce pixel size, even optimized HSI setups generally struggle to detect particles smaller than approximately 100 μm. This is a critical limitation, as a substantial proportion of environmentally relevant MPs fall below this size threshold. At present, non-destructive detection of sub-100 μm MP in soil remains largely unexplored, highlighting an important research gap.

5.3.15 Challenges in field deployment and subsurface detection

Most non-destructive MP detection studies have been conducted under controlled laboratory conditions, where factors such as illumination, background uniformity, and sample preparation can be carefully regulated. In contrast, investigations explicitly targeting in situ field detection are scarce. Real-world environments introduce additional complexities, including variable lighting, soil moisture, surface roughness, and organic matter content, all of which can influence spectral responses. The potential use of autonomous platforms, such as unmanned ground vehicles (UGVs) or unmanned aerial vehicles (UAVs), for non-destructive MP detection represents an emerging research frontier. Such systems could enable rapid, large-area surveys but would require robust calibration strategies and advanced data processing to account for environmental variability. Furthermore, existing studies predominantly focus on MPs located at or near the soil surface. Subsurface MPs, which may constitute a significant fraction of total soil contamination, remain largely unexplored due to limitations in optical penetration depth. Addressing below-surface detection will be essential for developing comprehensive assessments of soil MP pollution.

5.3.16 Limitations of laboratory-based simulation studies

Laboratory simulations using MP-spiked soils have been widely employed to evaluate detection methods under controlled conditions. While these studies provide valuable methodological insights, their ecological relevance is subject to debate. Artificially spiked samples may not adequately capture the complexity of real environmental systems, where MPs vary widely in polymer type, color, shape, size, and degree of weathering. Naturally contaminated field soils often contain mixtures of multiple polymers that have undergone physical, chemical, and biological aging processes. These characteristics influence spectral signatures and detection performance but are rarely replicated in laboratory simulations. The absence of studies using naturally contaminated soils therefore represents a significant gap in the current literature. Incorporating real-field samples into method development and validation would enable a more realistic assessment of non-destructive detection capabilities.

Additionally, systematic comparisons between non-destructive techniques and established destructive analytical methods are lacking. Destructive methods, such as chemical digestion followed by spectroscopic identification, could serve as reference standards for validating non-destructive approaches. External validation using naturally contaminated soils and reference methods is essential to assess model transferability, reduce bias, and enhance the practical applicability of non-destructive soil MP detection.

5.3.17 Conclusion

Non-destructive techniques, including pointed spectroscopy and hyperspectral imaging, have been increasingly applied for the detection of soil MPs across spectral regions spanning the visible to terahertz range. These approaches offer rapid and efficient alternatives to conventional destructive methods and enable direct analysis of soil samples with minimal preparation. Pointed spectroscopy provides high sensitivity and accuracy, particularly for low MP concentrations, but is limited by small sampling areas, uneven particle distribution, and challenges associated with complex soil matrices. Hyperspectral imaging, in contrast, benefits from extensive spatial coverage and the ability to analyze multiple samples simultaneously yet faces difficulties in identifying dark-colored MPs and very small particles. Future progress in this field will depend on the development of more transferable and resilient models, potentially through approaches such as transfer learning and expanded spectral libraries that incorporate diverse soil types and MP characteristics. Addressing these challenges will improve the reliability of non-destructive detection methods and support more realistic assessments of MP contamination in agricultural soils, ultimately enhancing our understanding of their environmental distribution and ecological implications.

5.4 Advances in field sampling and laboratory analysis

LDIR represents a paradigm shift in infrared microscopy for MP analysis, driven by innovations in quantum cascade laser (QCL) technology. Unlike conventional FTIR systems, which rely on broad spectrum thermal emitters and pixel by pixel spectral acquisition, LDIR employs a highly collimated, narrow linewidth infrared beam generated by QCLs. This eliminates the need for spectrometers and cryogenically cooled detectors, thereby simplifying instrumentation and reducing operational costs. LDIR systems integrate rapidly scanning optics with a thermoelectrically coated mercury cadmium telluride (MCT) detector, enabling selective spectral acquisition only at relevant sample locations. This targeted approach minimizes data redundancy and significantly reduces file sizes and processing times. The traditional FTIR imaging of a 10 mm sample may require over 3 h and generate >30 GB of data, LDIR can complete the same analysis in under on hour with superior spectral fidelity.

The dual-mode operation of LDIR comprising scan and sweep modes further enhances analytical efficiency. In scan mode, a single wavelength rapidly surveys the sample to identify particle morphology and distribution. Sweep mode that pauses over detected particles to acquire full infrared spectra, ensuring precise chemical characterization without wasting time on irrelevant regions. This workflow is highly automated and scalable, making LDIR particularly suitable for high-throughput environmental monitoring across water, sediment, air, and biota (Agilent, 2025).

Advancements in folding reflective optics and detector miniaturization have also contributed to LDIR's enhanced spatial resolution and analytical precision, supporting large-scale studies with minimal manual intervention. Compared to labor-intensive techniques like μ-FTIR and Raman spectroscopy, LDIR offers unmatched speed, reproducibility, and operational simplicity, positioning it as a preferred tool for MP research in complex ecosystem.

5.4.1 Light source: Quantum cascade laser

Quantum Cascade Lasers (QCLs) are semiconductor-based infrared sources engineered through band-structure manipulation of multiple quanta well heterostructures. These structures, typically composed of III-V materials such as GaGs-AIGaAs and InGaAs-InAIAs-InP, allow precise tuning of emission wavelength from 2.75 to 161 μm by adjusting quantum well depths during fabrication. LDIR systems utilize four multiplexed QCL modules to span the mid infrared fingerprint region (800–1900 cm-1), which is critical for biochemical and polymer specific spectral analysis. Compared to conventional globar sources, QCLs offer single-mode emission with high spectral resolution, enabling customizable and detailed imaging of MP particles. Their polarized, high-intensity output is particularly advantageous for advanced spectroscopic techniques such as vibrational circular dichroism and polarimetry, which enhance signal to noise rations and spatial discrimination. These attributes make QCLs integral to the performance of LDIR systems, supporting rapid, high resolution chemical imaging without the limitation of traditional thermal emitters. The integration of QCLs into LDIR platforms marks a significant advancement in the field of MP detection and environmental spectroscopy.

5.4.2 Detector: Mercury–Cadmium-Telluride

Mercury–Cadmium-Telluride (MCT) detectors are integrated to the performance of LDIR systems, offering exceptional sensitivity and spectral responsiveness in the mid-infrared (mid-IR) region. These thermoelectrically cooled detectors operate efficiently across wavelengths extending up to 18 μm, with bandgap energies reaching as low as 0.07 eV making them particularly suitable for detecting MP particles with high precision.

Single element MCT detector employed in LDUR platforms exhibit high detectivity, enabling rapid spectral acquisition and cost-effective operation. The integration of high refractive index microlenses further enhances their optical performance by expanding the effective collection area, thereby improving spatial resolution and signal intensity. This configuration allows LDIR systems to outperform conventional infrared techniques in terms of data acquisition speed, sensitivity, and spatial accuracy critical parameter for high throughput MP analysis in environmental samples (Lopez-Rosales et al., 2024).

5.4.3 Optical system design

The optical architecture of LDIR systems plays a pivotal role in directing infrared radiation from the quantum cascade laser (QCL) source to the sample and subsequently to the detector. To achieve diffraction-limited imaging, LDIR platforms incorporate custom designed refractive optics optimized for mid-IR transmission. While QCLs can be retrofitted into existing infrared microscopes, substantial redesigns of the optical pathway are required to maintain image fidelity and prevent aberration. Earlier LDIR configurations utilized reflective relay optics and Schwarzschild objectives, which, although functional, suffered from central obscuration and reduced image clarity. Contemporary systems have transitioned to advanced refractive optics that offer superior resolution and spatial localization, thereby enhancing the accuracy pf MP identification. In contrast, traditional IR methods often struggle with limited imaging resolution and prolonged analysis times, particularly heterogeneous environmental matrices (Ghanadi et al., 2024).

LDIR spectroscopy has demonstrated robust performance in inter-laboratory validation studies (Whiting et al., 2022). conducted ring tests using certified reference materials such as polystyrene and polyethylene, confirming LDIR's reproducibility and precision across multiple laboratories. Similarly (Ghanadi et al., 2024), reported high accuracy and automation in LDIR based MP analysis of sediment and water samples underscoring its reliability in collaborative research settings and its suitability for large-scale environmental monitoring.

5.4.4 Microplastic analysis using LDIR

Laser Direct Infrared Imaging (LDIR), exemplified by the Agilent 8700 platform, has emerged as a transformative analytical technique for the detection and characterization of MP in environmental matrices. Its non-destructive nature preserves both the chemical and morphological integrity of particles, eliminating the need for chemical pre-treatment and enabling rapid, high-resolution analysis. LDIR's automated workflow facilitates the acquisition of mid-infrared spectra for individual particles, which are then matched against comprehensive polymer libraries, covering particle sizes from 20–500 μm (Whiting et al., 2022).

LDIR's analytical efficacy has been validated across multiple studies (Whiting et al., 2022). developed an optimized protocol using 8 μm porosity polycarbonate filters, achieving recovery rates between 80%–100% for particles sized 200–500 μm across water, sediment, and biota samples. The total analysis time was reduced to approximately 3 h per sample, representing a substantial improvement over traditional spectroscopic method. In a separate investigation, Lopez-Rosales et al. (2022) applied LDIR to assess MP contamination in fish gastrointestinal tracts. The study compared digestion protocols and identified enzymatic oxidative digestion coupled with automated evaporation as the most effective. This approach yielded near complete recovery for particles $>$500 μm and approximately 75% recovery for fibers as thin as 10 μm, demonstrating LDIR's precision across a broad size spectrum.

LDIR has also been employed to quantify MP concentration in environmental samples, with reported values ranging from 8 to 132 particles/m^3. Notably, 96% of detected particles exceeded 100 μm in diameter, and polymer typer included polyurethane (PU), polyethylene terephthalate (PET), and polypropylene (PP), underscoring LDIR's sensitivity and compositional resolution (Hildebrandt et al., 2022)

Expanding its utility, microwave-assisted enzymatic-oxidative digestion with LDIR imaging to analyze MPs in diverse environmental matrices. This method identified 20 distinct polymer types, with PET and PU comprising 20% and 15% of the total, respectively. Moreover, 93% of particles were <100 μm, highlighting LDIR's capability to detect fine scale contamination and its adaptability for large scale environmental monitoring. These studies collectively affirm LDIR's superiority in MP analysis, offering rapid, accurate, and non-destructive identification across complex matrices. However, the reliability of LDIR is contingent upon optimized sample preparation, particularly digestion protocols, that isolate MPs without compromising their structural integrity. Inadequate digestion may result in false negatives or inaccurate quantification. Therefore, methodological rigor in sample processing is essential to fully leverage LDIR's advanced capabilities in environmental MP detection.

5.4.5 Sediment analysis in coastal and riverine systems

Zhao et al. (2021) employed LDIR to evaluate MP pollution in sediment samples from the Yellow River and Yellow Sea. Their findings revealed that MP concentration in the Yellow River (54,813 ± 19,355 particles/kg dry sediment) were approximately 2.9 times higher than those in the Yellow River (18,780 ± 9951.8 particles/kg). Dominant polymer types included silicone, fluoro rubber, and polypropylene (PP), with over 90% of particles measuring ≤100 μm. The ecological risk assessment indicated significant threats in both regions, underscoring the need for continuous monitoring and mitigation strategies.

***MP Residues in Sea Salt Processing*:** The impact of crushed and washed seas salt (CWSS) processing on MP contamination. Crude sea salt samples contained an average 256 ± 26 MP particles per 10 g, originating from seawater crystallization. The CWSS process achieved removal efficiencies ranging from 56.2% to 97.6%, yet both crude and commercial salts retained MP residues. Polyamide (PA) and polyurethane (PU) were the most prevalent polymers, while PP contamination was linked to packaging stages. These findings suggest that refining post-processing protocols could further minimize plastic contamination in edible salts (Li et al., 2022).

***Freshwater Reservoir Contamination*:** MP in sediments from the Rzeszow Dam Reservoir using LDIR. The study identified a predominance of irregular, jagged, and fibrous particles, with 91% exceeding 50 μm and 27% ranging between 3–6 μm. PU and PM were the dominant polymers. MP accumulation was higher near the reservoir inflow, where particle surface areas were also higher. These spatial patterns raise concerns about localized ecological impacts and highlight the broader implications of MP pollution in freshwater systems (Nowak et al., 2020).

***Agricultural Soil Contamination*:** LDIR to quantify MP and phtahalate esters (PAEs) in agricultural soils across China. MP concentrations ranged from 380 to 386 particles/kg, while PAE levels varied between 0.30 and 1.58 mg/kg. Greenhouse soil exhibited significantly higher MP loads compared to open field soils. The study concluded that MP contamination was primarily driven by external inputs, whereas PAE concentration were influenced by both input and degradation/removal processes. These findings emphasize the need for integrated management strategies to mitigate plastic pollution in agroecosystems (Xiang et al., 2022).

5.4.6 Standardization challenges in microplastic research

Recent advancements in MP research have introduced a suite of innovative techniques that enhances the efficiency, precision, and scalability of sampling, separation, and characterization processes. These developments include both refinements of conventional methods and transdisciplinary innovations, such as robotic sampling platforms, microfluidic separation systems, polarization-based classification, and purpose-built instruments like the Plastic Explorer (PLEX). This section highlights emerging methodologies for MP collection and analysis, highlighting their operational advantages and potential limitations.

5.4.6.1 Sampling innovations in freshwater environments

To address the limitations of traditional sampling approaches, several novel methods have been developed for MP collection in freshwater systems (Fig. 5.2). One such advancement involves the deployment of robotic vehicles equipped with surface water trawls. These autonomous platforms offer enhanced portability compared to conventional Manta trawls and are powered by watertight battery systems, eliminating the need for fuel combustion and thereby reducing environmental impact. Complementing this, large volume in situ filtration systems have been introduced as an evolution of submerging pumping techniques. These systems are optimized for sampling at specific depths within the water column, improving representativeness and reducing contamination risks. Additionally, the PLEX instrument has been designed to simultaneously collect bulk water samples and perform inline filtration of small volumes, enabling efficient multi-depth sampling and particle size segregation.

Aquatic drones represent another frontier in sampling technology. These compact, remotely operated devices are capable of collecting water samples while simultaneously recording metadata such as sampling speed, duration, distance, and geolocation. Their integration of real-time telemetry with sample acquisition enhances data quality and operational control. The use of cascade filtration assemblies in conjugation with these platforms has proven effective for capturing MP particles across a broad size spectrum (Herath et al., 2022).

Compared to traditional methods, such as manual Manta towing or static pumping systems, robotic vehicles and aquatic drones offer superior mobility, automation, and adaptability across diverse aquatic environments. While the PLEX instrument is particularly suited for small-volume, high-resolution sampling, all these technologies contribute to improved cost-efficiency over time by reducing labor demands and minimizing sampling duration.

5.4.7 Advancing sampling technologies

Innovative robotic platforms have been developed to enhance sediment sampling efficiency. One such system involves a medium-sized underwater legged robot equipped with a customized grab sampler mounted on an underactuated mechanism, enabling targeted collection of MP laden sediments. Additionally, magnetically actuated microrobots have demonstrated the capability to capture dispersed MPs in situ under the influence of external magnetic fields, offering a dynamic and responsive sampling strategy.

Despite the promise of automated water sampling technologies such as those deployed in the Dutch River Delta and Amsterdam Canal challenges persist regarding data reliability and statistical robustness. Studies have shown that low particle counts per sample, coupled with limited filtration volumes, can result in high variability and increased false negative rates (Fan et al., 2022). To address these limitations, future sampling strategies should integrate robotic automation with conventional methods to enhance stability, repeatability, and representativeness in complex aquatic systems. Emerging separation techniques or MP isolation are broadly categorized into passive and active modalities:

Passive methods: include hydrodynamics filtration, spiral microfluidic, electrostatic sieving, inertial microfluidics, and bacteria chemotaxis. These approaches exploit physical or biological properties to segregate particles without external energy input.

5.4.8 Emerging identification techniques

Recent advancements in MP detection have introduced a suite of innovative identification methods that prioritize speed, sensitivity, no-destructive analysis, and miniaturization. These techniques span optical, spectroscopic, electrochemical, and mass based domains, often integrated with machine learning to enhance classification accuracy and throughput.

5.4.8.1 Optical and spectroscopic innovations

Electrochemical impedance spectroscopy (EIS), utilizing graphene-based electrodes, has emerged as a rapid and sensitive method for MP quantification. This technique detects changes in impedance caused by polymer interactions at the electrode surface, offering a label-free and scalable approach for environmental monitoring. External cavity quantum cascade laser (EC-QCL) hyperspectral imaging systems represent a significant leap in infrared microscopy. These systems deliver 14-fold higher pixel density than conventional FTIR, expanding the field of view and drastically reducing acquisition time. Similarly, terahertz spectroscopy enables non-invasive detection by probing frequency-dependent absorption and refractive index variations. Surface-functionalized terahertz microfluidic metamaterials further enhance detection speed and sensitivity, enabling in situ analysis of suspended MP. Polarimetric techniques, such as Muller matrix polarimetry, provide insights into particle morphology and physiological states. When combined with machine learning, these methods can classify MPs based on shape, size, and refractive properties. Dense sampling of polarized light pulses has also shown promise for high-throughput characterization of MPs at sub-micron scales. Digital holography, and interferometric imaging technique, offers real-time, in-situ detection of transparent and semi-transparent MPs. It is highly sensitive to surface deformations and refractive index variations, making it suitable for complex aqueous environments.

5.4.9 Mass-based and elemental techniques

Emerging mass spectrometry platforms provide molecular and elemental resolution for MP identification:

- ToF-SIMS (Time of Flight Secondary Ion Mass Spectrometry) enables surface-level mapping of isotopic and molecular features with nanometer-scale resolution.
- MALDI-ToF/MS is effective for characterizing biopolymers and synthetic plastics, offering borad mass detection capabilities.
- LA-ICP-MS (Laser Ablation Inductively Coupled Plasma Mass Spectrometry) Detects metal adsorption/desorption on aged MPs, useful for environmental fate studies.

- Single Particle ICP-TOFMS allows real-time elemental fingerprinting of individual MP particles via carbon detection in complex suspensions.
- Quantitative Proton Nuclear Magnetic Resonance (qNMR) has been adapted for MP quantification at environmentally relevant concentrations.

While these techniques offer exceptional precision, they are often limited by high operational costs, instrument complexity, and polymer specificity, restricting their routine application in field settings.

5.5 Future directions in microplastic quantification

The quantification across studies have led to the development or rescaling strategies that correct for differences in unit expression and particle size ranges. MP abundance derived from surface trawling typically reported in items/km^2 can be converted to volumetric units (items/L or items/m^3) using the immersion depth of the tow net, thereby enabling cross study comparability (Maurizi et al., 2025). Additionally, size-rescaling frameworks have been proposed to normalize particle counts to a default size range (1–5000 μm) using correction factors informed by environmental distributions. Probability density functions (PDFs) representing environmental MP profiles allow researchers to estimate number, volume, and mass concentrations interchangeably. Incorporating particle morphology particularly distinguishing between fibrous and non-fibrous shapes enhances the accuracy of volume-based estimating and improves number to mass conversions (Zhao et al., 2023).

Despite these methodological advancements, the absence of globally standardized protocols for sampling, identification, and reporting remains a critical barrier to data interoperability. Although frameworks such as the UN Joint Group of Experts on the Scientific Aspects of Marine Environmental Protection (GESAMP) provide guidance for marine MP monitoring, no universally accepted standards exist for freshwater and terrestrial matrices.

Even widely adopted identification techniques, including FTIR and Raman spectroscopy, may yield inconsistent results due to ambiguities in spectral matching particularly for semi-synthetic or regenerated polymers. Harmonization is thus recommended as a precursor to standardization, involving the definition of procedural benchmarks, development of certified reference materials, and execution of interlaboratory comparisons under international quality assurance protocols (Whiting et al., 2022). Such harmonized datasets will serve as foundational references for further method development, enabling robust estimates of MP emissions, transport dynamics, exposure risks, and ecological impacts. They are also essential for informing policy decision and modeling trophic transfer and habitat degradation.

To address global disparities in analytical capacity, especially in low resource regions, there is a pressing need to develop cost-effective, scalable technologies capable of detecting MPs across the full-size spectrum (1 μm–5 mm) while capturing morphological and compositional attributes. The integration of particle based and mass-based techniques such as LDIR, ToR-SIMS, and qNMR offers a promising pathway toward comprehensive characterization. Automation and in situ detection technologies are also gaining traction. Although still in development stages, these systems have demonstrated feasibility in complex aquatic environments. Combining multiple optical and spectroscopic sensors can enhance detection accuracy and reduce analytical noise. Furthermore, artificial intelligence (AI) driven approaches are poised to revolutionize MP analysis by enabling real-time classification, quantification, and error correction. As spectral databases expand and low complexity algorithms mature, AI integration will become increasingly viable for routine environmental monitoring.

References

Atkinson, R., Arey, J., 2003. Atmospheric degradation of volatile organic compounds. Chem. Rev. 103 (12), 4605–4638. https://doi.org/10.1021/cr0206420.

Bolea-Fernandez, E., Rua-Ibarz, A., Velimirovic, M., Tirez, K., Vanhaecke, F., 2020. Detection of microplastics using inductively coupled plasma–mass spectrometry operated in single-event mode. J. Anal. Atom. Spectr. 35, 455–460.

Brunnbauer, L., Gajarska, Z., Lohninger, H., Limbeck, A., 2022. A critical review of recent trends in sample classification using laser-induced breakdown spectroscopy (LIBS). TrAC, Trends Anal. Chem. 159, 116859.

Caputo, F., Vogel, R., Savage, J., Vella, G., Law, A., Della Camera, G., Hannon, G., Peacock, B., Mehn, D., Ponti, J., Geiss, O., Aubert, D., Prina-Mello, A., Calzolai, L., 2021. Measuring particle size distribution and mass concentration of nanoplastics and microplastics: addressing some analytical challenges in the submicron size range. J. Colloid Interface Sci. 588, 401–417.

Chen, D., Wang, T., Ma, Y., Wang, G., Qunwei, K., Zhang, P., Li, R., 2020a. Rapid characterization of heavy metals in single microplastics by laser-induced breakdown spectroscopy. Sci. Total Environ. 743, 140850.

Chen, G., Fu, Z., Yang, H., Wang, J., 2020b. An overview of analytical methods for detecting microplastics in the atmosphere. TrAC, Trends Anal. Chem. 130, 115981.

Chicea, D., 2020. An artificial neural network assisted dynamic light scattering procedure for assessing living cell size in suspension. Sensors 20, 3425.

Deng, J., Guo, P., Zhang, X., Su, H., Zhang, Y., Wu, Y., Li, Y., 2020. Microplastics and accumulated heavy metals in restored mangrove wetland surface sediments at Jinjiang Estuary (Fujian, China). Mar. Pollut. Bull. 159, 111482. https://doi.org/10.1016/j.marpolbul.2020.111482.

Derwent, R.G., Jenkin, M.E., Saunders, S.M., Pilling, M.J., 2018. Characterization of volatile organic compounds in the atmosphere. Atmos. Environ. 193, 3–13.

Fan, Y., Zhang, F., Yang, L., Wang, J., Li, Y., 2022. Performance evaluation of laser direct infrared spectroscopy for microplastic detection in urban freshwater environments. Sci. Total Environ. 806, 150588.

Fischer, D., Kappler, A., Oberbeckmann, S., Schernewski, G., Labrenz, M., Eichhorn, K.-J., Voit, B., 2016. Analysis of environmental microplastics by vibrational microspectroscopy: FTIR, Raman or both? Anal. Bioanal. Chem. 408, 8377–8391.

Fu, W., Min, J., Jiang, W., Li, Y., Zhang, W., 2020. Separation, characterization and identification of microplastics and nanoplastics in the environment. Sci. Total Environ. 721, 137561.

Ghanadi, F., Akhbarizadeh, R., Dobaradaran, S., Schmidt, N., Fauvelle, V., 2024. High-resolution analytical approaches for microplastic characterization in complex environmental matrices. Sci. Total Environ. 906, 167512.

Gonzalez de Vega, R., Goyen, S., Lockwood, T.E., Doble, P.A., Camp, E.F., Clases, D., 2021. Characterisation of microplastics and unicellular algae in seawater by targeting carbon via single particle and single cell ICP-MS. Anal. Chim. Acta 1174. https://doi.org/10.1016/j.aca.2021.338737.

Gonzalez-Herrera, R., Doblado-Onieva, A., Montoro-Leal, P., López-Guerrero, M.M., 2026. Advancing microplastic and associated pollutants detection: a comprehensive review on high-sensitivity analysis using mass spectrometry techniques. TrAC, Trends Anal. Chem. 194, 118534.

Herath, D., Gunasekara, M., Egodawatta, P., Goonetilleke, A., 2022. Development of automated sampling platforms and cascade filtration systems for monitoring microplastics in aquatic environments. Sci. Total Environ. 806, 150588.

Hildebrandt, L., El Gareb, F., Zimmermann, T., Klein, O., Emeis, K., Profrock, D., Kerstan, A., 2020. Fast, automated microplastics analysis using laser direct chemical imaging: characterizing and quantifying microplastics in water samples from marine environments. Mar. Pollut. Bull. 190, 114880.

Hildebrandt, L., El Gareb, F., Zimmermann, T., Klein, O., Emeis, K., Profrock, D., Kerstan, A., 2022. Fast, automated microplastics analysis using laser direct chemical imaging: characterizing and quantifying microplastics in environmental samples. Mar. Pollut. Bull. 176, 113469.

Holmes, L.A., Thompson, R.C., Turner, A., 2020. In vitro avian bioaccessibility of metals adsorbed to microplastic pellets. Environ. Poll. 261. https://doi.org/10.1016/j.envpol.2020.114107.

Huang, S., Huang, X., Bi, R., Guo, Q., Yu, X., Zeng, Q., Huang, Z., Liu, T., Wu, H., Chen, Y., 2022. Detection and analysis of microplastics in human sputum. Environ. Sci. Technol. 56, 2476–2486.

Jorquera, A., Castillo, C., Murillo, V., Araya, J., Pinochet, J., Narváez, D., Pantoja-Gutiérrez, S., Urbina, M.A., 2022. Physical and anthropogenic drivers shaping the spatial distribution of microplastics in the marine sediments of Chilean fjords. Sci. Total Environ. 814, 152506.

Kosore, C.M., Ojwang, L., Maghanga, J., Kamau, J., Shilla, D., Everaert, G., Khan, F.R., Shashoua, Y., 2022. Microplastics in Kenya's marine nearshore surface waters: current status. Mar. Pollut. Bull. 179, 113710.

Laborda, F., Trujillo, C., Lobinski, R., 2021. Analysis of microplastics in consumer products by single particle-inductively coupled plasma mass spectrometry using the carbon-13 isotope. Talanta 221, 121486. https://doi.org/10.1016/j.talanta.2020.121486.

Lai, Y., Dong, L., Li, Q., Peng, L., Zhineng, H., Yu, S., Liu, J., 2021. Counting nanoplastics in environmental waters by single particle inductively coupled plasma mass spectroscopy after cloud-point extraction and in situ labeling of gold nanoparticles. Environ. Sci. Technol. 55 (8), 4783–4791. https://doi.org/10.1021/acs.est.0c06839.

Lam, T.W.L., Fok, L., Ma, A.T.H., Li, H.-X., Xu, X.-R., Cheung, L.T.O., Wong, M.H., 2022. Microplastic contamination in marine-cultured fish from the Pearl River estuary, South China. Sci. Total Environ. 827, 154281.

Law, K.L., Morét-Ferguson, S., Maximenko, N.A., Proskurowski, G., Peacock, E.E., Hafner, J., Reddy, C.M., 2010. Plastic accumulation in the North Atlantic subtropical gyre. Science 329, 1185–1188.

Le Juge, C., Gagnon, M.M., Tremblay, L., Juteau, P., 2021. Compound-specific stable carbon isotope analysis of volatile organic compounds associated with plastic materials. Environ. Sci. Technol. 55, 12540–12549.

Li, J., Zhang, K., Zhang, H., Thompson, R.C., 2022. Microplastics in sea salt: occurrence, removal efficiency during processing, and implications for food safety. Sci. Total Environ. 806, 150551.

Liu, S., You, H., Mu, H., Cheng, J., Kuang, S., Wang, F., Chen, H., Zheng, M., Xu, Y., Liu, T., 2023. Abundance, characteristics and risk assessment of microplastics in aquatic sediments: a comparative study in the Yellow River and Yellow Sea. Waste Manag. 172, 326–334.

Lopez-Rosales, B., Ferreiro, B., Andrade, J., Fernández-Amado, M., Gonzalez-Pleiter, M., Lopez-Mahía, P., Rosal, R., Muniategui-Lorenzo, S., 2024. A reliable method to determine airborne microplastics using quantum cascade laser infrared spectrometry. Sci. Total Environ. 913, 169678.

Maurizi, A., Fossi, M.C., Baini, M., Romeo, T., Campani, T., Panti, C., 2025. Harmonizing surface trawl data for microplastic monitoring: converting areal to volumetric concentration metrics. Mar. Pollut. Bull. 201, 116103.

Mu, H., Wang, Y., Zhang, H., Guo, F., Li, A., Zhang, S., Liu, S., Liu, T., 2022. High abundance of microplastics in groundwater in Jiaodong Peninsula, China. Sci. Total Environ. 839, 156318.

Muller, A., Osterlund, H., Marsalek, J., Viklander, M., 2020. The pollution conveyed by urban runoff: a review of sources. Sci. Total Environ. 709, 136125.

Nowak, B., Krawczyk, J., Czech, T., Sitarz, M., 2020. Microplastics in sediments of a freshwater reservoir: distribution, characteristics, and environmental implications. Sci. Total Environ. 741, 140329.

Penalver, R., Bolea-Fernandez, E., Vanhaecke, F., 2022. Advanced mass spectrometry-based strategies for the characterization and source attribution of microplastics. TrAC, Trends Anal. Chem. 150, 116569.

Pereira, L.G., Ferreira, G.V.B., Justino, A.K.S., de Oliveira, K.M.T., de Queiroz, M.T., Schmidt, N., Fauvelle, V., Carvalho, V.L., 2023. Exploring microplastic contamination in Guiana dolphins (Sotalia guianensis): insights into plastic pollution in the southwestern tropical Atlantic. Mar. Pollut. Bull. 194, 115407.

Pico, Y., Barceló, D., 2020. Pyrolysis gas chromatography–mass spectrometry in environmental analysis: focus on organic matter and microplastics. TrAC, Trends Anal. Chem. 130, 115964.
Pinto, J., Paiva, M.C., Bernardo, C.A., 2019. Release of volatile organic compounds from polymer-based consumer products. Polym. Test. 73, 1–9.
Prata, J.C., Silva, A.L.P., Walker, T.R., Duarte, A.C., Rocha-Santos, T., 2020. COVID-19 pandemic repercussions on the use and management of plastics. Environ. Sci. Technol. 54 (13), 7760–7765. https://doi.org/10.1021/acs.est.0c02178.
Primpke, S., Christiansen, S.H., Cowger, W., De Frond, H., Deshpande, A., Fischer, M., Holland, E.B., Meyns, M., O'Donnell, B.A., Ossmann, B.E., 2020. Critical assessment of analytical methods for the harmonized and cost-efficient analysis of microplastics. Appl. Spectrosc. 74, 1012–1047.
Priya, A.K., Jalil, A.A., Dutta, K., Rajendran, S., Vasseghian, Y., Karimi-Maleh, H., Soto-Moscoso, M., 2022. Algal degradation of microplastic from the environment: mechanism, challenges, and future prospects. Algal Res. 67, 102848. https://doi.org/10.1016/j.algal.2022.102848.
Rochman, C.M., Brookson, C., Bikker, J., Djuric, N., Earn, A., Bucci, K., Athey, S., Huntington, A., McIlwraith, H., Munno, K., 2019. Rethinking microplastics as a diverse contaminant suite. Environ. Toxicol. Chem. 38, 703–711.
Sadri, S.S., Thompson, R.C., 2014. On the quantity and composition of floating plastic debris entering and leaving the Tamar Estuary, Southwest England. Mar. Pollut. Bull. 81, 55–60.
Sakanupongkul, A., Sirisinha, K., Saenmuangchin, R., Siripinyanond, A., 2024. Analysis of microplastic particles by using single particle inductively coupled plasma mass spectrometry. Microchem. J. 199, 110016. https://doi.org/10.1016/j.microc.2024.110016.
Smith, D., Spanel, P., 2011. Ambient analysis of trace gases by selected ion flow tube mass spectrometry. Mass Spectr. Rev. 30, 236–267.
Smith, E.C., Turner, A., 2020. Mobilisation kinetics of Br, Cd, Cr, Hg, Pb and Sb in microplastics exposed to simulated, dietary-adapted digestive conditions of seabirds. Sci. Total Environ. 733, 138802. https://doi.org/10.1016/j.scitotenv.2020.138802.
Tamminga, M., 2017. Nile red staining as a subsidiary method for microplastic quantifica-tion: a comparison of three solvents and factors influencing application reliability. SDRP J. Earth Sci. Environ. Stud. 2 (2). https://doi.org/10.15436/JESES.2.2.1.
Whiting, J., Cowger, W., Primpke, S., De Frond, H., Fischer, M., Gerdts, G., Scholz-Böttcher, B.M., 2022. Interlaboratory comparison of laser direct infrared spectroscopy for the identification and quantification of microplastics using certified reference materials. Anal. Bioanal. Chem. 414, 4019–4033.
Xiang, X., Liu, X., Li, Q., Zhao, Y., Zhang, Y., Li, Y., 2022. Occurrence of microplastics and phthalate esters in agricultural soils: distribution patterns, influencing factors, and environmental implications. Sci. Total Environ. 838, 155881.
X-ray fluorescence (XRF) analysis of porcelain: background paper, 2017. Anal. Methods 9, 2371–2374.
Xu, J.L., Thomas, K.V., Luo, Z., Gowen, A.A., 2019. FTIR and Raman imaging for microplastics analysis: state of the art, challenges, and prospects. TrAC, Trends Anal. Chem. 119, 115629.
Yadav, S., Kataria, N., Khyalia, P., Rose, P.K., Mukherjee, S., Sabherwal, H., Chai, W.S., Rajendran, S., Jiang, J., Khoo, K.S., 2023. Recent analytical techniques and potential ecotoxicological impacts of textile fibrous microplastics (FMPs) and associated contaminants: a review. Chemosphere 326, 138495.
Zhang, K., Shi, H., Peng, J., Wang, Y., Xiong, X., Wu, C., Lam, P.K.S., 2018. Microplastic pollution in China's inland water systems: a review of findings, methods, characteristics, effects, and management. Sci. Total Environ. 630, 1641–1653.
Zhao, S., Zhu, L., Wang, T., Li, D., 2021. Microplastic pollution in sediments from the Yellow River and Yellow Sea: distribution, characteristics, and ecological risk assessment. Sci. Total Environ. 761, 143226.
Zhao, Y., Liu, X., Zhang, Y., Wang, J., Li, Q., 2023. Influence of particle shape on microplastic mass estimation and number-to-mass conversion in environmental samples. Sci. Total Environ. 858, 159964.

Zhou, Y., Liu, X., Wang, J., 2019. Characterization of microplastics and the association of heavy metals with microplastics in suburban soil of central China. Sci. Total Environ. 694, 133798. https://doi.org/10.1016/j.scitotenv.2019.133798.

Zhu, X., Qiang, L., Shi, H., Cheng, J., 2020. Bioaccumulation of microplastics and its in vivo interactions with trace metals in edible oysters. Mar. Pollut. Bull. 154, 111079. https://doi.org/10.1016/j.marpolbul.2020.111079.

CHAPTER

6 Potential health risks and toxicological implications of microplastics

Melvin S. Samuel

Department of Civil, Construction, and Environmental Engineering, Marquette University, Milwaukee, WI, United States

Highlights

- Microplastics enter the human body through food, water, air, and consumer products.
- Inhaled Microplastics can induce lung inflammation, oxidative stress, and fibrosis.
- Microplastics act as carriers for toxic chemicals and pathogenic microorganisms.
- Children are particularly vulnerable due to higher exposure and developing organs.
- Standardized detection methods and health-based regulatory limits are urgently needed.

6.1 Introduction

Plastics released into the environment do not remain intact indefinitely. Through prolonged exposure to ultraviolet radiation, mechanical abrasion, hydrolysis, and biological activity, larger plastic materials progressively fragment into smaller particles known as microplastics (MPs). These particles, typically defined as plastic debris with a diameter of 5 mm or less, originate either from the intentional manufacture of small plastic particles or from the secondary breakdown of larger plastic products. Owing to their persistence and widespread distribution, MPs have become an emerging class of environmental contaminants of global concern. Beyond their physical presence, MPs pose additional risks due to their strong affinity for environmental pollutants, their large surface area, and hydrophobic nature enable them to absorb organic contaminants, pharmaceuticals, and heavy metals from surrounding media, thereby increasing the mobility and bioavailability of these hazardous substances. In parallel, MPs can act as substrates for microbial colonization, including pathogenic microorganism, creating complex MP–pollutant–microbe assemblages that may amplify toxicity and infection risk in exposed organisms. Human exposure to MPs is now recognized as unavoidable. These particles enter the body through multiple routes, including ingestion, inhalation, and dermal contact. Among these pathways, ingestion has traditionally received the most attention, as MPs are readily consumed by planktons and progressively transferred through aquatic food webs to higher tropical levels. Elevated MP concentrations have been detected in marine organisms, including fish tissues and organs such as the liver, facilitating transfer to wildlife and ultimately to humans through seafood consumption. In addition, bottled drinking water has been shown to contain substantial quantities of MPs, many of

Microplastics in Aquatic Environments. https://doi.org/10.1016/B978-0-443-45619-0.00003-0

which are sufficiently small to penetrate biological barriers via the gastrointestinal tract or respiratory system.

Airborne MPs represent another important yet comparatively underexplored exposure route. MPs suspended in the atmosphere can be inhaled and deposited throughout the respiratory tract, where smaller particles may evade clearance mechanisms and accumulate in lung tissue. Experimental and observational evidence suggests that certain MPs are capable of translocating beyond the lungs, potentially crossing biological barriers and contributing to systematic effects. In addition to inhalation, MPs deposited on the skin may partially penetrate the epidermis, particularly when the skin barrier is compromised, further increasing internal exposure.

The global distribution of airborne MPs has been documented across a wide range of environments, including urban centers, remote regions, and uninhabited areas. These particles occur in diverse morphologies, including fibers, fragments, films, foams, and irregular particles, with their relative abundance varying across locations and environmental conditions. Differences in polymer composition have also been observed, with certain plastics predominating in specific settings, reflecting variations in local sources and human activities. Major contributors to airborne MPs include synthetic textiles, clothing, tire wear, plastic consumer products, waste incinerations, agricultural plastics, and road abrasion.

Indoor environments frequently exhibit higher MP concentrations than outdoor settings, largely due to the presence and degradation of synthetic textiles and household materials. Activities such as wearing, washing, and drying synthetic fabrics release substantial quantities of microfibers into both air and wastewater streams. Ventilation practices and human movement strongly influence indoor MP distribution. Once airborne, MPs can be transported over long distances by wind and atmospheric circulation, eventually depositing into terrestrial and aquatic ecosystems through dry and wet deposition processes. Repeated cycles of deposition and resuspension further promote fragmentation, leading over time to the formation of nanoscale plastic particles. Given their pervasive presence in air, inhalation of MPs is effectively unavoidable. While the respiratory systems possess natural defense mechanisms such as mucociliary clearance to remove inhaled particles, fine MPs particularly those smaller than 2.5 μm can penetrate deep into the lower airways. Excessive or prolonged exposure may overwhelm these defenses, resulting in particle retention and tissue injury. MPs have been detected in lung tissues of both humans and animals. Occupational studies indicate that individuals working in plastic-related industries face elevated risk of respiratory disorders.

Clinical and experimental evidence link chronic MP exposure to a range of pulmonary conditions, including inflammation, fibrosis, and neoplastic changes. Observations of increased microfiber accumulation in lung tissue associated with pathological lesions further support the potential role of MPs in respiratory disease development. Animal studies have demonstrated structural lung damage, inflammatory cell infiltration, surface lesions following exposure to specific plastic polymers, underscoring the biological plausibility of MP-induced pulmonary toxicity.

In light of these concerns, this chapter focuses on airborne MP and their adverse effects on the respiratory system, It systematically synthesizes existing research to elucidate the toxicological mechanisms through which MPs impact pulmonary health, with particular emphasis on oxidative stress, inflammatory responses, apoptosis and autophagy, microbial dysbiosis, disruption of lug surfactant function, inhibition of cellular repair, pulmonary fibrosis, and synergistic toxicity arising from interactions between MPS and coexisting environmental pollutants.

6.2 Human health implications of microplastic exposure

MPs under sunlight, pressure or some physical tears break down into smaller particles and released into the environment roughly around 70%–80% of the total plastic exposure. On the other hand, nanoplastics (NPs) often form larger plastics. People are exposed to these particles through drinking water, breathing air, through food or products they use daily with their skin as they are tiny and easily transported (Mariano et al., 2021). The amount of exposure a person receives depends on various factors such as age, diet, occupation, and mainly due to the lifestyle they live. MPs contaminants across aquatic, terrestrial, and atmospheric environments are considered universal (Table 6.1), as they occur in water, sediments, biota, and airborne dust. Although the ecological impacts of MPs on plants and animals are studied and analyzed, in recent years there is a serious concern about human exposure and potential health risks have been raised. Primary exposure to MPs in humans occurs through ingestion and inhalation while dermal contact is to lesser extent. The most available understandings are taken from studies on model organisms or in vitro systems, so there is no direct evidence on human effects. The emerging global concern is that MPs and airborne fibers can stay on the surface of the food and contaminate the drinking water.

The effects of MPs on potential health can be broadly categorized into physical, chemical, and biological effects. Physical interactions may include particle-induced inflammation or cellular disruption, while chemical risks derive from monomers, additives, and adsorbed environmental pollutants. Biological risks involve interactions with microbes or biofilms associated with MPs. The detailed study exposes various risk factors. Some studies reveal that oxidative stress will be induced when the human epithelial and neuronal cells are exposed to MPs or plastic-associated chemicals, but the impacts on cell viability were minimal.

Inflammatory responses will be produced when direct contact between polypropylene MPs happens, and human cells stimulate them to release cytokine and histamine. MPs also act as vectors for persistent organic pollutants (POPs) and other contaminants. Their hydrophobic surfaces adsorb chemicals such as Polychlorinated biphenyls (PCBs), Polycyclic Aromatic Hydrocarbons (PAHs), pesticides, antibiotics, and trace metals. MP accumulation in agricultural soil is contributed to by the use of polyethylene mulching films, leading to the potential transfer of contaminants into crops (Huang et al., 2022). MPs may transport these pollutants over long distances which is indicated by high affinity of polyamide for antibiotics in aquatic environments (Li et al., 2020a,b). Elevated concentrations of heavy metals are found to be edible oysters from China which revealed MPs in all tissues sampled. These studies reveal the raising concerns about combined exposure risks through seafood consumption (Zhu et al., 2020).

The airborne MP pollution is contributed to by the growing production of synthetic fibers, which is about ~16% of global plastic output. Higher deposition rates are found to be in urban environments. Studies show that atmospheric MP fallout of nearly 20 times higher was reported in London when compared with remote areas (Wright et al., 2013). Synthetic textiles are the main source of microfibers in airborne MPs, nearly 90% originating from them.

Even though the human respiratory tract can expel many inhaled fibers, some may persevere into the lungs causing airway irritation, inflammatory lesions, or deeper lung penetration (Prata et al., 2020). Even exposure through diet is also confirmed. In a preliminary human study MPs were detected in all fecal samples with dominant polymers such as polypropylene and PET Schwabl et al. (2019); Zhang et al. (2020) reported the contamination of MPs are found to be in foods such as seafood, fruits, vegetables, water, beer, and even table salt. MPs are being accumulated in fruits and vegetables through the contaminated soil or irrigation water. Recent research

Table 6.1 Bioaccumulation of microplastics on aquatic species.

Water sources	Fish species	Plastic types	MP concentration	Bioaccumulation profile	References
Freshwater	Oreochromis niloticus	PS-MPs	100 mg/L	Gut > Gills > Liver = Brain	Biginagwa et al. (2016)
Freshwater	Carassius auratus	PS-MPs	50 mg/L	Gut	Grigorakis et al. (2017)
Freshwater	Brachionus koreanus	PS-MPs	10 mg/L	Digestive tract	Jeong et al. (2016)
Marine water	Pomatoschistus microps	PE-MPs	18.4 and 184 mg/L	Neurotoxicity	Oliveira et al. (2013)
Marine water	Mytilus galloprovinicialis	PE- or PS-MPs	1.5 g/L	Neurotoxicity	Avio et al. (2015)

from animal models further highlights the chance for systemic effects. The alteration in fatty acid mechanism with the offsprings produced by pregnant mice which are exposed to polystyrene. Even though there is no change in the growth parameters. Human placental perfusion model demonstrated that polystyrene nanoplastics can overcome the placental barrier, suggesting possible fetal exposure under real-world conditions (Luo et al., 2019). The findings show the limited direct applicability to human health risk assessments (Wang et al., 2023a,b,c). Even though there are various studies that were conducted to understand the exposure pathways and potential mechanisms of toxicity, there is a significant knowledge gap remaining, particularly regarding long-term human exposure and clinical outcomes (Fig. 6.1).

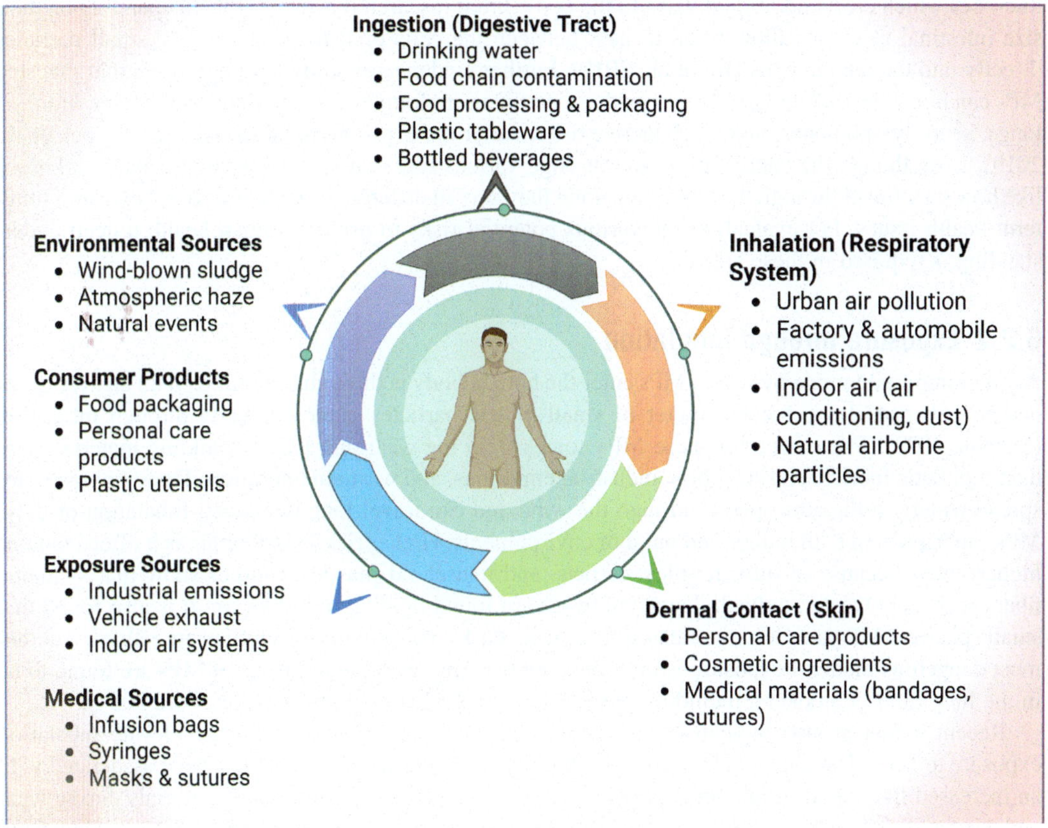

FIGURE 6.1

Overview of major human exposure pathways to microplastics. The diagram illustrates ingestion via the digestive tract, inhalation through the respiratory system, and dermal contact as key routes of exposure, originating from urban, natural, consumer, medical, and personal care sources. Contributions from food chains, drinking water, packaging materials, airborne emissions, and medical or cosmetic products collectively highlight the diverse and continuous pathways through which microplastics can enter the human body.

6.2.1 Exposure through ingestion

The most important means of MPs exposure that enters human body is through eating and drinking. The consumption of MP particles in everyday food is confirmed by the presence of particles in human blood, urine, stool, and even breast milk. These are contributed by Foods, beverages, and food-contact plastics. Depending on an individual diet and lifestyle the average person may ingest between 39,000 and 52,000 MP particles each year (Cox et al., 2019). Seafood, bottled water, table salt, beer, honey, milk, canned products, fruits, vegetables, some particles released from kitchen tools and packaging are identified to be the potential sources of MPs exposure (Danopoulos et al., 2020). When we consume these foods MPs interact with the digestive system, attached to the intestinal mucus, gets contact with gut cells, and influence gut bacteria. MPs from these foods may contain heavy metals, chemicals, and microbes which create additional harm to the body. From research it is evident that depending on their size intestinal cells can allow them to pass between the cells or it may take up the small particles directly into the human cells (Jin et al., 2019). Further evidence in animal studies prove that ingested MPs can move beyond the gut. Traces of these particles are found in organs such as the liver, kidneys, lungs, heart, lymph nodes, and reproductive tissues after the exposure to various sources (Stock et al., 2019). Even though the causes are found through various research many questions remain unknown like how they travel through the body, how long they stay in different organs, whether they cause long-term health issues. For evaluating the various potential risks to protect human health a deep understanding is required in these areas.

6.2.2 Exposure through inhalation

Another important mode by which MPs enter the human body is through breathing. With an average of nearly 9.80 particles per cubic meter of small plastic particles enter our body through inhalation (Statista, 2022). The source for these MPs comes from the wear and tear of building materials, tire dust, synthetic fabrics, plastic debris, industrial emissions, and waste incineration. These tiny particles spread widely in the atmosphere through the wind and can travel long distances. Inhalation of these MPs can happen in both indoor and outdoor environments. Higher risk in indoor air as it often contains higher levels because of dust, textiles, carpets, and household plastics. Studies show that synthetic fibers such as polyester, polyethylene, and nylon are found floating in indoor air. It is also found that small spherical MP particles are released from face masks, particularly when they are reused. Another most dangerous mode of exposure is through smoking. Higher concentrations of MPs are found to be in the lung fluid of smokers, including polyurethane and silicone particles (Ali et al., 2024).

Recent evidence also indicates that certain everyday behaviors may further elevate inhalation exposure to MPs. The use of face masks particularly when masks are reused has been associated with an increased likelihood of inhaling spherical MP particles released from mask materials during wear and mechanical (Li et al., 2021). This finding highlights an unintended exposure pathway arising from prolonged or repeated use of personal protective equipment composed of synthetic polymers. Smoking represents another important source of inhaled MPs. Combustion processes and contact with plastic-containing components in tobacco products can generate or release MP particles that are subsequently inhaled. In support of this, Lu et al. (2023a,b,c) reported substantially higher concentrations of MPs in the bronchoalveolar (11.34 MP/g) and silicone (1.15 MP/g) were particularly prominent. These findings suggest that smoking may significantly contribute to MP accumulation in the respiratory systems, potentially exacerbating pulmonary health risks associated with long-term exposure. The travel capacity of these particles into the lungs depends on the size of these particles. Particles around 2.5 μm or smaller can interfere with lung function by reaching the deepest airways. The effective measures to protect human health and improve air quality is mandatory at this level of the growing presence of airborne MPs which needs further research.

6.2.3 Exposure through dermal contact

Even though the primary routes of MP exposure are through ingestion and inhalation, a point of direct contact is made through skin. Synthetic textiles and personal care products which we use every day can release MPs that comes in contact with the skin. Clothes from materials such as polyester, nylon, and acrylic can shed microfibers during washing, wearing, and disposal. Synthetic textiles are estimated to be contributor of about 60% of microfibers in the environment (Xu et al., 2018). These fibers are accumulated in common areas such as indoor surfaces, in household dust, and in natural water bodies. Major contributions to the environmental contamination occur through bleaching and dyeing which releases microfiber in textile manufacturing processes. Microfibers into wastewater systems through domestic laundry cause serious concerns like long-term health effects. Chemicals such as flame retardants (PBDEs), in some MPs enter the skin through sweat or skin oils, arising concerns about potential toxic effects (Abafe et al., 2023). Even microbeads are found in personal care products including exfoliating scrubs, soaps, face washes, and sunscreens have direct access to the skin. A common polymer used in these beads is Polyacrylamide which breaks down into acrylamide, a serious compound found to affect the nervous system when certain amount got absorbed (Miraj et al., 2019). Regardless of various understandings and the studies of dermal exposure the exact amount of exposure remains unknown. For assessing the various health risks and safer design of products higher levels of understanding is needed about the amount of absorption and the other sources.

Many small plastic particles are found to be in shower gels, face scrubs, exfoliating cleansers, and sunscreens of in personal care products that act as polishing agents, stabilizers, or texture enhancers (Rahman et al., 2021). These products provide direct contact with MPs with skin. Polyester, nylon, and acrylic are another common source of synthetic fabrics. These sources are found to shed microfibers during daily wear, washing, and friction, which can adhere or settle on the skin. MPs are even released by certain medical products which have polymer-based materials, on routine use. Even wound dressings, catheters, surgical masks, and contact lenses come under this category. Reports from studies show that prolonged light exposure to contact lenses can cause degradation and release large number of MP fragments (Liu et al., 2023).

The particle size and the condition of the skin barrier determine the number of MPs penetrating the skin. Larger MPs tend to remain on the surface or become lodged within hair follicles or sweat glands as the diameter of Human skin pores typically range from 40 to 80 μm. However, the particles in very small size those in nanoscale are very likely to pass through damaged, inflamed, or compromised skin and reach deeper layers. When compared to the lungs or digestive tract, skin is generally less permeable but repeated or long-term contact with MPs may result in cumulative exposure especially from cosmetics, clothing, or medical materials (Larese Filon et al., 2016). Further research is needed in this area to analysis the particles behavior on the skin and the related long term health effects that may cause.

6.2.4 Microplastic exposure through the digestive tract

Humans are most exposed to MPs by Ingestion. These tiny particles are deeply spread through natural ecosystems as they have been found throughout the food web from plankton and shellfish to fish, seabirds, and marine mammals. This contamination of MPs has now become common across both aquatic and terrestrial foods as they are detected in crops such as rice, wheat, fruits, and vegetables. As a result, people consume MPs through agricultural products grown in polluted soil or irrigated with contaminated water along with seafood. Not only through food, but MPs also enter our body through various everyday products. When they are heated, scratched, or worn-down plastic packaging, storage containers, utensils, and drinking bottles releases particles which got ingested easily (Kogel et al., 2020). Another major source of MPs is indoor dust, synthetic fibers shed from clothing, carpets, and household items can settle on food surfaces. Studies highlights the homes with poor ventilation are likely to be exposed more. Nearly 100 particles per liter were found to be present in the bottled water. Billions of MP fragments are released into a cup of tea when steeping a single plastic tea bag in boiling water (Ali et al., 2023). These MPs are released from normal activities such as stirring hot soup with a plastic spoon or chewing from plastic tableware. Heat plays a major role in this process; fragmentation increases in the plastic structure when there is a rise in temperature. Many MP particles are released when PET plastics are exposed to hot water as they lose almost half of their molecular integrity in high temperature. Various experimentations on animals show that heat-generated MPs can accumulate in the body and the dietary exposure causes internal buildup which are excreted slowly overtime (Hernandez et al., 2019). Zhang et al. (2024) reported that long-term exposure to heat-treated disposable plastic foods significantly increased MP ingestion in mice. After 1 month of exposure, fecal MP concentrations reached 24.65 particles g^{-1}, indicating substantial gastrointestinal intake. Notable, following cessation of exposure, MP levels declined to 9.81 particles g^{-1} within the subsequent month, suggesting partial limitation once the exposure source was removed.

6.2.5 Variability in MP exposure pathways and bioaccumulation

Ingestion, inhalation, and skin contact are the three main pathways through which MPs enter the human body. Based on age, lifestyle, environmental pollution levels, and the types of products used the number of MPs entering the body through each route varies (Fig. 6.2). Estimation from the current studies shows particularly in the areas high indoor dust or heavy urban pollution approximately as many as 30 million MP particles per year were inhaled by the people. Another major source of exposure to humans is by drinking water. Studies reveal that individuals are subjected to around 9000 particles per year who depend on bottled water (Zuri et al., 2023). On the other hand, those who drink tap water were exposed to roughly half that number of particles. These exposure estimates vary depending on age and gender. Based on the drinking preferences, use of plastic products and the time spent indoors, the level of exposure to MPs in individual varies. On experiment it is found that bottled water consumption contributes to about 205 particles per day for male children, 174 for female children, 349 for adult men, and 255 for adult women. On the whole MP exposure depends on the multifaceted factors such as environmental conditions, product use, and human behavior. To reduce the MP intake, it is essential to understand the various pathways discussed (Cox et al., 2019).

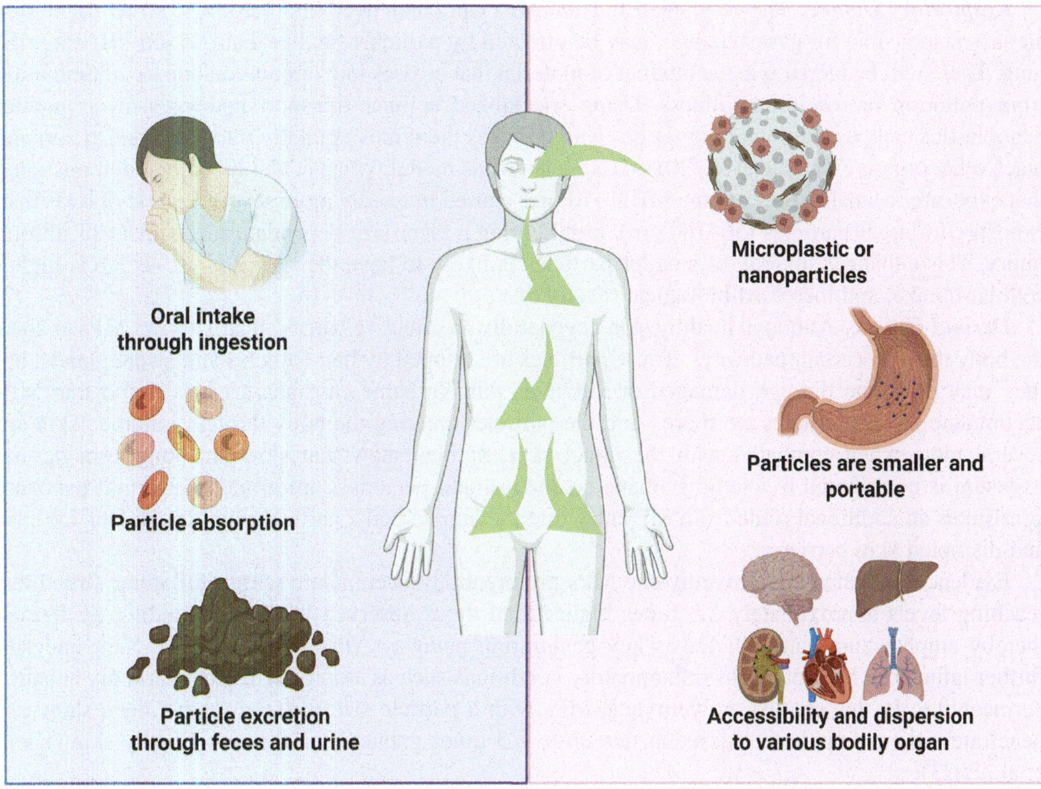

FIGURE 6.2

Schematic representation of the pathways and fate of microplastics and nanoplastics in the human body following oral exposure. The figure illustrates ingestion, gastrointestinal transformation, absorption into the bloodstream, distribution to major organs and tissues, and partial elimination via feces and urine. Potential particle size reduction under gastric conditions and systemic dispersion highlight key mechanisms underlying human exposure and health risk assessment.

6.2.6 Mechanisms of biological uptake across exposure routes

Ingestion, inhalation, and skin contact are the major ways by which MPs enter the human body. Factors such as particle size, surface charge, and the body's physiological state determine the ability of MPs to cross biological barriers and enter the human body. These particles move into deeper tissues when there is inflammation or tissue damage.

Gastrointestinal Uptake: MPs interact with the lining of the digestive tract when they are swallowed along with food or water. The natural cellular transport processes take up the particles into intestinal cells, especially very small particles and even some particles pass into the bloodstream. The positively charged particles bind strongly with the protective mucus layer in the intestines when compared with the neutral or negatively charged particles, as they are not absorbed quickly which makes surface charge an important role in MPs absorption (Liang et al., 2021).

Respiratory Uptake: The very small Inhaled MPs can reach deep into the lungs. Alveoli, the tiny air sacs responsible for gas exchange, may be affected by particles smaller than 2.5 μm that enter the lung. There will be increase in the amount of material that crosses into circulation due to inflammation from pollution or respiratory illness. Using specialized cellular transport pathways, even smaller nanoplastics can pass through lung tissues which makes them move quickly into the bloodstream and reach other organs (Vianello et al., 2019). Using a murine model, Wang et al. (2023a,b,c) demonstrated that exposure to smaller polystyrene MP (1–10 μm) caused markedly more severe hepatic fibrosis than exposure to larger particles (50–100 μm), highlighting a clear size dependent mechanism of fibrotic injury. This enhanced toxicity of smaller particles is likely to have their greater surface area, higher cellular uptake, and increased biological reactivity.

Dermal Uptake*:* Although healthy skin is generally an effective barrier, it still allows MPs to enter the body through certain pathways. These particles are trapped by hair follicles and sweat glands, but they may penetrate through damaged or inflamed skin. In some case studies it is found that MPs accumulate in hair follicles are more. And the particles entering the body through inflamed skin are several more when compared with the intact skin. Studies show that slow and long-term dermal exposure is contributed by clothing fibers, cosmetics, and personal care products. Dermal exposure constitutes an additional route from MP entry into the human body, particularly through hair follicles and disrupted skin barriers.

Evidence indicates that polyethylene MPs preferentially accumulate within follicular structures, reaching levels approximately 5.8 times higher than those observed in the surrounding epidermis, thereby emphasizing hair follicles as key penetration pathways (Patel et al., 2022). Skin integrity further influences MP uptake, as inflammatory conditions such as atopic dermatitis markedly enhance permeability. In this context, polyethylene MPs with a particle size of 0.5 μm have been shown to penetrate inflamed epidermal tissue at rats up to 8.3 times greater than through healthy skin (Klein et al., 2023)

Vulnerable Populations and Early-Life Exposure*:* Infants and young children are vulnerable groups prone to that MP exposure. The exposure of particles to children is more when compared to their body weight as they play closer, breathe faster and frequently place objects in their mouths. These particles move into their sensitive tissues as their biological barriers, such as the blood–brain barrier and intestinal lining, are also not fully developed. Some research shows the presence of MPs in placenta which indicates exposure may happen before birth even in womb.

Bioaccumulation and Tissue Distribution: The MPs which enter the body are not evenly distributed across the tissues, they will accumulate in specific organs, including the lungs, liver, kidneys, and intestines. These accumulations are found to be high in concentration in diseased or inflamed tissues in contrast with healthy ones. On comparison with normal tissue, colorectal tumor tissues seem to contain several times more MPs from the same area. This study shows the pattern that environment with altered tissues is more prone as they could worsen the inflammation and act with the disease processes.

Toxicological Uncertainties and Differences among Exposure Routes**:** Even though MPs have now been detected in human blood, lungs, placentas, and various organs, the actual concentration of particles which cause serious harm is still unknown even to scientists. Different biological effects are produced by different exposure pathways. Research shows that eating MPs affects one set of neurological pathways on the other hand, absorbing them from water causes some other disorders (Sripada et al., 2022). It is challenging to determine the long-term health risks in humans as they have different pathways in entering the tissues.

Overall Implications: Exposure to MPs is unavoidable as they are present in our environment, food, drinking water, household air, and common consumer products. The long-term effects on human health remain largely unknown, as these particles can enter and accumulate in various tissues. The risk assessments have become complicated by differences in particle type, size, surface chemistry, and exposure conditions. The other factors that influence the estimation include age, diet, and health status. For developing effective strategies to reduce exposure the knowledge gap in these areas must be closed. It is also essential for improving health guidelines and refining toxicity testing.

6.3 Microplastic actions and gastrointestinal toxicity

6.3.1 Microplastic actions

Humans encounter MPs mainly through Ingestion and these particles can intensely affect gut immunity, microbial communities, and epithelial barrier integrity shown by some studies. Studies from various experiments show that these MPs pledge initiate intestinal inflammation and alter microbiota composition (Table 6.2). Not only does it cause inflammation but also weakens the gastrointestinal tract immunological and structural defenses.

Immune Activation and Inflammatory Signaling: The innate immune pathways can be activated by MPs which trigger chronic gastrointestinal inflammation. Studies made by Li et al. (2020a,b) show that MPs, for example Polystyrene can stimulate Toll-like receptor 4 (TLR4) signaling which results in intensified immune activation and release inflammatory cytokine. Uptake often begins in Peyer's patches, where microfold (M) cells sample luminal particles and deliver them to submucosal lymphoid tissues. These MPs disturb the normal mucosal immune regulation, when adopted and distort the balance between tolerance and activation. Over the counter, this immune malfunction may result in reduced barrier function, stubborn inflammation and increased vulnerability to gastrointestinal disease.

Disruption of Gut Microbial Communities: Microbial dysbiosis in gut is mainly caused due to the increased MPs exposure. Studies made in animals shows that there is a shift in microbial composition, reduced beneficial commensals and mainly the increase in pathogenic taxa when they are ingested with chronic MPs. Experimental studies made by Lu et al. (2018) in mice shows that there is a significant downstream effect on host metabolism and the gut–liver axis due to altered intestinal microbial communities and are linked to hepatic lipid abnormalities caused by polystyrene MPs. Such dysbiosis can impair nutrient utilization, weaken immune defenses, and elevate the risk of metabolic and inflammatory disorders.

Damage to Intestinal Structure and Barrier Function: The epithelial lining of the gastrointestinal tract is damaged to the exposure of MPs. Villus swelling, epithelial erosion, reduced goblet cell numbers are the additional effects caused by MPs. It compromises the mucosal architecture which impairs nutrient absorption and metabolic homeostasis. Liang et al. (2021) found that Polystyrene particles in the range of 50–500 nm induce apoptosis of intestinal epithelial cells through ROS-mediated oxidative stress, leading to a more permeable barrier. The luminal toxins, microbial fragments, and MPs enter systemic circulation due to the increased intestinal permeability which highlights the metabolic burden and systemic inflammation.

Interaction with Preexisting Gastrointestinal Disorders: MP toxicity is vulnerable to the individuals who have preexisting gut conditions. An experiment with mice shows that MP exposure causes colitis and further disrupted barrier integrity, intensified inflammation, and worsened hepatic

Table 6.2 Combined toxicity of MPs and the attached pollutants on microorganisms.

Types of MPs	Attached pollutants	Affected organisms	Adverse effects	References
PE-MPs	Erythromycin	Escherichia coli	Inhibited expression of global regulator rpoS and DNA mismatch repair genes mutS and uvrD	Han et al. (2024)
PE-MPs, PVC -MPs	Dibutyl phthalate, dimethyl phthalate and bisphenol a	Anaerobe	Reduced abundance of important anaerobic bacteria such as acid-producing bacteria, protein-hydrolyzing bacteria	Jiang et al. (2022)
PS-MPs	CAP	Microcystis aeruginosa	Decreased expression of photosynthesis-related genes psaB and psbD, leading to induced membrane lipid peroxidation and oxidative stress damage	Zheng et al. (2024)
PS-NPs	PCB-44	Chlorella	Altered the cell cycle, inhibited growth, and reduced the hydrophobicity of the cell surface and the synthesis of chlorophyll *a* and chlorophyll *b*	Zhang et al. (2023)
PS-NH_2	B	Microcytosis aeruginosa	Increased CAT activity and disturbed the integrity of the thylakoid membrane	Li et al. (2023)

lipid metabolism when induced with dextran sulfate sodium (DSS) (Luo et al., 2022). These findings suggest that compromised mucosal defenses amplify MP-mediated injury by worsening microbial imbalance, reducing mucus secretion, and impairing epithelial repair. Resulting in chronic gastrointestinal disorders such as inflammatory bowel disease (IBD), irritable bowel syndrome (IBS), and colorectal inflammation.

6.3.2 Microplastic-induced gastric toxicity

MPs also impact the stomach, where they interact with gastric epithelial cells, immune mediators, and adsorbed chemical contaminants. These interactions activate proinflammatory pathways, increase oxidative stress, and compromise mucosal integrity.

Immune Activation and Cytokine Signaling: The inflammatory mediators in gastric and immune cells can be released by the stimulation of MPs, shown by some studies made in laboratories. IL-6 and TNF-α secretion are induced by Polystyrene MPs from the human peripheral blood mononuclear cells, which triggers histamine release from mast cells and amplifying inflammation (Hwang et al., 2019). These MPs retain in the stomach for 2–6 hours during digestion. It develops increased particle–epithelium contact due to the continuous mechanical mixing, increasing irritation and tissue injury.

Pollutant Desorption in Gastric Conditions: the release of chemicals and pollutants adsorbed onto MP surfaces are highly accelerated by the highly acidic gastric environment (pH 1.5–3.5). The experiment made by (Bakir et al., 2014) shows that hydrophobic contaminants separate under simulated gastric conditions, including persistent organic pollutants, plasticizers, and metals. Chemical exposure is increased due to this process which contributes to epithelial damage.

MPs produce gastric inflammation through two major pathways:

Chemical vector pathway: The high hydrophobicity acidic conditions (log K_ow > 5.2) are pollutants released from MP surfaces. It increases ROS production by ~1.8-fold and promoting oxidative stress-mediated inflammation (Valavanidis et al., 2013).

Physical mediator pathway: The gastric mucosa is irritated by highly porous MPs (≥68% porosity) which serve as substrates for colonization by gastric pathogens, activating TLR4/NF-κB signaling and increasing cytokine production (Valavanidis et al., 2013).

6.3.3 Microplastics as carriers of gastric pathogens

MPs can act as vectors for microorganisms, including gastric pathogens. Experimental studies in mice shows that Polyethylene MPs were shown to bind Helicobacter pylori, which improves bacterial colonization and inflammation got intensified (Tong et al., 2022). This suggests that preexisting gastric diseases are worsened by MPs and pathogen-induced to gastritis are increased.

Cellular Effects in Gastric Disease and Tumor Biology: Gastric cancer may also be developed by the influence of MPs. The morphology and the propagation are altered when the polystyrene particles readily enter gastric cancer cells and the expression of inflammatory genes, including IL-6 and IL-8 (Forte et al., 2016). The concerns raise about potential interactions between MPs and oncogenic pathways as these cytokines are associated with chronic gastric inflammation and tumor development.

Oxidative Stress and ROS-Driven Gastric Damage: Oxidative stress plays a crucial role in gastric MP toxicity. ROS are generated directly or promoted within epithelial cells by MPs which leads to mitochondrial dysfunction, cell injury, and mucosal inflammation. The gastric barrier function is weakened by sustained oxidative stress and inclines chronic gastritis or ulceration in tissues (Fig. 6.3).

6.3.4 Liver oxidative stress

Liver is the most important organ in our body, in charge for removing toxins, breaking down nutrients and supporting the immune system. As it is responsible for these essential roles, it is highly delicate to harmful substances like MPs. Starting with oxidative stress, a state where harmful molecules called free radicals overwhelm the liver's natural defenses, these MPs can cause serious problems in liver when they enter the human body. The developed stress could damage cell membranes as the antioxidant system gets weakened which affects the normal function of the liver. Further the metals and

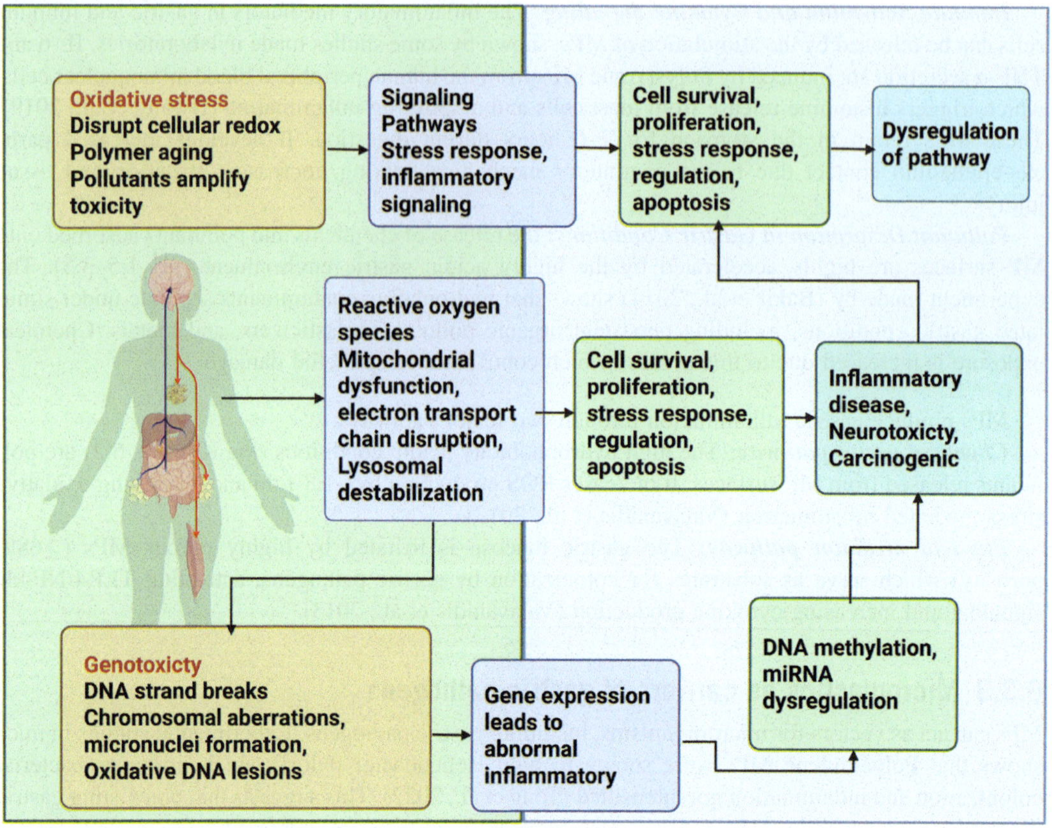

FIGURE 6.3

Conceptual illustration of the biological mechanisms underlying micro- and nanoplastic exposure in humans. The schematic shows how ingestion and inhalation trigger oxidative stress and reactive oxygen species (ROS) generation, leading to inflammation, genotoxicity, mitochondrial dysfunction, and altered gene expression. Disruption of signaling pathways, cytokine regulation, and programmed cell death may collectively contribute to cancer initiation and progression, highlighting potential long-term human health risks.

chemicals released from some plastics increase the production of harmful molecules. The strain caused by these molecules affects the powerhouse of the cell, the mitochondria resulting in increased cell injury and reduced energy. MPs are common markers of liver damage, as they activate immune cells in the liver which triggers inflammation. When these exposures are prolonged for a period it leads to chronic inflammation due to the fat buildup in the liver. Further it may pay a path to fibrosis, a condition where healthy tissue is replaced by scar tissue. The harmfulness of the MPs increases when they are combined with other pollutants. It is evident from the above studies that MPs cause serious liver issues. But it needs further studies to ensure the particle size, nanoplastics and chemical mixtures which intervene in long-term liver health.

Experimental evidence demonstrates that MP can disrupt hepatic redox balance and cellular homeostasis through multiple converging mechanisms. Exposure to 5 μm polystyrene microspheres has been shown to suppress endogenous antioxidant defenses by inhibiting the Nrf2/ARE signaling pathway, resulting in a 2,3-fold increase in malondialdehyde (MDA) levels and a 40% reduction in superoxide dismutase (SOD) activity in murine liver tissue (Hwang et al., 2019). In parallel, metal ions such as Co^{2+} and Sb^{3+} released from polyethylene terephthalate plastics have been reported to activate NADPH oxidase, thereby further amplifying intracellular reactive oxygen species (ROS) generation (Zhang et al., 2023).

Beyond oxidative damage, MPs have been shown to elicit pronounced immune and inflammatory responses in hepatic tissue. Zhao et al. (2021) reported enhanced infiltration of natural killer cells and macrophages into the liver following MP exposure, with subsequent activation of NF-kB-dependent inflammatory signaling. Consistently, Chen et al. (2021) observed that mice exposed to 2 μm polyvinyl chloride MPs for 60 days exhibited elevated oxidative stress, increased P13K/Akt-mediated apoptotic gene expression, and significantly higher serum alanine aminotransferase (ALT) and aspartate aminotransferase (AST) levels, indicating hepatocellular injury. Chronic MP exposure has also been associated with hepatic fibrosis and metabolic dysregulation. Lu et al. (2021) reported that MPs promote persistent liver inflammation and lipid accumulation in fish, disrupting energy metabolism and hepatic homeostasis. Mechanistically, activation of NF-kB signaling by polystyrene MPs has been shown to drive sustained inflammatory responses, with IL-6 and IL-1β expression increasing by approximately 3.2 and 2.8-fold, respectively (Liao et al., 2021). This inflammatory milieu is further reinforced by macrophages activation and the release of extracellular traps through the ROS/TGF-β/Smad2/3 axis, facilitating epithelial–mesenchymal transition and progressive hepatic remodeling. In addition, impairment of autophagic processes appears to exacerbate metabolic injury. Suppression of autophagy, indicated by a 40% reduction in the LC3-II/LC3-1 ratio, has been linked to enhanced lipid accumulation and accelerated progression of nonalcoholic fatty liver disease in Nrf2-deficient mice exposed to MPs (Chen et al., 2021). Collectively, these processes establish a self-perpetuating cycle of oxidative stress, inflammation, and metabolic dysfunction. Despite these advances, the precise mechanism underlying MP-induced hepatotoxicity, particularly those related to particle size and interactions with simultaneously occurring environmental stressors, remains insufficiently characterized.

6.3.5 Respiratory system

MPs can enter the human body through breathing. MPs of very small size can reach soft air sacs by deeply penetrating the lungs, through which the oxygen exchange happens. Usually, the larger particles are trapped in the upper airways and are cleared by normal clearing mechanisms. MPs of very small sizes under the range of 5 μm can easily stay in the lower respiratory tract over a longer period (Vianello et al., 2019). Research shows that stronger inflammation and greater stress to organs beyond the lungs are caused by these smaller particles. These inhaled MPs causes irritation and oxidative stress in lung cells, thereby weakening the protective lining and promoting inflammation. The lower concentration of MPs such as polystyrene and PVC causes cell damage in the lungs while the higher concentration may lead to the release of release inflammatory chemicals or undergo programmed cell death. MPs in people particularly with asthma or other preexisting lung conditions, activate immune cells in the lungs which releases additional inflammatory substances that triggers breathing difficulties (Wang et al., 2023a,b,c).

Pulmonary fibrosis, scarring in the lungs, is developed when the MPs are exposed for a longer period, as there is continuous irritation in the tissues by these particles. This scarring process is developed due to the involvement of certain signaling pathways in lung cells such as the Wnt/β-catenin pathway (Li et al., 2022). MPs in the atmosphere air are in small amount, but continuous inhalation for many years can cause serious health issues. The other particles like asbestos and fine dust inhalation can lead to cancer.

6.3.6 Reproductive system

Recent research shows the reproductive system of both men and women are affected by these MPs. The harmful molecules are created by these tiny plastic particles which create stress and disrupt the hormone levels in the human body thereby influencing the normal development of reproductive cells (Zhao et al., 2023). Experimental studies show that male fertility is affected as they found MPs in human testes and semen. MPs in animals show that they affect the structure of the tests, variation in levels of testosterone leading to the abnormal production of sperm. The internal framework for developing sperm cells is disturbed by the very small particles, as they are essential for their proper maturation. MPs in female reproductive system causes stress and damage the cells in the ovaries which reduces the quality and number of developing eggs. It leads to the early loss of follicles causing fertility related issues. Exposure to MPs during pregnancy has serious concerns, as they reach the developing fetus through the placenta (Xu et al., 2020).

Emerging evidence also demonstrates that MP exposure negatively affects female reproductive health. Xie et al. (2020) reported that polystyrene MPs trigger oxidative stress and apoptosis in ovarian tissue through activation of the p38 MAPK signaling pathway, resulting in follicular atresia and diminished oocyte quality in female mice. Consistent with these observations, Patel et al. (2022) found that MP exposure led to a reduction in sinus follicle numbers and impaired oocyte viability, indicating a potential decline in female fertility associated with MP accumulation. Beyond their intrinsic toxicity, MPs can amplify reproductive harm by acting as carriers for endocrine disrupting chemicals. Due to their high surface area and porous structure, MPs readily adsorb compounds such as bisphenol A and phthalates (logKow = 3.4–5.0). Studies made in animals show that exposure to MPs causes various alterations in metabolic processes of the offspring, raising concerns about long-term developmental effects. The scientists underline the need for more investigation regarding the influence of MPs in reproductive risks like fertility issues and pregnancy outcomes.

6.3.7 Nervous system

Due to their capacity to traverse the blood–brain barrier (BBB), accumulate within neural tissue, and disrupt central nervous system homeostasis, nanoscale plastic particles are considered as potential neurotoxicants. Nanoplastics can bypass or compromise BBB integrity due to their small size and surface reactivity. Which enables their entry into sensitive brain regions such as the hippocampus. These particles can initiate oxidative stress, activate microglia, and amplify neuroinflammatory signaling. Once they enter the neural systems, it causes neuronal injury. Experimental studies have shown that polystyrene MPs alter synaptic gene expression and disrupt glutamatergic receptor function, leading to measurable impairments in learning and memory (Lee et al., 2022). Nanoplastics can further induce cytotoxicity in human neuronal cells, including abnormalities in synapse formation, nuclear deformation, and leakage of intracellular contents which are demonstrated by various studies. These observations suggest that MPs may interfere with neuronal architecture, synaptic transmission, and neuroimmune regulation, raising concerns about their potential involvement in neuro-developmental and neurodegenerative processes (Ban et al., 2021). Comprehensive mechanistic and epidemiological studies are required to understand the substantial gaps regarding exposure thresholds and long-term accumulation to have a clear view on human neurological health caused by nanoplastics.

***Influence of Particle Properties on Neurotoxicity*:** Growing evidence shows that nanoplastics and the physicochemical properties of MPs play a pivotal role in determining their neurotoxic potential. The particle size remains the most important characteristics when compared with others. Due to the reduced immune recognition, slower clearance, and enhanced interactions with cellular membranes, smaller nanoplastics shows high perseverance in neural tissues when compared with larger particles. These particles tend to translocate across the blood–brain barrier due to their nanoscale dimensions and accumulate in the brain when exposed for longer period. The accumulation in brain cells causes oxidative stress and amplifies neuroinflammation, leading to neuronal dysfunction.

Another important determining factor of neurotoxicity is the surface charge. Nanoplastics are positively charged which interact strongly with neuronal membranes and intracellular organelles as they are negatively charged. For example, the neuronal reactive oxygen species production is increased more than double by positively charged polystyrene nanoplastics (+36.5 mV) which is accompanied by a major reduction in dendritic spine density which is an essential feature for synaptic plasticity and cognitive function (Wang et al., 2023a,b,c). Similarly, aminated nanoplastics demonstrate high uptake efficiency by microglia. Once internalized, they activate the NLRP3 inflammasome, producing a substantial elevation in IL-1β secretion. This proinflammatory cascade promotes chronic neuro-inflammation, which is closely linked to neurodegeneration. Toxicity profiles are further modified by the chemical composition and associated contaminants. Chemical additives such as di(2-ethylhexyl) phthalate (DEHP) are found to be in many plastics especially from PVC, causes leaching and induce additional cellular stress. DEHP has been shown to disrupt mitochondrial transport through activation of the PPARγ/RXRα signaling pathway, thereby impairing neuronal energy metabolism. The more concerning one is the absorption of toxic metals onto plastic surfaces (Wang et al., 2023a,b,c). Lead ions (Pb^{2+}) bound to polyethylene terephthalate (PET) fragments increase abnormal tau phosphory-lation more than three-fold, accelerating pathological events associated with Alzheimer's disease (Liu et al., 2023b). These results highlight that MPs act not only as physical particulates but also as vectors for a complex mixture of neurotoxic chemicals.

6.4 Toxicological studies and long-term health effects

MPs and nanoplastics (MPs–NPs) are increasingly recognized for their ability to interfere with im-mune function, raising significant concerns regarding their potential immunotoxicity. Evidence from both in vitro cell models and in vivo animal studies demonstrates that these particles can disrupt innate and adaptive immune responses through oxidative stress, apoptosis, metabolic reprogramming, and impaired cellular clearance (Fig. 6.4). Laboratory investigations have used particle sizes ranging from

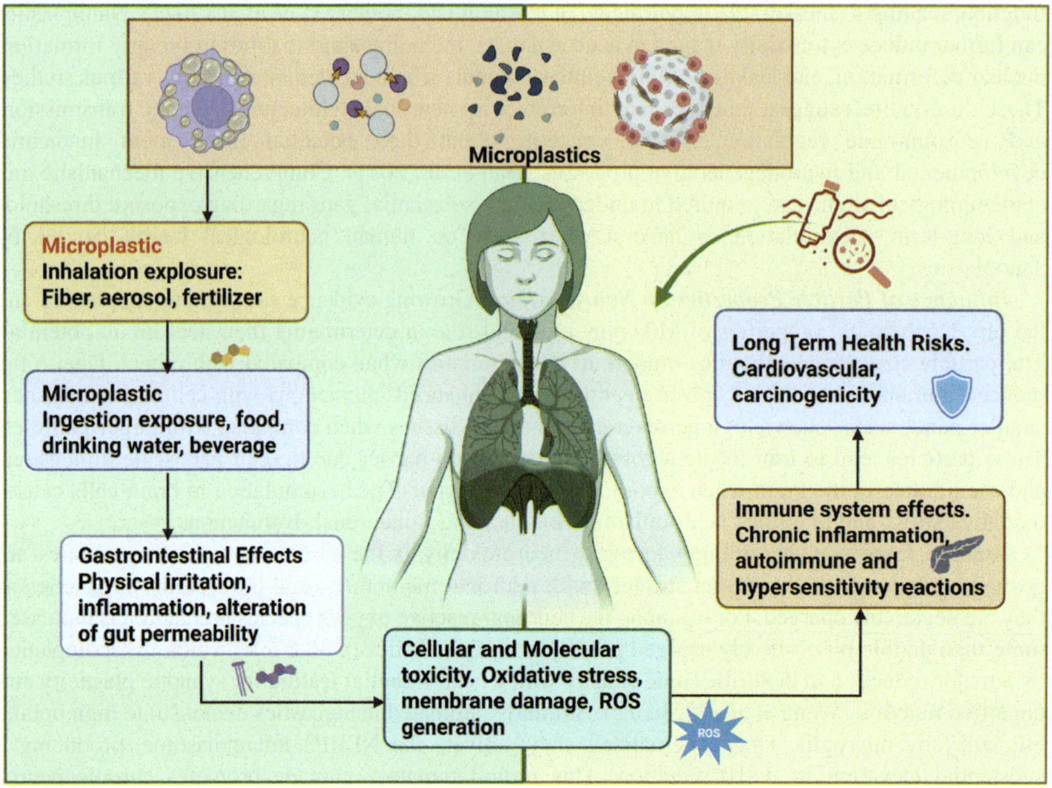

FIGURE 6.4

Schematic illustration of major sources and forms of microplastics, including fibers, beads, fragments, and foams originating from tires, textiles, paints, and plastic products. The figure highlights primary human exposure pathways through inhalation (airborne fibers and aerosols) and ingestion (food, drinking water, and beverages), leading to particle translocation within the body. Potential adverse health outcomes associated with microplastic exposure, such as oxidative stress, neurotoxicity, reproductive toxicity, carcinogenicity, and metabolic disruption, are also summarized.

approximately 50 nm–200 μm, with exposure periods extending from acute (1 h) to chronic (up to 49 days), providing a broad overview of their biological effects. Even though varies experimental studies show mechanistic insights, exposure to real-world human conditions remains an important question to be resolved. The most sensitive immune cell types affected by MP–NPs is macrophages. As primary mediators of the innate immune response, macrophages are responsible for recognizing, engulfing, and degrading foreign particles.

Studies show that 10 μm polystyrene MPs are efficiently internalized by mouse macrophages but resist breakdown by classical autophagy pathways. Microscopic analyses reveal partial colocalization with LC3, indicating disrupted autophagic flux and compromised phagocytic activity (Chen et al., 2021). These impairments are accompanied by a metabolic transition from oxidative phosphorylation to glycolysis, a metabolic reprogramming pattern typically associated with inflammatory activation. Despite these findings, long-term effects on macrophage function particularly with environmentally relevant doses—remain poorly understood due to limited chronic exposure studies. Systemic immune disturbances have also been documented in mammalian models. In C57BL/6 mice, exposure to 100 nm asymmetrical MPs at 600 μg/mL for 15 days resulted in marked hematological abnormalities, including erythrocyte injury, decreased red and white blood cell counts, and an elevated neutrophil-to-lymphocyte ratio an indicator of systemic inflammation (Abdel-Zaher et al., 2023). Similar experiments using 5 μm polystyrene MPs revealed reduced bone marrow activity and impaired leukocyte function. Transcriptomic analyses further showed altered expression of genes involved in immune regulation and cellular metabolism, including those in the JAK/STAT signaling pathway (Zhao et al., 2021). Although these studies offer mechanistic clarity, questions remain regarding the environmental relevance of the high exposure concentrations frequently used in laboratory settings.

Comparative studies highlight notable species-specific variability. In an investigation of polyvinyl chloride MPs ranging from 0.16 to 1.82 μm, human lymphocytes demonstrated pronounced sensitivity, exhibiting dose-dependent increases in oxidative stress, mitochondrial damage, lysosomal destabilization, and activation of caspase-3 (Salimi et al., 2022). In contrast, fish lymphocytes showed minimal cytotoxic responses under the same experimental conditions. These findings may reflect intrinsic differences in cellular physiology and immune architecture between species, underscoring the need for caution when extrapolating animal data to human health outcomes. Genotoxicity represents an additional concern. Human peripheral blood lymphocytes exposed to sterilized fluorescent polyethylene–methacrylic acid MPs (10–45 μm) exhibited increased frequencies of micronuclei, nuclear buds, and nucleoplasmic bridges well-established biomarkers of DNA damage and chromosomal instability (Çobanoglu et al., 2021). Although these findings demonstrate clear genotoxic potential, the implications for long-term mutation accumulation and disease risk have yet to be fully addressed.

6.4.1 Cell-type specific immune responses to MPs

Immune responses to MPs vary widely across cell types and are strongly influenced by particle size, polymer composition, and concentration. Studies examining secondary polypropylene (PP) particles are one of the most abundant forms of environmental plastic waste—demonstrate that particle characteristics critically shape cellular outcomes. At concentrations below 0.5 mg/mL, PP particles did not induce measurable cytotoxicity in human dermal fibroblasts (HDFs) or peripheral blood mononuclear cells (PBMCs). However, at higher doses, HDF viability decreased by approximately 40%, whereas PBMCs remained comparatively resistant, highlighting important cell-type differences in susceptibility. Similar trends were observed in work by Han et al. (2020), found that polyvinyl chloride (PVC) and acrylonitrile–butadiene–styrene (ABS) MPs (25–200 μm) did not produce over cytotoxicity in PBMCs, HMC-1 mast cells, HDFs, or HeLa cells. Despite the absence of cell death, immune activation was still evident: PVC exposure elevated TNF-α secretion from PBMCs, while ABS increased IL-6 production, indicating that MPs can stimulate inflammatory signaling even when cell viability is preserved (Han et al., 2020). These findings underscore that immunomodulation not only cytotoxicity is a critical aspect of MP-induced biological effects.

The immune activity within the central nervous system is also influenced by MPs. A noticeable sensitivity is seen in the Human HMC3 neural tissues, a primary defender when exposed to polystyrene MPs (0.2–2 μm). As a result of exposure there is alteration in morphological structure, increased apoptosis, and modulation of innate immune signaling pathways. Transcriptomic and microRNA profiling revealed broad changes in genes associated with inflammation, oxidative stress, and cell survival, suggesting that MPs may contribute to neuroimmune dysregulation and persistent neuroinflammatory states (Lu et al., 2023a). Collectively, these studies indicate that MP-induced immune responses are multifaceted, cell-type dependent, and capable of occurring independently of traditional cytotoxic endpoints.

6.4.2 Immune toxicity insights from aquatic organisms

Studies made on aquatic species shows an essential view on the immunotoxic potential of MPs and nanoplastics (NPs) affect nonmammalian immune systems and the broader ecological implications. The researchers evaluated the phagocytic uptake of both commercial microbeads and particles generated from consumer plastics, using a carp macrophage cell line. From the study it is evident that particle size strongly influenced phagocytic efficiency. The smaller microbeads (0.83 μm) were adopted more readily than larger fragments, results from 3-day period shows there is a maximum acceptance occur at approximately 16 hours before progressively returning to baseline. Comparable findings were reported in rainbow trout, where both myeloid cells and B cells exhibited substantial phagocytic activity toward MPs. The results make sure that different immune cell populations may respond divergently to MP exposure, as B cells demonstrated greater efficiency than myeloid cells in internalizing smaller microbeads (Zwollo et al., 2021). The interactions with environmental contaminants make the immune impacts of MPs in aquatic organisms further complicated. MPs and NPs alone can compromise plasma membrane integrity and enhance lysosomal activity, shown by the studies using mussel hemocytes. It produces greater immune dysfunction when they are coexposed with hydrophobic chemical pollutants such as benzo(a)pyrene, which indicates a synergistic toxic effect (Katsumiti et al., 2021). It is evident that MPs found in natural waters typically carry adsorbed pollutants, but the experiments made with pristine plastic particles show the significant true immunological hazard. So, in future ecological and immunotoxicity assessments it must involve MPs that are aged, weathered, or chemically contaminated to determine the actual neurological issues.

6.4.3 In vitro and in vivo gastrointestinal toxicity

The gastrointestinal (GI) tract acts as a primary interface for ingested MPs and NPs, and a growing body. To investigate the potential adverse effects research has employed both in vitro and in vivo models. To examine the cytotoxicity and inflammatory responses, in vitro studies frequently utilize human colorectal cancer cell lines (HT29, HCT116, SW480, SW620), colorectal adenocarcinoma models (Caco-2, HRT-18), and murine rectal carcinoma cells (CMT-93). There is a considerable gap in experimental validation at the organismal level as there are only two in vivo mouse studies that have systematically evaluated GI toxicity (Brynzak-Schreiber et al., 2024). The available studies show that MP sizes range from approximately 0.25–13 μm with the duration of exposure varying from 1 hour to nearly 4 weeks.

Species-Specificity, Chemical Transformations, and Gastrointestinal Alterations: Incorporation of MPs by colorectal epithelial cells raises important concerns regarding their potential to disrupt gut homeostasis and influence disease processes. Uptake across HT29, HCT116, SW480, and SW620 cell lines occur in a size- and dose-dependent manner, with HCT116 cells exhibiting the highest internalization rates. MPs enter cells through energy-dependent endocytic pathways and preferentially accumulate within lysosomes, suggesting potential interference with intracellular trafficking and degradation pathways. Importantly, evidence that 0.25 μm MP can enhance migratory behavior in colorectal cancer cells raises the possibility that MPs may influence tumor cell motility or metastatic potential (Brynzak-Schreiber et al., 2024).

Although these findings derive from transformed cell lines rather than healthy intestinal epithelium. Additional research is needed to determine whether noncancerous tissues would have similar effects.

Interspecies and Gastrointestinal Responses to Microplastics**:** Research shows that depending on the species and the type of cell examined the MPs can affect intestinal cells in different ways. When exposed to polystyrene MPs both human and mouse colorectal cells showed reduced survival. However, the clear signs of oxidative stress are found only in the mouse. The above findings highlight the different biological responses made by varies species and applying the laboratory findings to human health directly may cause a critical issue. The studies using intestinal epithelial cell models also show similar discrepancies. In experiments made with low-density polyethylene MPs although there is damage in cell membranes, mitochondria disruption and triggered oxidative stress, these effects are varied based on different factors such as particle size, chemical properties and mainly depends on experiment procedure. In laboratory studies using pristine particles, environmental conditions such as ultraviolet exposure increase the toxicity by altering surface chemistry. The gastrointestinal responses vary even in animal studies also. Few laboratory studies show the effect of oxidized polyethylene MPs in the human body causes tissue injury, promote inflammation, and disturb gut microbiota as they accumulate in the intestine which also reduces beneficial *Lactobacillus* species (Wang et al., 2023b). In contrast, some studies show that there is no toxicity to similar particles. This depends on the amount of exposure as smaller quantities did not reach the target tissues, but the overtime exposure is not considered. The above findings conclude that the MP toxicity in the gut is influenced by species differences, particle characteristics, and experimental conditions. It requires long term studies and more realistic experiments to determine the actual effect on human gastrointestinal health.

6.4.4 in vitro and in vivo lung toxicity

The potential impacts of airborne MPs and nanoplastics (NPs) on respiratory health has raised serious concerns as they are detected considerably in outdoor and indoor environments. Studies made in laboratories to characterize cellular responses to MP and NP exposure on lung toxicity depend on human respiratory cell lines most commonly adenocarcinomic alveolar epithelial cells (A549) and bronchial epithelial cells (BEAS-2B) (Jeon et al., 2023). In vivo studies have primarily used murine models, with a smaller number of long-term investigations conducted in chickens. The murine models are used in *in vivo* studies, a long-term study with a small number of chickens. In these experiments, the period of exposure varies from 24 hours, a short-term exposure to chronic treatments lasting several weeks with particle sizes typically range from 20 nm to 10 μm. The above studies give the understanding of interaction of MPs and NPs with the respiratory system. Although the major limitation lies in experimental designs, particle types, and exposure doses.

6.4.5 Cytoskeletal impairments, viral infection exacerbation, and epithelial-to-mesenchymal transition

The distraction in cytoskeletal organization in lung epithelial cells is found from the in vitro studies. When A549 cells are exposed to polystyrene MPs it causes reduced cell proliferation with remarkable changes in morphology and altered lamellipodia and filopodia formation. The disturbances in actin remodeling resulted from these effects, which makes it difficult to maintain epithelial barrier function and coordinating intracellular signaling. The studies made in experiments use pristine polystyrene particles, these particles do not capture the chemical complexity of environmentally aged MPs and therefore the actual toxicity remains underestimated. The potential of MPs to intensify respiratory viral infections is the major area of concern. The internalization and replication of influenza A virus seemed to be improved when the cultured A549 cells are exposed to fluorescent polystyrene MPs which lead to considerably higher viral titters. These effects reduce the important antiviral pathways, including the downregulation of IFITM3, IRF3, and IFN-β (Wang et al., 2023a). The results show that inhaled MPs may weaken innate antiviral defenses and increase vulnerability to respiratory infections. However the realistic approach of exposure to MPs are to be studied to find the actual effects, as these Results are made with concentrations that made exceed the amount in actual environmental settings. Epithelial-to-mesenchymal transition (EMT), a process associated with fibrosis, chronic inflammation, and cancer progression is also triggered by nanoplastics. Studies made with A549 cells show that smaller (20 nm) and positively charged nanoplastics made stronger EMT responses when compared with larger (50 nm) or negatively charged particles. From these findings it is evident that biological outcomes are shaped by the particle size and surface charge (Goodman et al., 2021). Still the results to whole-organ responses are not accurate as these findings are based on single cell models and high particle concentrations.

6.4.6 Charge-dependent toxicity, oxidative stress, and inflammation

The key determinant of MP toxicity in the respiratory system is the surface charge. The important cytotoxicity in BEAS-2B cells and mouse lung tissue is caused only by positively charged amino-modified polystyrene particles. While activating inflammatory pathways such as PERK–eIF2α and ATF4–CHOP, these particles induced oxidative stress, endoplasmic reticulum stress, and autophagic cell death (Jeon et al., 2023). The findings from these studies show the significance of particle surface chemistry, but there is a limitation in application of these results in human exposure as these animals are exposed to higher dosages. The notable amount of lung toxicity is caused by other polymer types, such as polyethylene terephthalate (PET). These PET MPs causes damage in mitochondria, increase in reactive oxygen species generation and stimulate inflammatory responses, in BEAS-2B cells. Cellular energy metabolism disorder and redox homeostasis are found in multiomics analyses. Conversely, these findings cannot be infused into *vivo* conditions as the in vitro models do not consider important factors in lung physiology such as mucus layer, surfactant presence and interactions with immune cells.

6.4.7 In vivo evidence: Chronic lung toxicity in chicken models

The potential chronic effects are found to be caused by MPs on the respiratory system on chickens during long – term exposure. The polystyrene MPs by suppressing the PI3K/Akt/mTOR pathway induces oxidative stress, reduces the antioxidant enzyme activity, and inhibits autophagy when exposed over a 42-day period of time. An activation of apoptosis-related MAPK signaling and structural abnormalities such as mitochondrial fragmentation and nuclear deformation are other effects observed by these MPs. Even though these findings provide various information regarding the exposure outcomes, it has limitations in human risk management. Due to the different airway structure and breathing patterns in avian and mammalian respiratory systems. However, the studies made in vivo studies are based on exposure with higher concentrations when compared with the actual amount present in the environment.

6.4.8 In vitro neurotoxicity

Owing to the concerns caused by the MPs and nanoplastics (NPs) to affect the nervous system, there is an increase in exploring the neurotoxic potential by in vitro studies. The studies made so far depend on human cerebral organoids, primary and immortalized neuronal cultures, astrocytes, neural stem cells, and rat-derived neural cell types, including neurons, oligodendrocytes, and astrocytes. There is a considerable gap in estimation of systemic and long-term neurological impacts, as there are no reports that were made in neurotoxicity by in vivo studies till date. The experiments made by in vitro experiments provide only systemic understanding and not the actual exposure levels. Because these studies were made with particle sizes between 30 and 500 nm, over a period of 24 hours to 7 days with exposure concentrations ranging from 0.0005 to 1 mg/mL.

6.4.9 Differential susceptibility among neural cell types and the role of astrocytes

From the studies it is evident that we receive different types of responses for micro- and nanoplastics from different types of brain cells. Demonstration of polystyrene nanoplastics from experiments shows that neurons are highly sensitive, as it may endure considerable cell death by absorbing the particles easily. Whereas the supporting cells in the brain, the astrocytes get inflamed and releases immune-related molecules when reacting with the particles. This demonstrates that when exposed to particles the astrocytes help brain to drive inflammation. On the other hand, some research shows that small nanoplastics tend to accumulate inside neural stem cells by crossing the blood brain barrier and activating strong apoptotic responses. But the results are not accurate because high concentration of particles is considered which may amplify the effects. The comparative study made in neurons, astrocytes, and oligodendrocytes shows that cell which is responsible for nanoplastic exposure is astrocytes, found by changes in cell structure, lipid metabolism, and inflammatory signaling. The experimental study made in human cerebral organoid models, the miniature version of brain like tissues discovered that mitochondrial energy production was damaged, and the normal neuronal development was affected by nanoplastics. Although the various damages in the nervous systems are detected by these experiments, the realistic approach must be analyzed to understand the real impact of nanoplastics on the human health.

6.5 Nephrotoxicity of microplastics and nanoplastics

Accumulation of MPs and nanoplastics (MPs–NPs) accumulate in the kidneys that entering through the bloodstream affects the normal filtration and cellular function. These particles visibly reduce cell viability, trigger oxidative stress, and disrupt mitochondrial activity when studied using various kidney cell models such as human proximal tubular cells, canine kidney epithelial cells, and human embryonic kidney cells. For instance, when polyethylene and polystyrene particles are exposed, it shows a pattern associated with cellular stress like impairment in cell health accomplished with increased metabolic activity. Some research shows that MPs exposure leads to elevated levels of reactive oxygen species and reduced expression of protective enzyme, as they localize near the cell nucleus and interfere with antioxidant defenses. The MPs in human kidney cell models compromise the structural and functional integrity of renal tissue found by the changes in cell shape and propagation effects.

Experiments made by *vivo* studies show that polystyrene MPs in mice leads to mitochondrial dysfunction, inflammation, activation of cellular stress pathways and mainly accumulation of particles in kidney tissues. Animal models activate signaling pathways such as MAPK, AKT, and mTOR when exposed for longer period which is associated with oxidative damage and impaired kidney function. Mitochondrial injury, oxidative stress, and inflammation are observed in the chicken studies mediated by activation of the NF-κB pathway. A regulated inflammatory form of cell death is observed in these models, through necroptosis. Though these effects are found by studies, it appears when they are subjected to high dosages. This reflects the need for future studies to be done in real world environment as the concentration in the atmosphere is minimal when compared with the study environment.

6.5.1 Toxic effects on epithelial, muscle, bone, and skin tissues

The available findings shows that the epithelial tissues, muscle, bone, and skin are also vulnerable to MP-induced stress, but the studies based on these tissues are limited when compared with the immune system, gut, and lungs. The exposure to polystyrene particles in epithelial cell models shows remarkable decrease in cell viability along with noticeable structural damage. The changes in mitochondrial dysfunction, elevated oxidative stress, and disruption of the mitochondrial membrane potential show these effects. Calcium imbalance and destabilization of cellular homeostasis are caused due to the Cellular injury, which are results of endoplasmic reticulum stress and activation of the PERK/eIF2α/CHOP pathway. Skeletal muscle health is also impaired by MPs. When studies are made in myoblasts and tibial muscle tissue, especially when exposed to larger particles, it shows muscle regeneration delay. Incomplete repair of muscle fibers and increased lipid accumulation are found when histological analyses are made with these tissues. In cell models, MPs reduce the ability of myoblasts to differentiate into mature muscle cells and activate stress-related pathways such as p38/MAPK and NF-κB. Even though these results show serious concerns, a realistic approach must be made to determine the actual health risks, as these findings are made with higher exposure than the typical level.

Bone cells are sensitive; in osteosarcoma cell lines the cell viability is reduced both in concentration and time dependent manner when exposed to MPs between 0.1 and 2 μm range. Particles detected in human tissues such as the placenta, lungs, and breast milk are around 2 μm in size causing serious concern. The dermal health affected by MPs are explained by Skin toxicity studies. MPs are readily internalized and can induce mitochondrial oxidative stress. This is found by experiments using

Table 6.3 Neurotoxic effects on the fish body.

Fish species	MP type	Organ effected	MP concentration	References
Cypriuns carpio	PE	Plasma	250, 500 μg/L	Jabeen et al. (2018)
Danio rerio	PS	Brain, liver	0, 100 μg/L	Umamaheshwari et al. (2021)
Pomatoschistus microps	PE	Brain	–	Pannetier et al. (2020)
Clarius gariepinus	PVC	Brain, gills	0.5%, 1.5%, 3.0% in pellet	Jabeen et al. (2018)
Carassius auratus larvae	PS	–	10,100, 1000 μg/L	Yang et al. (2020)

skin carcinoma cells and normal keratinocytes. The mitochondrial DNA is released due to the stress which disturbs the mitochondrial permeability and activates the NLRP3 inflammasome a key driver of inflammation. This activation encourages the rapid increase in cancer cells, while in normal keratinocytes it triggers inflammation and in pyroptosis causes cell death. MPs in skin form aggregates that stick to cell membranes, which act as a carrier to transport pathogens or toxic chemicals across the skin surface. The results show that MPs influence inflammatory skin conditions and weaken the protective functions of the skin.

6.5.2 In vitro and in vivo multiorgan toxicity

Several organ systems are affected simultaneously by MPs and nanoplastics (MP–NPs). It encourages the development of experimental models designed to examine their broader systemic impacts. The multiple tissue toxicity studies in parallel are only limited up to date. The published studies show in vitro multiorgan model exposing human intestinal, lung, immune, and skin cells to MPs for 24 hours and in vivo studies with exposure durations ranging from a few hours to several months have used mice, chickens, and a range of aquatic organisms including gastropods, crustaceans, shrimp, carp, amphipods, fish (Table 6.3), and microalgae. Polystyrene, polyethylene, and plastic foam fragments are commonly tested particles with the size ranging from 100 nm to 300 μm.

6.6 Human and animal models to study systemic MP–NP effects

6.6.1 Multicellular human in vitro systems

MPs response to different tissues under comparable exposure conditions is understood through multicell culture systems.

In a study, reductions in cell viability were observed when human intestinal (Caco-2), lung (A549), immune (Jurkat, THP-1, U397), and skin (HaCaT) cells were exposed to polyethylene MPs of size 6 and 30 μm. The results depend greatly on dosage, and greater vulnerability found to be in intestinal and lung epithelial cells. Oxidative stress is triggered by particles in immune cells such as dendritic cells, macrophages, and T lymphocytes. In keratinocytes and THP-1 monocytes, exposure promoted inflammatory activation, underscoring that MPs can elicit tissue-specific stress responses even under short-term exposure.

***Systemic Toxicity in Mouse Models*:** Studies made in animals provide further evidence that MPs can accumulate in several organs and disrupt physiological function. Experiment made in mice shows that accumulation was detected in the brain, kidneys, and gastrointestinal tract after only 7 days when orally exposed to fluorescent polystyrene particles (0.2–10 μm). Noticeable effects are found by histological examinations such as inflammatory lesions, immune cell infiltration, and increased cytokine expression. Exposure to polyethylene MPs for over 90-day causes weight loss, organ injury, elevated IgA levels, and an increase in splenic lymphocyte populations. The chronic exposure effects immune regulation, development, and possibly reproduction. Studies show that polystyrene MPs exposure in mice for 28 days causes evident organ damage. Yet, change in behavior and shifts in immune markers were observed in the liver and brain. The effects suggest that children or older individuals are subjected to more vulnerability when exposed to MP exposure.

***Avian Multiorgan Toxicity*:** Systemic risks are found in avian models which provide additional understandings about MPs through evidence. A study in chickens for period of 6-week exposure showed that MPs induced neurological damage, oxidative stress, and liver injury including loss of Purkinje cells and the formation of microthrombi. The blood–brain barrier is also weakened by the exposure, indicated by reduced expression of claudin-3, occludin, and ZO-1.

These structural disruptions were accompanied by impaired antioxidant signaling through the Nrf2 pathway, as well as activation of ferroptosis, apoptosis, and autophagy-related stress responses. The above studies shows that complex multi organ toxicity is developed in vertebrates when exposed to MPs involving oxidative stress, inflammation, and cell death.

***Current Trends, Challenges, and Future Directions*:** There is a rapid growth of impact in both biological and ecological structure due to the widespread distribution of MPs and nanoplastics across air, water, soil, and food which leads path to the increased research in this area. From studies it is evident that humans are exposed to these particles by various means such as inhalation, ingestion, and dermal contact based on how MPs are absorbed, distributed, metabolized and eliminated once they enter the body. The major limitations in these studies are due to the conclusions made by theoretical assumptions rather than the direct experiment with the actual number of particles exposed.

So, the outcome remains unresolved as the questions on toxicokinetics, long-term accumulation, dose thresholds, and interactions with cocontaminants remain untouched.

Moreover, the results depend on experiments done with high particle concentrations and do not reflect the real-world exposure which doesn't give the actual human health risks.

For better understanding, the dosage and chronic exposure models must be prioritized while studying the effects of MPs–NPs on human and ecological health.

6.7 Environmental distribution, size heterogeneity, and chemical composition of particles

A careful consideration of the environmental distribution, size heterogeneity, and chemical complexity are required to understand the MP–NP toxicity. The MPs and NPs samples are found in multiple physical forms such as fibers, fragments, films, pellets, granules, and foams of which fibers are dominant. But synthetic and commercially manufactured particles are most involved in laboratory studies. Primarily PS (33 in vitro, 31 in vivo) are used and to a lesser extent PE ($n = 3$ in vitro; $n = 9$ in vivo), PVC, ABS, PP, oxidized PE, PS with DEHP, P–F biofoams, or PET are involved. The usage of these particles causes limitations as polymers exposed to humans may be varied in environment when compared with laboratory generated ones.

6.7.1 Size and bioavailability considerations

In general, particles within the range 1–10 μm is used in the experimental studies, some studies found to include particles >11–100 μm but this does not match the environment available particles which are mostly in the range of 90–100 μm.

Some points to be noted:

- MPs >10 μm are typically excreted as they have limited capability to cross the intestinal barrier.
- NPs <1 μm exhibit higher bioavailability and toxicity because they can cross biological membranes.

Due to differences in particle size, surface area, reactivity, and biological persistence, MPs and NPs therefore exhibit distinct toxicological profiles. Larger MPs contribute chronic inflammation and fibrosis as they are prone to accumulate in the gut, liver, brain, and reproductive organs. These MPs can cause physical obstruction and particularly in aquatic organisms they can alter nutrient absorption and disrupt the microbiome. The toxic effects of MPs are increased by other environmental contaminants such as metals and persistent organic pollutants.

On the other hand, NPs can penetrate the intestinal epithelium causing blood–brain barrier (BBB), and penetration into placenta therefore enabling systemic distribution. It supports the formation of protein corona, alteration in cellular internals and interact with immune modulation due to their large surface-area-to-volume ratio. Furthermore, they can generate noticeable oxidative stress, mitochondrial dysfunction, DNA damage, apoptosis, and inflammation.

6.7.2 Polymer composition and surface chemistry

- Based on the polymer type, surface charge, and functional groups toxicity also varies
- Symptoms such as oxidative stress, apoptosis, neurotoxicity, and inflammation are caused by Polystyrene (PS) due to leaching.
- In zebrafish lipid peroxidation and neurodevelopmental disruption are promoted by PVC.
- Even though PE and PP are less reactive, oxidative stress and behavioral abnormalities are induced in vivo.
- In vitro, hazardous by-products are released by ABS, which escalate xenobiotic accumulation, and increase necrosis and inflammation.
- Enhanced reactivity is exhibited by Oxidized PE (ox-PE) which induces mitochondrial dysfunction.
- ROS generation is promoted in marine invertebrates because P–F biofoams leach more chemicals than typical plastics.
- Carcinogen, antimony is released when exposed to PET which is linked to ocular, pulmonary, and liver disorders.

- Pregnancy complications, neurotoxicity, and immune disruption are often linked to soft plastics, DEHP.

  MP–NP toxicity is further modulated by surface properties. The studies which demonstrated these properties are:
- Positive surface charge
- Oxygen-containing functional groups

  To understand environmental aging, pollutant interactions, and degradation behavior, systematic investigation of these properties is urgently required.

6.7.3 Major limitation

The global concern regarding MPs and NPs has been heated up due to the widespread distribution of these particles and their chronic health effects on humans and animals. Even though MP–NP toxicology research was done widely, a certain knowledge gap prevails. Although the studies made till now explain the accumulation of these particles in biological tissues and their chronic effects in various organs, the pharmacokinetics, absorption, distribution, metabolism, and excretion of these particles are not mentioned clearly which are essentials to be noted. For the advancement in human health risk assessment and to develop the accurate model for exposure these points to be addressed clearly. The major limitations in the current research are due to the high particle doses and short exposure durations, which do not match the real environmental exposure conditions. The results from these approaches do not provide the actual health risk, as the outcome may be overstating or amplified long term effects. In future studies, it is better to simulate real-world conditions and to include chronic, low-dose, environmentally relevant exposures to expose the long-term toxicity. Another important challenge is the consideration of small division of polymers such as polystyrene (PS) and polyethylene (PE). As there are more variety available which may also cause effects on ecosystems. Even after their wide range of presence in the daily usage polymers such as polypropylene (PP), polyvinyl chloride (PVC), acrylonitrile–butadiene–styrene (ABS), polyethylene terephthalate (PET), and P–F biofoams remain underexplored. The factors such as particle behavior, reactivity, and toxicity are more likely to be affected by surface charge, functional groups, environmental weathering, and chemical aging which are not considered. To understand the MPs and NPs interaction with biological systems the above-mentioned characteristics must be investigated completely. Toxicity is mostly influenced by the size, the NPs which are smaller in size possess greater risk as they can penetrate deep into the tissues by crossing biological barriers including the gut epithelium, blood–brain barrier, and placenta.

In contrast, the experimental studies are made with MPs of size (1–10 μm) which are mostly not adsorbed into the tissues as they get excreted before entering the circulation. This raises concerns about the real-world environmental exposure with laboratory studies. The evidence from the current research shows that the reproductive, immune, gastrointestinal, respiratory, renal, hepatic, and cardiovascular systems are affected by MPs and NPs. The effects in these systems are due to the oxidative stress, inflammation, apoptosis, ER stress, and autophagy, involving pathways such as ROS/MAPK, Nrf2/HO-1, NF-κB, JAK/STAT, PERK, PI3K/Akt/mTOR, and NLRP3.

The studies made so far depend on microscopy, gene expression analysis, and biochemical assays. Although these approaches give valuable information, by incorporating integrating multiomics technologies, machine learning, high-resolution imaging, and nanotechnology-based analytical methods the deeper knowledge can be obtained. To identify biomarkers of exposure and support the development of therapeutic or preventive interventions these advanced techniques may be useful. Overall, considerable progress in the research is achieved, but more detailed and multidisciplinary research must be done. In future studies, realistic exposure scenarios, a broader diversity of polymers, environmentally relevant transformations, and advanced analytical tools are to be incorporated. To address the fast-growing risk caused by plastic pollution, efforts must be made to fill the knowledge gap about the exposure with evidence-based environmental policies and public health strategies.

6.8 Toxicological mechanisms of lung damage caused by microplastics

6.8.1 Oxidative stress

The body's natural balance between reactive oxygen species (ROS) harmful oxygen molecules and the antioxidants that normally protect our cells can be disturbed when MPs enter the lungs (Fig. 6.5). MPs can develop reactive chemical groups on their surfaces under sunlight, which generate long-lasting free radicals. Damaging molecules such as superoxide and hydroxyl radicals are produced in water or lung fluids, when these radicals react with oxygen. MPs first come into contact with lung surfactants when inhaled, which is the protective fluid lining the air sacs. When this interaction happens the formation of ROS is further triggered causing the lungs' natural antioxidant defenses to be overwhelmed. More ROS are produced when stimulated by MPs which enter lung cells directly. Antioxidant enzymes are increased temporarily in the cells at the low dosages but when they are exposed to higher dosage or repeated exposed exhausts these defenses quickly. Harmful oxidizing enzymes increase when antioxidants level fall which creates an imbalance leading to oxidative stress. DNA, fats and proteins are damaged due to this stress causing inflammation and chronic diseases such as chronic obstructive pulmonary disease (COPD). Body's metabolic processes are also interfered with by MPs (Chikuma et al., 2020). Studies reveal that oxidative stress is worsening when MPs ingestion reduces important antioxidants which disturb normal energy production in mitochondria and increase lipid peroxidation. Any interference with mitochondria function can rapidly intensify cell injury as they are the main energy producers and a major source of ROS inside cells. Mitochondrial membrane stability is reduced by MPs by which the electron transport chain is blocked, and the release of free iron (Fe^{2+}) is increased. ROS production is further amplified by these iron particles which on chemical reaction create harmful cycles of oxidative damage. Furthermore, ROS formation is promoted by chemicals on plastic surfaces, plastic pollution which damages the antioxidant systems causing serious damage. Oxidative stress is further intensified by the mechanical stress endured during normal breathing as more MPs are absorbed into the lung cells. The lung cells are pushed toward inflammation, premature aging, and long-term respiratory injury due to the combined effects of MPs, their additives, and enhanced ROS production (Wei et al., 2025). Overall, MPs harm the lungs through oxidative stress.

6.8.2 Inflammation

Inflammation is triggered whenever MPs enter the lungs, this is the natural defense response to harmful substances from human body. Studies show that MPs can activate immune cells and stimulate the release of inflammatory molecules. These include cytokines such as IL-1β, IL-6, and TNF-α, which are commonly elevated in lung fluid and tissue after exposure. One major pathway involved in this process is the NLRP3 inflammasome, a protein complex that helps regulate inflammation and is known to play

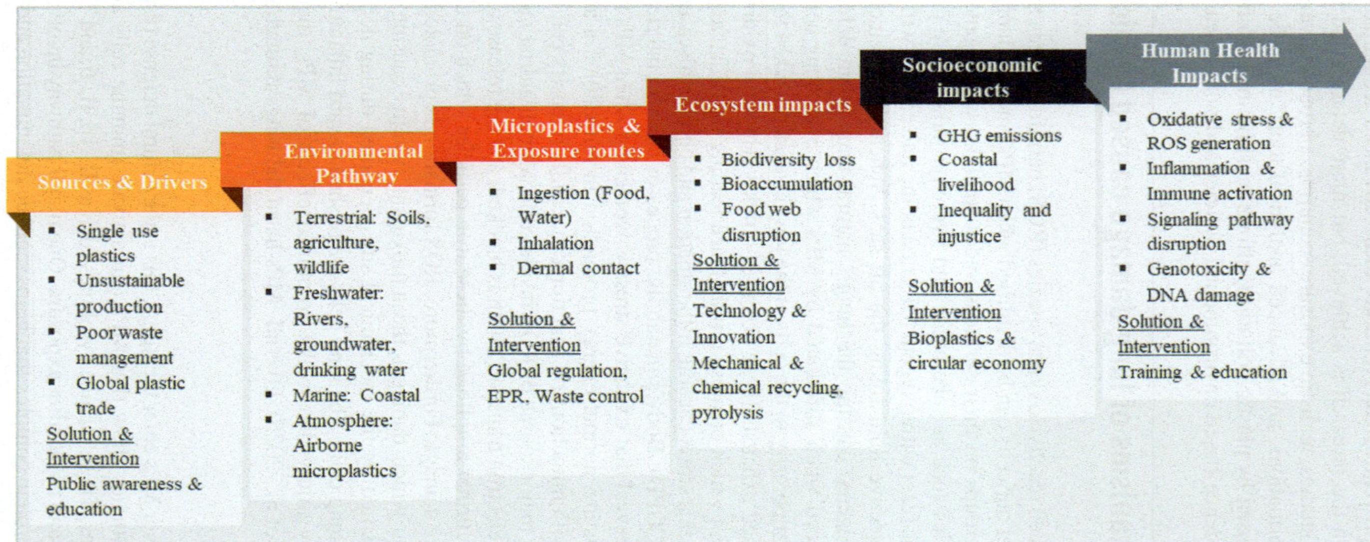

FIGURE 6.5

Conceptual framework illustrates the progression of microplastics from sources and driving factors through environmental pathways and human exposure routes. The figure links ingestion and inhalation exposure to ecosystem-level effects, including biodiversity loss and bioaccumulation, and extends to socioeconomic consequences such as greenhouse gas emissions and impacts on livelihoods. Associated human health risks, including oxidative stress, inflammation, immune activation, and ROS generation, are highlighted along this continuum.

a role in conditions like asthma, chronic obstructive pulmonary disease (COPD), and lung injury. MPs, especially polystyrene (PS) and polypropylene (PP) can activate NLRP3, leading to inflammation and tissue irritation. Another important pathway is NF-κB, a central regulator of inflammation (Lu et al., 2024a,b). When MPs enter the airway, they can stimulate receptors on lung cells, such as Toll-like receptors (TLR2 and TLR4), which then activate NF-κB and cause the release of more inflammatory molecules. These processes can worsen allergic reactions, asthma symptoms, and airway sensitivity (Xia et al., 2024).

The protective barrier of the lungs is weakened by MPs. The immune reactions such as higher IgE and Th2-type responses are triggered by MPs and cause severe inflammation. These effects are caused by pollutants and allergens which penetrate deeper into the tissue, when the proteins that keep lung cells tightly connected are disrupted. The research done recently shows that inflammation in the lung cells gets worse further due to the release of additional inflammatory substances known as SASP factors, caused by the MPs in the lung cells to undergo senescence, premature aging. The MP induced inflammation is also contributed by small regulatory RNA molecules such as microRNAs, long noncoding RNAs, and circular RNAs when changes occur due to the injury cell response. In the site of exposure, MPs attract immune cells such as macrophages, neutrophil, eosinophil, and dendritic cells, at the tissue level. These cells release tissue damaging chemicals to remove foreign particles. Experiments made in animals show that after MP exposure the effects such as thickened alveolar walls, higher levels of inflammatory factors and increased inflammatory cells around blood vessels are found to be in the body. The above effects show that due to the buildup of immune cells, especially neutrophils, there is a similar inflammatory response developed as there in chronic lung diseases (Fan et al., 2022). In general, MPs contribute to lung damage and respiratory problems by triggering inflammation through multiple pathways at the molecular, cellular, and tissue levels.

6.8.3 Apoptosis and autophagy

The lung cells are harmed by MPs by triggering two major cellular processes. By programmed cell death, apoptosis and another way autophagy, a mechanism that helps in cell breakdown and recycle damaged components. The cell's energy-producing structures, mitochondria, are damaged when MPs, especially polystyrene (PS–MPs), enter the lungs. Molecules like cytochrome *c* are released due to this damage causing apoptosis by activating proteins called caspases. MPs can set another pathway for cell death by activating death receptors on the cell surface. Oxidative stress is the major driver of this process, the normal cell functions are interfered with by the accumulation of excess reactive oxygen species (ROS), especially protein folding inside the endoplasmic reticulum (ER). The unfolded protein response (UPR) is activated as the ER becomes stressed, when there is buildup of many misfolded proteins. The UPR triggers apoptosis and cell damage, if the stress becomes overwhelming. In addition to apoptosis, the balance of calcium inside cells is also disturbed by MPs. The organelles are damaged, and the cell death is amplified further when too much of calcium enter both the ER and mitochondria. At the instance, autophagosomes are formed to remove damaged proteins and organelles due to the activation of autophagy as the cells try to protect themselves. The key levels of autophagy proteins such as LC3-II, p62, and Beclin one are increased due to the exposure of MPs, which indicates the pathway through which lung cells attempt to clear the damage. The autophagy-regulating pathways, including PI3K/Akt/mTOR, AMPK, and MAPK are disrupted by oxidative stress, occasionally turning this protective mechanism into another driver of cell injury (Lu et al., 2023b). Overall, it is evident that by various factors such as by damaging mitochondria, disturbing calcium balance, overwhelming the ER, and generating excessive ROS MPs push lung cells toward apoptosis. Depending on the severity of cellular stress, they can activate or disrupt autophagy at the same time. Collectively, these effects play a major role in the contribution of lung tissue injury and the development of respiratory diseases.

6.8.4 Disruption of microbial balance

The delicate balance of microbes in the lungs is also harmed by the MPs which normally protect the respiratory system. The amounts and the types of bacteria living in the nose or lungs are altered by MPs once they enter the body. The beneficial microbes like Bacteroides and Eubacterium are decreased while the harmful bacteria such as Helicobacter, Klebsiella, and Acinetobacter tend to increase causing the airways more vulnerable to inflammation, infection, and other lung problems due to the imbalance. Scientists found the influence of lung health on gut known as "gut–lung axis", in which that MPs disrupt not only the microbes in the lungs but also those in the gut. Lactic acid–producing species are increased in the gut when MPs build up in the intestine, which can alter gut bacteria. This leads to lung injury when the gut lining is damaged, the lactic acid in the gut leaks into the bloodstream and travel to the lungs (Xuan et al., 2023). The microbes are affected differently by different types of MPs. Nonbiodegradable plastics like polypropylene (PP) and polyethylene terephthalate (PET) raise the risk of lung infections by encouraging the growth of disease-causing bacteria as they reduce the microbial diversity. On the other hand, some biodegradable plastics show very small effects on microbial communities. Prominently, MPs introduce pathogens into the respiratory system upon inhalation as they carry their own biofilms clusters of bacteria and fungi. Bacterial toxins such as lipopolysaccharide (LPS) enter the bloodstream and even reach the brain due to the MPs exposure causing the lung blood and blood brain barriers weakening causing severe damage. These studies show that MPs can contribute to lung disease by disturbing the respiratory microbiome.

6.8.5 How microplastics affect cell growth in the lungs

A steady cycle of cell growth to replace old or damaged cells and maintain normal tissue function are required for healthy lungs. Research shows that natural renewal processes are slowed down by polystyrene MPs (PS–MPs). The cell multiplication and the cell growth such as Ki67 is reduced when these tiny particles enter the lung cells. Premature aging in lung tissue, including higher levels of p21, p16, and p27 are signs of long-term exposure to PS–MPs (Wu et al., 2023). These molecules are associated with cell cycle arrest. Aging in the cells causes stop dividing and begin releasing inflammatory substances which develops harmful environment that contributes to chronic lung problems such as asthma, chronic obstructive pulmonary disease (COPD), and fibrosis (Wei et al., 2025). Research shows that normal cell cycle is disturbed by the MPs causing DNA damage, pushing cells into abnormal "S-phase arrest," and stopping them from completing division. Stronger effects appear in the cells due to the smaller particles, as well as UV-aged MPs. These responses are found to be caused by certain RNAs such as circular RNAs and microRNAs, found by research. For instance, after PS-MP exposure the molecule circ_kif26b becomes elevated and interferes with miR-346-3p, a type of microRNA that normally helps control the senescence gene p21 (Fan et al., 2022). MPs promote cellular aging and disrupt lung tissue repair, by blocking this protective pathway. Overall, the findings shows that the lungs ability to regeneration is interfered in addition to inflammation caused by MPs, which contribute to long-term respiratory disease.

6.8.6 Microplastics and the development of pulmonary fibrosis

Pulmonary fibrosis is a serious condition in which lung tissue becomes stiff and scarred, making it difficult to breathe. One way this process begins is through epithelial–mesenchymal transition (EMT), a biological shift where lung epithelial cells lose their normal structure and take on a more mobile, fibroblast-like form that promotes scarring. Studies increasingly show that MPs can trigger EMT in lung cells. When exposed to PS-MPs or other types of MPs, lung cells begin to express higher levels of mesenchymal markers such as vimentin, N-cadherin, α-SMA, and EMT-related transcription factors like Snail, ZEB1, and Twist1. At the same time, the key epithelial marker E-cadherin decreases, signaling the early steps of fibrotic change. MPs also appear to enhance cell migration, further driving the tissue remodeling that ultimately results in fibrosis (Song et al., 2024).

In animal studies, MPs have been shown to reach deep into lung tissue, where they trigger chronic inflammation, oxidative stress, and mitochondrial damage. This combination activates profibrotic signals especially TGF-β that encourage fibroblasts to produce collagen and other scar-forming proteins. Some studies reveal that MPs can even cross the placental barrier and affect fetal lung development, suggesting long-term consequences across generations. Beyond EMT, newer research highlights additional pathways contributing to fibrosis, such as ferroptosis, a form of iron-dependent cell death that is stimulated by MP-induced oxidative stress through pathways like cGAS/STING (Goodman et al., 2021). MPs can worsen fibrosis indirectly as well. They may elevate lactic acid levels in the lungs and intestines, disturb the microbial balance, activate the NF-κB and Wnt/β-catenin pathways, and promote ongoing inflammation and apoptosis all processes known to accelerate fibrotic disease (Cao et al., 2023). While some compounds, such as melatonin, have shown early promise in reducing this oxidative damage, the overall picture remains concerning. Together, these findings indicate that MPs can promote pulmonary fibrosis through several interconnected mechanisms, highlighting the urgent need for strategies to reduce exposure and protect lung health.

6.8.7 How microplastics and attached pollutants harm the lungs

MPs can act like tiny sponges in the environment, picking up pollutants such as heavy metals, toxic chemicals, pharmaceuticals, and even harmful microbes. They attach to these substances through simple chemical interactions, which make the pollutants easier to transport and more likely to enter living organisms. When these contaminated particles are inhaled, the lungs are exposed not only to the MPs themselves but also to the pollutants they carry. In many cases, this combined exposure creates a stronger toxic effect than either substance alone. For example, some studies show that MPs make metals like arsenic or mercury more harmful by helping them enter tissues more easily, leading to greater damage to lung cells and reduced breathing efficiency in exposed animals. In other situations, MPs may slightly reduce the toxicity of certain chemicals, although such cases are far less common. Recent research also shows that MPs found in lung tumors often contain attached metals, which may activate pathways linked to cancer growth and immune disruption. Airborne particles such as polypropylene MPs mixed with chromium can also weaken lung surfactants, the substance that keeps the air sacs from collapsing making breathing more difficult. The respiratory tissues are further damaged by MPs when they combine dyes or proteins which increase oxidative stress and inflammation in lung cells, studies prove this. Even though it is proven that MPs can carry pollutants deep into the lungs and trigger harmful biological responses, some of the questions remain unknown even for scientists. The questions such as how different types of plastics interact with specific chemicals, what levels of combined exposure are most dangerous, and how long-term, low-dose inhalation affects lung health (Yu et al., 2025). For assessing the real risks of MPs in the air we breathe understanding these interactions are essential.

6.9 Inhalable microplastics in urban air: Emerging health risks

Over the decades, it is understood that air pollution was mainly due to the well-known contaminants such as particulate matter (PM_{10} and $PM_{2.5}$), carbon monoxide, lead, sulfur oxides, nitrogen oxides, and ozone. In recent years, a new group of pollutants called respirable emerging contaminants (RECs) is recognized by the scientists. Some of them include volatile organic compounds (VOCs), persistent organic pollutants (POPs), pharmaceuticals, inhalable MPs (iMPs), potentially toxic elements (PTEs), and personal-care product residues (PPCPs), bioaerosols, and engineered nanomaterials (Wei et al., 2022). Activities such as waste handling, industrial processes, abrasion, diffusion, and volatilization cause these substances to enter the atmosphere, and many remain suspended at human breathing height about 1.2–1.8 m. RECs have become an urgent area of environmental research, as their health effects are still poorly understood. AMPs create a complex mixture of biological and chemical hazards as they act as carriers for microorganisms and chemical pollutants. Studies made on viruses such as SARS-CoV-2 show that they can survive on plastic surfaces. Depending on particle shape, weathering, humidity, and temperature common airborne bacteria and fungi including Staphylococcus aureus, Klebsiella pneumoniae, and Aspergillus species can attach to MP surfaces.

The microbial survival, pathogenicity are increased by these interactions and resistance to antimicrobial drugs. Universally, airborne MPs have been detected including in Paris, London, and several Indian cities. However, only larger particles or only a limited number of polymers are detected by many sampling methods, the nanoplastics (<1 μm) remains dangerous. Concerns have raised about systemic exposure as these ultrafine particles can overcome biological barriers such as the lungs, placenta, blood–brain barrier, and mammary glands. In India, recent measurements have revealed that extremely high AMP concentrations are found in megacities especially in Delhi, Kolkata, Mumbai, and Chennai. Severe $PM_{2.5}$ pollution affects the cities which is home to more than 80 million and the public health burden is further intensified by addition of MPs (Air Quality Life Index (AQLI), 2021).

The research done recently provide the first extensive evidence on inhalable MPs (iMPs) especially airborne plastic particles smaller than 10 μm that can enter the human respiratory system. The measurements made from studies shows the iMP concentrations at human breathing height across major Indian cities. It reveals that higher levels of concentration are found in winter evenings due to increased crowd density, widespread use of synthetic clothing, and inadequate waste management. The highest concentration is recorded in Kolkata 14.23 $\mu g/m^3$, followed by Delhi with concentration 14.18 $\mu g/m^3$. On the other hand, because of better air circulation, milder winters, and reduced reliance on heavy synthetic apparel, coastal cities such as Chennai (4 $\mu g/m^3$) and Mumbai (2.65 $\mu g/m^3$) showed significantly lower levels (Biswas et al., 2025). The principal component analysis supports further these regional differences.

Eleven polymer types of samples are identified by chemical analysis using pyrolysis–GC–MS. Most dominant polymers mainly released from textiles is Polyester (PET), followed by polyethylene (PE) and styrene–butadiene rubber (SBR) which are originated from single-use plastics, packaging materials, vehicle components, and footwear. A rough estimation of lung burden of nearly 2.9 g per person is found over a lifetime if on average of about 8.8 μg/m^3, iMPs in urban Indian air. The study highlights the concern of raising chronic inhalation exposure as the presence of harmful MP-associated chemicals, including diethyl phthalates and lead. Overall, this is the first work to demonstrate that airborne MPs can act as carriers for ultrafine particulate matter, potentially toxic elements, persistent organic pollutants, pharmaceuticals and personal-care product residues, and even microorganisms. Pathogenic species like *Aspergillus fumigatus* and newly emerging bacteria and fungi with antibiotic-resistance and virulence-factor genes and other microbial communities are found to be carried with airborne MPs during periods of high exposure such as autumn festivals. The inhalable MPs may enhance the spread and survival of multidrug-resistant organisms as these microbes showed increased pathogenic potential and resistance to common antibiotics such as tetracycline (Biswas et al., 2025).

6.10 Conclusion and future perspective

MPs are readily ingested by aquatic organisms across multiple trophic levels, including plankton, invertebrates, and fish. Once internalized, these particles accumulate in digestive tissues and can translocate to other organs, impairing physiological functions. Experimental and field studies consistently demonstrate that MP ingestion disrupts feeding behavior, induces oxidative stress, alters immune responses, and compromises growth and reproduction in aquatic species. Moreover, MPs absorb environmental contaminants and microorganisms, increasing their bioavailability and facilitating trophic transfer to higher organisms, including humans. Humans are exposed to MPs through three primary routes: ingestion, inhalation, and dermal contact. Dietary intake represents a major pathway, with MPs detected in seafood, meat, dairy products, fruits, vegetables, drinking water, beverages, and processed foods. Inhalation exposure arises from airborne MPs released from textiles, urban dust, and industrial activities, allowing particles to deposit in the respiratory tract. Dermal exposure, although considered less significant, may occur through personal care products and contaminated water. The diversity in particle size, shape, and polymer type influence organ distribution and toxicological outcomes.

MPs exert toxic effects through multiple, often interconnected mechanisms. In the respiratory system, inhaled particles vade clearance and accumulate in lung tissues, triggering oxidative stress through excessive reactive oxygen species generation. This imbalance damages cellular macromolecules and activates inflammatory signaling pathways, leading to chronic inflammation. MPs also interfere with lung surfactant function, disrupt alveolar integrity, and suppress cellular repair processes. Prolonged exposure has been linked to pulmonary fibrosis, asthma-like conditions, and potentially carcinogenic outcomes.

Beyond the lungs, MPs interact with gastrointestinal, hepatic, immune, endocrine, nervous, and reproductive systems. Cellular uptake occurs via phagocytosis, endocytosis, or passive diffusion, particularly for nanoscale particles capable of crossing biological barriers such as the blood-brain and placenta barriers. Chronic exposure is associated with metabolic dysregulation, hormonal imbalance, immune suppression, and epigenetic alterations. Importantly, MPs amplify toxicity by transporting copollutants, resulting in synergistic effects that intensify tissue damage.

Despite rapid advances, significant uncertainties remain regarding exposure thresholds, biodistribution, doe-response relationships, and long-term health outcomes of MP exposure. Current toxicological studies often rely on acute, high-dose models that do not reflect realistic environmental conditions. Additionally, the lack of standardized sampling, analytical protocols, and reporting frameworks limits comparability across studies. Regulatory efforts remain fragmented, with insufficient enforceable limits for MP in food, water, and consumer products.

6.11 Knowledge gaps and future directions

Even though studies from experiments shows that MPs and nanoplastics can overcome the blood–brain barrier and influence neuroinflammatory, oxidative, and metabolic pathways, the neurological consequences on humans for long term remains unknown. The current evidence is based primarily on animal models or in vitro systems as the exposure levels in humans are difficult to quantify. The question on effects on human body is induced by whether environmentally relevant concentrations is unknown. The details on particle aging, bio-corona formation, and the factors altering neurotoxicity in real-world conditions is known less. Research must be done on low exposure models, developmental susceptibility and their interactions with other environmental pollutants to clarify the risks to neurodevelopment and neurodegeneration. To link the MP exposure with clinical neurological outcomes epidemiological studies are also essential. To assess the public health implications of MPs and to guide effective regulatory and mitigation strategies, a deeper mechanistic understanding is urgently needed.

Future research must prioritize chronic-low-dose exposure studies, size, and polymer-specific toxicity assessments, and synergistic effects with cocontaminants. Equally important is the development of harmonized analytical methods and evidence based regulatory frameworks. Strengthening international collaboration, advancing safer material alternatives, and implementing precautionary public health strategies will be critical to mitigating MP related hazards and safeguarding and environmental health.

References

Abafe, O.A., Harrad, S., Abdallah, M.A., 2023. Novel insights into the dermal bioaccessibility and human exposure to brominated flame retardant additives in microplastics. Environ. Sci. Technol. 57 (29), 10554–10562. https://doi.org/10.1021/acs.est.3c01894.

Abdel-Zaher, S., Mohamed, M.S., Sayed, A.E.D.H., 2023. Hemotoxic effects of polyethylene microplastics on mice. Front. Physiol. 14, 1072797. https://doi.org/10.3389/fphys.2023.1072797.

Ali, T., Habib, A., Muskan, F., Mumtaz, S., Shams, R., 2023. Health risks posed by microplastics in tea bags: microplastic pollution – a truly global problem. Int. J. Surg. 109, 515–516. https://doi.org/10.1097/JS9.0000000000000055.

Ali, N., Katsouli, J., Marczylo, E.L., Gant, T.W., Wright, S., De La Serna, J.B., 2024. The potential impacts of micro-and-nano plastics on various organ systems in humans. EBioMedicine 99, 104901. https://doi.org/10.1016/j.ebiom.2023.104901.

Avio, C.G., Gorbi, S., Milan, M., Benedetti, M., Fattorini, D., d'Errico, G., Pauletto, M., Bargelloni, L., Regoli, F., 2015. Pollutants Bioavailability and Toxicological Risk from MPs to marine mussels. Environ. Pollut. 198, 211–222. https://doi.org/10.1016/j.envpol.2014.12.021.

Bakir, A., Rowland, S.J., Thompson, R.C., 2014. Enhanced desorption of persistent organic pollutants from microplastics under simulated physiological conditions. Environ. Pollut. (Amsterdam, Neth.) 185, 16–23. https://doi.org/10.1016/j.envpol.2013.10.007.

Ban, M., Shimoda, R., Chen, J., 2021. Investigation of nanoplastic cytotoxicity using SHSY5Y human neuroblastoma cells and polystyrene nanoparticles. Toxicol. Vitro 76, 105225. https://doi.org/10.1016/j.tiv.2021.105225.

Biginagwa, F.J., Mayoma, B.S., Shashoua, Y., Syberg, K., Khan, F.R., 2016. First evidence of microplastics in the african great lakes: recovery from lake victoria nile perch and nile tilapia. J. Great Lakes Res. 42 (1), 146–149. https://doi.org/10.1016/j.jglr.2015.10.012.

Biswas, A., Saini, N., Chivukula, N., Samal, A., Jansari, M.R., Bhadury, P., Darbha, G.K., 2025. The dawn of a new air pollutant: inhalable microplastics as emerging vectors of hazardous contaminants and their implications for human health. Environ. Int. 205, 109897.

Brynzak-Schreiber, E., Schogl, E., Bapp, C., Cseh, K., Kopatz, V., Jakupec, M.A., et al., 2024. Microplastics role in cell migration and distribution during cancer cell division. Chemosphere 353, 141463. https://doi.org/10.1016/j.chemosphere.2024.141463.

Cao, J., Xu, R., Geng, Y., Xu, S., Guo, M., 2023. Exposure to polystyrene microplastics triggers lung injury via targeting toll-like receptor 2 and activation of the NF-κB signal in mice. Environ. Pollut. 320, 121068.

Chen, Q., Zhang, Y., Wang, J., 2021. Combined effects of microplastics and heavy metals on hepatic oxidative stress in zebrafish. Sci. Total Environ. 784, 147203. https://doi.org/10.1016/j.scitotenv.2021.147203.

Chikuma, K., Arima, K., Asaba, Y., Kubota, R., Asayama, S., Sato, K., Kawakami, H., 2020. The potential of lipid-polymer nanoparticles as epigenetic and ROS control approaches for COPD. Free Radic. Res. 54, 829–840.

Çobanoglu, H., Belivermis¸, M., Sıkdokur, E., Kılıç, O., Çayır, A., 2021. Genotoxic and cytotoxic effects of polyethylene microplastics on human peripheral blood lymphocytes. Chemosphere 272, 129805. https://doi.org/10.1016/j.chemosphere.2021.129805.

Cox, K.D., Covernton, G.A., Davies, H.L., Dower, J.F., Juanes, F., Dudas, S.E., 2019. Human consumption of microplastics. Environ. Sci. Technol. 53, 7068–7074. https://doi.org/10.1021/acs.est.9b01517.

Danopoulos, E., Jenner, L.C., Twiddy, M., Rotchell, J.M., 2020. Microplastic contamination of seafood intended for human consumption: a systematic review and meta-analysis. Environ. Health Perspect. 128 (12). https://doi.org/10.1289/ehp7171.

Fan, Z., Xiao, T., Luo, H., Chen, D., Lu, K., Shi, W., Sun, C., Bian, Q., 2022. A study on the roles of long non-coding RNA and circular RNA in the pulmonary injuries induced by polystyrene microplastics. Environ. Int. 163, 107223.

Forte, M., Iachetta, G., Tussellino, M., Carotenuto, R., Prisco, M., De Falco, M., Laforgia, V., Valiante, S., 2016. Polystyrene nanoparticles internalization in human gastric adenocarcinoma cells. Toxicol. Vitro 31, 126–136. https://doi.org/10.1016/j.tiv.2015.11.006.

Goodman, K.E., Hare, J.T., Khamis, Z.I., Hua, T., Sang, Q.A., 2021. Exposure of human lung cells to polystyrene microplastics significantly retards cell proliferation and triggers morphological changes. Chem. Res. Toxicol. 34, 1069–1081.

Grigorakis, S., Mason, S.A., Drouillard, K.G., 2017. Determination of the gut retention of plastic microbeads and microfibers in goldfish (*Carassius auratus*). Chemosphere 169, 233–238. https://doi.org/10.1016/j.chemosphere.2016.11.055.

Han, S., Bang, J., Choi, D., Hwang, J., Kim, T., Oh, Y., et al., 2020. Surface pattern analysis of microplastics and their impact on human-derived cells. ACS Appl. Polym. Mater. 2 (11), 4541–4550.

Han, X., Fu, L., Yu, J., Li, K., Deng, Z., Shu, R., Wang, D., You, J., Zeng, E.Y., 2024. Effects of erythromycin on biofilm formation and resistance mutation of *Escherichia coli* on pristine and UV-aged polystyrene microplastics. Water Res. 256, 121628.

Hernandez, L.M., Xu, E.G., Larsson, H.C.E., Tahara, R., Maisuria, V.B., Tufenkji, N., 2019. Plastic teabags release billions of microparticles and nanoparticles into tea. Environ. Sci. Technol. 53, 12300–12310. https://doi.org/10.1021/acs.est.9b02540.

Huang, D., Wang, X., Yin, L., Chen, S., Tao, J., Zhou, W., Chen, H., Zhang, G., Xiao, R., 2022. Research progress of microplastics in soil-plant system: ecological effects and potential risks. Sci. Total Environ. 812, 151487. https://doi.org/10.1016/j.scitotenv.2021.151487.

Hwang, J., Choi, D., Han, S., Choi, J., Hong, J., 2019. An assessment of the toxicity of polypropylene microplastics in human derived cells. Sci. Total Environ. 684, 657–669. https://doi.org/10.1016/j.scitotenv.2019.05.071.

Jabeen, K., Li, B., Chen, Q., Su, L., Wu, C., Hollert, H., Shi, H., 2018. Effects of virgin microplastics on goldfish (*Carassius auratus*). Chemosphere 213, 323–332. https://doi.org/10.1016/j.chemosphere.2018.09.031.

Jeon, M.S., Kim, J.W., Han, Y.B., Jeong, M.H., Kim, H.R., Sik Kim, H., et al., 2023. Polystyrene microplastic particles induce autophagic cell death in BEAS-2B human bronchial epithelial cells. Environ. Toxicol. 38 (2), 359–367. https://doi.org/10.1002/tox.23705.

Jeong, C.B., Won, E.J., Kang, H.M., Lee, M.C., Hwang, D.S., Hwang, U.K., Lee, J.S., 2016. Microplastic size-dependent toxicity, oxidative stress induction, and p-JNK and p-p38 activation in the monogonont rotifer (Brachionus koreanus). Environ. Sci. Technol. 50 (16), 8849–8857. https://doi.org/10.1021/acs.est.6b01441.

Jiang, C., Ni, B.-J., Zheng, X., Lu, B., Chen, Z., Gao, Y., Zhang, Y., Zhang, S., Luo, G., 2022. Changes in microplastic behavior during adsorption and anaerobic digestion of waste activated sludge induced by hydrothermal pretreatment. Water Res. 221, 118744.

Jin, Y., Liang, L., Tu, W., Luo, T., Fu, Z., 2019. Impacts of polystyrene microplastic on the gut barrier, microbiota and metabolism of mice. Sci. Total Environ. 649, 308–317. https://doi.org/10.1016/j.scitotenv.2018.08.353.

Katsumiti, A., Losada-Carrillo, M.P., Barros, M., Cajaraville, M.P., 2021. Polystyrene nanoplastics and microplastics can act as Trojan horse carriers of benzo (a) pyrene to mussel hemocytes in vitro. Sci. Rep. 11 (1), 22396. https://doi.org/10.1038/s41598-021-01938-4.

Klein, M., Fischer, E.K., Fischer, M., Scholz-Bottcher, B.M., Gerdts, G., 2023. Dermal absorption of microplastic fibres from synthetic textiles by human ex vivo skin. Environ. Int. 171, 107706. https://doi.org/10.1016/j.envint.2023.107706.

Kogel, T., Bjoroy, O., Toto, B., Bienfait, A.M., Sanden, M., 2020. Micro- and nanoplastic toxicity on aquatic life: determining factors. Sci. Total Environ. 709, 136050. https://doi.org/10.1016/j.scitotenv.2019.136050.

Larese Filon, F., Bello, D., Cherrie, J.W., Sleeuwenhoek, A., Spaan, S., Brouwer, D.H., 2016. Occupational dermal exposure to nanoparticles and nano-enabled products: Part I—factors affecting skin absorption. Int. J. Hyg Environ. Health 219, 536–544. https://doi.org/10.1016/j.ijheh.2016.05.009.

Lee, C.-W., Hsu, L.-F., Wu, I.-L., Wang, Y.-L., Chen, W.-C., Liu, Y.-J., Yang, L.-T., Tan, C.-L., Luo, Y.-H., Wang, C.-C., Chiu, H.-W., Yang, T.C.-K., Lin, Y.-Y., Chang, H.-A., Chiang, Y.-C., Chen, C.-H., Lee, M.-H., Peng, K.-T., Huang, C.C.-Y., 2022. Exposure to polystyrene microplastics impairs hippocampus-dependent learning and memory in mice. J. Hazard Mater. 430, 128431. https://doi.org/10.1016/j.jhazmat.2022.128431.

Li, B., Ding, Y., Cheng, X., Sheng, D., Xu, Z., Rong, Q., Wu, Y., Zhao, H., Ji, X., Zhang, Y., 2020a. Polyethylene microplastics affect the distribution of gut microbiota and inflammation development in mice. Chemosphere 244, 125492. https://doi.org/10.1016/j.chemosphere.2019.125492.

Li, Z., Zhu, S., Liu, Q., Wei, J., Jin, Y., Wang, X., Zhang, L., 2020b. Polystyrene microplastics cause cardiac fibrosis by activating Wnt/β-catenin signaling pathway and promoting cardiomyocyte apoptosis in rats. Environ. Pollut. 265, 115025. https://doi.org/10.1016/j.envpol.2020.115025.

Li, L., Zhao, X., Li, Z., Song, K., 2021. COVID-19: performance study of microplastic inhalation risk posed by wearing masks. J. Hazard Mater. 411, 124955. https://doi.org/10.1016/j.jhazmat.2020.124955.

Li, X., Zhang, T., Lv, W., Wang, H., Chen, H., Xu, Q., Cai, H., Dai, J., 2022. Intratracheal administration of polystyrene microplastics induces pulmonary fibrosis by activating oxidative stress and Wnt/β-catenin signaling pathway in mice. Ecotoxicol. Environ. Saf. 232, 113238. https://doi.org/10.1016/j.ecoenv.2022.113238.

Li, N., Zeng, Z., Zhang, Y., Zhang, H., Tang, N., Guo, Y., Lu, L., Li, X., Zhu, Z., Gao, X., 2023. Higher toxicity induced by co-exposure of polystyrene microplastics and chloramphenicol to Microcystis aeruginosa: experimental study and molecular dynamics simulation. Sci. Total Environ. 866, 161375.

Liang, B., Zhong, Y., Huang, Y., Lin, X., Liu, J., Lin, L., Hu, M., Jiang, J., Dai, M., Wang, B., Zhang, B., Meng, H., Lelaka, J.J.J., Sui, H., Yang, X., Huang, Z., 2021. Underestimated health risks: polystyrene micro- and nanoplastics jointly induce intestinal barrier dysfunction by ROS-mediated epithelial cell apoptosis. Part. Fibre Toxicol. 18, 20. https://doi.org/10.1186/s12989-021-00414-1.

Liao, Z., Ji, X., Ma, Y., Lv, B., Huang, W., Zhu, X., Fang, M., Wang, Q., Wang, X., Dahlgren, R., Shang, X., 2021. Airborne microplastics in indoor and outdoor environments of a coastal city in Eastern China. J. Hazard Mater. 417, 126007. https://doi.org/10.1016/j.jhazmat.2021.126007.

Liu, S., Guo, J., Liu, X., Yang, R., Wang, H., Sun, Y., et al., 2023. Detection of various microplastics in placentas, meconium, infant feces, breastmilk and infant formula: a pilot prospective study. Sci. Total Environ. 854, 158699.

Liu, S., Guo, J., Liu, X., Yang, R., Wang, H., Sun, Y., et al., 2023b. Detection of various microplastics in placentas, meconium, infant feces, breastmilk and infant formula: a pilot prospective study. Sci. Total Environ. 854, 158699.

Lu, L., Wan, Z., Luo, T., Fu, Z., Jin, Y., 2018. Polystyrene microplastics induce gut microbiota dysbiosis and hepatic lipid metabolism disorder in mice. Sci. Total Environ. 631–632, 449–458. https://doi.org/10.1016/j.scitotenv.2018.03.051.

Lu, K., Lai, K.P., Stoeger, T., Ji, S., Lin, Z., Lin, X., Chan, T.F., Fang, J.K.-H., Lo, M., Gao, L., Qiu, C., Chen, S., Chen, G., Li, L., Wang, L., 2021. Detrimental effects of microplastic exposure on normal and asthmatic pulmonary physiology. J. Hazard Mater. 416, 126069. https://doi.org/10.1016/j.jhazmat.2021.126069.

Lu, C., Liang, Y., Cheng, Y., Peng, C., Sun, Y., Liu, K., et al., 2023a. Microplastics cause reproductive toxicity in male mice through inducing apoptosis of spermatogenic cells via p53 signaling. Food Chem. Toxicol. 179, 113970. https://doi.org/10.1016/j.fct.2023.113970.

Lu, H., Yin, K., Su, H., Wang, D., Zhang, Y., Hou, L., Li, J.B., Wang, Y., Xing, M., 2023b. Polystyrene microplastics induce autophagy and apoptosis in birds lungs via PTEN/PI3K/AKT/mTOR. Environ. Toxicol. 38 (1), 78–89. https://doi.org/10.1002/tox.23663.

Lu, W., Li, X., Wang, S., Tu, C., Qiu, L., Zhang, H., Zhong, C., Li, S., Liu, Y., Liu, J., Zhou, Y., 2023c. New evidence of microplastics in the lower respiratory tract: inhalation through smoking. Environ. Sci. Technol. 57 (23), 8496–8505. https://doi.org/10.1021/acs.est.3c00716.

Lu, H., Guo, T., Zhang, Y., Liu, D., Hou, L., Ma, C., Xing, M., 2024a. Endoplasmic reticulum stress-induced NLRP3 inflammasome activation as a novel mechanism of polystyrene microplastics (PS-MPs)-induced pulmonary inflammation in chickens. J. Zhejiang Univ. Sci. B 25, 233–243.

Lu, S., Feng, Q., Chen, M., Zeng, X., Wei, H., Chen, Q., Guo, H., Su, L., Yan, B., Wu, Y., Yang, X., Ma, P., 2024b. Mechanisms underlying Th2-dominant pneumonia caused by plastic pollution derivatives (PPD): a molecular toxicology investigation that encompasses gut microbiomics and lung metabolomics. J. Hazard Mater. 480, 136326.

Luo, T., Zhang, Y., Wang, C., Wang, X., Zhou, J., Shen, M., Zhao, Y., Fu, Z., Jin, Y., 2019. Maternal exposure to different sizes of polystyrene microplastics during gestation causes metabolic disorders in their offspring. Environ. Pollut. (Amsterdam, Neth.) 255, 113122. https://doi.org/10.1016/j.envpol.2019.113122.

Luo, T., Wang, D., Zhao, Y., Li, X., Yang, G., Jin, Y., 2022. Polystyrene microplastics exacerbate experimental colitis in mice tightly associated with the occurrence of hepatic inflammation. Sci. Total Environ. 844, 156884. https://doi.org/10.1016/j.scitotenv.2022.156884.

Mariano, S., Tacconi, S., Fidaleo, M., Rossi, M., Dini, L., 2021. Micro and Nanoplastics identification: classic methods and innovative detection techniques. Front. Toxicol. 3. https://doi.org/10.3389/ftox.2021.636640.

Miraj, S.S.A., Parveen, N., Zedan, H., 2019. Plastic microbeads: small yet mighty concerning. Int. J. Environ. Health Res. 31 (7), 788–804. https://doi.org/10.1080/09603123.2019.1689233.

Oliveira, M., Ribeiro, A., Hylland, K., Guilhermino, L., 2013. Single and combined effects of MPs and pyrene on juveniles (0þ group) of the common goby Pomatoschistus microps (Teleostei, Gobiidae). Ecol. Indic. 34, 641e647. https://doi.org/10.1016/j.ecolind.2013.06.019.

Pannetier, P., Morin, B., Le Bihanic, F., Dubreil, L., Cl'erandeau, C., Chouvellon, F., Van Arkel, K., Danion, M., Cachot, J., 2020. Environmental samples of MPs induce significant toxic effects in fish larvae. Environ. Int. 134, 105047. https://doi.org/10.1016/j.envint.2019.105047.

Patel, S., Clark-Ginsberg, A., Declet-Barreto, J., 2022. Human exposure pathways to microplastics: a global perspective. Environ. Sci. Technol. 56 (15), 10976–10988. https://doi.org/10.1021/acs.est.2c02981.

Prata, J.C., Da Costa, J.P., Lopes, I., Duarte, A.C., Rocha-Santos, T., 2020. Environmental exposure to microplastics: an overview on possible human health effects. Sci. Total Environ. 702, 134455. https://doi.org/10.1016/j.scitotenv.2019.134455.

Rahman, A., Sarkar, A., Yadav, O.P., Achari, G., Slobodnik, J., 2021. Potential human health risks due to environmental exposure to nano- and microplastics and knowledge gaps: a scoping review. Sci. Total Environ. 757, 143872. https://doi.org/10.1016/j.scitotenv.2020.143872.

Salimi, A., Alavehzadeh, A., Ramezani, M., Pourahmad, J., 2022. Differences in sensitivity of human lymphocytes and fish lymphocytes to polyvinyl chloride microplastic toxicity. Toxicol. Ind. Health 38 (2), 100–111. https://doi.org/10.1177/07482337211065832.

Schwabl, P., Koppel, S., Konigshofer, P., Bucsics, T., Trauner, M., Reiberger, T., Liebmann, B., 2019. Detection of various microplastics in human stool. Ann. Intern. Med. 171 (7), 453–457. https://doi.org/10.7326/m19-0618.

Song, S., van Dijk, F., Vasse, G.F., Liu, Q., Gosselink, I.F., Weltjens, E., Remels, A., de Jager, M.H., Bos, S., Li, C., Stoeger, T., Rehberg, M., Kutschke, D., van Eck, G., Wu, X., Willems, S.H., Boom, D., Kooter, I.M., Spierings, D., Wardenaar, R., Cole, M., Nawijn, M.C., Salvati, A., Gosens, R., Melgert, B.N., 2024. Inhalable textile microplastic fibers impair airway epithelial differentiation. Am. J. Respir. Crit. Care Med. 209, 427–443.

Sripada, K., Wierzbicka, A., Abass, K., Grimalt, J.O., Erbe, A., Rollin, H.B., et al., 2022. A children's health perspective on nano-and microplastics. Environ. Health Perspect. 130 (1), 015001.

Stock, V., Bohmert, L., Lisicki, E., Block, R., Cara-Carmona, J., Pack, L.K., Selb, R., Lichtenstein, D., Voß, L., Henderson, C.J., Zabinsky, E., Sieg, H., Braeuning, A., Lampen, A., 2019. Uptake and effects of orally ingested polystyrene microplastic particles in vitro and in vivo. Arch. Toxicol. 93 (7), 1817–1833. https://doi.org/10.1007/s00204-019-02478-7.

Tong, X., Li, B., Li, J., Li, L., Zhang, R., Du, Y., Zhang, Y., 2022. Polyethylene microplastics cooperate with *Helicobacter pylori* to promote gastric injury and inflammation in mice. Chemosphere 288, 132579. https://doi.org/10.1016/j.chemosphere.2021.132579.

Umamaheswari, S., Priyadarshinee, S., Bhattacharjee, M., Kadirvelu, K., Ramesh, M., 2021. Exposure to polystyrene microplastics induced gene modulated biological responses in zebrafish (*Danio rerio*). Chemosphere 281, 128592. https://doi.org/10.1016/j.chemosphere.2020.128592.

Valavanidis, A., Vlachogianni, T., Fiotakis, K., Loridas, S., 2013. Pulmonary oxidative stress, inflammation and cancer: respirable particulate matter, fibrous dusts and ozone as major causes of lung carcinogenesis through reactive oxygen species mechanisms. IJERPH 10, 3886–3907. https://doi.org/10.3390/ijerph10093886.

Vianello, A., Jensen, R.L., Liu, L., Vollertsen, J., 2019. Simulating human exposure to indoor airborne microplastics using a Breathing Thermal Manikin. Sci. Rep. 9, 8670. https://doi.org/10.1038/s41598-019-45054-w.

Wang, C., Wu, W., Pang, Z., Liu, J., Qiu, J., Luan, T., et al., 2023a. Polystyrene microplastics significantly facilitate influenza A virus infection of host cells. J. Hazard Mater. 446, 130617. https://doi.org/10.1016/j.jhazmat.2022.130617.

Wang, J., Tian, H., Shi, Y., Yang, Y., Yu, F., Cao, H., et al., 2023b. The enhancement in toxic potency of oxidized functionalized polyethylene-microplastics in mice gut and Caco-2 cells. Sci. Total Environ. 903, 166057. https://doi.org/10.1016/j.scitotenv.2023.166057.

Wang, Y., Xu, X., Jiang, G., 2023c. Microplastics exposure promotes the proliferation of skin cancer cells but inhibits the growth of normal skin cells by regulating the inflammatory process. Ecotoxicol. Environ. Saf. 267, 115636. https://doi.org/10.1016/j.ecoenv.2023.115636.

Wei, W., Ho, W.-C., Behringer, M.G., Miller, S.F., Bcharah, G., Lynch, M., 2022. Rapid evolution of mutation rate and spectrum in response to environmental and population-genetic challenges. Nat. Commun. 13, 4752. https://doi.org/10.1038/s41467-022-32353-6.

Wei, Y.Y., Chen, T.T., Zhang, D.W., Zhang, Y., Li, F., Ding, Y.C., Wang, M.Y., Zhang, L., Chen, K.G., Fei, G.H., 2025. Microplastics exacerbate ferroptosis via mitochondrial reactive oxygen species-mediated autophagy in chronic obstructive pulmonary disease. Autophagy 21, 1717–1743.

Wright, S., Thompson, R.C., Galloway, T.S., 2013. The physical impacts of microplastics on marine organisms: a review. Environ. Pollut. 178, 483–492. https://doi.org/10.1016/j.envpol.2013.02.031.

Wu, D., Zhang, M., Bao, T.T., Lan, H., 2023. Long-term exposure to polystyrene microplastics triggers premature testicular aging. Part. Fibre Toxicol. 20 (1), 35. https://doi.org/10.1186/s12989-023-00546-6.

Xia, Q., Wei, Y., Hu, L., Zeng, F., Chen, Y., Xu, D., Sun, Y., Zhao, L., Li, Y., Pang, G., Peng, W., He, M., 2024. Inhalation of microplastics induces inflammatory injuries in multiple murine organs via the toll-like receptor pathway. Environ. Sci. Technol. 58, 18603–18618.

Xie, X., Deng, T., Duan, J., Xie, J., Yuan, J., Chen, M., 2020. Exposure to polystyrene microplastics causes reproductive toxicity through oxidative stress and activation of the p38 MAPK signaling pathway. Ecotoxicol. Environ. Saf. 190, 110133. https://doi.org/10.1016/j.ecoenv.2019.110133.

Xu, X., Hou, Q., YingangXue, Y.J., Wang, L.P., 2018. Pollution characteristics and fate of microfibers in the wastewater from textile dyeing wastewater treatment plant. Water Sci. Technol. 78, 2046–2054. https://doi.org/10.2166/wst.2018.476.

Xu, J.L., Lin, X., Wang, J.J., et al., 2020. Microplastics in aquatic environments: occurrence, accumulation, and biological effects. Sci. Total Environ. 703, 134699. https://doi.org/10.1016/j.scitotenv.2019.134699.

Xuan, L., Xu, Z., Luo, J., Wang, Y., Yan, Y., Qu, C., Xie, Z., Skonieczna, M., Zhou, P., Huang, R., 2023. Lactate exacerbates lung damage induced by nanomicroplastic through the gut microbiota–HIF1a/PTBP1 pathway. Exp. Mol. Med. 55, 2596–2607.

Yang, H., Xiong, H., Mi, K., Xue, W., Wei, W., Zhang, Y., 2020. Toxicity comparison of nano-sized and micron-sized microplastics to goldfish Carassius auratus Larvae. J. Hazard. Mater. 388, 122058. https://doi.org/10.1016/j.jhazmat.2020.122058.

Yu, H., Tiao, M., Huang, S., Sheu, J.J., Tain, Y., Sheen, J., Lin, I.C., Tsai, C., Huang, L., Hsu, C., Tsai, C., Lin, Y., Lee, P., Su, Y., 2025. Impact of maternal microplastic exposure on offspring lung structure and function: insights into transcriptional misregulation and the TGF-β/α-SMA pathway. Ecotoxicol. Environ. Saf. 299, 118397.

Zhang, Q., Xu, E.G., Li, J., Chen, Q., Ma, L., Zeng, E.Y., Shi, H., 2020. A review of microplastics in table salt, drinking water, and air: direct human exposure. Environ. Sci. Technol. 54, 3740–3751. https://doi.org/10.1021/acs.est.9b04535.

Zhang, C., Lin, X., Gao, P., Zhao, X., Ma, C., Wang, L., Sun, H., Sun, L., Liu, C., 2023. Combined effects of microplastics and excess boron on Microcystis aeruginosa. Sci. Total Environ. 891, 164298.

Zhang, X., He, X., Pan, D., Shi, L., Wu, Y., Yang, Y., Zhu, Y., Wang, Y., Wang, H., Pu, L., He, S., Meng, L., Li, J., 2024. Effects of thermal exposure to disposable plastic tableware on human gut microbiota and metabolites: a quasi-experimental study. J. Hazard Mater. 462, 132800. https://doi.org/10.1016/j.jhazmat.2023.132800.

Zhao, L., Shi, W., Hu, F., Song, X., Cheng, Z., Zhou, J., 2021. Prolonged oral ingestion of microplastics induced inflammation in the liver tissues of C57BL/6J mice through polarization of macrophages and increased infiltration of natural killer cells. Ecotoxicol. Environ. Saf. 227, 112882. https://doi.org/10.1016/j.ecoenv.2021.112882.

Zhao, Q., Zhu, L., Weng, J., Jin, Z., Cao, Y., Jiang, H., Zhang, Z., 2023. Detection and characterization of microplastics in the human testis and semen. Sci. Total Environ. 877, 162713. https://doi.org/10.1016/j.scitotenv.2023.162713.

Zheng, Q., Wu, H., Yan, L., Zhang, Y., Wang, J., 2024. Effects of polystyrene nanoplastics and PCB-44 exposure on growth and physiological biochemistry of Chlorella vulgaris. Sci. Total Environ. 918, 170366.

Zhu, K., Jia, H., Sun, Y., Dai, Y., Zhang, C., Guo, X., Wang, T., Zhu, L., 2020. Enhanced cytotoxicity of photoaged phenol-formaldehyde resins microplastics: combined effects of environmentally persistent free radicals, reactive oxygen species, and conjugated carbonyls. Environ. Int. 145, 106137. https://doi.org/10.1016/j.envint.2020.106137.

Zuri, G., Karanasiou, A., Lacorte, S., 2023. Microplastics: human exposure assessment through air, water, and food. Environ. Int. 179, 108150. https://doi.org/10.1016/j. envint.2023.108150.

Zwollo, P., Quddos, F., Bagdassarian, C., Seeley, M.E., Hale, R.C., Abderhalden, L., 2021. Polystyrene microplastics reduce abundance of developing B cells in rainbow trout (*Oncorhynchus mykiss*) primary cultures. Fish Shellfish Immunol. 114, 102–111. https://doi.org/10.1016/j.fsi.2021.04.014.

CHAPTER

Microplastics in fisheries and food security: Implications for human health

7

Melvin S. Samuel

Department of Civil, Construction, and Environmental Engineering, Marquette University, Milwaukee, WI, United States

Highlights

- Microplastics are persistent contaminants widely distributed across food systems.
- Food packaging, processing, and environmental exposure are major microplastic entry pathways.
- Microplastics can accumulate in human tissue and promote inflammation and chronic effects.
- Detection of microplastics in food remains methodologically challenging and nonstandardized.
- Strong mitigation strategies and regulatory frameworks are needed.

7.1 Introduction

Tiny plastic particles of size ranging from 1 μm to 5 mm are known as MPs, which originate from a wide variety of sources. These are manufactured purposefully by industries at microscopic sizes, such as resin pellets, textile fibers, and exfoliating microbeads used in cosmetics, based on their usage. Some MPs originate from larger plastic objects by degradation through sunlight, scraps of mechanical parts, and various environmental changes. Nowadays, these MPs are found throughout marine, freshwater, and terrestrial environments because of the various sources of origin. Detection of MP fragments and fibers in human feces shows that these particles are entering the human body and that people are regularly exposed to these through food and water consumption (Jin et al., 2021). Evidence confirms that several interconnected pathways are involved in human exposure to these MPs. One of the most commonly reported dietary sources is seafood. Many aquatic species, such as fish, shellfish, oysters, and lobsters, ingest MPs directly from polluted waters and then pass them on to humans. It may also be ingested by terrestrial animals and livestock through contaminated feed, soil, or water. Not only are animals vulnerable, but also crops. The plants absorb micro and nanosized plastics from polluted soils and irrigation supplies, and these MPs enter our body from various sources through the food chain (Aboobakri et al., 2023).

Food-contact plastics such as packaging films, bottles, storage containers, and processing equipment introduce an additional layer of contamination, which can shed MPs during usage, especially under heat, friction, or prolonged storage. One of the main sources of overall MP intake is drinking water, including both bottled and tap water (Zhang et al., 2020). Another major contributor is ready-to-eat meals, which may also acquire MPs not only during processing, preparation, or packaging, but also

Microplastics in Aquatic Environments. https://doi.org/10.1016/B978-0-443-45619-0.00012-1

because the raw ingredients themselves contain few or no particles. Thermal processing either increases particle release or breaks down existing particles, depending on temperature and various cooking conditions (Kwon et al., 2020). Through various routes, modern food systems have multiple forms of contamination, which reflects the complexity of MP movement. As these studies show, seafood, processed foods, drinking water, table salt, beverages, and various processed products contain MPs (Fig. 7.1). For an effective approach, it's essential to understand the way MPs enter our food and how they move through agricultural, industrial, and domestic environments. This chapter examines the presence of MP in various food categories and estimates the influential contamination pathways, concentrations, and polymer types in different foods. Finally, it addresses the potential health risks of dietary exposure.

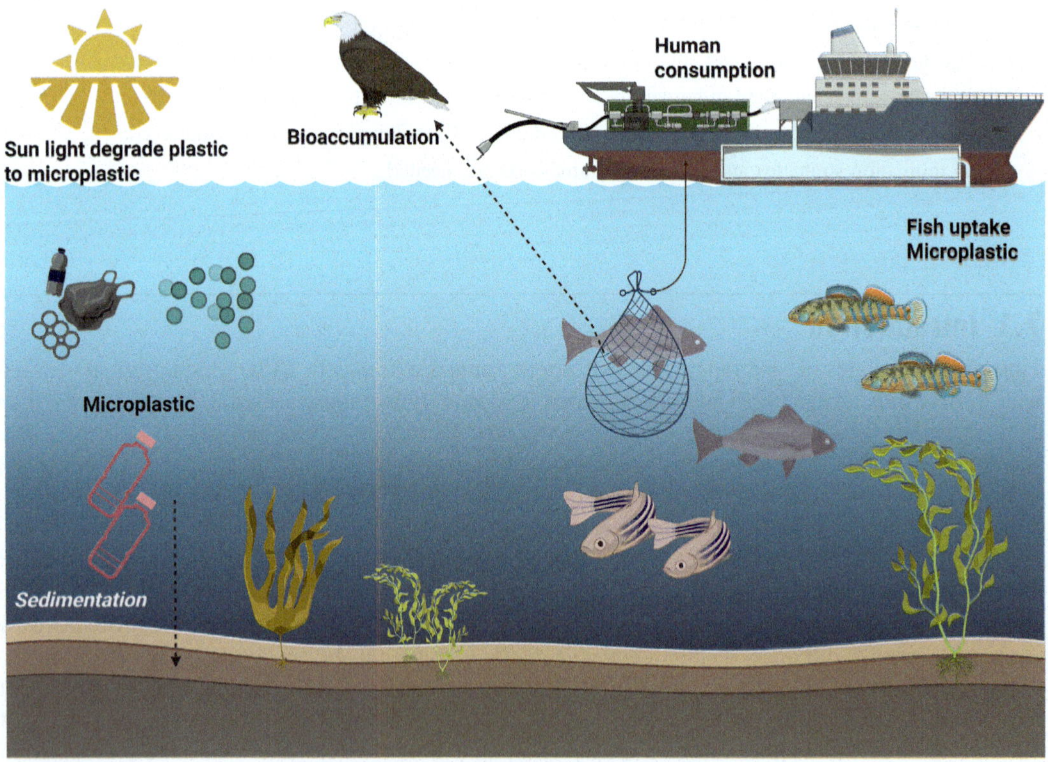

FIGURE 7.1

Schematic representation of the transformation of plastics from production and consumer use into microplastics within aquatic environments and food webs. The figure illustrates the fragmentation of macroplastics into microplastics, their interaction with co-occurring contaminants (e.g., heavy metals and hydrophobic organic pollutants) through adsorption, and the formation of pollutant-bound microplastics. Subsequent trophic transfer from plankton to higher-level consumers highlights bioaccumulation processes and associated ecological risks, with implications for seafood safety and human exposure.

7.2 Microplastic contamination in fisheries and food products

7.2.1 Microplastics in aquatic foods

The widespread microplastic (MP) contamination in marine environments is due to the rapid growth of global plastic production and inadequate waste management practices, creating serious ecological and public health concerns (Table 7.1). MPs found in aquatic systems are from various sources, such as stormwater discharges, poor waste management, and surface runoff. Also, from other marine activities, including fishing, aquaculture, and shipping. Seafood is one of the major sources of MPs' entry to the human digestive system. In marine ecosystems, plankton, shellfish, crustaceans, echinoderms, and other small organisms are the major carriers of these particles into the food web. MPs move upward through the food chain as the contaminated prey is consumed by the predators, commonly in fish, shrimp, crabs, and other seafood consumed by humans. As a result, oxidative stress, inflammation, or other physiological disruptions will occur once the MPs migrate into our tissues and cells from the ingested aquatic organisms.

Studies show that commercially important fish and shellfish have MPs in the gastrointestinal tract, gills, liver, muscle, and skin. A measurable amount of MP levels is found in the most traded marine species, as per the report by the Food and Agriculture Organization. Ingestion of up to 11,000 MP particles can be found in individuals with high shellfish consumption, while average consumers ingest roughly as per the report. The level of contamination varies depending on the species and regions. Approximately three to five particles per 10 g of tissue have been reported in Belgian commercial mussels. The edible bivalves in China were found to range from 2 to 11 particles per gram of MPs and 4–57 particles per individual, based on the species and their location. MPs have been detected in the stomachs and livers of most commonly consumed fish, such as small pelagics, anchovies, and sardines (Compa et al., 2018). During the cleaning process, MPs are removed from the gastrointestinal tract. However, some seafood and fish, such as mollusks, crustaceans, juvenile fish, and dried fish, are consumed whole, which corresponds to greater exposure to MPs. Additionally, recent studies show that eviscerated dried fish, such as Greenback mullet and Belanger's croaker, still contain MPs at higher concentrations because they are absorbed into the muscle tissue. Therefore, evisceration does

Table 7.1 Occurrence of microplastics in aquatic food.

Examples	MP-level particles	MP size range (μm)	Types of MPs	References
Umbonellid moon snail	6	10–5000	Fibers, fragments, pellets, film	Naji et al. (2018)
Asiatic clam	9.22	5–5000	Fibers, fragments, pellets	Li et al. (2015)
Green tiger prawns	7.8	<100 – >1000	Fibers	Abbasi et al. (2018)
Atlantic Pomfret	3; 33%	217–4810	Fibers	Neves et al. (2015)
European seabass	40; 23%	≤1000–5000	Fibers, fragments	Bessa et al. (2018)

not guarantee the absence of MPs (Karami et al., 2017). Studies have shown that strengthening the mechanistic and toxicological research is required in some areas, as the MPs are found in a wide range of aquatic foods, which raises concerns in dietary exposure and impacts on human health.

7.2.2 Microplastics in processed foods

Beyond seafood, fruits, vegetables, and rice, MPs are detected in a wide range of various packaged food items such as bottled water, table sugar, honey, beer, milk, edible seaweed, table salt, and various beverages. Various studies show that products from the manufacturing units possess higher levels of MPs when compared with the unprocessed food products (Kwon et al., 2020). This happens during various steps like processing, handling, or packaging in the manufacturing units. Some research suggests that MPs interact with human intestinal cells and combine with dietary proteins and lipids during digestion. The amount of daily human intake by inhalation requires further study. Although environmental issues and polymer risks are extensively analyzed, other concerns, such as chemical hazards, pollutants, and biological problems, remain understudied and require further analysis. Furthermore, exposure to MPs causes gastrointestinal disturbances and reproduction-related issues. Gastrointestinal disturbances, changes in liver metabolism, and behavioral alterations are likely to be found in mice, as studied by the mammalian toxicology department. The most frequently detected polymers in the food processing analysis are polyethylene (PE), polyethylene terephthalate (PET), and polypropylene (PP), which are mainly associated with food contamination. Plastic packaging can contribute to MPs not only during processing, but also through tearing, cutting, or twisting (Venkatesan et al., 2023; Ruiz-Pozo et al., 2024). Research shows that repeated plastic bottle opening can also add particles due to the interface with the cap. Sea salt, a raw ingredient, also contains MPs due to contamination, which contributes to the processing industries. A significant number of particles are detected in the analysis of human feces, as well as in the polymers present in packaged products. PP and PET are commonly found in these analyses. Additionally, beverage contamination is a major concern in bottled water, soft drinks, juices, and beer. This contamination not only affects the source water but also results from production, filtration, bottling, and storage practices (Tan et al., 2020).

7.2.3 Fruits and vegetables

Essential nutrients for good health come from fruits and vegetables. At least 400 g of fruit and vegetables should be taken daily as recommended by the World Health Organization. But these foods are contaminated by MPs, as they were present in soil, water, and soil (Table 7.2). Studies found that MPs are present in larger amounts; fruits had a median level of about 223,000 particles per gram, while vegetables contained around 97,800 particles per gram (Jin et al., 2021). Around 190,000–200,000 particles per gram are found to be in apples and pears. In vegetables such as broccoli and carrots, exceeding 100,000 particles per gram of MPs were found. MPs can attach to and accumulate on the root surfaces of crops, including potatoes, beets, radishes, and wheat, which cause a severe threat to humans. As a result, MPs move from the roots to the various edible parts of fruits and vegetables and enter the human body through daily intake, providing global exposure across the food chain.

Table 7.2 Occurrence of microplastics in other food products.

Food types	MP level	MP size range	Types of MP	References
Fruits and vegetables	176,500–223,000 (fruits) 52,050–97,800 (vegetables)	1990–2170 1510–2520	–	Jin et al. (2021)
Milk	6500	0.1–5 mm	Polysulfone	Pironti et al. (2021)
Honey	54	0.013–0.25 mm	PP, PE, PAAm	Pironti et al. (2021)
Sugar	217 ± 123	–	–	Liebezeit and Liebezeit (2013)
Soft drinks	40 ± 24.53	100–3000	PA, PEA, ABS	Shruti et al. (2020)

7.2.4 Seafood and packaged foods

Concerns about MP contamination are rising as seafood becomes a larger part of the global diet. MPs are widespread in the food web, from tiny zooplankton to crabs, mussels, oysters, and many fish species. These MPs reach humans through various levels of the food chain, from seawater to tiny organisms to larger fish and other aquatic animals. Seafood is not the only source of MP intake; packaged and ready-to-eat foods also contribute to MP intake worldwide. Due to their various advantages, such as being cheap, durable, and versatile, nearly 40% of plastics are used for packaging. However, when exposed to heat, friction, or long storage times, these materials can release MPs into food in notable quantities. This issue has been exacerbated by the rapid growth of takeout culture. MPs accumulate when hot food is packed and delivered in plastic containers. Research reveals that people who frequently order takeout meals from these food containers can consume roughly 12 to 203 MP particles per week. Exposure increases when these containers are reused. Abrasion or environmental stress can increase the number of particles during the manufacturing, sealing, and transport of plastic packaging. For instance, one of the most consumed packaged foods is milk. Studies show that milk contains MPs, most commonly sulfone-based particles, revealing contamination occurring during dairy processing. Other factors, such as vibration and temperature changes during transportation, can release more MPs into food products. Overall, these studies reveal that exposure to MPs in seafood and packaged food is greater, highlighting the need for standardization in packaging and other food handling practices.

7.2.5 Salt and sugar

Sea salt, which is produced through the evaporation of seawater, is an essential part of the human diet. It gets contaminated when plastic pollution enters marine ecosystems, which is a significant source of MP exposure. Micro- and nanoplastics (MNPs) have been widely reported in edible salts derived from diverse sources, including sea salt, rock salt, and lake salt (Table 7.3). Reported concentrations span a broad range, from nondetectable levels up to 39.8 particles g^{-1}, with most particles exhibiting fibrous

Table 7.3 Occurrence of microplastics in salt.

Location	Type of salt	MP levels	Types of MP	References
Pacific Ocean	Rock salt	0–14	PE, PET, Teflon	Kim et al. (2018)
Atlantic Ocean	Sea salt (coarse)	95–140	PET, PE, PP	Gundogdu (2018)
Indian Ocean	Sea salt	56–103	PES, PET, PA, PE, PS	Gundogdu (2018)
Mediterranean Sea	Sea salt (fine)	13,500–19,800	PE, PP	Kim et al. (2018)
Black Sea	Sea salt	16–84	PET, PE, PP, PU, PMMA, PVC, PA-6	Gundogdu (2018)

or irregular fragmented morphologies (Peixoto et al., 2019). Based on current consumption patterns, the estimated annual intake of MPs through dietary salt alone ranges from approximately 36 to over 36,000 particles per adult, underscoring salt as a nonnegligible contributor to human MP exposure (Lee et al., 2021). In addition to salt, sweeteners such as sugar and honey have also been identified as carriers of MNPs. A study conducted in Bangladesh revealed the presence of MPs in all analyzed sugar samples, with an average concentration of 344 ± 32 particles g^{-1}, indicating widespread contamination during processing, packaging, or storage (Afrin et al., 2022). In contrast, honey generally contains lower levels of MPs, with reported concentrations ranging between 22 and 114 particles L^{-1} (Diaz-Basantes et al., 2020). The lower abundance observed in honey may be related to differences in production pathways and filtration practices.

As per studies, up to 19,000 MP particles per year may be consumed by people through sea salt alone. Not only sea salt, but salt obtained from various sources, such as lake salt, well salt, and rock salt, also contains a significant number of MPs. Reports show that concentrations ranging from almost none to over 1600 particles per kilogram can be found. It shows the widespread contamination of salt in MPs and their contribution to human exposure. Apart from salt, another most consumed ingredient is sugar. Due to increased consumption of sweets, baked goods, and sugary beverages, global sugar consumption has increased drastically. Researchers found that MPs are even found in the sugar. When various samples of sugar are tested, they show traces of MPs in the form of fibers. These fibers are contaminated by various sources, as they vary in colors such as red, black, and white and are larger than 500 μm. The most common polymer types detected are PE and PP. It is estimated that a common adult might ingest up to 23,660 MP particles per kilogram of sugar, which raises a long-term health impact. It is evident that serious quality control and hygiene must be improved in the food production and packaging processes.

7.2.6 Drinking water and beverages

Drinking water, an important source of daily exposure, has also been detected with MPs in both tap water and bottled water (Table 7.4). Studies show that, on average, about five MP particles are found in every liter of water, and around 81% of water samples contain these particles. Bottled water generally contains higher levels of MPs than tap water. A large study found that about 93% of different bottled water brands from around the world contained MPs ranging in size from about 6 to 100 μm. This proves that plastic bottles, caps, and packaging may contribute directly to contamination. Even after treatment, tap water can still contain MPs, which we consume day to day. These particles come not

Table 7.4 Occurrence of microplastics in drinking water.

Water source	MP levels (particles/ individual)	MP size range	Types of MP	References
Surface water	266 ± 56	1–10 μm	PA, PVC, PP, PET, PE, others	Shen et al. (2021)
Ground water	13.23	>10 μm	PEST, NY, PS	Chu et al. (2022)
Raw water	0.7	50–150	PES, PV, PP, PET, PE, others	Mason et al. (2018)
Tap water	0–61	100–5000	–	Kosuth et al. (2018)
One-way plastic bottles	14	>100	PEST, PET, PES, PP, PE, PA	Schymanski et al. (2018)

only from the original water entering the treatment plant, but also from plastic materials in the water distribution system. MPs are released into the water from the commonly used Polyvinyl chloride (PVC) and Polyethylene terephthalate (PET) pipes over time and from the degradation of epoxy resin coatings inside storage tanks (Winkler et al., 2019). MPs can easily transfer into drinks such as soft drinks, energy drinks, tea, and beer because water is used as the base ingredient in many beverages. In Mexico, MPs over 84% were found in 27 brands when analyzed in 57 beverage samples. Beer products are the highest levels, averaging 28 particles per liter (Shruti et al., 2020a,b). MPs were shed into beverages during production and from the packaging materials, such as PET and polyamide. Even MPs were found in tea leaves, likely due to processing and packaging steps.

Obmann et al. (2018) reported exceptionally high concentrations in bottled water, with levels reaching up to 2649 ± 2857 particles L^{-1} in single-use PET bottles and 6292 ± 10,2521 particles L^{-1} in glass bottles. These values represent some of the highest MP abundances reported for bottled water to date. The elevated concentrations observed in glass bottles indicate that MP contamination cannot be attributed solely to plastic packaging; additional sources, such as bottle cleaning procedures, filling operations, storage conditions, and transportation processes, likely contribute to contamination. Particle size analysis has shown that most MPs detected in bottled water fall within 1–5 μm range, which raises particular concern due to their increased potential for biological interaction and uptake (Akhbarizadeh et al., 2020). In contrast, MP concentrations in the tap water are generally lower. Current studies report a maximum abundance of approximately 930 particles L^{-1} in tap water (Kirstein et al., 2021). Importantly, common household practices such as boiling tap water prior to consumption have been shown to be ineffective in reducing the concentration of micro- and nanoplastics (Munno et al., 2018). Based on available exposure data, adults may ingest up to 4.59×10^5 MP particles per year from tap water consumption, whereas intake from bottled water could reach 3.57×10^7 particles annually, highlighting bottled water as a potentially more significant exposure pathway (Danopoulos et al., 2020). Beyond drinking water, micro-nanoplastics have also been identified in a variety of beverages, including beer, tea, soft drinks, and milk (Da Costa Filho et al., 2021). Since water is the primary component of most beverages, contamination of the water source directly influences the MP content of these products. Li et al. (2022) analyzed 15 beer brands from different countries and reported MP concentrations ranging from 1.2×10^4 to 9.7×10^4 particles L^{-1}. The large discrepancy between studies is primarily attributable to differences in analytical methodologies.

Tea has also been identified as a notable MP source, particularly when brewed using commercial tea bags. During infusion, approximately 50 plastic particles per tea bag may be released into the beverage because of thermal and mechanical stress on polymer-based packaging materials (Afrin et al., 2022). Similarly, milk and dairy products, key dietary sources of protein and calcium, have been shown to contain MPs. Reported MP concentrations in milk vary widely across regions, ranging from 3 to 11 particles L^{-1} in Brazil (Kutralam-Muniasamy et al., 2020), 164–427 particles L^{-1} in India (Kiruba et al., 2022), and up to 2.04×10^3–1.0×10^4 particle L^{-1} in Switzerland (Da Costa Filho et al., 2021). Furthermore, environmental exposure also contributes to MPs. Drinks may collect airborne MP fibers and fragments when they encounter open air. These studies show that both drinking water and packaged beverages are major contributors that require additional treatment processes, improved manufacturing controls, and safer packing materials.

7.2.7 Kitchen devices and utensils

Apart from food sources, many everyday kitchen tools are sources of MP contamination in the food, as they are made from plastic, which we use to prepare and eat. When exposed to high heat during cooking, common kitchen tools such as plastic spatulas, spoons, ladles, and strainers can release small plastic particles, particularly when scraped against cookware surfaces or in high temperature. MPs are released into foods, drinks, water from plastic containers, wraps, bags, cutting boards, and cleaning sponges due to wear down over the long run. Even microscopic fragments can be shed from nonstick cookware as its coating breaks down with repeated use. In due course, larger kitchen appliances such as blenders, coffee machines, toasters, and dishwashers that contain plastic parts can also contribute to MP release during normal operation. For instance, scraping of the blender blades against the plastic cup and breakdown of plastic utensils and containers in the dishwasher due to heat, detergent, and water pressure.

MPs can easily enter our prepared meals without any notice on a daily basis because these utensils and appliances are used most in daily cooking. Eventually, the accumulation of these particles in our body raises various health concerns. These studies highlight areas to minimize the contamination in the home environment, like the plastic use in kitchen tools, waste management, and the emergence of safer and more durable materials. When compared with bottled water or beer, beverages may pose a higher chemical risk caused by the processing-related introduction of MPs.

There are dual approaches suggested for decreasing the MP risks in food processing:

1. Contamination in the raw materials must be reduced.
2. Contamination caused by the external inputs during processing, manufacturing, and packaging must be minimized.

7.2.8 Microplastics in meat and livestock products

Regardless of various sources such as air, soil, and water, MPs are found everywhere. Due to this widespread availability, these particles break down into even smaller particles and easily enter living organisms. These MPs enter the direct food chain through many species, including animals, plants, and aquatic organisms, which humans consume. In this way, these tiny plastics and the harmful MPs pose a threat not only to ecosystems but also to human health. Through the mulch films, irrigation pipes, construction materials, contaminated sludge, and polluted water, agricultural soils are most frequently exposed to plastic waste. Plastic fragments are mixed with this soil, so the livestock grazing the fields eventually are exposed to MPs. Cattle, buffalo, sheep, and goats often ingest plastic due to poor waste management practices. These plastics accumulate in their digestive systems and mix with the bloodstream, milk, and meat. Studies on various livestock found traces of toxic metals in the tissues, liver, kidneys, and rumen fluids of affected animals. MPs are consumed through contaminated feed or soil by poultry such as chickens, ducks, and turkeys, leading to inflammation and reduced nutrient absorption. These MPs then enter the human body through meat, milk, and eggs. They can intensify long-term health risks for both animals and humans, as MPs are carriers of other pollutants like heavy metals and organic chemicals.

7.2.9 Microplastics in plants and agricultural soil

MPs in soil act as a long-term reservoir, particularly in agricultural fields. The plastic mulch and other materials degrade into small particles and penetrate deep into soil layers during plowing, irrigation, and soil organisms, so that these particles are difficult to remove. Once these MPs go deep into the soil, they interact with plant roots and alter the soil structure, nutrient availability, and microbial communities. PVC fragments smaller than 10 μm enter the root tissues and disrupt the plant growth by damaging the roots. The exposure to MPs in crops like tomatoes and onions results in a significant reduction in fruit yield and root health. Studies show that important nutrients and antioxidant levels are also depleted. Not only in the yield, but MPs also show severe interference in nutrient absorption, seed germination, and metabolic processes. These MPs create another pathway to enter the human diet through vegetables, fruits, and grains as these particles move from roots to edible plant parts.

7.2.10 Microplastics in seafood and aquatic foods

Marine and freshwater environments are also prone to MP pollution. MPs are easily consumed by plankton because of their small size, which is the base of aquatic food webs. MPs move upward through the food chain with fish, crustaceans, and shellfish as they eat contaminated plankton. Researchers found higher numbers of MPs in mussels, fish, shrimp, and other seafood species. These microparticles can reach consumers with seafood species, as many of them are eaten whole or only partially cleaned. These MPs disrupt natural food webs by interacting with nutrients and pollutants that are essential for ecosystem stability. Health risks like inflammation and endocrine disruption occur when humans consume contaminated seafood. The toxic chemicals associated with MPs may cause potential long-term effects such as increased cancer risk.

***Overall Impacts and Need for Action*:** MPs are deeply embedded in global food systems, as they are found in meat, plants, and seafood. Significant concerns for food safety and human health arise as they can carry harmful chemicals, have the ability to accumulate in tissues, and disturb biological processes. To focus on this issue, there should be strict controls on agricultural and industrial pollution, stronger waste management, better monitoring of MPs in food production, and most importantly, reduced usage of single-use plastics. For betterment, a wide range of solutions is to be adopted to protect animal health and to safeguard ecosystems. This will reduce the health risk and burden of MPs on human populations.

7.3 Potential risks to global food security and nutrition

7.3.1 Exposure routes and concentrations of microplastics in food

Starting from farming and processing to packaging, transport, and storage, plastics are used at every stage of the modern food system. This makes it impossible to avoid MPs in our diets. These tiny particles can enter raw foods such as vegetables, grains, fish, and poultry through various sources, such as contaminated soil, water, or feed. They can also enter food accidentally during processing, be released from machinery, or leak from plastic packaging during storage, heating, or cooking. These MPs from the food chain can build up in human tissues, arising various concerns about long-term health effects. MPs in different foods react in different ways, as the exact measure is not known. For evaluating dietary exposure, exact information about the number of particles present and the source of MPs is essential to protect both ecosystems and human health. There are three major pathways in an MP area survey: marine environments, freshwater and agricultural systems, and food processing and packaging (Fig. 7.2).

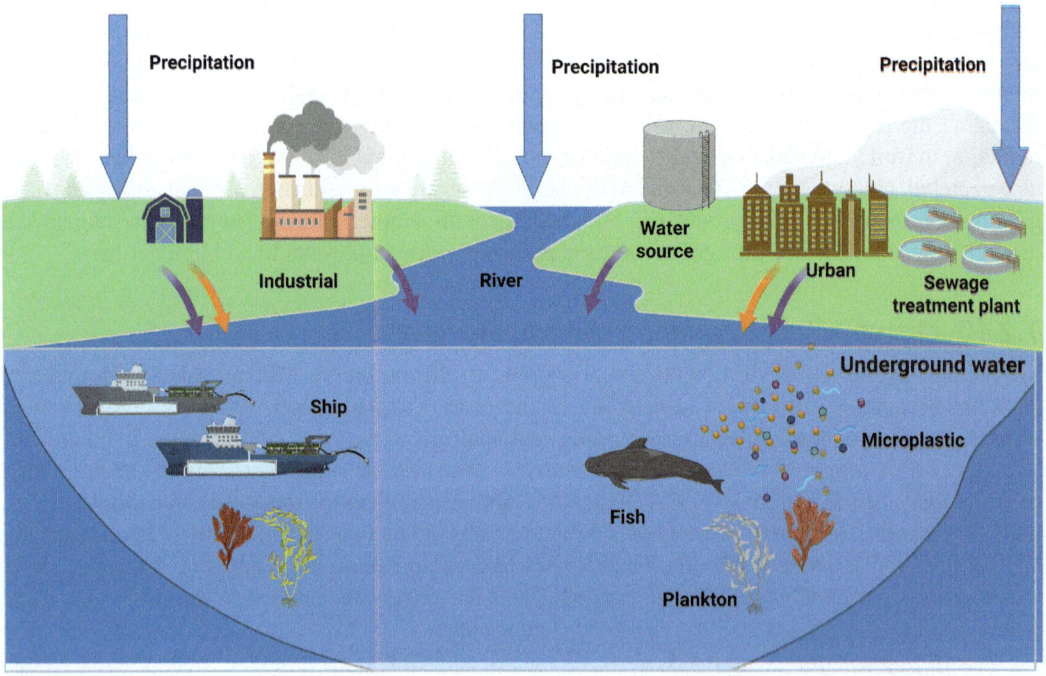

FIGURE 7.2

Conceptual illustration of microplastic circulation within aquatic environments. The schematic depicts major sources of microplastics, including urban runoff, sewage treatment facilities, maritime activities, and agricultural inputs, and their transport through rivers, lakes, groundwater, and coastal waters. Atmospheric precipitation, surface runoff, infiltration, and hydrological connectivity facilitate the redistribution of microplastics across freshwater and marine systems.

7.3.2 Bioaccumulation and retention of MPs in aquatic organisms

Bioaccumulation occurs when organisms gradually accumulate MPs within their bodies over time, which happens when the rate of intake exceeds the rate of excretion. The most common pathway of MPs in aquatic environments is through exposure:

- Filter-feeding organisms, such as zooplankton, mussels, and some fish species, are particularly vulnerable because they continuously strain large volumes of water with minimal particle selectivity.
- Various studies, particularly from the East China Sea, show that zooplankton ingest MPs dominated by fibers, with omnivorous species demonstrating especially high uptake due to their broad feeding habits.

The absorption and retention of MPs within tissues are influenced by their hydrophobicity and physicochemical properties, including particle size, shape, and polymer composition. Particles smaller than 500 μm are found in the gills, liver, and muscle after moving beyond the digestive tracts. Contaminants such as polychlorinated biphenyls (PCBs) and polycyclic aromatic hydrocarbons (PAHs), carried by MPs, can attach in the digestive tract, increasing the toxicity on exposed organisms. Additional concerns arise because of fiber-shaped MPs having elongated structures, causing entanglement in gills or gut linings, causing disruption in feeding, inflammation, or extended retention.

***Role of Polymer Density in Environmental Distribution and Organism Exposure*:** The major role in determining MPs' environmental fate depends on the density of an MP relative to water. PE and PP are low-density polymers that float, but the remaining MPs are suspended in the water column, which are easily swallowed by pelagic fish, zooplankton, and other surface-dwelling species. On the other hand, polyvinyl chloride and polyester are higher-density plastics that sink and accumulate in sediments. Benthic organisms such as worms, crustaceans, sea cucumbers, and bottom-feeding fish are most likely exposed to these fibers as they ingest sediment particles and contact the surface. The density, transport pathways, and physical form of MPs control their movement through aquatic food webs and influence which organisms are most vulnerable to retention and bioaccumulation.

***Density-Dependent Bioaccumulation of MPs Across Aquatic Species*:** Evidence shows that the movement of MPs through aquatic environments strongly depends on the density of MP polymers and the exposure patterns of different species. Various studies clearly demonstrate this density-dependent distribution. Demersal fish in Campeche Bay have higher concentrations of polymers than pelagic species, reflecting their contact with plastics in deeper waters. Similarly, observations along the Indian coastline show PE and PP in benthic mollusks, indicating habitat-specific exposure. Experiments in the controlled microcosm also show the same results as more polyester are ingested by sediment-dwelling organisms; in contrast, buoyant polymers such as PP are more likely to be consumed by mobile algae grazers. In sorting MPs, the sediment grain size also plays a major role; denser plastics concentrate in finer sediments. In coastal areas of Thoothukudi, India, surveys show that benthic species carried significantly higher MP loads than pelagic fish, highlighting the sedimentary risks of dense polymers. Even though risks are found in seawater, some studies on freshwater species show only the ingestion of certain selective polymers and not others. This proves the MPs' exposure depends on various factors such as preference, feeding mode, sensory cues, and shape. MP bioaccumulation across species is determined by factors such as habitat depth, feeding strategies, and polymer density.

***Retention and Depuration Dynamics Across Taxa*:** The retention times of MPs vary considerably among aquatic organisms. European seabass show the presence of MPs in their gills, liver, blood, and gastrointestinal tissues. These MPs disturb immune and metabolic functions. In these species, MPs are distributed in multiple organs, so rapid elimination from tissues is not possible. However, Antarctic krill have efficient depuration mechanisms that eliminate most ingested particles within 5 days. While some species have effective depuration mechanisms, prolonged MP pollution overwhelms the natural clearance processes due to continuous exposure. This continuous retention causes physiological stress, an impaired energy balance, and a modified nutrient intake.

***Ecosystem-Level Impacts of MP Accumulation*:** The MP bioaccumulation consequences not only affect the individual organisms, but they can also affect whole ecosystem functions. Mussels and oysters are filter feeders that help in removing suspended particles to regulate water quality. But their filtration efficiency is reduced by MP ingestion, and so the ability to maintain water clarity and nutrient balance is disrupted. Even coral reefs and mangrove forests are affected, as they are sensitive habitats. MPs intensify bleaching events in corals as they carry adsorbed pollutants and harmful microorganisms, and there is a possibility of sediment structure alteration. It also affects mangrove forests, which are vulnerable to climate change.

***Technological Advancements in Studying MP Accumulation*:** The ability to detect, track, and understand MPs within biological systems has greatly improved with rapid progress in analytical technologies. The precise identification of polymer types within animal tissues is done by Fourier transform infrared (FTIR) spectroscopy and Raman spectroscopy. As demonstrated in zebrafish embryos, the movement of particles can be visualized using fluorescently labeled MPs. Modern technologies, such as machine learning and artificial intelligence models, have been used to predict MP distribution patterns. These technologies identify the species or habitats at greatest risk, offering an effective tool for ecological forecasting.

***Environmental and Seasonal Factors Shaping Bioaccumulation*:** Exposure and accumulation of MPs are strongly influenced by environmental conditions. MP concentrations look highly elevated in the coastal regions with high human activity. In the East China Sea, there is a visible impact on the zooplankton communities. MP dynamics are also modified by the seasonal changes. The feeding rates and metabolic activity are affected by the variations in temperature, salinity, and food availability, which result in the probability of MP ingestion and a reduced depuration rate. The reduced gut transit time can prolong MP retention during colder seasons. Seasonal plankton blooms may dilute MP concentrations in the water column.

***Interactions With Other Environmental Stressors*:** MPs most frequently carry co-contaminants, such as heavy metals, antibiotics, and hydrocarbons, which are rarely found in isolation. These chemical interactions amplify the biological impacts. The increased accumulation of the antibiotic roxithromycin in fish is due to polystyrene MPs, which increases its toxic effects. Further complications develop due to these climate change interactions. The surface charge of MPs is modified by ocean acidification, increasing their attraction to chemical pollutants. Metabolic activity and pollutant uptake are accelerated by the rising temperature. Deeper analysis is required in the MPs within multipollutant and climate-relevant frameworks to overcome these issues.

***Trophic Transfer and Accumulation Across Food Webs*:** The primary way that MPs spread throughout aquatic food webs is through trophic transfer, which is the movement of MPs from prey to predator. This process begins with primary consumers, such as zooplankton. They readily ingest MPs present in the water column, which then reach higher-level consumers through consumption of prey. A study conducted in tropical regions, particularly the Sundarbans estuary, revealed an increase in MP concentration from 0.56 particles in primary consumers to 4.17 particles in quaternary consumers. There is a major concern about biomagnification in top predators, including species consumed by humans. This highlights the contamination of MPs in aquatic ecosystems.

***Trophic Transfer Pathways and Evidence Across Aquatic Food Webs*:** Many research studies and case reports clearly show that MPs can move from prey to predator. These particles spread throughout aquatic food webs via trophic transfer, a major pathway for this movement. The clearest mechanistic evidence is provided by some of the controlled experiments. In an experimental field study, it was found that mussels exposed to MPs were consumed by crabs. These particles persevered in the crab's hemolymph and digestive organs for up to 3 weeks. Although each species has different retention times, this study shows that MPs can survive digestion and accumulate in predators, increasing biomagnification. In a field study of North Sea food webs, benthic invertebrates were first detected with contaminated MPs; then, they were found in pelagic fish. Ultimately, seabirds feed on those fish. Although seabirds carried relatively low concentrations of MPs, more than half of the sampled fish contained them. This indicates widespread, multiple-level trophic movement. Accumulation patterns are further influenced by age. A study of Alaska pollock found MP particles between 100 and 500 μm in individuals. Similar trends have been found in Arctic polar cod documentation, in which habitat type and feeding strategy influence ingestion patterns. There is growing concern about direct dietary exposure, given the number of MPs found in commercially important fish species.

7.3.3 Direct and indirect exposure routes in aquatic species

There are two primary pathways by which aquatic organisms acquire MPs. The first method is by direct intake of MPs from the surrounding water. This includes active ingestion when particles resemble food items, passive filtration in species such as bivalves, and water intake during routine physiological processes. Filter-feeding organisms, including mussels, oysters, and tunicates, are particularly efficient at capturing small MPs suspended in the water column, making them important indicators of local contamination. The second major route is through the trophic transfer indirect process, in which MPs travel to predators through prey found mostly in omnivorous and carnivorous species. Experimental studies in the laboratory show that gobies and zebrafish are found to have an increased number of MP accumulation as they mainly feed on contaminated prey, such as mussels or zooplankton. Although there are two different pathways, most aquatic ecosystems are likely to be involved in both direct intake and tropical transfer. The habitat characteristics and feeding behavior decide the pathway. The difficulties of MP exposure in aquatic environments were highlighted in the above studies. The overall accumulation of MPs in the organisms is due to various multiple and overlapping mechanisms, which make it difficult to discriminate the original pathway of exposure. To estimate the potential human health risks and contamination pressures, specific species' feeding habits, trophic interactions, and mainly habitat use are to be considered during ecological risk measurements.

***Ecosystem-Specific Variability in Trophic Transfer of Microplastics*:** Various studies in different ecosystems show that trophic transfer of MPs varies phenomenally. This is due to the variations in environmental conditions, species composition, local pollution pressures, and feeding strategies. The results show that the movement of MPs through food webs depends not only on trophic transfer. Ecological context plays a major role in exposure patterns.

***Deep-Sea Food Webs*:** The MP transfer in deep-sea ecosystems shows unique patterns. The observations made in Monterey Bay show that higher-order predators did not depend on whether they contain greater amounts of MPs even though MPs were found across all five major trophic levels. The results show that apart from trophic position, MP retention is greatly controlled by other physiological traits such as gut morphology, feeding efficiency, and digestion time. Despite their position in the top food web, the deep-sea organisms modulate or reduce the MP accumulation in their body due to the slow metabolism or unique feeding adaptations.

***Tropical and Subtropical Lakes*:** In contrast, dynamic patterns are observed in the ecosystems with high biodiversity and rapid predator–prey turnover. Driven by dense trophic interactions and fast ecological cycling, rapid and substantial MP accumulation across species is found in Baiyangdian Lake in China. Some studies show that due to the retention delay and the constant transfer from the prey, MP concentrations rise by quadruple times in the top predators over time (Ma & You, 2021). The results highlight that the movement of MPs through freshwater food webs is magnified by the fast energy flow and high biological activity.

***Coastal Marine Food Webs*:** A distinct trophic transfer behavior is observed in the coastal ecosystems. Studies found that consistent biomagnification is not observed in the species exposed to MPs in British Columbia. The alternate feeding strategy appeared to be the strongest predictor of MP load. The higher concentrations of MPs are found in the suspension feeders and small planktivorous fish species, which are constantly exposed to the particles (Yıldız et al., 2022). Observations show that instead of trophic level, web complexity and feeding mode determine the intake of MPs.

***Terrestrial Aquatic Ecotones*:** Another pattern was found in the ecosystems situated between terrestrial and aquatic environments. A unique polymer signatures were introduced into resident fish and bird species where informal e-waste recycling sites were formed. From these sites, MPs enter the adjacent streams and wetlands, causing terrestrial pollution (Zheng et al., 2023). These findings show the relation between land-based pollution sources and aquatic ecosystem damage through hydrological flows, causing MP distribution.

***Ecological Consequences of Trophic Transfer*:** The predator–prey relationship is altered by the MP movement, and the ecological interactions are disturbed. The changes, such as impaired mobility, reduced feeding efficiency, and weakened physiology, make predators more vulnerable when they feed on prey organisms exposed to MPs. When midges consume contaminated mosquito larvae, they ingest more MPs, which alters predation rates and potentially reshapes food web dynamics (Cuthbert et al., 2019). The adsorbed contaminants, such as PAHs and heavy metals, are also carried by the MPs. In some Arctic ecosystems, MPs reduced the biomagnification of PCBs by binding them strongly, yet simultaneously increased predator exposure to additives leached from the plastics themselves. Depending on factors such as environmental conditions, pollutant types, and biological interactions, MPs can both suppress and enhance contaminant transfer.

7.4 Isolation and characterization of microplastics in fruits and vegetables

7.4.1 Extraction of microplastics

Research on MPs in fruits and vegetables is still developing, and standardized methodologies are limited. Larger plastic pieces (>1 mm) can sometimes be identified visually, but smaller particles typically require laboratory extraction for detection and verification. Extraction is necessary because MPs are often embedded in complex plant tissues containing cellulose, pigments, lipids, and other organics that can obscure or trap particles. Although methods vary, most procedures follow a sequence of organic matter digestion, density-based separation, and filtration, followed by further characterization. Digestion methods for plant samples are generally classified as acidic, alkaline, oxidative, or enzymatic.

Acidic digestion is commonly used because strong mineral acids efficiently break down protein and carbohydrates, adding MP recovery. Hydrochloric acid (HCl) and nitric acid (HNO_3) are frequently applied under heated conditions, with HNO_3 being particularly effective for removing protein- and carbohydrate-rich material. However, harsh acid treatments, especially at high temperatures, may damage or alter some polymers, potentially reducing MP recovery or biasing polymer identification. Alkaline digestion is less common for fruits and vegetables because plant tissue contains high levels of cellulose and hemicellulose, substances that resist base degradation. Although potassium hydroxide (KOH), including 10% solutions, is often used for animal tissues, resistant materials such as humins, cellulosic residue, and chitin may remain, hindering separation. PET and polycarbonate (PC) undergo alkaline treatments.

Oxidative digestion is considered practical for plant-based matrices because it reduces organic material while preserving the integrity of the polymers. Enzymes have minimal impact on MPs compared to aggressive chemical digestion. Hydrogen peroxide (H_2O_2) is widely used at various concentrations and temperatures to remove fats, greases, and fibrous plant components. Wet peroxide (WPO), which uses H_2O_2 with Fe^{2+} as a catalyst, can accelerate digestion by producing reactive radicals. Careful control of pH and temperature is necessary, as suboptimal conditions can cause iron hydroxide precipitation, leaving residues that interfere with filtration and microscopic analysis.

Enzymatic digestion is preferred when preserving polymer integrity is important. Enzymes have minimal impact on MPs compared to aggressive chemical digestion. Hydrolytic enzymes such as protease, proteinase-K, cellulase, and chitinase are used individually or in combination to degrade specific plant polymers. Although enzymatic digestion can take longer than oxidative methods, it often results in cleaner filters and better preservation of MPs. Hybrid protocols combining enzymes with mild oxidation have improved digestion efficiency in complex matrices, including soils and plant tissues. Pancreatic enzyme-based methods have also shown effective organic matter removal with minimal polymer damage in bivalves, highlighting multienzyme strategies for biological samples.

Method choice depends on sample type, organic content, and tissue properties like fiber density and lignification; harder or more resistant tissues may require stronger digestion than soft leafy samples. Studies suggest mineral acids can fully dissolve hard calcareous tissue, while oxidative methods are less effective, and alkaline treatments achieve only limited organic reduction. Acid or peroxide digestion, often combined with density separation, remains common for plant matrices, adjusted according to tissue hardness and target polymer.

After digestion, density separation is used to isolate MPs from residues. This relies on the density difference between most plastics (~0.94–1.46 g/m^3) and heavier mineral or denser matter sinks. Sodium chloride (NaCl) is widely used due to availability and cost, but its maximum density (~1.2 g/mL) limits recovery of denser polymers, making salts like zinc chloride ($ZnCl_2$) or sodium iodide (NaI) necessary for polymers like PET. Recovery also depends on particle size and morphology. Fibers may be recovered less efficiently due to buoyancy issues, adhesion to containers, or density overlap with NaCl, while smaller particles are harder to isolate, causing size-dependent biases. Ionic composition can affect aggregation: divalent cations (e.g., Ca^{2+}) can induce organic flocculation that traps MPs, and some high-density solutions can interact with filters, complicating observation.

The limited number of studies in agroecosystems reflects the absence of a standardized analytical workflow. Plant tissue poses challenges because pigments, fibrous structures, and heterogeneous organic matter interfere with recovery and identification. To address this, studies typically combine visual inspection with instrumental confirmation. Stereomicroscopy allows preliminary classification by size and shape, while spectroscopic and mass spectrometric techniques confirm polymer identity and reduce misidentification, particularly for irregular or colorless particles. Common tools include Scanning Electron Microscopy (SEM)/Transmission Electron Microscopy (TEM) for morphology and FTIR or Raman spectroscopy for polymer identification, with mass spectroscopy increasingly used for complex samples.

7.4.2 Soil preparation

Soil preparation is a crucial initial step for detecting MPs in terrestrial environments. Before extraction, samples are usually dried and sieved to lower moisture content, enhance particle separation, and remove extraneous materials. A review of current studies shows considerable variability in soil preparation protocols, indicating a lack of standardization. Variations occur in drying methods, temperatures, durations, and sieve sizes, but the primary goal remains consistent: introducing a homogenous, dry soil matrix that enables efficient MP recovery while minimizing contamination and analytical bias. Drying approaches include passive air-drying, oven-drying, and freeze-drying. Large debris, such as stones, plant material, and visible litter, is typically removed manually before drying. Air-drying is simple and cost-effective but prone to airborne MP contamination. Oven-drying provides more control, though high temperatures can alter polymer structures or cause thermal degradation. Freeze-drying preserves soil structure and limits thermal stress on MPs, though it requires specialized equipment and can lead to MP fragmentation during sublimation.

After drying, soils are sieved to remove coarse material and homogenize particle size. Stainless steel or iron sieves are preferred to minimize contamination. Mesh sizes vary by study objectives, with 5 mm sieves commonly used, though smaller meshes or sequential sieving are also reported. Coarser sieves may allow fine debris to remain, whereas finer sieves can trap MPs within soil aggregates. Therefore, sieve selection balances recovery efficiency with matrix cleanliness.

7.4.3 Microplastic isolation

Isolation of MPs from soil involves density separation, removal of organic matter, and filtration, allowing physical separation of MPs from the soil matrix. Once isolated, MPs are analyzed using microscopic or spectroscopic techniques.

7.4.3.1 Microplastic separation

Density separation, or flotation, is the most widely used method for extracting MPs from soils. This method relies on differences in density between plastics and mineral soil particles. Typically, the process involves three steps: suspending and agitating soil in a flotation solution, allowing mineral particles to settle while MPs float, and filtering the supernatant to recover floating MPs.

The effectiveness of density separation depends on the solution's density relative to the plastics. Polymers less dense than the solution float, while heavier soil components sink. Selecting a suitable flotation solution requires knowledge of the target polymers. Agricultural soils often contain multiple polymer types from various sources, making polymer-specific flotation impractical. High-density solutions, such as magnesium chloride ($MgCl_2$), are often used for comprehensive recovery but may leave crystalline residues that interfere with FTIR analysis, highlighting the need for careful solution selection or complementary tools. Recovery efficiency is commonly assessed using spiked soil experiments, where known polymer amounts are introduced and subsequently recovered.

7.4.3.2 Filtration

Filtration retains MPs recovered from the supernatant after flotation. Efficiency depends on filter materials, pore size, and chemical compatibility. Quartz and glass fiber filters can degrade or shed particles, causing contamination or loss. Quantitative filter papers are prone to mechanical failure and are often unsuitable for vacuum filtration. Polytetrafluoroethylene (PTFE) and nylon membranes are mechanically strong and chemically resistant but have limitations: PTFE is hydrophobic, slowing filtration, while nylon can obscure transparent MPs due to its translucence. Mixed cellulose ester (MCE) membranes are preferred for manual counting because their gridded structure provides contrast, reducing visual fatigue. PC membranes are recommended for fluorescence-based detection because they produce lower background signals. The absence of standardized filtration protocols limits compatibility across studies, emphasizing the need for consistent specifications regarding pore size, material, and compatibility.

7.4.3.3 Organic material removal

Soils contain substantial organic matter that can obscure MPs and reduce extraction efficiency. Residual organics may encapsulate MPs during flotation, leading to underestimation. Therefore, digesting organic matter prior to MP identification is strongly recommended. Common approaches include acidic, alkaline, oxidative, or enzymatic digestion. Oxidative digestion with H_2O_2 is highly effective for organic-rich soils and often outperforms acid or alkali treatments, which may fail to fully remove organics and can damage polymers, particularly under heated conditions. Enzymatic digestion is gentler and preserves polymer integrity, but it is typically more costly, time-consuming, and operationally complex. Some studies improve efficiency through repeated digestion cycles or by combining multiple methods. Sequential oxidative or enzymatic treatments can substantially enhance organic removal. Preliminary trials are advised to determine the most suitable method for a given soil type and organic content. Standardized, matrix-specific digestion protocols are essential to increase reliability and compatibility in soil MP studies.

7.5 Impacts on aquaculture and fisheries economics

One of the main sources of MPs is the marine environment because a significant amount of the world's plastic waste eventually ends up in the oceans. There, sunlight, wave action, and seawater gradually break the plastic waste into MPs and nanoplastics. Many marine organisms, including fish, crustaceans, and mollusks, ingest these MPs as they spread through the water. Even sea salt produced from evaporated seawater contains MPs. A considerable amount of MP exposure can be seen in the areas where seafood is consumed frequently. In Europe, the number of shellfish consumers is higher, so that those people might have ingested thousands of plastic particles each year. From marine systems, MPs can enter the human diet through several routes:

Ingestion by Plankton: Planktons are small and are mostly available in seawater and can easily consume tiny plastic particles. Once ingested by the food chain, these MPs can easily enter humans as we consume various shrimp, crabs, shellfish, and small fish species, which eat planktons.

Trophic Transfer and Bioaccumulation: The accumulation of MPs in larger predators occurs through smaller plankton. Commercially important fish and marine mammals carry these particles in their digestive systems, increasing their exposure to people.

Contamination of Sea Salt: Another nonnegligible source includes sea salt. Studies have found tens of thousands of MP particles in a person's annual intake contributed by sea salt.

MPs as Carriers of Microbes: The surface of MPs can host bacteria, including marine pathogens and antibiotic-resistant strains. When these contaminated particles enter edible species, they may pose additional food safety risks.

Most marine MP pollution begins on land, where plastics degrade in soils, rivers, lakes, and even the air. Because water is central to both farming and food processing, freshwater systems serve as an important link between environmental contamination and the human diet.

Freshwater Contamination: Even rivers, lakes, groundwater, and drinking water supplies are contaminated with MPs. Concentrations of MPs vary largely based on the size and flow of the water body. Research from the Great Lakes shows MP levels are larger when sampled at deeper points. MPs in bottled water and processed beverages are higher due to contamination during bottling or packaging, which sometimes shows in higher numbers than tap water.

Impacts on Agriculture: MP concentrations in soils are even higher when compared to water bodies. Buildup of MPs in agricultural practices includes the breakdown of plastic mulches, sewage sludge application, and more often irrigation with polluted water. Once MPs enter the soil, they can affect plant growth, absorption of nutrients, and even root development. MPs enter the human diet from these plants and crops as they reach the edible part through contaminated soils.

Livestock and Poultry Exposure: Ingestion of MPs in animals happens through grazing on contaminated land. Various studies show that terrestrial food sources are vulnerable, as MPs are found in dairy products, meat, and poultry. These contaminated diets from animals added another source of MPs to the food web. Studies show the occurrence of MPs in marine and freshwater systems, agricultural land, livestock, and plants, and how these particles are deeply rooted in global food production. MPs travel through different ecosystems and into food webs and reach consumers through multiple routes: seafood, drinking water, processed foods, agricultural produce, and animal products. MPs' source of occurrence, path of entry, and levels in different sources are to be assessed to determine health risks and to develop preventive measures for maintaining the healthy ecosystem.

7.5.1 Health risks of microplastics in fish and implications for food safety

Seafood is an increasingly important component of global diets, yet it has emerged as a major route of human exposure to MPs. These particles, typically defined as plastic debris smaller than 5 mm, are readily ingested by fish and shellfish and can subsequently enter the human food chain. Beyond their physical presence, MPs raise significant toxicological and microbiological concerns because they act as vectors for a wide range of hazardous substances, including persistent organic pollutants (POPs), heavy metals, and pathogenic microorganisms (Unuofin and Igwaran, 2023). A growing body of evidence suggests that MPs have a strong capacity to sorb environmental contaminants, complicating the assessment of pollutant bioavailability in seafood. Seminal work by Rochman et al. (2013) demonstrated that POPs readily adsorb onto the surface of MPs and are subsequently transferred into fish tissues following ingestion. This vector effect enhances the internal exposure of aquatic organisms to toxic chemicals that may otherwise be less bioavailable. In addition to chemical transport, MPs exert direct biological effects on marine organisms. Experimental studies have shown that exposure to MPs can induce oxidative stress, inflammation, and immune suppression in fish, resulting in cellular and tissue damage (Kim et al., 2021). These physiological disturbances may further facilitate the accumulation and trophic transfer of contaminants along the food chain. Enhanced uptake of PCBs and brominated flame retardants has been reported in seabass exposed to contaminated MPs (Granby et al., 2018), while a similar increase in pollutant bioaccumulation has been observed in zebrafish (Rainieri et al., 2018).

MP also provides a favorable substrate for microbial colonization, forming complex biofilm communities collectively referred to as the plastisphere (Fig. 7.3). This microbial assemblage often includes opportunistic and pathogenic bacteria, raising additional concerns for seafood safety. Studies have shown that MP-associated biofilms can harbor harmful microorganisms that may be transferred to higher trophic levels upon ingestion (Zettler et al., 2013). Of particular concern is the frequent detection of Vibrio species on MP surfaces, as these bacteria are known to cause seafood-related illnesses (Tavelli et al., 2022). Beyond pathogenicity, the plastisphere has been identified as a potential reservoir for antibiotic resistance genes (ARGs). Wang et al. (2021) demonstrated that MPs collected from wastewater and natural environments can carry antibiotic-resistant bacteria and ARGs, potentially facilitating the dissemination of antimicrobial resistance. The ingestion of MPs containing resistant microorganisms represents a potential risk for human health, as it may contribute to a gastrointestinal infection or the broader spread of antimicrobial resistance (Hong et al., 2023). Overall, MPs in seafood represent a multifaceted hazard, combining physical exposure with chemical toxicity and microbial risks. Fig. 7.4 illustrates the pathways through which MPs interact with environmental contaminants and microbial communities, highlighting their role in pollutant transport and food chain transfer.

7.5.2 Regional evidence of microplastic contamination in seafood and implications for food safety

Most consumed marine species, including fish, mussels, and shellfish, are contaminated by MP, which is evident from the global seafood market analysis. The results pinpoint serious dietary exposure to seafoods and emphasize the need for improved monitoring of seafood safety.

***Mussels from the North Sea and Baltic Sea*:** The most sensitive bioindicators of MP pollution are mussels, due to their filter-feeding behavior. Studies found that contamination levels ranging from 0.2 to 1.6 MP particles per gram of tissue are found along the North Sea and Baltic Sea coasts, with elevated concentrations in mussels, especially near urbanized and industrial shorelines (Lenz et al., 2016; Sainio et al., 2021). The observations show that these serve as a direct route of MP ingestion in populations with high shellfish consumption, as mussels are often consumed whole, where most of MP's accumulation is in the gastrointestinal tract.

***Farmed Fish and Shrimp in Southeast Asia*:** An emerging hotspot in MP contamination is the aquaculture systems in Southeast Asia. After some experimental studies near the Chao Phraya River

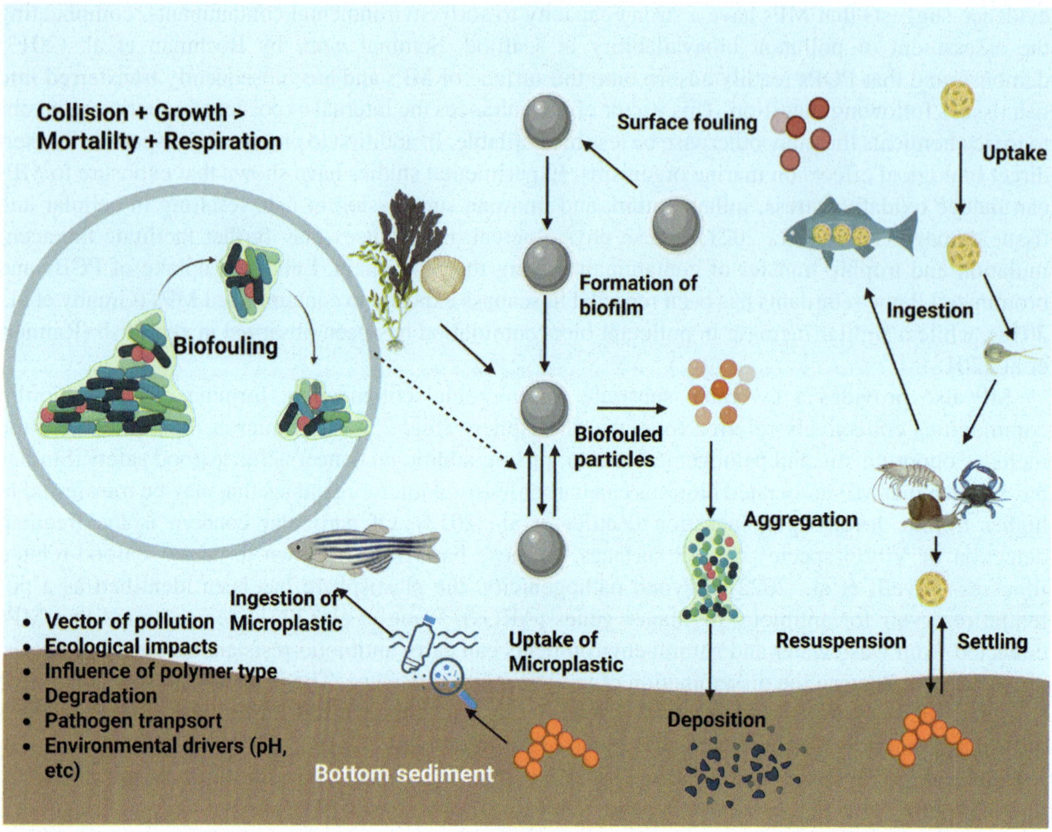

FIGURE 7.3

Conceptual illustration of the key processes governing the distribution and fate of microplastics in aquatic environments. The schematic depicts biofouling and biofilm formation on microplastic surfaces, aggregation with organic matter and marine snow, and subsequent sinking, resuspension, and deposition in bottom sediments. Biological interactions, including ingestion and uptake by aquatic organisms across trophic levels, are shown alongside physical processes such as oscillatory motion, settling, and resuspension, highlighting the dynamic cycling of microplastics between the water column and sediments.

Estuary in Thailand, it was found that there is a link in MP contamination to both environmental inputs and farming practices throughout the aquaculture zones (Ta and Babel, 2020). Moreover, substantial quantities of MPs, primarily fragments and fibers, are found in farmed white-leg shrimp, which clearly indicates the contamination of feed, water, or infrastructure used in aquaculture systems (Vitheepradit and Prommi, 2023). Results from various studies depict serious concerns in one of the world's largest aquaculture-producing regions about seafood safety.

Wild-Caught Fish from the Mediterranean Sea: A higher rate of MP ingestion was seen in small pelagic fish, including anchovies (*Engraulis encrasicolus*) and sardines (*Sardina pilchardus*). Analysis shows that MPs in 14%–15% of specimens are mostly PET (Compa et al., 2018). On further

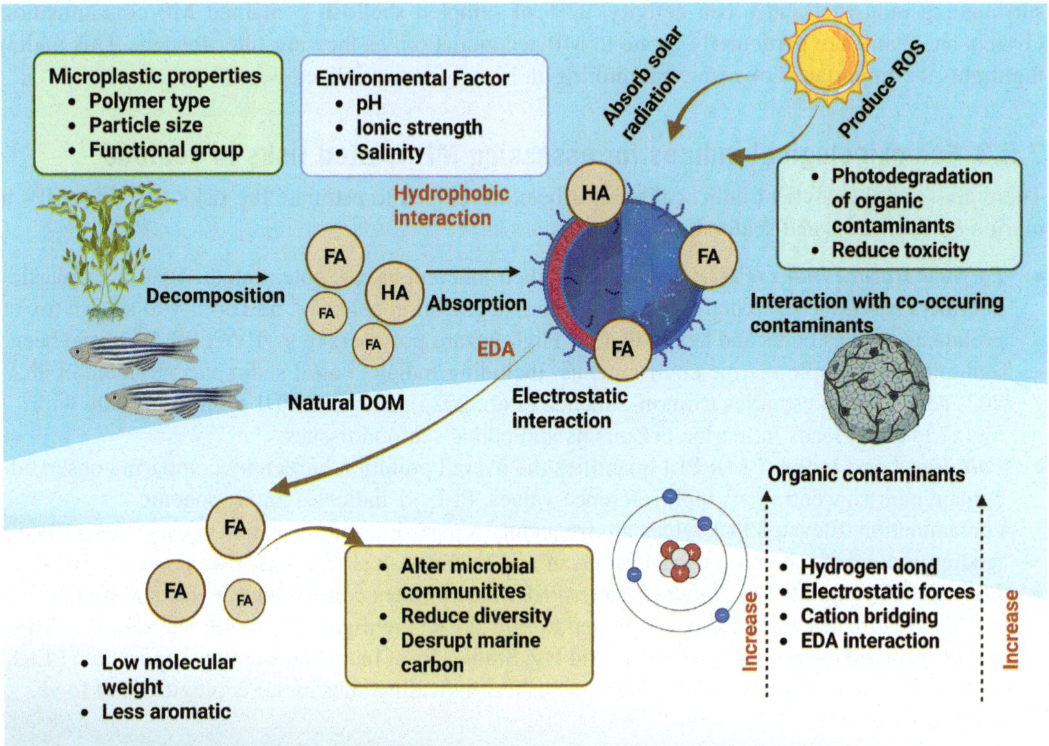

FIGURE 7.4

Conceptual illustration of the interactions between microplastics (MPs), dissolved organic matter (DOM), and co-occurring contaminants in aquatic environments. The figure highlights how MP properties (e.g., polymer type, size, surface functionality) and environmental conditions (pH, salinity, ionic strength) govern hydrophobic, electrostatic, and hydrogen-bonding interactions with natural and microplastic-derived DOM. These processes influence contaminant adsorption, photochemical reactions under solar radiation, reactive oxygen species (ROS) formation, and subsequent effects on microbial communities, carbon cycling, and contaminant fate in aquatic systems.

observation, MP ingestion in 58% of sardines and 60% of anchovies has risks in the reproductive phase and various health conditions (Pennino et al., 2020). The exposure from these fish is higher for consumers when compared with filleted fish, as they are consumed whole, including the guts.

Seafood from Japanese Markets: Not even highly regulated seafood markets are exempt. MPs are detected in commonly sold species, including mackerel, squid, and sardines, especially in mackerel, where it is found to be up to two MP particles per gram (Tanaka and Takada, 2016). This indicates that even in nations with stringent food safety systems, contamination can occur across harvesting, processing, and transport stages, making it a global issue.

Shellfish from the Gulf of Mexico: Globally, the highest contamination was found to be in the oysters and clams collected along the Gulf of Mexico. Ding et al. (2020) found that due to the surrounding industrial and urban activity, 92% of sampled shellfish contained MP contamination. Oysters and clams are particularly prone to MP accumulation, as they are filter feeders. This finding highlights the importance of severe monitoring in heavily industrialized coastal zones.

7.5.3 Ecotoxicological indices for assessing MP-related risks in seafood

There are several analytical indices that have been developed to estimate the risks posed by MPs in marine environments and seafood products:

- ***Polymer Hazard Index (PHI)*:** According to their inherent hazard potential, the PHI ranks plastic polymers based on their chemical composition, degradation behavior, and ability to adsorb toxic pollutants such as POPs and heavy metals. High-hazard polymers (e.g., PVC, PS, PA) have been identified in multiple marine environments, including Indian coastal sediments (Ranjani et al., 2022) and Danish estuaries (Simon-Sanchez et al., 2024). Elevated PHI scores correlate with higher toxicant loads in marine organisms and edible seafood tissues.
- ***Pollution Load Index (PLI)*:** PLI quantifies the overall pollution burden by comparing observed contaminant concentrations with reference values. PLI >1 indicates anthropogenic contamination. Elevated PLI values are frequently reported in sediment-rich ecosystems, critical feeding grounds for benthic fish in Mexican and Indian waters (Duenas-Moreno et al., 2024).
- ***Potential Ecological Risk Index (PERI)*:** PERI incorporates contaminant levels and toxicity factors to estimate ecological risk. Applied to MPs, PERI highlights the combined hazard of MPs and coadsorbed metals such as Cd, Pb, and Hg. Studies from India and Vietnam report high PERI values in bivalves and sediment-dwelling species, indicating substantial ecological and food safety concerns.
- Integrating PHI, PLI, and PERI provides a more comprehensive risk picture, linking environmental MP contamination to potential human exposure.

Implications for Food Safety and Public Health: Safety challenges imposed by MPs on food are unique compared with traditional biological and chemical hazards. Their impacts include:

- Chemical risks: Transfer of sorbed pollutants (POPs, heavy metals).
- Microbiological risks: MPs act as vectors for pathogenic bacteria and ARGs.
- Physiological risks: MPs can induce inflammation, oxidative stress, and immune disruption in fish, effects that may translate to human health through chronic exposure (Fig. 7.5).

7.6 Summary and conclusion

MPs have emerged as a complex environmental and public health challenge that extends well beyond issues of pollution and toxicology. Evidence consolidated in this chapter demonstrates that the ingestion of MP-contaminated food, particularly seafood, drinking water, beverages, and other widely consumed products, represents a tangible health concern. MPs not only persist as particulate contaminants but also function as carriers for hazardous substances, including POPs, heavy metals, pathogenic microorganisms, and ARGs. Long-term dietary exposure to such combined stressors may contribute to chronic health outcomes, including endocrine disruption, immune dysregulation, gastrointestinal inflammation, and metabolic disturbances.

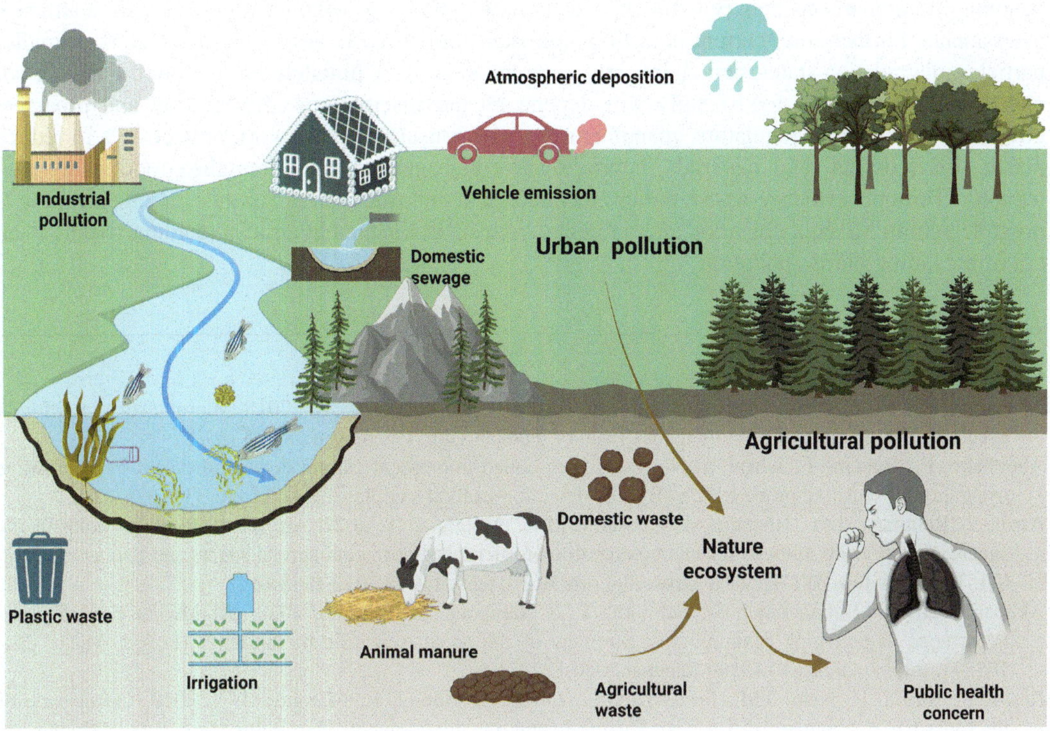

FIGURE 7.5

Schematic overview of the major pathways through which microplastics are introduced into natural environments. The diagram illustrates contributions from urban sources (industrial emissions, domestic sewage, plastic waste, and vehicle-related inputs), agricultural activities (plastic mulching, irrigation, and manure application), and atmospheric deposition. These inputs facilitate the transfer of microplastics into terrestrial and aquatic ecosystems, where they circulate through natural systems and may ultimately reach humans via environmental exposure and food-chain pathways.

This chapter highlights clear differences in MP contamination between wild-caught and aquaculture-derived food products, reflecting variations in environmental exposure, rearing practices, feed composition, and infrastructure materials. These findings underscore the need for targeted improvements in aquaculture management, stricter controls on plastic usage, and broader environmental interventions aimed at reducing plastic inputs into aquatic and terrestrial ecosystems. Importantly, MPs are not confined to seafood; they are now documented across a wide range of foods, including fruits, vegetables, salt, beverages, and processed products, reinforcing the notion that human exposure is cumulative and multi-pathway.

Despite the increase in global research activity, there are still knowledge gaps regarding sustainability. Variability in reported MP concentrations across studies is driven not only by differences in environmental contamination but also by inconsistencies in sampling, extraction, identification, and quantification methods. The absence of standardized analytical protocols continues to hinder meaningful comparisons between studies and limits the ability to conduct robust human health risk assessments. Furthermore, current detection approaches often fail to adequately characterize smaller particles, degradation states, and chemically complex systems. Ultimately, mitigating the risks associated with MPs in the food supply will depend on interdisciplinary collaboration and proactive governance. Scientists, regulators, manufacturers, and consumers must work collectively to reduce plastic use, redesign safer materials, improve waste management, and strengthen food safety protections. Through evidence-based policy, technological innovations, and public engagement, it is possible to limit MP contamination and safeguard both environmental integrity and human health in an era of escalating plastic pollution.

References

Abbasi, S., Soltani, N., Keshavarzi, B., Moore, F., Turner, A., Hassanaghaei, M., 2018. Microplastics in different tissues of fish and prawn from the Musa Estuary, Persian Gulf. Chemosphere 205, 80–87.

Aboobakri, E., Heidari, T., Jahani, M.N., 2023. S co-doped fluorescent carbon dots synthesized by microwave irradiation: a sensitive probe for Pb (II) ions detection in food samples. Carbon Lett. 33, 1629–1638.

Afrin, S., Rahman, M.M., Hossain, M.N., Uddin, M.K., Malafaia, G., 2022. Are there plastic particles in my sugar? A pioneering study on the characterization of microplastics in commercial sugars and risk assessment. Sci. Total Environ. 837, 155849. https://doi.org/10.1016/j.scitotenv.2022.155849.

Akhbarizadeh, R., Dobaradaran, S., Schmidt, T.C., Nabipour, I., Spitz, J., 2020. Worldwide bottled water occurrence of emerging contaminants: a review of the recent scientific literature. J. Hazard. Mater. 392, 122271. https://doi.org/10.1016/j.jhazmat.2020.122271.

Bessa, F., Barría, P., Neto, J.M., Frias, J.P.G.L., Otero, V., Sobral, P., Marques, J.C., 2018. Occurrence of microplastics in commercial fish from a natural estuarine environment. Mar. Pollut. Bull. 128, 575–584.

Chu, X., Zheng, B., Li, Z., Cai, C., Peng, Z., Zhao, P., Tian, Y., 2022. Occurrence and distribution of microplastics in water supply systems: in water and pipe scales. Sci. Total Environ. 803, 150004.

Compa, M., Ventero, A., Iglesias, M., Deudero, S., 2018. Ingestion of microplastics and natural fibres in Sardina pilchardus (Walbaum, 1792) and Engraulis encrasicolus (Linnaeus, 1758) along the Spanish Mediterranean coast. Mar. Pollut. Bull. 128, 89–96. https://doi.org/10.1016/j.marpolbul.2018.01.009.

Cuthbert, R.N., Al-Jaibachi, R., Dalu, T., Dick, J.T.A., Callaghan, A., 2019. The influence of microplastics on trophic interaction strengths and oviposition preferences of dipterans. Sci. Total Environ. 651, 2420–2423. https://doi.org/10.1016/j. scitotenv.2018.10.108.

Da Costa Filho, P.A., Andrey, D., Eriksen, B., Peixoto, R.P., Carreres, B.M., Ambuhl, B.M., Descarrega, J.B., Dubascoux, S., 2021. Detection and characterization of small-sized microplastics (5 μm) in milk products. Sci. Rep. 11, 24046. https://doi.org/10.1038/s41598-021-03458-7.

Danopoulos, E., Twiddy, M., Rotchell, J.M., 2020. Microplastic contamination of drinking water: a systematic review. PLoS One 15, e0236838. https://doi.org/10.1371/journal.pone.0236838.

Diaz-Basantes, M.F., Conesa, J.A., Fullana, A., 2020. Microplastics in honey, beer, milk and refreshments in Ecuador as emerging contaminants. Sustainability 12, 5514. https://doi.org/10.3390/su12145514.

Ding, J., Li, J., Sun, C., Jiang, F., He, C., Zhang, M., Ju, P., Ding, N.X., 2020. An examination of the occurrence and potential risks of microplastics across various shellfish. Sci. Total Environ. 739, 139887. https://doi.org/10.1016/j.scitotenv.2020.139887.

Duenas-Moreno, J., Mora, A., Capparelli, M.V., Gonzalez-Domínguez, J., Mahlknecht, J., 2024. Potential ecological risk assessment of microplastics in environmental compartments in Mexico: a meta-analysis. Environ. Pollut. 361, 124812. https://doi.org/10.1016/j.envpol.2024.124.

Granby, K., Rainieri, S., Rasmussen, R.R., Kotterman, M.J.J., Sloth, J.J., Cederberg, T.L., Barranco, A., Marques, A., Larsen, B.K., 2018. The influence of microplastics and halogenated contaminants in feed on toxicokinetics and gene expression in European seabass (*Dicentrarchus labrax*). Environ. Res. 164, 430–443. https://doi.org/10.1016/j.envres.2018.02.0.

Gundogdu, S., 2018. Contamination of table salts from Turkey with microplastics. Food Addit. Contam. Part A Chem. Anal. Control Expo. Risk Ass. 35, 1006–1014.

Hong, J.-K., Oh, H., Lee, T.K., Kim, S., Oh, D., Ahn, J., Park, S., 2023. Tracking the evolution of microbial communities on microplastics through a wastewater treatment process: insight into the "plastisphere". Water 15 (21), 3746. https://doi.org/10.3390/w15213.

Jin, M., Wang, X., Ren, T., Wang, J., Shan, J., 2021. Microplastics contamination in food and beverages: direct exposure to humans. J. Food Sci. 86, 2816–2837.

Karami, A., Golieskardi, A., Choo, C.K., Larat, V., Galloway, T.S., Salamatinia, B., 2017. Corrigendum: the presence of microplastics in commercial salts from different countries. Sci. Rep. 7, 46838.

Kim, J.-S., Lee, H.-J., Kim, S.-K., Kim, H.-J., 2018. Global pattern of microplastics (microplastics) in commercial food-grade salts: sea salt as an indicator of seawater MP pollution. Environ. Sci. Technol. 52, 12819–12828.

Kim, J.-H., Yu, Y.-B., Choi, J.-H., 2021. Toxic effects on bioaccumulation, hematological parameters, oxidative stress, immune responses and neurotoxicity in fish exposed to microplastics: a review. J. Hazard. Mater. 413, 125423. https://doi.org/10.1016/j. jhazmat.2021.125423.

Kirstein, I.V., Gomiero, A., Vollertsen, J., 2021. Microplastic pollution in drinking water. Curr. Opin. Toxicol. 28, 70–75. https://doi.org/10.1016/j.cotox.2021.09.003.

Kiruba, R., Preethi, M., Aganasteen, R., Rithick Raj, M., Hannah Thabitha, C., Monica, P., Sakthivel, J., Levince, C., 2022. Identification of microplastics as emerging contaminant in branded milk of Tamil Nadu State, India. Asian J. Biol. Life Sci. 11, 181. https://doi.org/10.5530/ajbls.2022.11.25.

Kosuth, M., Mason, S.A., Wattenberg, E.V., 2018. Anthropogenic contamination of tap water, beer, and sea salt. PLoS One 13, e0194970.

Kutralam-Muniasamy, G., Perez-Guevara, F., Elizalde-Martínez, I., Shruti, V.C., 2020. Branded milks – are they immune from microplastics contamination? Sci. Total Environ. 714, 136823. https://doi.org/10.1016/j.scitotenv.2020.136823.

Kwon, J.H., Kim, J.W., Pham, T.D., Tarafdar, A., Hong, S., Chun, S.-H., Lee, S.-H., Kang, D.-Y., Kim, J.-Y., Kim, S.-B., et al., 2020. Microplastics in food: a review on analytical methods and challenges. Int. J. Environ. Res. Publ. Health 17, 6710.

Lee, H.J., Song, N.S., Kim, J.S., Kim, S.K., 2021. Variation and uncertainty of microplastics in commercial table salts: critical review and validation. J. Hazard. Mater. 402, 123743. https://doi.org/10.1016/j.jhazmat.2020.123743.

Lenz, R., Enders, K., Beer, S., Sørensen, T.K., Stedmon, C.A., 2016. Analysis of microplastic in the stomachs of herring and cod from the North Sea and Baltic Sea. Technical University of Denmark. DTU Aqua National Institute of Aquatic Resources. https://doi.org/10.13140/RG.2.1.1625.1769.

Li, J., Yang, D., Li, L., Jabeen, K., Shi, H., 2015. Microplastics in commercial bivalves from China. Environ. Pollut. 207, 190–195.

Li, Y., Peng, L., Fu, J., Dai, X., Wang, G., 2022. A microscopic survey on microplastics in beverages: the case of beer, mineral water and tea. Analyst 147, 1099–1105. https://doi.org/10.1039/D2AN00083K.

Liebezeit, G., Liebezeit, E., 2013. Non-pollen particulates in honey and sugar. Food Addit. Contam. Part A Chem. Anal. Control Expo. Risk Ass. 30, 2136–2140.

Ma, Y., You, X., 2021. Modelling the accumulation of microplastics through food webs with the example Baiyangdian Lake. China. Sci. Total Environ. 762, 144110. https://doi.org/10.1016/j.scitotenv.2020.144110.

Mason, S.A., Welch, V.G., Neratko, J., 2018. Synthetic polymer contamination in bottled water. Front. Chem. 6, 407.

Munno, K., Helm, P., Jackson, D.A., Rochman, C.M., Sims, A., 2018. Impacts of temperature and selected chemical digestion methods on microplastic particles. Environ. Toxicol. Chem. 37, 91–98. https://doi.org/10.1002/etc.3935.

Naji, A., Nuri, M., Vethaak, A.D., 2018. Microplastics contamination in molluscs from the northern part of the Persian Gulf. Environ. Pollut. 235, 113–120.

Neves, D., Sobral, P., Ferreira, J.L., Pereira, T., 2015. Ingestion of microplastics by commercial fish off the Portuguese Coast. Mar. Pollut. Bull. 101, 119–126.

Obmann, B.E., Sarau, G., Holtmannspötter, H., Pischetsrieder, M., Christiansen, S.H., Dicke, W., 2018. Small-sized microplastics and pigmented particles in bottled mineral water. Water Res. 141, 307–316. https://doi.org/10.1016/j.watres.2018.05.027.

Peixoto, D., Pinheiro, C., Amorim, J., Oliva-Teles, L., Guilhermino, L., Vieira, M.N., 2019. Microplastic pollution in commercial salt for human consumption: a review. Estuar. Coast Shelf Sci. 219, 161–168. https://doi.org/10.1016/j.ecss.2019.02.018.

Pennino, M.G., Bachiller, E., Lloret-Lloret, E., Albo-Puigserver, M., Esteban, A., Jadaud, A., Bellido, J.M., Coll, M., 2020. Ingestion of microplastics and occurrence of parasite association in Mediterranean anchovy and sardine. Mar. Pollut. Bull. 158, 111399. https://doi.org/10.1016/j.marpolbul.2020.111399.

Pironti, C., Ricciardi, M., Motta, O., Miele, Y., Proto, A., Montano, L., 2021. Microplastics in the environment: intake through the food web, human exposure and toxicological effects. Toxics 9. https://doi.org/10.3390/toxics9090224.

Rainieri, S., Conlledo, N., Larsen, B.K., Granby, K., Barranco, A., 2018. Combined effects of microplastics and chemical contaminants on the organ toxicity of zebrafish (Daniorerio). Environ. Res. 162, 135–143. https://doi.org/10.1016/j.envres.2017.12.019.

Ranjani, M., Veerasingam, S., Venkatachalapathy, R., Jinoj, T.P.S., Guganathan, L., Mugilarasan, M., Vethamony, P., 2022. Seasonal variation, polymer hazard risk and controlling factors of microplastics in beach sediments along the southeast coast of India. Environ. Pollut. 305, 119315. https://doi.org/10.1016/j.envpol.2022.11931.

Rochman, C.M., Hoh, E., Kurobe, T., Teh, S.J., 2013. Ingested plastic transfers hazardous chemicals to fish and induces hepatic stress. Sci. Rep. 3 (1), 3263. https://doi.org/10.1038/srep03263.

Ruiz-Pozo, V.A., Guevara-Ramírez, P., Paz-Cruz, E., Tamayo-Trujillo, R., CadenaUllauri, S., Frias-Toral, E., Zambrano, A.K., 2024. The role of the Mediterranean diet in prediabetes management and prevention: a review of molecular mechanisms and clinical outcomes. Food Agric. Immunol. 35, 2398042.

Sainio, E., Lehtiniemi, M., Setal, O., 2021. Microplastic ingestion by small coastal fish in the northern Baltic Sea. Finland. Mar. Pollut. Bull. 172, 112814. https://doi.org/10.1016/j.marpolbul.2021.1.

Schymanski, D., Goldbeck, C., Humpf, H.-U., Fürst, P., 2018. Analysis of microplastics in water by micro-Raman spectroscopy: release of plastic particles from different packaging into mineral water. Water Res. 129, 154–162.

Shen, M., Zeng, Z., Wen, X., Ren, X., Zeng, G., Zhang, Y., Xiao, R., 2021. Presence of microplastics in drinking water from freshwater sources: the investigation in Changsha, China. Environ. Sci. Pollut. Res. Int. 28, 42313–42324.

Shruti, V.C., Fermín, P.-G., Elizalde-Martínez, Gurusamy, K.-M., 2020a. First study of its kind on the microplastic contamination of soft drinks, cold tea and energy drinks - future research and environmental considerations. Sci. Total Environ. 726, 138580.

Shruti, V.C., Pérez-Guevara, F., Elizalde-Martínez, I., Kutralam-Muniasamy, G., 2020b. First study of its kind on the microplastic contamination of soft drinks, cold tea and energy drinks - future research and environmental considerations. Sci. Total Environ. 726, 138580.

Simon-Sanchez, L., Vianello, A., Kirstein, I.V., Molazadeh, M.-S., Lorenz, C., Vollertsen, J., 2024. Assessment of microplastic pollution and polymer risk in the sediment compartment of the Limfjord. Denmark. Sci. Total Environ. 950, 175017. https://doi.org/10.1016/j.scitotenv.2024.175017.

Ta, A.T., Babel, S., 2020. Microplastics pollution with heavy metals in the aquaculture zone of the Chao Phraya River estuary. Thailand. Mar. Pollut. Bull. 161, 111747. https://doi.org/10.1016/j.marpolbul.2020.111747.

Tan, H., Yue, T., Xu, Y., Zhao, J., Xing, B., 2020. Microplastics reduce lipid digestion in simulated human gastrointestinal system. Environ. Sci. Technol. 54, 12285–12294.

Tanaka, K., Takada, H., 2016. Microplastic fragments and microbeads in digestive tracts of planktivorous fish from urban coastal waters. Sci. Rep. 6 (1), 34351. https://doi.org/10.1038/srep34351.

Tavelli, R., Callens, M., Grootaert, C., Abdallah, M.F., Rajkovic, A., 2022. Foodborne pathogens in the plastisphere: can microplastics in the food chain threaten microbial food safety? Trends Food Sci. Technol. 129, 1–10. https://doi.org/10.1016/j.tifs.2022.08.02.

Unuofin, J.O., Igwaran, A., 2023. Microplastics in seafood: implications for food security, safety, and human health. J. Sea Res. 194, 102410. https://doi.org/10.1016/j. seares.2023.10241.

Venkatesan, R., Sana, S.S., Ramkumar, V., Alagumalai, K., Kim, S.C., 2023. Development and characterization of poly (butylene adipate-co-terephthalate)(PBAT) composites with N, P-doped carbons for food packaging. Carbon Lett. 33, 1679–1687.

Vitheepradit, A., Prommi, T.-O., 2023. Microplastics in surface water and tissue of white leg shrimp, Litopenaeus vannamei, in a cultured pond in Nakhon Pathom Province. Central Thailand. AIMS Environ. Sci. 10 (4), 478–503. https://doi.org/10.3934/environsci.202.

Wang, Z., Gao, J., Zhao, Y., Dai, H., Jia, J., Zhang, D., 2021. Plastisphere enrich antibiotic resistance genes and potential pathogenic bacteria in sewage with pharmaceuticals. Sci. Total Environ. 768, 144663. https://doi.org/10.1016/j.scitotenv.2020.144663.

Winkler, A., Santo, N., Ortenzi, M.A., Bolzoni, E., Bacchetta, R., Tremolada, P., 2019. Does mechanical stress cause microplastic release from plastic water bottles? Water Res. 166, 115082.

Yıldız, D., Yalçın, G., Jovanovic, B., Boukal, D.S., Vebrova, L., Riha, D., Stankovic, J., Savic-Zdrakovic, D., Metin, M., Akyurek, Y.N., Balkanlı, D., Filiz, N., Milosevic, D., Feuchtmayr, H., Richardson, J.A., Beklioglu, M., 2022. Effects of a microplastic mixture differ across trophic levels and taxa in a freshwater food web: in situ mesocosm experiment. Sci. Total Environ. 836, 155407. https://doi.org/10.1016/j.scitotenv.2022.15540.

Zettler, E.R., Mincer, T.J., Amaral-Zettler, L.A., 2013. Life in the "Plastisphere": microbial communities on plastic marine debris. Environ. Sci. Technol. 47 (13), 7137–7146. https://doi.org/10.1021/es401288x.

Zhang, D., Cui, Y., Zhou, H., Jin, C., Yu, X., Xu, Y., Li, Y., Zhang, C., 2020. Microplastic pollution in water, sediment, and fish from artificial reefs around the Ma'an Archipelago, Shengsi. China. Sci. Total Environ. 703, 134768.

Zheng, X., Wu, X., Zheng, Q., Mai, B.-X., Qiu, R., 2023. Transfer of microplastics in terrestrial and aquatic food webs: the impact of E-waste debris and ecological traits. Environ. Sci. Technol. 57 (3), 1300–1308. https://doi.org/10.1021/acs.est.2c06473.

CHAPTER

8 Microplastic contamination in diverse aquatic environments: Insights from regional case studies

Melvin S. Samuel

Department of Civil, Construction, and Environmental Engineering, Marquette University, Milwaukee, WI, United States

Highlights

- Microplastics are present across all studied aquatic ecosystems, from tropical rivers to Arctic Sea ice.
- Freshwater systems act as critical but underrecognized pathways for marine Microplastic pollution.
- Polar environments functions as temporary sinks, releasing Microplastics during seasonal ice melt.
- Urbanization, hydrology, and ecosystem residence time strongly influence Microplastic accumulation.
- Case-study comparison reveals the need for region-specific monitoring and mitigation approaches

8.1 Introduction

Microplastic (MP) contamination has evolved from a localized pollution concern into a truly global environmental issue, affecting aquatic ecosystem across continents, climate zones, and levels of economic development. Evidence from Africa, Asia, Europe, North America, and the Arctic collectively demonstrates that MPs are no longer confined to densely populated or industrialized regions but are now embedded within natural hydrological and biogeochemical cycles. This global distribution underscores the need to interpret MP pollution not as isolated incidents, but as a systemic outcome of modern material use, environmental transport process, and ecosystem connectivity.

Observations from African and Asian aquatic systems highlights the dominant role of riverine transport in MP dispersal. Rapid urbanization, insufficient waste management infrastructure, and high population densities contribute to substantial plastic inputs into rivers, lakes and coastal water. These systems frequently act as primary entry points through which MPs move from land to sea. Seasonal hydrology, monsoonal rainfall, and flooding further amplify transport efficiency, leading to pronounced spatial and temporal variability in MP concentrations. Such findings emphasize the importance of freshwater systems as active drivers of global MP fluxes rather than passive conduits. In contrast, Europe aquatic ecosystems often reflect the long-term accumulation and retention of MPs within managed landscape. Studies from rivers, lakes, and coastal zones reveal that even regions with advanced wastewater treatment and regulatory frameworks experience persistent MP contamination. High residence times in lakes and reservoirs promote sedimentation and accumulation, while urban runoff and legacy plastic sources sustain continuous inputs. These observations demonstrate that

Microplastics in Aquatic Environments. https://doi.org/10.1016/B978-0-443-45619-0.00011-X

technological controls alone are insufficient to eliminate MP pollution once it becomes embedded with aquatic systems.

Evidence from North American freshwater and marine environments bridges these perspectives by illustrating strong linkages between inland waters, coastal ecosystems, and human exposure pathways. Large lakes and estuaries functions simultaneously as sinks and secondary sources, redistributing MPs through resuspension, biological interactions, and downstream transport. Importantly, studies documenting MPs in commercially relevant species highlight the convergence of ecological and human health concerns, reinforcing the need for ecosystem scale assessments. The Arctic region provides some of the most compelling evidence for the long-range transport and delayed release of MPs. Despite minimal local plastic production, MPs are consistently detected in sea ice, snow, and remote water, confirming atmospheric and oceanic transport from lower latitudes. Cryospheric systems act as temporary storge reservoirs, trapping MPs during ice formation and releasing them during seasonal and climate driven melt. This process introduced a time lagged contamination pathway that is likely to intensify under ongoing climate warming, with implications for polar food webs and downstream marine environments.

When viewed collectively, these continues sale observational reveal that MP contamination is governed by three interconnected processes: (i) land-based generation and freshwater transport, (ii) ecosystem-specific retention and transformation, and (iii) long-range redistribution driven by oceanic and atmospheric circulation. Case-study integration across Africa, Asia, Europe, North America, and the Arctic thus provides critical insights into how MPs behave differently across ecosystems while remaining globally interconnected. By adopting a comprehensive, cross-regional perspective, this chapter advances beyond isolate concentration measurements to emphasize system-level understanding. Such integration is essential for developing harmonized monitoring strategies, improving risk assessment frameworks, and designing mitigation approaches that are responsive to regional environmental realties while addressing the global nature of MP pollution.

8.2 Technological interventions for mitigating microplastic pollution in Malaysia's marine ecosystems

MP pollution has recently become a significant environmental issue in Malaysia's coastal and marine waters. Studies have documented measurable levels of MP pollution in various locations, including Kuala Nerus and Kuantan Port. Field surveys reported 1713 from 0.13 to 0.69 particles per liter and 0.14 to 0.15 per liter respectively (Kurniawan et al., 2023b). Most of these pollutants stem from everyday disposable plastics and insufficient waste management practices. Frequently identified sources include plastic bags, cigarette butts, food packaging and beverage bottles. The MP particles detected are primarily composed of polymers such as polyester, polystyrene, polyamide, PVC, polypropylene and polyethylene. Malaysia's economy is closely linked to healthy marine ecosystem, which support fisheries and tourism. MPs pose risks to fish populations, harm coastal habitats and diminish the esthetic value of beaches, potentially degrading tourism. Financially, mismanaged plastics incur losses exceeding USD 13 billion annually from land filling alone while marine plastic pollution contributes to natural capital losses ranging between USD 3300 to USD 33,000 per ton of waste each year (Kazour et al., 2019). In addition, reduced sea food quality and availability can impact global food security, given that 1/5th of the world's population relies on sea food as a primary source of animal protein. Therefore, safeguarding Malaysia's marine resources is vital for the livelihoods of communities engaged in fishing, aqua culture and tourism. Efforts to reduce MP pollution also align with international sustainability objectives, including the United Nation's sustainable development goal 14 (Life below water). In response to this growing concern, the Malaysian government introduced the "***Road map to zero single-use plastics 2018–2030***", which outlines steps to progressively eliminate disposable plastic items. These initiatives include restrictions on plastic straws, cutlery and packaging alongside the promotion of environmentally friendly alternatives. Despite these measures, challenges persist, such as low recycling rates, limited public awareness and the higher cost of biodegradable material (Lamberti et al., 2020).

Long-term and effective mitigation of MP pollution requires collaboration among government agencies, industry stakeholders, NGOs, academic institutions and the public. A coordinated strategy can help Malaysia transition from a linear, plastic-dependent economy to a circular and sustainable model, reducing environmental damage and protecting marine biodiversity. Proposed interventions include enhancing waste management policies, leveraging technological innovations, strengthening regulatory enforcement and fostering community participation. While MP pollution is a global problem, understanding Malaysia's specific environmental, cultural and policy circumstances provide important regional insights (Kumar et al., 2020). Emerging technologies, including advanced monitoring systems, sensor networks, and artificial intelligence present promising opportunities to detect, manage, and remove MPs from marine environments.

8.2.1 Waste treatment and filtration technologies

Wastewater treatment is a key strategy for preventing MPs from entering rivers, lakes and oceans. The process begins with primary treatment, which removes large solids, sediments and floating debris. While this step eliminates some MPs, it is not specifically designed for their removal. Secondary treatment, relying on biological processes, can further reduce MP levels, but its efficiency is limited (Spano et al., 2022). Higher removal rates are generally achieved during tertiary or advanced treatment, where modern filtration techniques, such as membrane bioreactors (MBRs), advanced filters, and reverse osmosis, are employed. These systems can achieve very high MP removal; for instance, full-scale MBR installations have reported up to 99.7% elimination of MPs (Kurniawan et al., 2022c). Additional steps, including activated carbon treatment or biological adsorption, can further enhance MP capture. However, challenges remain advanced systems often demand substantial energy and operational costs (Goh et al., 2022), and very fine MP particles may still pass through treatment barriers. Highlighting the need for ongoing technological improvements. Wastewater treatment plants also serve as valuable monitoring points to track environmental MP contamination trends. Current research is focussed on developing more cost effective, efficient, and scalable solutions, including improved filtration materials, optimized treatment workflows, and biological degradation approaches. Although treatment technologies are essential. Tackling plastic pollution at the source through better waste management and reduced plastic use is equally critical.

Malaysia's wastewater sector illustrates both achievements and persistent challenges. The country operates over 8000 public sewage treatment plants and a wide-reaching sewerage network (Lusher et al., 2019), treating approximately six million tons of sewage annually. Most plants rely on preliminary, primary and secondary treatment to remove solids, grease and organic pollutants. Persistent issues include low treatment fees that do not cover operational costs, high volumes of oil and grease entering the system, and limited public awareness regarding proper sewage disposal. To enhance performance, Malaysia is upgrading treatment facilities by implementing technologies such as Moving

Bed Biofilm Reactors and MBRs, which improve efficiency and produce cleaner effluent. Digital monitoring, automation, and real-time water quality tracking further optimize plant operations, supporting environmental protection, regulatory compliance, and alignment with international standards (Gatidou et al., 2019). Community engagement also plays an important role, For Examples, a UNEP supported pilot project in Lok Urai introduced low energy, decentralized sewage treatment units to provide sanitation solutions to remote settlements. At the policy level, Malaysia has established a regulatory framework through the Water Services Industry Act 2006, supervised by the National Water Services Commission (SPAN), which governs the water and wastewater sector (Ryu, 2019). Despite progress, additional improvements in enforcement, technology, private education, and financial sustainability are needed. Ongoing collaboration among government agencies, private companies, researchers, and local communities will be vital to strengthen wastewater management and protect Malaysia's freshwater and marine environments (Guo et al., 2021).

8.2.2 Cleanup and remediation technologies

Malaysia has undertaken proactive measures to mitigate MP pollution through a combination of technological advancements, policy interventions, and community-based initiatives, although notable challenges remain. Across the country, many wastewater treatment plants have incorporated advanced processes, including membrane bioreactors (MBR) and ultrafiltration systems. These enhanced treatments have demonstrated high efficiency, with evidence indicating that tertiary can elimination approximately 80%–90% of MPs from treated effluent (Elkhatib and Oyanedel-Craver, 2020). Government-led initiatives have been equally influential. The national Roadmap to Zero Single-Use plastic items such as straws, utensils, and packaging materials. Regulatory authorities also promote the adoption of improvized wastewater management technologies and best practices among industries and private operators (Ahmed et al., 2020).

Community engagement further reinforces these efforts. One illustrative case is the UNEP-supported pilot project in Lok Urai, Sabah, where compact sewage treatment units were installed beneath houses and schools in a floating village setting. These energy-efficient systems can treat domestic wastewater generated by a five-member household within approximately 7 hours prior to safe discharge, highlighting the potential of decentralized treatment solutions for underserved communities (Fu et al., 2023). Nevertheless, Malaysia's wastewater sector continues to face economic and operational constraints. Treatment expenses frequently surpass revenue obtained from sewerage tariffs, and public awareness regarding appropriate wastewater management particles remain limited. With a population of about 27 million generating nearly six million tonnes of swage annually much of which ultimately enters rivers untreated or inadequately treated wastewater poses serious risks to water quality. This concern is amplified by the fact that roughly 98% of Malaysia's freshwater supply is derived from surface water sources, including rivers and reservoirs (Satya et al., 2021).

Freshwater systems are central to understanding the dynamics of plastic pollution. Rivers, in particular, function as major conduits, transporting plastics, from terrestrial environments to marine systems. Simultaneously, they serve as storage zones, retaining plastics within sediments, vegetation, floodplains, dam reservoirs. This dual role has both transport pathways and sink enable rivers to accumulate substantial quantities of MP debris over extended periods (Wang et al., 2024). Despite their importance, freshwater plastic pollution has attracted considerably less scientific attention than marine pollution. With research coverage remaining uneven across geographic regions. Urban stormwater runoff, agricultural drainage, industrial effluents, and recreational activities all contribute to the presence of MPs in freshwater environments. Once introduced into rivers, plastics may accumulate along banks, settle on riverbeds, or become trapped within wetlands and debris zones. Over time, physical and chemical degradation can fragment these materials into smaller particles, complicating removal efforts and heightening ecological risks (Yuan et al., 2023). In addition, the absence of standardized approach for sampling, classification, and monitoring across studies restricts data comparability and creates obstacles for effective policy formulation (Hurley et al., 2023). Addressing freshwater plastic pollution, therefore requires comprehensive knowledge of pollution sources, transport mechanism, and accumulation areas. Strengthened waste management system, enhanced public education, stricter regulatory enforcement, and harmonized monitoring protocols can enable more targeted and effective interventions. Collaboration among governmental bodies, industry stakeholders, researchers, and local communities remain essential for developing sustainable solutions.

Malaysia has also demonstrated commitment through financial investment and international collaboration. For instance, USD 85 million was allocated under the 2009 National Budget to expand rural water supply infrastructure and improve public awareness related to sanitation and clean water initiatives (Selamat et al., 2024). Engagement in international environmental programmes, including UNEP—supported wastewater projects, further enhances national capacity to address MP pollution. Overall, Malaysia's response reflects an integrated approach that combines technological innovations, regulatory planning and community participation. Sustained progress will depend on securing long term funding, advancing treatment technologies, strengthening waste management systems, and improving public understanding to ensure durable protection of freshwater and marine ecosystem.

8.2.3 Policy instruments for microplastic control

8.2.3.1 Existing regulatory framework

Malaysia has implemented a range of policies and national initiatives aimed at reducing plastic pollution, including MPs. One of the earliest measures was the "***No Plastic Bag Day*(NPBD)**" campaign introduced in 2011, which imposed an RM 0.20 levy on plastic bags at supermarkets and retail outlets to discourage reliance on disposable plastics (Ma et al., 2020). This initiative was later complemented by the Roadmap Toward Zero Single-Use Plastics 2018–2020, which sets out long-term actions such as pollution charges, extended producer responsibility, and the gradual elimination of single-use plastic products. Legal support for household-level waste separation is provided under the Solid Wate and Public Cleansing Management Act 2007 (Act 672), reflecting efforts to institutionalize structured waste management systems (Usman et al., 2022). Recycling is further promoted through the National Recycling Program (NRP), launched in 1993, which emphasizes the principles of Reduce, Reuse, and Recycle. Despite these initiatives, recycling rates remain low, estimated at only 3%–5%, indicating a persistent gap between policy objectives and on-the-ground implementation (Kamaruddin et al., 2022).

Plans to expand waste-to-energy facilities have been proposed; however, progress has been constrained by public opposition and environmental concerns. To address the influx of foreign plastic waste, the government has tightened import regulations, halted the issuance of new permits for overseas waste processing operations, and shut down 148 illegal plastic recycling factories in 2019. Malaysia's adherence to the Basel Convention, particularly following amendments related to plastic waste, has further reinforced controls on transboundary waste movements. A transition toward a circular economy has been identified as a long-term strategy, emphasizing recovery, reuse, and conversion of plastic waste into new materials rather than disposal. The government recognizes that stronger enforcement mechanisms, economic incentives, infrastructure improvements, and broader public participation are necessary to facilitate this transition (Fauziah et al., 2021). Although biodegradable plastics are promoted as alternatives, challenges persist due to limited consumer awareness, weak market demand, and unclear end-of-life management pathways. Despite these collective actions, ongoing issues such as insufficient waste collection coverage, fragmented recycling systems, and low levels of public engagement remain evident. In 2018 alone, more than 0.94 million tonnes of plastic waste were mismanaged nationwide (Li et al., 2020), underscoring the urgency for sustained improvements in policy implementation and public awareness.

8.2.3.2 Zero single use plastic road map

Malaysia continues to pursue a multi-dimensional strategy to address MP pollution by integrating legislation, nationwide campaigns, technological innovation and international cooperation. Initiatives such as the NPBD campaign, the Zero Single-Use Plastics Roadmap, and expanded recycling programs signal increasing political commitment. Public awareness efforts, including those conducted under the NRP, seek to improve recycling rates; however, progress has been slow due to the low cost of virgin plastics, limited recycling infrastructure, and insufficient environmental literacy (Kamaruddin et al., 2022). Looking ahead, Malaysia aims to accelerate its shift toward a closed-loop plastic economy through stronger collaboration among government agencies, industries, researchers, and communities, investments in green technologies, enhanced waste collection systems, and improved product design will be critical for achieving a sustainable, low plastic future (Pathak et al., 2023). Engagement in international agreements such as the Basel Convention further affirms Malaysia's commitment to reducing plastic pollution, including MPs at both national and global levels. While meaningful progress has been achieved, continued effort, adequate financial support, technological advancements, and deeper public engagements will be essential for long term success. Malaysia's evolving policy framework reflects its broader dedication to environmental protection, resource efficiency, and sustainable development (An et al., 2020).

8.2.4 International best practices

Drawing lessons from international experiences offers Malaysia important guidance foe addressing plastic and MP pollution. A few effective policy approaches implemented elsewhere could be adapted to the Malaysian context, although their success would depend on local economic conditions, social structures, and cultural norms. Irelands widely cited plastic bag levy, for instance, resulted in a 90% reduction in plastic bag usage, clearly illustrating the ability of fiscal measures to influence consumer behavior. Likewise, California's statutory ban on single-use plastic bags led to a 50% decrease in consumption, providing a potential model for strengthening Malaysia's existing "No Plastic Bag Day" initiative (Wang and Karasik, 2022). Economic incentives have also proven effective in improving recycling outcomes. Sweden's deposit-refund scheme for beverage containers consistently achieves recycling rates exceeding 80%, while Norway's extended Producer Responsibility (EPR) framework-requiring manufactures to oversee plastic packaging throughout its lifecycle-reported a 90% rate of recycling or energy recovery in 2019 (Maitre-Ekren et al., 2021). These examples highlight the benefits of reallocating responsibility to both producers and consumers. Education campaigns and pricing mechanisms, such as France's eco-tax on single-use water bottles, have reduced sales of targeted products by nearly 30% within 2 years' time. At the same time, China's Plastic Restriction Order underscores the need for strong recycling systems and waste management infrastructure to support plastic reduction policies. Regional collaboration, similar to coordinated policy making within the European union, could also offer advantages for ASEAN countries, including Malaysia. Community lead initiatives further demonstrated their values; in Kenya, a nationwide plastic bag ban resulted in an 80% reduction in use, reflecting the effectiveness of community engagement reinforced by enforcement and awareness measures. To assess outcomes, many countries increasingly reply on data-driven policy evaluation, regular reporting and standardized monitoring frameworks to inform decision and refine interventions (Heller et al., 2020). In applying international best practices, Malaysia must carefully consider national development priorities, socioeconomic diversity, existing waste management capacity, and prevailing attitudes toward consumption and recycling. An incremental, evidence-based strategy-starting with pilot initiatives and expanding proven programs-may offer the most practical route toward sustained reduction in plastic pollution while supporting broader global sustainability objective (Kurniawan et al., 2023a,b)

8.2.5 Public participation and awareness

8.2.5.1 Role of public awareness campaigns

Public involvement plays a vital role in mitigating MP pollution, and Malaysia has introduced multiple awareness-oriented initiatives aimed at encouraging behavioral change. Among the most prominent national efforts in the *Roadmap to Zero Single-Use Plastics 2018–2030,* which proposes the gradual phase-out of disposable plastic products such as straws, food packaging, and cutlery by promoting environmentally responsible alternatives (Kurniawan et al., 2023a,b). At the state level, community-based measures including "*No Plastic Bag Campaigns*" and weekly "*No Plastic Bag Days*" in Penang, Selangor, Kedah, and Pahang have contributed to reducing routine reliance on plastic bags. Survey finding suggests that while Malaysians generally recognize plastic pollution as a serious issue, many remain insufficiently informed about alternative materials and the long-term environmental and health consequences of plastic use. Nevertheless, most respondents indicate a willingness to support initiatives aimed at reducing MPs emissions (Baensch-Baltruschat et al., 2020).

The global significance of MPs has also been highlighted by organizations such as the World Health Organization, which has emphasized the need for expanded research, stricter regulatory controls, and enhanced international cooperation. Malaysia's experience mirrors broader global pressures, as rapid urbanization, rising waste generation, and a legacy of imported plastic waste have placed additional strains on existing waste management system (Wen et al., 2021). Research indicates that nearly half of Malaysia's 0.94 million tonnes of inadequately managed plastic waste eventually enters rivers or costal environments. MPs have been identified in drinking water, marine organisms, and seafood from regions such as Kuala Nerus and Kuantan, raising concerns related to food safety and public health. Although recent polices and awareness campaigns have improved public understanding and stimulated positive behavioral changes, sustained progress will require long-term commitments. Enhancing education initiatives, strengthening policy enforcement, expanding recycling and recovery infrastructure, and integrating circular economy principles will be critical. Ultimately, coordinated participation by household, industries, governments and communities will shape Malaysia's capacity to protect ecosystems, safeguard public health, and reduce MP pollution over time.

8.2.5.2 Case studies on integrated efforts

Effective response to MP pollution typically relies on a combination of policy measure, technological solution, and active public participation. Malaysia offers a clear illustration of this integrated approach. The Roadmap to Zero Single-Use Plastic 2018–2030establishes a national strategy to progressively eliminate disposable plastics, a major source of MP contamination. This framework is reinforced through community outreach initiatives, public education programs, and local actions that promote reduced plastic consumption and improved waste disposal practices. In parallel, Malaysia's Fourth Industrial Revolution (IR4.0) Policy supports the advancements of digital technologies, smart infrastructure, and innovative waste management systems, linking environmental objectives with long-term economic development (Ma et al., 2020).

At the global level, the European Union's Plastic Strategy illustrate how harmonized legislation, advanced recycling technologies, and strong public engagement can drive systemic change. The EU's target of achieving a 90% recycling rate for plastic beverage bottles by 2025 is expected to substantially reduce MP releases into the environment. Progress toward this goal is supported by public communication campaigns, deposit-refund schemes and improvements in sorting and recycling infrastructure. Public participation also plays a critical role internationally. The International Coastal Cleanup, coordinated by Ocean Conservancy, mobilizes volunteers across more than 70 countries annually. These activities have led to the removal of millions of plastic items from shorelines and waterways, while simultaneously generating valuable datasets for researchers, policymakers, and governments (Accinelli et al., 2019). Such initiatives demonstrate how citizens involvement can effectively complement policy and technological interventions.

UNEP highlights the global scale of the issue, estimating that approximately eight million tonnes of plastic enter the ocean each year, including significant quantities of MPs in Malaysia, integrating environmental priorities into education and workforce development such as through TVET sector influenced by IR4.0 may help prepare future professionals to address MP related challenges (Tahir and Abdullah, 2023). Collectively, these case studies indicate that isolated actions are insufficient. Instead, long-term success depends on aligning government policy, technological innovation, and community engagement to form a comprehensive framework capable of reducing plastic use, improving waste management, and ultimately mitigating MP pollution.

8.2.6 Implications of this work

Prioritizing technological interventions has the potential to drive advancements in environmental monitoring, waste treatment, and MP removal technologies. Such developments could enhance Malaysia's role as a regional leader in green technology and encourage the application of innovative solutions across other environmental domains. Broad adoption of these technologies may also foster long-term sustainability practises while stimulating investment in research and innovation. In parallel, this contributes to the refinement of policy instruments that more effectively govern plastic production, consumption, and disposal. Outcomes could include stronger legislation, more effective enforcement mechanisms, and clearer regulatory frameworks. Should these approaches prove successful, Malaysia's experience could offer a valuable reference for other countries facing similar plastic pollution challenges, thereby supporting wider regional and global policy efforts to reduce plastic waste (Mitrano and Wohlleben, 2020).

8.3 Co-occurrence of microplastics and trace metals in seawater: Case studies from North America with transatlantic perspectives

8.3.1 Sampling plan

Morici et al., 2025 carried out through two independent oceanographic sampling campaigns that together covered more than 17,000 nautical miles across the Mediterranean Sea and the Atlantic Ocean, reaching both the eastern and western coastlines of the United States. Overall, 34 locations were sampled for trace metals (TM) and metalloids (Ms), while approximately 60 sites were analyzed specifically for MPs. Sampling areas included Sicily, Sardina, the Balearic Islands, southern Spain, the African Atlantic coast, the Canary Islands, Cape Verde, Suriname, Guyana, Venezuela, Colombia, Panama, Costa Rica, Nicaragua, El Salvador, Mexico and Baja California extending to Los Angeles. Additionally, samples were collected in waters surrounding Portugal, the United Kingdoms, the Faroe Islands, Iceland, Greenland, Canada, and along the U.S. East coast up to New York. The analytical framework for MPs, trace metals, and metalloids comprised sample collection, pretreatment, laboratory analyses, and data interpretation, and was implemented over the period from 2019 to 2022.

8.3.2 Results overview

MPs were identified in all seawater samples analyzed. Concentrations showed substantial variability, ranging from 230 to 3320 items per liter of seawater. Box-and-whisker plots revealed clear differences in data spread and distribution (Morici et al., 2025). During the 2019 transect from Palermo to New York, approximately 71% of samples were below the mean concentration, whereas a lower proportion (57%) fell below the mean during the Palermo-to-Los Angeles transect. The dataset collected in 2022 exhibited an extended upper whisker, indicating increased variability at higher concentration levels, although the most pronounced outlier was recorded during the 2019 campaign. Unexpectedly high MP concentrations were measured near the Faroe Islands, particularly in the vicinity of Torshavn in southern Streymoy, an area characterized by intensive fisheries and marine-related activities. These elevated values maybe linked to local oceanographic conditions, including upwelling processes and their interaction between warm Gulf Stram waters and colder Arctic currents (Vage et al., 2018).

A notable outlier in the 2022 dataset was observed in Colombia's Magdalena Sea; however, its concentration was comparable to levels reported in other regions such as the Strait o Gibraltar. Overall, the results indicate that anthropogenic pressure-including shipping activities, costal waste inputs, agricultural runoff, and petroleum-related operations-are key drivers of MP contaminations (Galindo Montero et al., 2023). In general, MP concentrations were consistently higher in coastal waters than in offshore environments, underscoring a strong relationship between pollution levels and proximity to population centers and industrial zones. Repeated sampling along shared transects, such as from Palermo to the Strait of Gibraltar, suggests a possible increase in MP accumulation over time, although seasonal variable and oceanographic dynamics may also influence observed concentrations.

8.4 Microplastic contamination in the Southern United States coastline: Occurrence, sources, and distribution

8.4.1 Sources, distribution, and transport pathways of microplastics in southeastern US coastal systems

MPs are now ubiquitous contaminating costal and marine environments globally. This chapter integrates field observations with oceanographic modeling to investigate the distribution, sources, and transport pathway of MPs across National Park Service (NSP) costal sites in the southeastern United States and the U.S. Virgin Islands. Eighteen costal locations were sampled using standardized surface sand collection techniques, and MP were extracted through density separation procedures. Metrics of urbanization and simulations of ocean circulation using the Regional Ocean Modeling System (ROMS) were evaluated to determine factors influencing spatial variability. MPs were detected at all sampling locations, including relatively remote areas. Their abundance showed a moderate relationship with nearby urban development and was strongly shaped by regional hydrodynamic conditions and riverine inputs. The combined use of empirical data and predictive modeling emphasizes the interaction between human activities and oceanographic processes in controlling MP distribution within coastal systems. The coastal plain of the Southeastern United States is characterized by a wide, low-relief shoreline and a variety of ecological habitats that support abundant wildlife as well as important fisheries and tourism sectors. Despite its environmental and economic significance, information regarding the occurrence and spatial patterns of MPs in this region remains scare. While macroplastic debris has been extensively documented along these coastlines (Viehman et al., 2011), the proximity of many locations to densely populated urban areas underscore the need for systematic assessment of MP contamination.

To represent this broad geographic area, several units of the U.S. National Park Service (NPS) were selected as sampling sites. These locations were chosen based on accessibility, logistical support from park staff, and their distinctive combination of relative protection from direct anthropogenic disturbance and exposure to common coastal stressors such as marine debris. Because NPS units vary considerably in their distance from urbanized area, they provide an effective setting for examining the influence of local development of MP inputs. Wastewater discharge associated with urban expansion are widely recognized as significant sources of both macroplastics and MPs in aquatic systems (Cesa et al., 2017). Consequently, investigations within these protected environments can yield important insights into regional contamination dynamics. Although previous studies have reported the presence of large plastic debris along these coasts (Viehman et al., 2011), comprehensive evaluations of MP abundance and transport processes remain limited. This chapter presents an integrated assessment of MP contamination across selected NPS coastal sites, combining standardized field sampling, laboratory analyses, and numerical oceanographic modeling. The primary objectives are to:

- Quantify MP abundance across a range of costal environments
- Evaluate the effects of local urbanization and riverine contributions
- Identify regional transport pathways using Regional Ocean Modeling System (ROMS) simulation

This section provides both a methodological and conceptual framework for understanding MP behavior in coastal ecosystem.

8.4.2 Field sampling collection and potential contributing cities

Yu et al. (2018) in his study, 18 coastal sites across 16 National Park Service (NPS) units in the southeastern United States were selected. Sixteen sites were distributed along the coastline from North Carolina to Texas, with two additional locations situated in the U.S. Virgin Islands. Within each park, sampling locations were identified by NPS personnel in areas where large marine debris had been repeatedly observed. To determine potential urban sources of MPs, 26 coastal cities were included in the modeling framework. Of these, 12 cities—Havana, Tampico, New Orleans, Gulfport, Mobile, St. Petersburg, Tampa, Orlando, Jacksonville, Savannah, Charleston, and Wilmington—were predicted to release particles that reached at least one of the sampled coastal locations. Model outputs showed agreement with field measurements. Based on drifter counts at day 300 of the simulation, the number of modeled particles arriving at individual sites exhibited a statistically significant positive correlation with MP abundances observed in situ ($R^2 = 0.55$, $P = .01$) (Yu et al., 2018). This relationship indicates that the model effectively captured the broad spatial patterns of MP transport.

8.4.3 Predicted contributions to samples sites

Model simulations indicated that the number if virtual MPs at each sampling site increased progressively over time, although a portion of particles was dispersed away by dominant ocean currents. Distinct regional source patterns were evident among individual cities. Along the Atlantic seaboard, Jacksonville was identified as a major contributor. Despite its greater distance compared with Wilmington, particles released from Jacksonville reached northern sites such as S1-CAHA and S2-CALO within approximately 200 modeled days. Savannah and Charleston were also identified as contributing sources, although their influence within the 50 km radius fluctuated seasonally throughout the simulation period. Orlando was predicted to supply virtual MPs to sites including S5-CUIS, S6-TIMU, S7-FOMA, and S8-CANA. Drifters arriving at S3-FOSU and S4-FOPU originated primarily from Jacksonville and Orlando, consistent with the northward transport associated with the Gulf stream. Within the Gulf of Mexico, Gulfport emerged as the dominant contributor to the Gulf Islands National Seashore (GUIS) sites, which also exhibited relatively high MP concentrations in field observations. Whilst other Gulf Coast cities contributed particles to Gulf sites, they did not deliver drifters to Atlantic sampling locations within the 300-day simulation period. Although longer transport times cannot be ruled out, these results suggest that Guld coast urban centers exert a weaker influence on Atlantic sites compared with cities located closer to those areas.

Several Florida locations, including S12-BICY and S10-EVER, were primarily influence by particles released from Tampa and St. Petersburg, although transport to these sites required approximately 100 modeled days. These two cities were also predicted to contribute to S9-BISC. Drifters originating from Havana were transported exclusively to S11-DRTO and were not detected until the August phase of the simulation. This finding is notable given Havana's proximity to the Gulf Stream, which might suggest potential influence on sites such as S9-BISC or S8-CANA; however, localized circulation patterns and seasonal variability limited such transport in the model. Notably, 11 modeled cities including Miami, Houston, Beaumont, Corpus Christi, Matamoros, Tuxpan, Coatzacoalcos, Merida, Campeche, Cancun, and Nassau did not contribute any particles to sampled sites within the 50 km radius (Yu et al., 2018). This outcome implies that MPs originating from these urban centers either disappear offshore before reaching coastal sampling areas or require transport durations exceeding the 300-day simulation window.

8.4.4 Conclusion

MPs were detected at all National Park service coastal sites included in the study, including those located in relatively remote areas. Fourier-transform infrared (FTIR) spectroscopy confirms the polymeric nature of the recovered particles, with polyethylene terephalate (PET) identified as the most prevalent polymer. In general, MP abundance increased with the degree of surrounding urbanization, and sites influenced by large river systems also exhibited elevated contamination levels. These patterns are consistent with the widely recognized role of urban and industrial wastewater discharges as major sources of MPs to marine environments. Ocean circulation at both local and regional scales exerted a strong control on MPs transport and accumulation. Model results indicated that particles tend to remain within distinct oceanographic domains; for example, MPs released from Texas and Louisiana predominantly circulated within the Gulf of Mexico rather than entering the Atlantic or Caribbean, while particles originating along the Atlantic coast were transported northward by the Gulf stream. Overall, the findings demonstrate that MP presence in coastal environments is governed by the combined effects of urban development, riverine inputs and oceanographic processes. Even well-protected or geographically isolated coastal areas remain vulnerable to MP contamination due to the interconnected nature of marine systems and the persistent movement of plastics within coastal currents.

8.5 Microplastic dynamics across pelagic and benthic compartments of San Pedro Bay, California, USA

8.5.1 Study region

Singh et al. (2025), monitoring of MPs in this study was conducted in the coastal waters of Southern California with particular emphasis on the San Pedro shelf. This shelf is part of the Southern California bight, a large embayment covering approximately 94,000 km^2(Sutula et al., 2021) that extends from point conception south to Baja, California. The San Pedro shelf is a gently inclined marine platform measuring roughly 40 km in width and 100 km in length, with depths approaching 900 m (Hickey, 1992). The seabed is dominated by soft sediments interspersed with low-relief rocky features and scattered boulders. Oceanographic conditions in the region are highly dynamic. Circulation is influenced by the southward-flowing California current, which transports subarctic Pacific waters, and the northward flowing California under current, carrying water masses from the Eastern tropical North Pacific (Kessouri et al., 2021). These major currents converge within the southern California bight, where coastal topography and atmospheric forcing further modify circulation patterns. Consequently, exhibits pronounce variability across inter annual, seasonal and shorter temporal scales.

The surrounding region experiences a Mediterranean-type climate, characterized by warm, dry summers with average temperatures of 20–24°C and mild wetter winters ranging from 8–14°C. Annual rainfall is highly variable, typically ranging between 635 and 1524 mm (Kauffman, 2003). San Pedro Bay is also strongly influenced by freshwater discharges from the Los Angeles and San Gabriel rivers, which together account for nearly 70% of total riverine input to the bight (Sutula et al., 2021). Ecologically the southern California bight supports high levels of marine biodiversity and delivers essential ecosystem services including climate regulation, recreation and productive fisheries. However, its proximity to one of North America's most densely populated and industrialized coastal regions increase its susceptibility to environmental stressors. These include overfishing, intense sport and shipping activities, climate change, invasive species, and widespread pollution, notably the accumulation of plastic debris and MPs.

8.5.2 Microplastic concentrations

MP concentrations within the water column of San Pedro Bay ranged from 0.03 to 8.71 particles/m^3. During the dry season, nearly all sampling stations exhibited higher MP levels than those recorded during wet season. Average concentrations were 1.29 particles/m^3 in the dry season and 0.90 particles/m^3 in the wet season; however, this seasonal difference was not statistically significant. The highest concentrations were observed at near shore locations, particularly in proximity to the outfalls of the Los Angeles, San Gabriel, and Santa Ana rivers. Mean near shore concentrations reached 1.78 particles/m^3 which was significantly greater ($P = .01$) than offshore values, where the mean concentration was 0.55 particles/m^3. Overall, MP abundance generally decreased with increasing distance from shore except at one site located just beyond the San Pedro shelf where concentrations were elevated related to other offshore stations. Distinct vertical patterns were also evident in the water column. Surface waters exhibited significantly higher MP concentrations (mean = 2.06 particles/m^3, $P = .02$) compared with mid water samples (0.53 particles/m^3) and epibenthic layers (0.75 particles/m^3). These trends suggest buoyancy-driven accumulation near the surface with progressively low concentrations at greater depths (Singh et al., 2025).

8.5.3 Microplastics in benthic sediments: Comparison with Pacific Ocean case studies

MP abundance in sediments collected from the San Pedro shelf ranged from 500 to 3000 particles/kg dry weight. No statistically significant seasonal difference was detected in sediment concentrations ($P = 0.72$). During the wet season, near shore sites exhibited higher MP levels whereas offshore sediments showed greater concentrations during the dry season. On average, MP abundance in nearshore sediments (1102 particles/kg) exceeded that of offshore sediments (416 particles/kg), although this difference was statistically significant ($P = 0.18$) the observed spatial distribution reflects the combined influence of riverine inputs, local hydro dynamics, and sediment transport and deposition processes on MP accumulation (Singh et al., 2025).

The arial concentration of MPs in San Pedro Bay measured at $8.64 \times 10^5 \pm 7.60 \times 10^5$ particles/km^2 using 333 μm mesh Manta nets, ranks among the highest reported for Pacific coastal and open-ocean sites. Comparable concentrations have been documented in San Fransico Bay (7×10^5 particles/km^2 (Sutton et al., 2016), highlighting the influence of densely urbanized regions. Lower concentrations have been reported in:

- the North Pacific Central Gyre (3.34×10^5 particles/km^2; Moore et al., 2001),
- the broader North Pacific Ocean (1.05×10^5 particles/km^2; Eriksen et al., 2014),
- the North Pacific sub-tropical Gyre (3.31×10^4 particles/km^2; Law et al., 2014) and
- the South Pacific sub-tropical Gyre (2.69×10^4 particles/km^2; Eriksen et al., 2013).

Volumetric concentrations in San Pedro Bay averaged 3.19 ± 2.96 particles/m^3, consistent with Lattin et al. (2004) who reported 3.29 particles/m^3. Surface-specific concentrations (2.06 ± 2.41

particles/m^3) where substantially higher than values reported in the Northeast Pacific (0.004 particles/m^3; Doyle et al., 2011) and mid-north Pacific 0.51 particles/m^3 (Pan et al., 2022) indicating elevated contamination in urbanized coastal zones.

8.6 Microplastic entrapment and redistribution in Alaskan Arctic landfast sea ice

The Arctic region is experiencing rapid environmental transformation, with surface temperatures rising between 3.1°C between 1971 and 2019 approximately 3 times the global average (AMAP, 2021; IPCC et al., 2021) warning has accelerated ice sheet loss, permafrost thaw an altered sea ice and snow cover, affecting biogeochemical cycles and ecosystem functional (Schuur et al., 2022; Hopwood et al., 2020). MPs now recognized as emerging pollutants, are increasingly reported throughout Arctic Sea water, sediments, sea ice, and snow (Zhang et al., 2023). Sea ice functions as both a temporary sink and transport medium for contaminants with MP concentrations ranging from 1.1×10^6 to 1.2×10^7 items/m^3 in the Farm strait and central Arctic (Peeken et al., 2018). Sources include atmospheric deposition, riverine inflows and ocean circulation (Bergmann et al., 2022) with ice drift modeling confirming basin scale transport (Kim et al., 2021). Melting sea ice releases stored MPs into marine systems contributing an estimated 5.1×10^{18} items to the Arctic Ocean (Zhang et al., 2023). At Point Barrow, Alaska, land fast ice dominates seasonal coastal processes, breaking as grounded ridges fail, and drifting westward in the Southern Beaufort Sea due to local North South pressure gradients and seasonal winds (Stabeno et al., 2018). MP contamination in coastal landfast sea ice and Lagoon Lake ice remains understudied, yet understanding its distribution is crucial for assessing ecological impacts in Arctic coastal systems under climate change. Ice core sampling near Point Barrow conducted between April and May 2019 provided insights into MP sources and transport pathways in this region.

8.6.1 Study area and ice sampling

Monitoring MPs in Arctic environments presents substantial challenges due to remoteness, extreme weather, and limited accessibility (Provencher et al., 2023). This study focussed on the coastal zone near Point Barrow, northern Alaska, where prevailing northeast winds and currents dominate local circulation. The orientation of shoreline features combined with seasonal ice variability creates a highly dynamic environment. In the southern Beaufort Sea, north-northeast winds contribute to a predominantly westward drift of sea ice (Jones et al., 2016). Between 23 April and May 15, 2019, research collected four landfast sea ice cores and three lagoon lake ice cores, with individual core depths ranging from 30 to 72.5 cm. Each core was sectioned at 15–20 cm intervals, yielding 10 landfast ice samples and nine lagoon lake ice samples. These intervals were chosen following methodological recommendations (Bergmann et al., 2023) to ensure sufficient meltwater (~1000 mL per section) for MP pretreatment and subsequent analysis.

8.6.2 Microplastic abundance, size and shape

MPs were detected in all sea ice and lake ice samples. In landfast sea ice near Point Barrow, concentrations ranged from 38.4 to 485.5 items/L, with a mean of 220.6 ± 140.1 items/L. Lagoon Lake ice from Elson Lagoon exhibited concentrations ranging from 23.4 to 478 items/L, with a mean of 148.9 ± 141.8 items/L (YuLan et al., 2025). Across both ice types, smaller MPs dominated. In sea ice, the 10–50 μm size class comprised an average of 66.7% of all particles (range: 30.7%–96.3%), with nearly all particles (<100 μm) making up 94.7 ± 6.8% of the total. Fragments in the 10–50 μm range were the most abundant particle type.

Lake ice showed a comparable pattern, with small particles (<50 μm) accounting for 40%–80% of all MP depending on depth. In sea ice, the middle layer had the highest proportion of 10–50 μm particles (92.6%), though fragment fractions were lower than those in top and bottom layers. In lake ice, the top layer contained fewer small particles, whereas the bottom layer approached 80%. Across all samples, fragments comprised 87.8% of MPs, with fibers representing 12.2%; in total, fragments accounted for approximately 95% of particles (YuLan et al., 2025). Fibers were most abundant in the middle sea-ice layer (16.8%), with no significant variation among lake-ice layers.

8.6.3 Chemical polymer composition

LDIR analysis revealed a diverse array of polymer types in all ice samples. Landfast sea ice was dominated by

- polyethylene (PE, 20.1%),
- polyamide (PA, 12.8%),
- rubber (12.7%),
- polyurethane (PU, 8.6%),
- polyethylene terephthalate (PET, 8.8%), and
- polypropylene (PP, 6.1%).

In contrast, lagoon lake ice primarily contained PE (21.8%) and rubber (44.3%), suggesting different polymer sources or environmental influences between the two ice samples. No consistent relationship was observed between polymer type and particle size or depth layer in either ice type (YuLan et al., 2025), indicating that polymer composition was not a primary factor determining MP occurrences in the sampled Arctic ice.

8.6.4 Abundance, origins, and seasonal release of microplastics in Arctic sea ice

Sea ice can incorporate MPs originating from surrounding seawater, atmospheric deposition, and surface-related inputs (Kim et al., 2021). With ongoing Arctic warming, ice melt facilitates the release of entrapped MPs back into marine systems, potentially affecting trophic interactions and food-web dynamics (Bergmann et al., 2023). Evidence of MP accumulation within ice-associated algae further suggests that algal communities may act as transfer pathways, conveying MPs to surface feeding organisms and potentially to deeper marine environments (Bergmann et al., 2023). MP concentrations observed in this study were several orders of magnitude higher than those reported from central Arctic Ocean ice cores (2–17 items/L; Kanhai et al., 2020), although values documented in Arctic snow ($\approx 9.8 \times 10^3$ items/L) remain considerably greater (Bergmann et al., 2019). Collectively, these results demonstrate the pervasive presence of MPs in northeastern Alaskan sea ice and emphasize the complexity of processes governing their transport, deposition, and retention. Regional variability across the Arctic likely reflects differences in oceanographic circulation, atmospheric inputs, and methodological approaches among studies.

Arctic coastal sea ice can receive MPs through a combination of long-distance transport mechanisms and local inputs, including ocean currents, atmospheric pathways and coastal human activities. Marine MPs are redistributed by surface circulation and wave dynamics (Shaw et al., 2023), while localized sources such as wastewater effluent, coastal infrastructure, recreational use, and fishing practices may further contribute to MP loading. Declining sea ice extent had coincided with increased shipping, fishing, and tourism across the Arctic, potentially enhancing plastic inputs to the region (Eriksen et al., 2020). For instance, discarded fishing gear constitutes approximately 65% of total plastic debris by mass in the Barents Sea (Novikov et al., 2021). Similarly, the Svalbard region receives significant MP inputs via the West Spitsbergen Current, which transports material from heavily fished areas of the North Atlantic (Bao et al., 2022).

Broad-scale Ocean connectivity also plays a critical role in shaping regional MPs distributions. MPs are widely documented in both the Berling Sea and the northeastern Pacific Ocean. Annual MP fluxes entering the Chukchi Sea from the Pacific Ocean are estimated to be 5.5 times greater than the standing stock of MPs withing Chukchi Sea waters (Ikenoue et al., 2023). Given that inflowing water masses through the Bering Strait strongly influence Chukchi Sea hydrography, a substantial proportion of MPs detected in this region likely originates from Pacific Ocean sources (Mu et al., 2019). Polymer compositions observed in both sea ice and lake ice samples indicate contributions from multiple source pathways. In sea ice, the predominate polymers included PA, PE, PET, PVC and rubber. PA is frequently linked to fishing-related materials (GESAMP, 2015). On a globe scale, PE and PP account for the largest share of plastic production (PlasticsEurope, 2023) and are commonly reported in the Arctic seawater and sea ice environments (Gong et al., 2023). PET is likewise prevalent in Arctic marine systems, while PVC has been identified as a significant component of MPs in both Arctic Sea ice and seawater (D'Angelo et al., 2023). The polymer profiles identified in Point Barrow Sea ice only partially reflect global production patterns, implying the influence of selective deposition mechanisms controlled by particle density, atmospheric transport, and nearshore hydrodynamic processes. Furthermore, similarities between Polymer assemblages at Point Barrow and those reported from the western Arctic Ocean and central Arctic Basin suggest strong connectivity with the Pacific-derived waters entering the Arctic through the Bering Strait (Kim et al., 2021).

8.6.5 Conclusions and future research

Accelerated Arctic warming is modifying sea ice formation, persistence and melt processes, thereby reshaping the pathways through which MPs accumulate, are transported, and are redistributed across the region. This study presents a comprehensive evaluation of MPs in landfast sea ice and lagoon lake ice from the coastal Alaskan Arctic.

The principal findings are as follows:

- MPs were prevalent in both landfast sea ice (220.6 ± 140.1 items/L) and lagoon lake ice (148.9 ± 141.8 items/L)
- Peak concentrations were detected in the basal layers of landfast ice and withing the middle to lower layers of lagoon lake ice.

- Fine size MPs (<50 μm) dominated both ice types, frequently accounting for more than 80% of total particle abundance.
- PA, PE, and PET were the most common polymers in sea ice, whereas PE and rubber predominated in lagoon lake ice.
- Comparisons with other Arctic investigations revealed pronounced regional variability, like driven by differences in ocean circulation, deposition mechanisms, and analytical methodologies (YuLan et al., 2025).

Collectively, these results demonstrate that MPs are restricted to surface waters but are distributed throughout the water column and embedded within ice matrices, underscoring the need for integrated, multi-compartment monitoring strategies. Substantial knowledge gaps remain concerning the relative contributions of oceanic transport, atmospheric deposition, and ice dynamics to observed MP distributions.

- Coordinated, multi-compartment sampling encompassing water, ice, sediments, and the atmosphere.
- High-resolution temporal monitoring to resolve seasonal and interannual variability.
- Source attribution approaches linking polymer types to Arctic and non-Arctic origins.
- Modeling efforts to assess connectivity between the Pacific inflow water and Arctic basin systems.
- Adopting a more systematic and collaborative research framework will be essential for advancing understanding of MP fate and transport within the rapidly evolving Alaskan Arctic environments.

8.7 Spatial distribution and risk assessment of microplastics in benthic fish of a major North American Estuary

8.7.1 Overview of microplastic pollution in benthic aquatic ecosystems

Benthic ecosystems, which occupy the bottom zones of rivers, lakes, estuaries, and coastal waters, function as major repositories for environmental contaminants, including MPs and persistent organic pollutants (Lenaker et al., 2021). These environments are ecologically significant, supporting diverse biological communities and numerous fish species that inhabit or forage near the sediment-water interface. Owing to their continuous interaction with sediments and overlying waters, benthic organisms may experience greater exposure to MPs than pelagic species (Keerthika et al., 2023). Despite growing global concern, substantial gaps persist in understanding the occurrence, spatial distribution, and environmental fate of MPs in benthic fish from freshwater and estuarine systems. The transport and behavior of MPs in aquatic environments are governed by particle-specific properties, including polymer composition, density, shape and size, as well as by environmental factors such as fluid density, hydrodynamic conditions, biofouling, aggregation, tidal influence, and wave action (Chubarenko et al., 2018). For instance, high-density MPs (>1.0 g/cm^3) tend to settle more rapidly and accumulate within benthic compartments (Chubarenko et al., 2016), thereby shaping exposure pathways for bottom-dwelling organisms, assessing MPs ingestion in benthic fish can therefore provide critical insight into the particle types posing the greatest ecological risks. Such knowledge is essential for environmental contamination assessments, ecological risk analyses, and the formulation of targeted management and policy responses.

8.7.2 Study area and environmental setting of the St. Lawrence River and Estuary

This investigation was carried out in the St. Lawrence River and Estuary (SLRE), Canada, a critical ecological corridor that connects the Laurentian Great Lakes with the Gulf of St. Lawrence. The estuarine system extends approximately 655 km from Lake Saint-Pierre to Pointe-des-Monts in Quebec and represents one of the largest and most dynamic aquatic environments in North America. The passage of MPs in SLRE waters and sediments has been documented in previous studies (Rowenczyk et al., 2022). Nevertheless, MP contamination in benthic fish remains insufficiently characterized, particularly within the higher energetic estuarine setting. Improved understanding of MP exposure in the SLRE is therefore essential for robust ecological risk assessments and development of effective management strategies.

Catfish (*Ictalurus punctatus*, $n = 26$) and tomcod (*Microgadus tomcod*, $n = 16$) were collected from the SLRE in the vicinity of Quebec City during a research cruise abroad the R/V Lampsilis conducted between 22 September and October 4, 2022. Sampling stations were selected to capture gradients in land-use influence, including upstream, downstream, and more distant locations, to evaluate the potential effects of urban activities on MP contamination patterns.

Catfish were sampled from three principal regions:

- Cap-Santé (CC1, $n = 10$),
- Québec City (CC2, $n = 1$; CC3, $n = 2$; CC4, $n = 4$),
- East of the ile d'Orleans (CC5, $n = 5$; CC6, $n = 3$; CC7, $n = 1$).

Tomcod were collected at two stations in Pointe-au-Pic (AT1, $n = 5$; AT2, $n = 10$) and one site at Rivière-du-Loup (AT3, $n = 1$).

Fish were captured using a bottom trawl deployed at depths ranging from 10 to 50 m, depending on local bathymetric conditions. Following retrieval, specimens were immediately wrapped in aluminum foil, placed in sealed containers, and stored at $-20°C$ for transport to the laboratory, minimizing the risk of contamination and degradation prior to analysis (Kabir et al., 2025).

8.7.3 Microplastic properties: Morphology, size distribution, color, and polymer type

8.7.3.1 Shape

Evaluation of MP morphology revealed pronounce differences in particle shapes among samples, with fibers constituting the most abundant form, followed by films and fragments. The prevalence of fibers is commonly attributed to their extensive release into aquatic environments from sources such as synthetic textiles, fishing gear, household laundering, and industrial discharges (GESAMP, 2016). Distinct species- and tissue-specific patterns were also evident. Catfish contained fibers, films, and fragments in both gastrointestinal tracts (GITS) and gill tissues. In contrast, tomcod was largely contaminated by fibers, with films detected only occasionally in GIT samples and fragments entirely absent from all tomcod specimens. These differences nicely reflect species-species anatomical and behavioral characteristics. The relatively larger mouth gape and gill structures of catfish may increase their exposure to a broader range of particle morphologies, a pattern consistent with observations reported by(Siddique et al., 2024).

Spatial variability was additionally observed across the SLRE. At upstream locations (e.g., CC1), catfish samples exclusively fibers, whereas downstream stations (CC2-7) exhibited a mixture of fibers, films and fragments. This pattern parallels earlier findings by Crew et al. (2020), who reported fiber-dominated contamination upstream near CC1 and higher proportions of fragments and films in downstream areas near Québec City. Such downstream enrichment likely reflects increased urban influence, wastewater inputs, and industrial activity. For tomcod, fibers remained the dominant shape across the dominant shape across all sampling sites, with films appearing only sporadically. The observed species- and location specific differences likely arise from combined effects of land-use patterns, hydrodynamic conditions, and habitat characteristics within the SLRE. Overall, the strong predominance of fibbers, together with marked spatial and biological variability, provides valuable insight into MP exposure pathways and potential ecological consequences. Future research should prioritize identifying the sources, transport pathways, and degradation processes of different MP shapes to better evaluate the ecological significance.

8.7.3.2 Size

MPs identified in this study were largely small in size, spanning a range of 26–809 μm. particles smaller than 1 mm, classified as small MPs Naji et al. (2019), dominated the assemblage detected in both intestinal trats (GITs) and gills of catfish and tomcod. This pattern is consistent with previous studies reporting a predominance of small-sized MP in benthic fish species (Lin et al., 2023). Particles within the 26–809 μm size range appear to be readily ingestible by both species, likely due to several interacting factors:

- First, high environmental availability may play a role, as the SLRE likely contains abundant MPs with size fractions, emphasizing the importance of simultaneous monitoring of sediments and overlying waters.
- Second, particles of the size closely resemble natural prey items such as zooplankton, phytoplankton, insect larvae, or contaminated prey organisms, increasing the probability of unintentional ingestion (Koongolla et al., 2020).
- Third, mechanical breakdown of larger plastic debris during feeding and digestive processes can generate smaller particles, further increasing their abundance (Zhang et al., 2021).
- Finally, small MPs often exhibit longer residence time within the GIT, which may enhance their detectability relative to larger particles (Roch et al., 2021).

Notably, differences in particle-size distribution were observed in gill tissues between the two species. Tomcod exhibited a higher occurrence of larger particles (>809 μm) in the gills, potentially reflecting specie-specific differences in respiratory filtration, gill morphology, or habitat use. The dominance of small MPs indicate that these particles are widely distributed within SLRE benthic environments as a result of fragmentation, hydrodynamic sorting, and secondary transport processes (Lin et al., 2023). Spatial variability in particle size across sampling locations further suggests the influence of localized pollution source, hydrodynamic conditions, and species-specific feeding behaviors. Small MPs are of ecological concern because of their enhanced potential for bioaccumulation, tissue retention, tropic transfer, and increased sure relativity (Zheng et al., 2020). Their widespread occurrence in benthic fish highlights the need for management measures aimed at limiting the input of small MPs into aquatic ecosystems.

8.7.3.3 Color

MPs recovered from catfish and tomcord exhibited a broad spectrum of colors, with blue particles representing the most abundant category, followed by transparent and white particles. Other colors, include red, black, violet, yellow, green, and gray, were detected at comparatively lower frequencies. This distribution mirrors global observations in which blue and transparent MPs commonly dominate benthic fish samples. The marked predominance of blue particles may be attributed to their visual similarity to natural prey or food associated cues. Foraging fish are known to ingest colored particles that resemble zooplanktons, algae, or small invertebrates, either directly or through secondary ingestion of contaminated prey (Ory et al., 2017). Such visual mimicry likely contributes to the frequent occurrence of blue MPs in both species examined. Color distribution also varied between tissues. Blue particles were more commonly detected in the GIT, consistent uptake through feeding, whereas transparent and white particles were more prevalent in gill tissues, possibly reflecting passive entrapment during respiration. These patterns suggest distinct exposure mechanisms for MP within fish. Spatial differences in color composition were also evident across the SLRE. Upstream locations (CC1) were characterized primarily by blue particles, while downstream sites (CC2-7) displayed greater color diversity, indicative of multiple pollution sources such as urban runoff, wastewater effluents, and industrial discharges. In tomcord, blue and black particles dominated most sampling stations, consistent with their benthic feeding behavior and habitat preferences. Although the lack of sediment color data in previous SLRE studies (Crew et al., 2020) limits direct comparisons, the observed patterns underscore the importance of characterizing sediment-associated MP to better link environmental contamination with ingestion profiles. Elucidating color-related ingestion patterns may aid in identifying MP types that pose elevated ecological risks and support the development of targeted mitigation strategies focused on the most visually attractive or environmentally relevant particles.

8.7.3.4 Polymer composition

A total of 11 polymer types were identified in catfish and tomcord, highlighting the heterogeneous nature of MP contamination within the SLRE. The dominant polymers- PET, PE, PP, and PS- reflect global plastic production patterns and are among the most frequently reported polymers in aquatic organisms. In addition, several less commonly detected polymers, including PCL, PMMA, PVC, ABS, PUR, and PA, were also present with higher diversity observed in catfish samples. Certain polymers such as EVA, PUR, and PA, were detected exclusively in catfish gill, indicating species-specific exposure pathways potentially related to differences in habitat use, feeding behavior, or respiratory filtration (Kabir et al., 2025). These observations align with findings reported by Rowenczyk et al. (2022), who documented polymers including PET, PP, PS, PUR, EVA and PVC in SLRE surface water and benthic bivalves. The overlap in polymer types identified in water, sediments, and fish tissues suggest that benthic organisms are primarily exposed to MPs originating from the immediate water-sediment interface.

The observed polymer diversity is likely governed by two principal factors. First, the environmental behavior of polymers-such as density, degradation processes, and hydrodynamic transport-strongly influence their partitioning among surface waters, sediments and biological compartments. Second, localized exposure to pollution sources within SLRE, including urban runoff, industrial effluents, wastewater discharges, maritime activities, and shoreline infrastructure, may introduce distinct polymer types, particularly in downstream areas. Collectively, the complex mixture of polymers underscores the need for enhanced monitoring of MP sources and transport pathways in the SLRE and highlight the importance of evaluating polymer-specific risks within benthic food web (Kabir et al., 2025).

8.7.4 Conclusion

This study provides a comprehensive baseline assessment of MP contamination in benthic fish from the St. Lawrance River and Estuary (SLRE), offering new insight into their occurrence, spatial heterogeneity, and associated ecological risk. Clear species-specific patterns were evident, with catfish exhibiting higher MP loads than tomcord, likely reflecting difference in feeding strategies, anatomical traits, and exposure routes. While tomcord appeared to be exposed predominantly through ingestion, catfish accumulated MPs via both gastrointestinal and gill pathways, emphasizing the relevance of multiple exposure mechanism in benthic predators. Spatial trends revealed elevated MP abundance at downstream sites near Quebec City, consistent with the influence of urban land use and intensified anthropogenic activity. Despite the predator-prey relationship between catfish and tomcord, there was no indication of MP biomagnification. Suggesting that tomcord consumption does not substantially enhance MP accumulation in catfish.

Across all samples, fibers represented the dominant MP shape, with small particles (<806 μm) and colors such as blue, red and black occurring most frequently. PET and PE were the most prevalent polymers, accompanied by a range of additional polymers-including PMMA, PUR, ABS, and PS-indicating inputs from diverse point and non-point sources. Fragment-shaped MPs were more common at downstream locations, further linking their occurrence to urban influences. High-density polymers (e.g., PVC, PMMA, PUR, ABS, PS) likely originated from sediment deposition followed by resuspension, thereby increasing exposure for benthic organisms. Ecological risk assessment results indicate that small-sized MPs (<809 μm) and high hazard polymers (e.g., PVC, PUR, PMMA, ABS, and PS) were the primary contributors to elevated ecological risk within the SLRE. The highest risk levels were observed downstream of Quebec City, underscoring the need for targeted management interventions aimed at reducing MP inputs from urban watersheds. Addressing remaining knowledge gaps will require an integrated, multi-compartment research approach encompassing sediments, water, and biota, along with expanded spatial and temporal monitoring. Future studies should also focus on hydrodynamic controls on MP transport and the long-term effects of MP exposure on fish health, trophic dynamics, and ecosystem functioning. Such efforts are critical for informing evidence-based policies to safeguard aquatic biodiversity and maintain the ecological integrity of the SLRE and comparable freshwater-estuarine systems worldwide.

8.8 Microplastic occurrence in mudflat sediments and varnish clams (*Nuttallia obscurata*) of British Columbia Estuaries

8.8.1 Study area and background: Microplastics in British Columbia's estuarine environments

MPs are increasingly recognized as emerging contaminants in British Columbia (BC), particularly in marine and estuarine environments (Alava, 2019). Costal studies along BC have reported elevated MP concentrations in surface waters, especially in nearshore areas affected by urbanization and industrial activities (Mahara et al., 2022). Evidence shows that zooplankton ingest MP, raising concerns about exposure of high tropic level species. For instance, adult pacific salmon may consume substantial quantities of MPs accumulation in BC marine food webs is influenced more by species-specific interaction than by tropic position alone (Covernton et al., 2022), emphasizing the need to understand habitat-and behavior-driven exposure pathways. Estuaries in BC are ecologically rich and provide essential habitats for diverse marine species, serving as critical spawning, nursery, and migratory areas. However, these systems are increasingly threatened by urban expansion, industrial discharge, agriculture runoff, and climate-related changes such as sea-level rise and altered water temperatures (Ross and Morales-Caselles, 2015).

8.8.2 Cowichan-Koksilah Estuary

The Cowichan-Koksilah Estuary is located on the east coast of Vancouver Island, between Victoria Nanaimo, formed by the confluence of the Koksilah River, originating from Waterloo Mountain, and the Cowichan River, which flows ~40 km from Cowichan Lake. Together, these rivers drain a watershed of nearly 8000 km^2, making it the eighth-largest estuary in BC. The area lies within the traditional territory of the Salish Peoples, particularly the Cowichan tribes, who have historically relied on the estuary's resources, including shellfish, salmon, herring roe, and seaweed (Schuerholz et al., 2006). The village of Cowichan Bay, near the estuary head, has ~2394 residents. Industrial activities including a metal manufacturing facility, sawmill, and marine shipping terminal-contribute potential pollutants such as untreated stormwater, mill effluent, sewage, and agriculture nutrient runoff (Alava et al., 2021), raising concerns about cumulative anthropogenic impacts.

8.8.3 Komoks Estuary

The komoks Estuary, historically known as the Courtenay River Estuary, lies downstream of the Puntledge and Tsolum river confluence. It provides critical habitat form numerous bird and fish species, with hydrology and ecological health largely determined by freshwater inflows from the Courtenay River, similar to the Cowichan Estuary System (Mackas et al., 2001). As the second most important estuary in BC after the Fraser River, Komoks supports 145 bird species (>70,000 individuals), 218 plant species, 29 fish species-including all five Pacific salmon species- and numerous intertidal organisms such as clams, worms, and microbes. However, industrial development, urban expansion, and shoreline construction (e.g., hospitals, long-term care facilities, resorts) have caused habitat loss, pollution, and long-term ecological degradation (Walters et al., 2019).

8.8.4 Sample collection and Cowichan Estuary

In July 2023, a comprehensive sediment sampling program was conducted in both estuaries to assess MP contamination and associated sediment characteristics. Bulk sediments were collected during low tide to maximize the exposure of intertidal mudflats. A total of 156 sediment samples were collected from 26 sites, selected to capture spatial variability of MP deposition on either side of the Westcan Terminal causeway, which divides the intertidal marsh and mudflats. Sampling locations included:

- South of the causeway (Koksilah River influence): S1–S4
- North of the causeway (Cowichan River influence): N1–N4 and C1–C6
- Additional sites: terminal sites T1, T2, T3, T5; Cowichan River outfalls CO1–CO3; north fork CO4-A, CO4-B; south fork CO5, CO6; Koksilah River K1-a, K1-B

This design enables a detailed assessment of MP distribution relative to multiple freshwater inputs and industrial activity (Zoveidadianpour et al., 2025).

8.8.5 Observed microplastic contamination in sediments and varnish clams

Analyses revealed substantial MP contamination across both estuaries. In Cowichan, sediments ($n = 78$) contained 1121 suspected particles, of which 74.4% (835 particles) were confirmed as MPs via spectroscopic analysis. Komoks sediments ($n = 27$) contained 836 suspected particles, with 52.08% (483 particles) confirmed. Varnish clams also showed significant exposure. Cowichan clams ($n = 57$) contained 112 suspected particles, with 67.40% (75 particles verified as MPs, whereas Komoks clams ($n = 103$) had 231 suspected particles, with 60.17% (139 particles) confirmed. Condition index measurements suggested difference in overall clam health between estuaries, with komoks clams showing a higher mean condition index (38.79 g/cm^3) than Cowichan clams (29.10 g/cm^3), indicating potentially better physiological status in the komoks population (Zoveidadianpour et al., 2025).

8.8.6 Microplastic abundance in sediments

MPs were detected in all sediment samples across both estuaries, though concentrations varied considerably.

8.8.6.1 Komoks Estuary

Sediments from the Komoks Estuary exhibited roughly double the MP concentration varied considerably:

- **Mean abundance:** 30.96 ± 14.58 particles/kg
- **Range:** 8–55 particles/kg

Concentrations differed significantly among sampling sites ($P < .05$). the highest levels were recorded at Komoks Marina (CM: 44 ± 9.65 particles/kg), followed by Goose Spit Park (GP) and Trent River (TR). Sites characterized by greater human activity, tourism, and population density showed elevated MP loads, highlighting the influence of anthropogenic pressure.

8.8.6.2 Cowichan Estuary

Sediments from the Cowichan Estuary contained lower MP levels:

- **Mean abundance:** 14.37 ± 11.57 particles/kg
- **Range:** 2–153 particles/kg

Although averages were lower, some sites-including T3 and central locations exhibited higher concentrations (22.11 ± 16.99 particles/kg and 20.17 ± 13.52 particles/kg respectively). One-way ANOVA revealed no significant differences among sites ($P = 0.2068$), expect for T3, which differed significantly in post hoc tests ($P < 0.05$). The lowest MP count (7 ± 2.898 particles/kg) was observed at K1.A in the Koksilah River (Zoveidadianpour et al., 2025).

8.8.6.3 Microplastic abundance in varnish clams

MPs of varying shapes and sizes were present in varnish clams from both estuaries.

- **Cowichan clams:** 3.62 ± 2.58 particles/g
- **Komoks clams:** 2.24 ± 1.96 particles/g

Interestingly, although Cowichan sediments had fewer MPs, clams from this estuary contained higher particle concentrations per gram of tissue, indicating differences in exposure pathways, bioavailability, and retention dynamics (Zoveidadianpour et al., 2025).

8.8.7 Characteristics of microplastic

8.8.7.1 Shape distribution

MP morphology varied between estuaries and sample types. In Cowichan sediments and clams, fibers predominated, accounting for 53.34% and 53.50% of sediments were dominated by fragments (53.70%), whereas fibers remained most abundant in clams (57.7%). The prevalence of fragments in komoks sediments likely reflects the breakdown of larger plastic debris, as indicated by irregular shapes, rough edges, and weathered surfaces typical of mechanical degradation (Andrady, 2011). Many fragments consisted of polyethylene (PE) and polypropylene (PP), common in consumer products, confirming that larger plastics are key contributors to fragment formation (Duis and Coors, 2016). High-density polyethylene (HDPE), known for its environmental persistence, likely contributes to fragment accumulation in komoks sediments, although targeted polymer-specific analyses are needed to confirm this (Kazmiruk et al., 2018).

8.8.7.2 Comparison with other sediment studies

The MP concentrations measured in Cowichan (14.37 $\pm$ 11.75 particles/kg) and komoks (30.96 $\pm$ 14.58 particles/kg) sediments are consistent with values reported globally in similar environments, including coastal BC, the Great Lakes, the Baltic Sea, San Francisco Bay, the English Channel, Dutch coasts, and Arctic regions. This demonstrates the widespread nature of MP contamination in estuarine sediments.

8.8.7.3 Comparison with clams elsewhere

MP concentration in Cowichan varnish clams (3.64 particles/kg) were lower than those reported in razor clams (*Siliqua patula*) from Oregon, Daya Bay (China), the Persian Gulf, and Yantai (China), but higher than levels observed in clams from the French English Channel, South Korea, Puducherry (India), Baynes Sound (BC), South Brazil, and other bivalve species in BC, Oregon, and Thailand. Komoks Estuary contained 2.92 ng/g, exceeding some prior BC and Hong Kong measurements, while remaining within the broader reported range (1.84–60.05 particles/g) for marine bivalve along the BC coast (Covernton et al., 2022).

8.8.7.4 Considerations and limitations

Although results align with global datasets, comparisons should be interpreted cautiously due to:

- Variations in sampling and extraction methods
- Differences in particle size detection threshold
- Environmental heterogeneity among habitats
- Species-specific feeding and filtration behaviors
- Temporal and seasonal variability
- Lack of standardized monitoring protocols

Such factors create variability that complicates direct global comparisons (Ding et al., 2022).

8.8.8 Summary of microplastic characteristics

This study confirms widespread MP contamination in both sediments and varnish clams from the Komoks and Cowichan estuaries, highlighting their integration into benthic habitats, food webs, and potential seafood pathways. Fibers were the most abundant from across samples, reflecting ongoing environmental degradation of larger plastics. Five major polymer types were identified: polyester (PES), polyethylene (PE), polypropylene terephthalate (PET), and polyamide (PA), with PES and PE being dominant, indicative of strong anthropogenic inputs. Elevated MP loads are linked to urbanization, wastewater discharge, tourism, and fishing activity. Spatial variation in polymer composition, particle size, and morphology demonstrated that estuarine MP contamination is dynamic and strongly influenced by human pressure and environmental processes. These findings underscore the need for continuous monitoring and targeted management strategies to mitigate MP pollution and protect the ecological health of BC estuarine environments.

8.9 Surface water contamination by MPs and anthropogenic microparticles in Great Slave Lake, Canada

8.9.1 Study site

Great Slave Lake, located in the northwest territories of Canada, is one of the largest freshwater systems in Mackenzi River Basin. Covering over 27,000 km^2, it is Canada's deepest lake, reaching 614 m in the east arm, and ranks among the deepest lakes in the world (Evans, 2000). The lake was formed through extensive glacial activity during the Pleistocene. Despite its size, it has active hydrology with a mean water residence time of around 16 years, largely due to the Slave River, which provides aboutn75% of total river inflow. Additional inflows come from tributaries such as the Taltson, Lockhart, and Hay rivers (Gibson et al., 2006). Compared with the Laurentian Great Lakes, the watershed around Great Slave Lake is sparsely populated and minimally industrialized. Yellowknife, with a population near 20,000, is the primary urban center. Smaller communities such as Dettah, Behchokoo, Lutsel Ke, Fort Resolution, Hay River, and Kattodeeche are scattered along the shoreline. To the west, the Mackenzie River flows out of Great Slave Lake to the Beaufort Sea, with Fort Providence near this outlet. These areas fall within the traditional lands of the Akitcho Dene First Nations, Dehcho First Nations, Tlicho Nation, and Northwest Territory Metis nation. Great Slave Lake holds significant cultural, spiritual and economic value, supporting transportation, subsistence harvesting and commercial fisheries (Stenekes et al., 2020).

Yellowknife Bay lies on the Northern shore of the lake, covering roughly 20 km^2 and stretching 11.3 km in length (Pelletier et al., 2020). It receives inflow from the Yellowknife River at the northern end, while Latham Island partly divides the bay, creating a sheltered section called Yellowknife Back Bay. Westwater from Yellowknife is treated a Fiddler's lagoon, an enhanced natural lagoon system typical of northern municipalities. Treated effluent flows annually through about 13 km of interconnected wetlands before reaching Great Slave Lake northwest of the bay, meaning there is no direct municipal wastewater discharge into the bay and its not a major MP source. Stormwater runoff, however, may act as a localized input. Yellowknife is divided into eight drainage catchments, two of which flow directly into Yellowknife Bay. These stormwater outlets can carry MPs and other human-derived particles, such as tire-wear debris into the bay.

8.9.2 Distribution and concentration of anthropogenic microparticles in surface waters

Anthropogenic MPs were detected in all 30 surface water samples from the Yellowknife River, Yellowknife Back Bay, and Yellowknife Outer Bay. Total concentrations (all particle types) ranged from 0.03 to 2.04 particles per cubic meter. Excluding fibers, concentrations ranged from 0 to 0.29 particles/m^3. the overall mean concentration was 0.42 ± 0.040 particle/m^3 for all particles, and 0.07 ± 0.08 particles/m^3 when fibers were excluded (Madelaine et al., 2024).

Average concentrations by region were:

- Yellowknife River: 0.51 ± 0.65 particles/m^3
- Yellowknife Back Bay: 0.25 ± 0.14 particles/m^3
- Yellowknife Outer Bay: 0.52 ± 0.10 particles/m^3

Sites near the Yellowknife River were extracted from all surface water samples. Fibers were dominant, accounting for 87% ($n = 883$), followed by fragments (11%, $n = 112$), films (1% $n = 7$), and foams (1%, $n = 10$). No paint particles were observed, suggesting sampling equipment did not contribute contamination.

8.9.3 Particle characteristics

A total of 1012 particles were extracted from all surface water samples. Fiber's dominant, accounting for 87% ($n = 883$), followed by fragments (11%, n = 112), films (1%, n = 7), and foams (1%, $n = 10$). No paint particles were observed, suggesting sampling equipment did not contribute contamination. ATR-FTIR spectroscopy was applied to about 10% of suspected particles. Of 136 particles analyzed, 79 particles (7.8% of the total) with match score >60% were used for polymer identification. Excluded particles were highly likely anthropogenic based on visual characteristics, such as blue, pink, or red fibers.

Fifteen polymer types were confirmed among identified particles. The most frequent were

- Low-density polyethylene (LDPE): 34%
- Polyethylene terephthalate (PET): 34%

Other polymers included anthropogenic cellulosics, polypropylene (PP), ethylene copolymers, nylon, polystyrene, and several polyacrylate-or isocyanate-based polymers. Polymer diversity was highest in Yellowknife Back Bay (9 types), followed by Outer Bay (7 types), and Yellowknife River (6 types). Fiber lengths ranged from 306 to 17,247 μm, with 70% below 5 mm. fragment, film, and foam surface areas ranged from 0.00,067 to 9.57 mm^2, demonstrating considerable morphological variation (Madelaine et al., 2024).

8.9.4 Spatial trends in microparticle distribution

High microparticle concentrations at river sites near the Territorial Park boat launch were almost entirely fibers (98%), likely reflecting boating, shore activities, or transient hydrodynamic effects. Fibers were present a all sites, including upstream river locations with minimal direct human influence, suggesting atmospheric decomposition is key pathway for microfibers to reach remote northern freshwater systems. Previous studies document long-range atmospheric transport of plastics to Arctic regions, with deposition occurring within days to weeks or dry fallout. Non-fiber particles (fragments, foams, films) were more frequently in Back Bay than in other regions, suggesting localized sources. While municipal wastewater is not directly discharged into the bay, two stormwater outlets covey runoff from urban catchments, potentially acting as point source for non-fiber microparticles. Further monitoring of these inputs is warranted (Madelaine et al., 2024).

8.9.5 Potential influence of sedimentation

Non-fiber particles in outer bay were present at lower concentrations than in Back Bay, possibly due to sedimentation before reaching the outer bay, plastics with densities $\geq$ water, or those accumulating biofilms, can sink into sediments. This sedimentation process has been observed in Arctic marine sediments, including deep-sea environments. Examining Yellowknife Bay sediments would clarify whether the benthic environment functions as a major sink for anthropogenic microparticles.

8.9.6 Comparison with other freshwater and arctic systems

In the Laurentian Great Lakes, mean MP concentrations reported since 2013 range from 0.02 to 0.66 particles/m^3, with fibers typically dominating. Yellowknife Bay concentrations were similar to Lake Erie (0.03–2.91 particles/m^3) but slightly higher than open water areas of the other Great Lakes, likely due to the bay's smaller, semi-enclosed nature. Lake Winnipeg, Canada's sixth-largest lake, recorded higher MP levels (0.58–8.31 particles/m^3). Differences among studies are influenced by sampling methods, mesh sizes, and whether fibers are included in counts. Globally, Yellowknife Bay's MP concentrations are lower than heavily polluted lakes but align with other sub-Arctic freshwater systems. Lake Kallavesi, Finland, reported 0.27 ± 0.18 particles/m^3, also fiber dominated. Arctic marine surface waters (e.g., Canadian Arctic Archipelago, Svalbard, Greenland, Fram Strait) typically range from 0.06–0.34 particles/m^3 and show fiber dominance when comparable sampling methods are applied (Madelaine et al., 2024).

8.10 Distribution of microplastics in coastal sediments and mussels along the British Columbia Coast, Canada

MP contamination has been increasingly documented along the British Columbia Coast, where complex Oceanographic conditions, urbanized shorelines, and diverse ecosystem interact to influence particle distribution and biological exposure. Studies conducted across this region consistently demonstrate the presence of MPs in both coastal sediments and mussels, highlighting the close coupling between environmental reservoirs and benthic-feeding organisms. Sediments along the British Columbia coast act as important sinks foe MPs, particularly in sheltered bays, estuaries, and nearshore environments characterize by low hydrodynamic energy. Fine fragments and fibers dominate sediment samples, reflecting the breakdown of larger plastic debris and the deposition of particles transported via riverine discharge and coastal circulation. Spatial variability in sediment MP abundance is strongly associated with proximity to urban centers, wastewater outfalls, and river mouths, emphasizing the role of land-based sources.

Mussels, as sessile filter-feeding organisms, provide valuable insight into MP bioavailability within coastal waters. MPs detected in mussels are typically small and largely composed of fibers and fragments, consistent with particles suspended in the surrounding water column, Variations in MP burden among sampling locations suggest that local environmental conditions, including water circulation patterns and suspended particle concentrations, strongly influence biological uptake. Comparative analysis of sediments and mussels indicates that while sediments serve as long term repositories for MPs, mussels reflect more immediate exposure conditions. This sediment-biota linkage underscores the importance of integrated monitoring approaches that asses both abiotic and biological compartments when evaluating coastal MP contamination. The presence of MPs in mussels also raises concerns regarding trophic transfer and potential implications for higher-level consumers, including species of commercial and ecological importance. Overall, evidence from the British Columbia coast highlights the persistence and widespread distribution of MPs in coastal environments. Integrating sediment and biomonitoring data provides a more comprehensive understanding of MP pathways, exposure dynamics, and ecological relevance in temperate coastal ecosystems.

8.11 Microplastics in British Columbia Coastal sediments: Results and interpretation

In this investigation, researchers visually identified 162 suspected MPs (SMPs) from 36 sediment samples, of which 52.5% underwent spectroscopic analysis. Only 36 of these particles were confirmed as plastics, representing 43% of the analyzed portion. In mussel samples ($n = 29$), 483 SMPS were visually observed, with 67.6% (328 particles) analyzed, and 47 particles verified as plastics (9.7%). The considerable gap between visually detected SMPs and spectroscopically verified plastics highlights the limitations of virtual assessment alone (Noel et al., 2022). These findings underscore the importance of FTIR or Raman spectroscopy for accurately classifying MPs, particularly in complex environmental matrices.

8.11.1 Microplastics in sediment along the British Columbia Coast

Sediments are recognized as long-term repositories for MPs and other pollutants because they accumulate particles through settling and hydrodynamic processes (Nakki et al., 2019). MPs were detected in 32 out of 36 sediment samples, with concentrations ranging from 0 to 141.6 particles/kg and an overall average of 32.6 ± 5.3 particles/kg:

- Spatial differences in sediment contamination
- MPs concentrations differed significantly between regions ($P < 0.05$)

Victoria Harbor had the highest levels (102.5 ± 23.6 MP/kg) followed by the North Coast (68.2 ± 8.2 MP/kg). Both areas include harbor environments (Victoria Harbor and Prince Rupert Harbor) where limited water exchange can enhance MP retention (Noel et al., 2022). Comparable patterns have been observed elsewhere, as harbors and enclosed coastal zones tend to accumulate more MPs due to restricted flushing (Claessens et al., 2011). Additionally, oceanographic conditions near the North Coast, such as circulation in Queen Charlotte Sound and around Haida Gwaii, favor the retention of floating plastics, which contributes to higher sediment concentrations.

8.11.2 Microplastic shapes and particle size in sediments and mussels

- Sediments: 70.3% fibers
- Museels: 73.1% fibers

Fragments were the second most frequent form (21.6% in sediments; 25% in mussels). Sheets accounted for 8.1% of sediment MPss, while microbeads (1.9%) appeared only in mussels. Foam, granules and pellets were occasionally present but not confirmed as plastics via FTIR. The widespread occurrence of microfibres (<1000 μm) in both sediments and mussels aligns with global observations. Fibers commonly originate from:

- The washing and degradation of synthetic textiles,
- Fishing and aquaculture gear,
- Atmospheric deposition, and
- Fragmented maritime ropes and lines

Their high mobility, buoyancy, and ability to remain suspended facilitate broad dispersal sampling sites. The uniformity of MP shapes indicates:

- Multiple diffuse sources,
- Large-scale atmospheric and coastal transport processes and consistent exposure pathways across regions.
- Consistent exposure pathways across regions (Noel et al., 2022).

8.11.3 Ecological implications of microplastic contamination in sediments and mussels: Insights from polymer composition

The predominance of fibers in both sediments and mussels indicates that filter-feeding organisms interact closely with the most abundant environmental particle types. Mussel contamination likely reflects:

- MP availability in nearby sediments,
- Resuspension of fine particles during tidal changes, and
- Direct ingestion from the water column.

Since mussels are frequently used as bioindicators of coastal pollution, these results reinforce their value for long-term monitoring the BC coast. However, the relatively low confirmation rate of plastics in mussels (9.7%) compared to sediments (43) emphasizes the need for more rigorous analytical approaches and standardized protocols, especially when studies rely on visual particle selection.

Polymer analysis revealed that polyester was the most common polymer in both sediments and mussels, accounting for 54.1% and 63.5% of confirmed MPs, respectively. Polyethylene was the second most frequent polymer, representing 16.2% of sediment MPs and 11.5% of those found in mussels. In sediments, the remaining MP polymers were mainly nylon and polystyrene, each contributing 13.5 of confirmed particles. In mussels, however, nylon (19%) and acrylic (7.7%) were significant additional components. A single rubber particle was detected in one sediment sample, though such occurrences were rare. No clear spatial patterns were observed in polymer type distribution across sampling regions. This may be partly due to the analytical approach: only spectroscopically confirmed MPs from a randomized subset of virtually observed SMPs were analyzed, meaning the dataset may not fully capture geographical variation in polymer types (Noel et al., 2022).

8.11.4 Conclusion

The Pollution Tracker monitoring program enabled, for the first time, the analysis of MPs in sediment and mussel samples along the entire BC coast using standardized methods. Victoria harbor and the North Coast (Prince Rupert area) were identified as MP hotspots, likely due to a combination of local sources and oceanographic factors. Although only a subset of SMPs was analyzed with FTIR due to cost and time constraints, limiting detailed geographic assessment of polymer patterns, the study provided insights into the polymer composition of MPs in sediment and mussels along the coast, identified polyester and polyethylene as dominant. This research demonstrated the widespread distribution of MPs in BC sediments and highlighted textiles as a major source of microfibres entering the marine environment.

8.12 Role of ecological traits in shaping microplastic accumulation in commercial estuarine crabs from the Southeastern Gulf of Mexico

8.12.1 Study site. Crab collection and tissue processing

This research focussed on evaluating MP accumulation in tissues of intertidal crab species of ecological and commercial importance: the stone crab (Menippe merceneria) and the blue crab (*Callinectes sapidus*), inhabiting Lagina de Terminos, a major coastal lagoon in southwestern Gulf of Mexico. MP concentrations were measured in the gills, digestive tract (DT), and muscle, and compared with MP types (shape, color polymer) present in surrounding water and sediments. The hypothesis was that MP abundance and traits in crab tissues would mirror the characteristics of the environmental matrices they most frequently contact (D'Costa, 2022).

Laguna de Terminos is Mexico's largest coastal lagoon (~7050 km^2) (INEGI, 1997) and is recognized as a Ramsar Site and Wildfire Protection Area (RSIS, 2006). It receives large freshwater inflows from the Paizada, Chumpan and Candelaria rivers, which carry sediments, nutrients, and land-based pollutants from agriculture, livestock, and other human activities. Sampling occurred at the Candeleria River inlet, a major freshwater entry, which drains a ~13,000 km^2 watershed extending across southern Campeche and parts of Guatemala. This extensive hydrological network increases vulnerability to land-based contaminants, including MPs (Capparelli et al., 2022). Forty adult crabs (20 per species, mixed sexes) were collected with site water. Upon arrival at the lab, they were cryo-anesthetized on crushed ice for 10 min before dissection. Gills, muscles and digestive tract (esophagus, stomach, intestine) were carefully removed and stored at −20°C until the MP extraction.

8.12.2 Microplastic content in water and sediment

MPs were found in all environmental matrices, with sediments showing much higher concentrations than water. Mean MP abundance in surface water was 1.74 ± 3.65 MP/L, whereas sediments contained 586.35 ± 29.32 MP/g, confirming the lagoon's role as a major MP sink. In water, 82% of particles were retained by a 100-μm mesh, indicating dominance of small particles. Sediments exhibited a bimodal size distribution: 32.8% of MPs >2 mm and 32.7% between 0.063–0.125 mm, reflecting fragmentation of large plastics and deposition of finer particles. Transparent MPs predominated in both matrices: 56.17% in water and 91.76% in sediments. These clear particles, likely from packaging, resist photodegradation, supporting persistence in aquatic systems.

Shape distribution differed between matrices. Sediments were overwhelmingly fiber-dominated (98.5%), whereas water contained mostly fragments (74.5%) and fewer fibers (25.5%). This reflects transport dynamic: low density fragments float and accumulate in water, whereas fibers, with higher surface-area-to-volume ratios, entangle with suspended solids and settle via turbidity-driven processes. Laguna de Terminos, a shallow lagoon (~2.4 m depth), retains plastics due to low flushing and continuous urban/agricultural inputs. Fragments likely originate from degrading macroplastics, while fibers may derive from untreated or inadequately treated wastewater (Capparelli et al., 2022).

8.12.3 Microplastic accumulation in crab tissues

Both crab species contained MPs in all tissues analyzed. Overall, *M.mercenaria* had nearly had nearly double the MP abundance compared to *C. sapidus*, indicating species-specific differences in feeding, habitat or filtration behavior. Tissue-specific differences showed *C. sapidus* had the highest MP load in gills (37.9 MP/g). while M. merceneria accumulated most MPs in the digestive tract (76 MP/g). Other tissues contained lower amounts:

- *C. sapidus*: DT = 8.62 MP/g, muscle = 6.89 MP/g
- *M. mercenaria*: gills = 37.91 MP/g, muscle = 20.25 MP/g

No significant differences were observed among tissues within each species ($P > 0.25$). Particle sizes varied by species: M. mercenaria ingested smaller MPs (0.9–1.7 mm), while C. sapidus had larger items (2.0–3.8 mm), likely reflecting feeding strategies or sediment contact.

Fragments dominated MPs in both species, particularly in *M. mercenaria*:

- *M. mercenaria*: 102 fragments/g; 18 fibers/g.
- *C. sapidus*: 41 fragments/g, 12 fibers/g.

Fragment differences between both species were significant ($P = 0.008$), whereas fiber differences were not ($P > 0.99$). In *M. mercenaria*, fiber versus fragment differences in the digestive tract were significant ($P < 0.001$), but not in other tissues or in *C. sapidus* (Capparelli et al., 2022). These results suggest species-specific exposure pathways: *C. sapidus* likely accumulates MPs through gill filtration, while *M. mercenaria*, a benthic forager, ingests MPs during sediment feeding, resulting in higher digestive tract accumulation.

8.12.4 Microplastics: Chemical composition and environmental samples

Across all MP samples, 37 chemical compounds were identified. Both crab species predominantly contained Silopren (SI), a silicone-based elastomer: 53.1% in *C. sapidus* and 47.8% in *M. mercenaria*. Polyethylene (PE) was the second most abundant polymer (12%–13%), consistent with common packaging materials:

- Water: dominated by polyethyl cyanoacrylate) (PECA, 23.5%) and poly (butadiene: styrene (POPS, 13.7%).
- Sediments: exclusively Silopren (100%), indicating localized industrial input

Silopren prevalence is plausible due to the high density of petrochemical industries nearby. Silicone-based industrial products may therefore represent a major MP source to the lagoon (Capparelli et al., 2022).

8.13 Microplastic contamination in cultured oysters: Seasonal and spatial patterns in Baja California, Mexico

8.13.1 Study rationale, hypotheses and area

This study measured MPs in Pacific Ocean (*Magallana gigas*) cultured in Todos Santos Bay (TSB) and San Quintin Bay (SQB) during winter and summer 2019, to evaluate human exposure through oyster consumption. The hypothesis was that TSB oysters would contain higher MP loads than SQB oysters, due to TSB's greater population density (≈10x higher) and more intensive coastal activities contributing anthropogenic inputs (Eduardo et al., 2021). Eight oyster aquaculture sites were sampled in Baja California during late winter (March) and summer (August) 2019: two in TSB and six in SQB.

8.13.2 Todos Santos Bay and San Quintin Bay

Todos Santos Bay (TSB) is in northwestern Baja California and is a semiconductor embayment covering roughly 330 km^2. The bay has an average depth of ~50m, with a submarine canyon descending to nearly 400 m. To the southeast, it connects to the Punat Banda Estuary through a sandbar with a single tidal inlet. Bay hydrodynamics are influenced by the California Current System, wind forcing, and tidal exchange, resulting in an estimated water residence time of 10–16 days (Cervantes-Audelo, 2013). The nearby city of Ensenada (~500,000 residents) (INEGI, 2017) exerts a significant anthropogenic influence, Key inputs include treated effluent from three wastewater plants—El Gallo (WTPG), El Naranjo (WTPN), and EL Sauzal (WTPS)- as well as agricultural runoff. Freshwater flows from Ensenada and El Gallo creeks occur mainly during the winter rainy season (December–February), while flows are minimal for most of the year. TSB supports multiple human activities, including fishing, aquaculture, and commercial port operations.

San Quintin Bay (SQB) lies ~200 km south of TSB along Baja California's Pacific coast. It is a shaloow, Y shaped coastal lagoon with two main arms connected to the ocean through a narrow (~1 km wide) inlet. The eastern arm extends ~11.5 km, and the western arm ~5.6 km. Covering ~42 km^2, SQB has a mean depth of 2 m, deepening to ~15 m near the entrance. Unlike TSB, SQB lacks consistent freshwater inflows; rainfall and runoff during winter are limited, and no wastewater is discharged directly. Water residence time varies from ~16 days in the upper lagoon to just a few hours near the mouth (Zertuche-Chanes, 2016). Extensive Zostera marina seagrass meadows cover up to 45% of the lagoon, peaking in summer and declining in winter. The surrounding region supports ~66,000 residents and is important for fishing, oyster farming and agriculture.

8.13.3 Preliminary microplastic assessment and concentrations in oysters

Across all oyster samples, 1945 particles were initially identified as suspected MPs. Fibers dominated (93.3%), while fragments comprised 6.7%; no pellets or films were detected. Seasonal and spatial patterns varied: in TSB, 359 suspected MPs were observed in winter and 195 in summer, whereas SQB recorded 909 in winter and 482 in summer. Approximately 60% of winter and 70% of summer samples were analyzed using μ-FTIR-ATR to determine polymer types. Procedural blanks indicated minimal laboratory contamination (~1% polyamide fibers, remainder non-synthetic), confirming low interference from lab processes (Eduardo et al., 2021). MPs were detected in oyster sample from TSB and SQB during winter and summer. In TSB, winter concentrations were significantly higher (0.07 ± 0.03 MP/g wet tissue), while summer values were lower. SQB showed a similar seasonal trend, with higher MP levels in winter (0.04 ± 0.02 MP/g wet tissue). Nested ANOVA indicated significant differences between bays during winter ($P < 0.01$), with TSB showing the highest wintertime concentrations. Seasonal differences in SQB were not statistically significant.

8.13.4 Microplastics in digestive system and soft tissue

MP distribution between digestive system (DS) and soft tissue (ST) was assessed only in summer-collected oysters. In TSB, oysters had the highest MP loads in both tissues:

- DS: 0.22 ± 0.20 MPs/organism
- ST: 0.38 ± 0.14 MPs/organism

In SQB, the highest MP concentration in soft tissue was 0.31 ± 0.10 MPs/organism.

8.13.5 Polymer composition of MPs

Eight polymer types were identified: seven synthetic and one semi-synthetic. The synthetic polymers included:

- Polyester/Polyethylene terephthalate (PET)
- Polyacrylonitrile (PAN)
- Polyethylene (PE)
- Polypropylene (PP)
- Polystyrene (PS)
- Polyamide (PA)
- Thermoplastic elastomer (T. elastomer)

The semi-synthetic polymer was rayon. Across both bays and seasons, PET, PAN, and rayon together accounted for $\geq$80% of all MPs. Analysis of aquaculture gear indicated that PP and PE were the dominant polymers, suggesting these as potential environmental sources (Eduardo et al., 2021).

8.13.6 Morphology, size, color of microplastics and conclusions

Only fibers and fragments were detected. Fibers accounted for 92% of MPs in TSB and 93% in SQB, consistent across both seasons. MPs were divided into five size classes: over 75% of fibers exceeded

250 μm, while all fragments were <500 μm, indicating extensive fragmentation or environmental degradation. Ten colors were observed, with black, transparent, and blue bing most common, matching typical textile fibers and degraded fishing or aquaculture materials (Eduardo et al., 2021). This study provides the first evidence of MP contamination in *Magillana gigas* cultivated in TSB and SQB. MPs were detected in all oysters, but concentrations were relatively low, suggesting current oyster consumption poses minimal human exposure risk. The results provide a baseline for MP occurrence, potential sources, and spatial distribution in Baja, California aquaculture. This information will support long-term monitoring and inform management strategies. Further research is recommended to:

- Identify sources and transport mechanism of MPs in both bays.
- Assess biological impacts of MPs on oysters and other local marine species.

8.14 Distribution of microplastics in abiotic and biotic compartments of two freshwater ecosystems in Pakistan

8.14.1 Environmental sample collection: Air, water, plankton, and biota

Sample collection: Abiotic and biotic components from each tropic level were examined for MP contamination samples of air, water, sediment, plankton, fish, and birds (n = 5 for each) were collected at the river site (Qaiser et al., 2023).

Air sampling: Airborne MPs were captured using petri dishes lined with 45 μm cellulose membrane filters (90 mm diameter), moistened with distilled water. Plates were left exposed for 30 min at 1 m above the ground (Wesch et al., 2017). Filters were rinsed with deionized water and checked microscopically to ensure no pre-existing contamination. Plates were sealed except during sampling to avoid external contamination.

Water and sediment sampling: Surface water samples (1L) were collected in sterile glass bottles, while top-layer sediments (1 kg) were gathered from shallow sites using pre labeled glass containers.

Plankton sampling: Plankton were collected by towing plankton nets across the water surface for 5 min from a boat. The material was then transferred into clean glass jars for laboratory processing.

Fish sampling: *Catla catla*, *Labeo rohita* and *Cirrhinus mrigala* ($n = 5$ each) were obtained from both the river and aquaculture ponds. All specimens were stored at −4°C in iceboxes until examination.

Avian sampling: The white-breasted kingfisher (*Halcyon smyrnensis*) was selected as the top trophic organism because fish constitutes more than 60% of its diet (Naher and Sarker, 2014).

8.14.2 Sample processing and analysis

Air samples: Airborne MPs were collected on moistened cellulose filters and examined under a stereomicroscope. This method minimized sample handling and reduced the risk of contamination. Water and sediment samples: Samples were sieved using a stacked system with mesh sizes of 0.2 mm, and 0.05 mm to separate particles by size. The retained material was dried in an oven and weighed. Organic matter was digested by incubating samples in 10% KOH at 60°C overnight (Rochman et al., 2015), followed by filtration, drying, and density separation to enhance MP recovery (Masura et al., 2015).

***Fish and avian samples*:** fish were dissected to separate flesh, gills, and the gastrointestinal tract (GIT). Each tissue was digested with 10% KOH, filtered, dried and examined under a stereomicroscope at 40× magnification. Avian tissues, including flesh, lungs, and intestines, were processed following the same procedure.

***Identification and classification*:** MPs were usually visually identified under a stereomicroscope based on Hidalgo-Ruz et al. (2012). Particles with certain identity were confirmed using the hot needle test. MPs were classified as fibers or fragments depending on their physical appearance. Further chemical analysis was not possible due to COVID-19 related lab closures.

***Contamination Control*:** To Prevent contamination, all equipment (Forceps, petri plates, funnels, beakers) was rinsed thrice with deionized water before and after use. Samples were handles in a controlled environment to reduce airborne contamination. MPs were detected in nearly all samples, with river samples showing the highest mean concentrations. Key observations included:

- Airborne MPs: River Ravi – 15.20 ± 3.35 items; Fishponds – 11.20 ± 1.89 items.
- Water and sediment: River samples had higher MP loads than pond samples across all size categories.
- Plankton: MPs were found in both environments (River – 2.8 ± 1.79n items; Ponds – 3.0 ± 1.58 items).
- Abiotic samples: Micro fragments accounted for over 80%, while microfibers were more common in GIT samples.

***Biotic Samples*:** Three freshwater fish species and one avian species were examined:

- MPs were more concentrated in the GIT than in gills or lungs.
- Flesh samples contained no MPs.
- MPs were absent in some gill (6/15) and lung (5/10) samples but present in all GIT samples (Qaiser et al., 2023).

MPs were identified in almost all samples collected from both study locations, with riverine samples showing the highest average concentration. Among the environmental matrices samples showing the highest average concentration. Among the environmental matrices, the mean airborne MP concentration at River Ravi was 15.20 ± 3.35 items, whereas a lower value of 11.20 ± 1.89 items was recorded at the UVAS fishponds. In addition, MP concentrations in water and sediment from River Ravi were markedly greater than those observed in the controlled pond system, irrespective of particle size.

MPs were also detected in plankton from both settings with average abundance of 3.0 ± 1.58 items in pond plankton and 2.8 ± 1.79 items in river plankton. Across all abiotic samples, micro fragments were the predominant form (>80%), while micro fibers occurred more frequently in gastrointestinal tract (GIT) samples. Two-way ANOVA results demonstrated that sampling location had a significant effect on MP concentrations in both water and sediment.

Overall, MPs were found at all trophic levels, with concentrations decreasing along the food chain. River samples consistently had higher MP loads than pond samples. Two-way ANOVA indicated that the sampling location significantly influences MP concentrations in water and sediment.

8.14.3 Conclusion

MPs are widespread contaminants in aquatic ecosystems. This study compared MP occurrence in a wild river system (Ravi) and a controlled pond environment in Punjab. River samples had significantly higher MP concentrations, suggesting potential irrigation water contamination and ecological risks. MPs were primarily detected in the GIT of fish and birds, indicating ingestion as the main exposure route. Although MPs were found across multiple trophic levels, their decreasing trend along the food chain suggests biomagnification under current conditions. This study highlights the need for further research on:

- Bioaccumulation and biomagnification
- Threshold exposure levels causing adverse effects
- Implications for food safety and ecosystem health.

Further studies should also identify MP sources an develop strategies to reduce plastic emissions into aquatic systems.

8.15 Occurrence and distribution of microplastics in Indian freshwater environment

8.15.1 Introduction

India exhibits a high susceptibility to MP contamination owing to its extensive network of rivers, lakes, wetlands, and reservoirs. These freshwater systems play a critical role in maintaining biodiversity, supplying portable water, and sustaining agricultural production (Lim, 2021). Despite their fundamental importance, the extent, spatial variability, and impacts of MP pollution within Indian freshwater environments remain inadequately characterized (Thacharodi et al., 2024). To date, most MP-related studies in India have concentrated on marine domains, including coastal waters, sediments and marine biota, resulting in a relative neglect of inland freshwater ecosystems (Zhou et al., 2019). Consequently, expanding scientific investigations into freshwater systems is vital for developing a comprehensive understanding of MP sources, transport pathways, and associated risks in these essential environments.

Recent research has identified multiple key sources contributing to MP contamination in India, such as industrial effluents, improper plastic waste disposal practices, and runoff linked to agricultural activities (Borah et al., 2024). Urban rover systems frequently function as conduits, transporting MPs to downstream lakes, reservoirs, and wetlands where plastic particles tend to accumulate. In rural settings, the widespread application of plastic materials in agricultural practices further exacerbates MP inputs. Although these pollution sources are increasingly acknowledged, systematic and large-scale datasets on MP abundance, spatial distribution, and ecological consequences across Indian freshwater systems remain scarce. In the absence of coordinated mitigation strategies, MP pollution is projected to intensity due to elevated plastic consumption and persistent shortcomings in waste management infrastructure (Sharma et al., 2021).

Addressing MP pollution effectively necessitates improvements in plastic waste management, particularly within urban areas, along with stricter enforcement of industrial discharge regulations and enhanced public awareness regarding the environmental implications of plastic use. MP concentration in freshwater ecosystems constitutes a significant environmental concern with direct consequences for ecosystem functioning and human health. Although recognition of the issue has increased, substantial knowledge gaps persist with respect to MP source, transport processes, and long-term effects, especially in rapidly industrializing and densely populated nations such as India (Yang et al., 2020). In the

Indian context, accelerated industrial growth, urbanization, and ineffective waste management practices have collectively contributed to escalating MP contamination in freshwater environments (Gupta et al., 2024). While global research on MP pollution is advancing rapidly, investigations centered on Indian freshwater systems-particularly those examining pollution sources, ecological consequences, and integrated evaluations on MP toxicity, interactions with co-existing contaminants, and the socioeconomic factors that perpetuate pollution.

By undertaking a comprehensive assessment of MP pollution in Indian freshwater systems, identifying dominant contributors such as urban wastewater, industrial discharges, and agricultural runoff, and outlining key research priorities, this study seeks to address these deficiencies. The work aims to establish a scientifically robust framework for MP mitigation by emphasizing multidisciplinary strategies, including enhanced monitoring programs, economically viable remediation approaches, and focussed policy measures. This assessment provides actionable insights for researchers, policymakers, and environmental professional, aligning with national development priorities and global sustainability goals.

8.15.2 Microplastics in Indian freshwater systems

Substantial volumes of plastic waste continue to enter aquatic environments annually as a result of improper disposal practices and inadequate waste management systems. At the global scale, approximately 8–12 million tonnes of plastic waste are estimated to reach the oceans each year, with freshwater systems functioning as major transport routes for this material (Pandey et al., 2023). In India, rapid urbanization, population increase, and industrial expansion have significantly amplified plastic consumption and waste production. According to data from the Central Pollution Control Board (CPCB), the country generates nearly 3.3 million metric tonnes of plastic waste each year, a considerable fraction of which is discarded in open dumping sites or poorly managed landfills (Gupta et al., 2023).

The growing issue of plastic pollution has led to the introduction of various policy and regulatory initiatives in India. Early actions included the prohibition of single-use plastics on ships in 2019, reflecting heightened concern regarding plastic-related environmental hazards. The state of Sikkim notably implemented of ban on single-use plastics as early as 1998, establishing a model subsequently followed by other regions. Although nationwide restrictions on single-use plastics were scheduled for implementation by 2022, progress has been constrained by the substantial economic role of the plastics sector, which has expanded considerably since its inception in the mid-twentieth century (All India Plastics Manufacturers Association, 2019). While this industry contributes meaningfully to economic development, persistent issues related to waste management inefficient and environmental degradation remain unresolved.

8.15.3 Microplastics in Indian lakes

Freshwater lakes across India have increasingly been recognized as significant accumulation zones for MPs (MPs), primarily due to ineffective waste disposal, discharge of untreated sewage, and surface runoff from urban and agricultural landscapes. MPs enter lake environments through diverse pathways, including stormwater inflows, agricultural drainage, and domestic wastewater, thereby posing threats to aquatic organisms and human health (Sarkar et al., 2020). The hydrodynamic properties of lake systems often promote the retention of MPs. Limited water circulation and weal current regimes facilities particle settling and long-term accumulation within sediments, effectively rendering lakes persistent sinks for plastic debris (Sathish et al., 2024). Moreover, MPs possess a strong ability to adsorb hazardous substances such as heavy metals and persistent organic pollutants, thereby enhancing their toxicity and promoting contaminant transfer through aquatic food webs.

Numerous investigations have reported MP contamination in lakes across India. Sediment analyses from Anchar Lake in the northwestern Himalayan region revealed MP concentrations ranging from 233 to 1533 particles per kilogram, with fragments, pellets, and fibers dominated by polystyrene (PS), polypropylene (PP), and polyamide (PA) (Neelavannan et al., 2022). In Red Hills Lake ear Chennai, mean MP abundances of approximately 27 particles per kilogram of sediment were observed, with polyethylene (PE) and PP identified as the predominant polymers, likely originating from adjacent urban waste inputs (Gopinath et al., 2020). Recreational activities also represent a notable source of MPs. At Kodaikanal Lake, sediment samples contained an average of 28.31 ± 5.29 MP items per kilogram, largely composed of PE, PP, and PS, indicating the combined effects of tourism pressure and inadequate waste management infrastructure (Laju et al., 2022). Comparable MP levels have been reported in Veeranam Lake, Tamil Nadu where PVC, PE, and PP constituted the principal polymer types (Srinivasalu et al., 2021).

Notably MP contamination extends beyond urbanized areas. Remote, high-altitude, and ecologically sensitive lakes in the Ladakh region—including Pangong, Tsomoriri, and Tsokar—have exhibited sediment MP concentrations between 160 and 1000 particles per kilogram. The detection of PE, PP, and PET in these isolated environments underscores the long-range transport potential and pervasive distribution of plastic pollution (Tsering et al., 2021, 2022). Vembanad Lake in Kerala, one of the largest freshwater bodies in India and a critical resource for fisheries and local livelihoods, has also been reported to square meter, with fragments, fibers and foams predominantly composed of PE, PP, and high-density polyethylene (HDPE) (Srinivasalu et al., 2021). These observations raise significant concerns regarding trophic transfer and potential human exposure through fish consumption.

8.15.4 Environmental and health implications, outlook and management needs

MPs represent a serious threat to ecological integrity and public health within freshwater systems. A wide range of aquatic organisms, including plankton, invertebrates and fish, are capable of ingesting MPs, which can result in physical damage, inhibited growth, reduced reproductive capacity, and, in extreme cases, mortality. Through trophic transfer, MPs can ultimately reach human populations via contaminated drinking water and aquatic food resources. Although the full extent of health impacts remains under investigation, chronic exposure has been linked to inflammation, endocrine disruption, and other adverse health outcomes (Nazir et al., 2024). The growing MP burden observed in Indian lakes mirrors broader challenges associated with rapid urban growth, extensive plastic use, and insufficient wastewater treatment facilities. Many freshwater bodies that are essential for drinking water supply and livelihoods are currently experiencing substantial pollution stress (Adeel et al., 2024).

The increasing prevalence of MPs in India's freshwater lakes highlights an urgent need for targeted intervention. Enhancing solid waste management frameworks, implementing stricter controls on plastic usage, and modernizing wastewater treatment infrastructure are critical measures for minimizing MP inputs (Sathish et al., 2024). Concurrently, sustained monitoring initiatives and focussed research efforts are necessary to accurately assess MP distribution, elucidate ecological effects, and deign effective mitigation strategies. Addressing these issues is essential for protecting India's freshwater ecosystems and ensuring long-term environmental sustainability and public health security.

8.15.5 Outlook and management needs s in Indian rivers

MPs have become a prominent category of pollutants within Indian riverine systems, which serve as vital transport routes conveying plastic debris from inland domestic, industrial, and urban sources to coastal and marine environmental (Amrutha and Warrier, 2020). Rapid population increase, accelerated urbanization, and insufficient management of solid waste and wastewater have collectively resulted in large volumes of plastic waste entering rivers throughout the country. These pressures are particularly intense in highly populated river basins, as outlined in Table 8.1 Recent studies consistently indicate severe MP contamination in major Indian rivers, including the Ganges and Brahmaputra, primarily driven by untreated sewage inputs and inefficient plastic waste management practices (Pandey et al., 2023).

The Ganges River, among the most culturally, economically, and environmentally significant rivers in India, has been identified as a major global source of plastic discharge to the oceans. Estimates by Lebreton et al. (2017) suggest that the Ganges alone contributes approximately 1.05×10^5 tonnes of plastic waste annually to the Indian Ocean. Subsequently field-based MP particles may be released each day from the river into marine environments. This exceptionally high plastic burden reflects the combined effects of dense population centers, urban surface runoff, informal waste disposal practices, and the widespread discharge of untreated or partially treated municipal wastewater along the river corridor (Farooq et al., 2023). MPs have been detected throughout both the water column and sediment compartments of the river, raising serious concerns regarding water quality and riverine ecosystem health.

Table 8.1 Scientific studies reported the abundance of microplastics in Indian waste treatment plants facilities.

S. no	Location	Sample	Polymer type	Extraction method	Detection	References
1	Bihar	Sewage	LDPE, PE, PVC, rayon	Digestion and density separation	Optical microscopy, SEM and FTIR	Parashar and Hait (2022)
2	Haryana	Sewage	PP, PE, and PS	Digestion and density separation	Optical microscopy, SEM and FTIR	Parashar and Hait (2022)
3	Uttar Pradesh (ganga stretch)	Sewage	PP, PE, and PS	Digestion and density separation	Optical microscopy, SEM and FTIR	Parashar and Hait (2023)
4	Nagpur	Sewage	LDPE, PE, PVC, rayon	Digestion and density separation	FTIR	Patil et al. (2023)

Similar patterns of MP pollution have been observed in other major river systems. In the lower reaches of the Ganges in easter India, Sarker et al. (2019) reported MP concentrations ranging from 380 to 684 particles per 1000 m^3 of surface water, with fibers and films identified as the predominant morphotypes. Polymer analyses revealed a dominance of polyethylene (PE), polypropylene (PP), and polystyrene (PS), materials commonly linked to household use and industrial activities (Das et al., 2023). Likewise, documented extensive MP contamination in the form of fibers, fragments, and pellets, demonstrating that the MP pollution extends into upstream and comparatively less industrialized regions (Chauhan and Trivedi, 2023).

The Brahmaputra River basin in northeastern India also exhibits considerable MP contamination. Sediment-based studies conducted by Tsering et al. (2022) reported MP films primarily composed of PE, PET, and PP. Seasonal monsoonal runoff, together with increasing urbanization and limited waste management infrastructure, appears to strongly influence the accumulation of MPs in rover sediments.

Rivers in southern India display comparable contamination trends. In the Netravathi River in Karnataka, MP concentrations between 9.44 and 253.27 particles per kilogram of dry sediment have been reported, largely attributed to urban runoff and the discharge of untreated wastewater from nearby urban centers (Amrutha et al., 2023). In the Kaveri River, Maheswaran et al. (2021) documented sediment MP abundances reaching upto 699 particles per kilogram, highlighting the magnitude of contamination in rivers that are crucial for irrigation, potable water supply, and fisheries (Majeed and Sharma, 2024).

Across Indian river systems, MPs primarily originate from daily-use plastic items, synthetic textile fibers, industrial effluents, and stormwater runoff. Once introduced to riverine environments, MPs accumulate within sediments or remain suspended in the water column, enabling downstream transport from terrestrial sources to estuaries and marine ecosystems. However, effective evaluation and control of MP pollution in Indian rivers are hindered by several challenges. Notably, the lack of standardized methodologies for sampling, analysis, and reporting limits comparability across studies and river basins (Neelavannan and Sen, 2023). In addition, conventional wastewater treatment facilities are not specifically designed to eliminate MPs, allowing these particles to persist in treated effluents and re-enter freshwater environments.

Establishing robust baseline datasets on MP concentrations in major Indian rivers is therefore critical for understanding pollution dynamics and informing targeted mitigation strategies (Veerasingam et al., 2020). Given their function as direct links between land-based activities and marine systems, rivers such as the Ganges, Brahmaputra, and Kaveri warrant immediate attention. Enhancing waste management practices, upgrading wastewater treatment technologies, and implementing standardized monitoring frameworks represent essential measures for reducing MP inputs, protecting riverine ecosystems, and safeguarding public health.

8.15.6 Transport of microplastics through the water cycle

India is both a leading producer and consumer of plastic materials, and the scale of plastic utilization has generated significant challenges for effective waste management. Consequently, MP– plastic particles smaller than 5 mm—are now widely detected across diverse aquatic environments in India, including rivers, lakes, coastal waters, and increasingly, groundwater systems (Robin et al., 2020). The country generates more than 3.5 million tonnes of plastic waste each year, a substantial portion of which is either inadequately collected or improperly disposed of. This mismanaged plastic waste undergoes fragmentation and dispersal, enabling MPs to circulate persistently within the hydrological cycle.

River networks play a pivotal role in the redistribution of MPs across landscapes. Large river systems such as Ganges and Yamuna act as major conduits, transporting plastic debris and MPs from densely populated urban centers, industrial zones, and agricultural regions toward the ocean (Rymbai and Jha, 2012). Urban stormwater runoff, untreated sewage effluents, and agricultural drainage—particularly from areas employing plastic mulching—introduce substantial MP loads into surface water. Upon discharge into the Arabian Sea and the Bay of Bengal, oceanic currents further disperse MPs along India's coastline and into offshore marine environments (Robin et al., 2020).

Ecologically sensitive regions, including the Sundarbans mangrove ecosystem, are especially vulnerable due to their hydrological connectivity with heavily polluted river systems and intense anthropogenic pressure. Beyond surface waters, MPs are increasingly detected in subsurface environments. Groundwater contamination has been reported in regions characterized by high plastic waste accumulation and extensive use of agricultural plastics. This issue of particular concern in rural India, where groundwater constitutes the primary source of drinking water for millions of residents. Although investigations into groundwater MPs remain limited, emerging evidence indicates that these particles can mitigate through soil profiles, raising serious concerns regarding long-term drinking water safety.

Atmospheric transport represents an additional and increasingly recognized pathway in the MP cycle. Wind-driven erosion of plastic debris and open burning of plastic waste release MPs into the atmosphere, allowing long-range transport before deposition through precipitation. In northern India, where ambient air pollution levels are already elevated, atmospheric deposition of MPs via rainfall imposes an added environmental burden.

The textile industry constitutes a major source of both airborne and aquatic MPs. The laundering of synthetic textiles releases substantial quantities of microfibers into wastewater streams, which subsequently enter rivers and coastal waters (Akyildiz et al., 2024). Prominent textile manufacturing centers such as Surat and Tirupur further intensify the issue. Exacerbating the problem, many urban wastewater treatment plants lack the technological capacity to efficiently remove MPs, permitting their continuous discharge into natural water bodies. Metropolitan areas such as Mumbai and Delhi release significant MP loads into adjacent rivers as a result of incomplete wastewater treatment.

The widespread occurrence of MPs in India's freshwater and marine environments presents serious ecological and public health concerns. Aquatic organisms, including fish, birds, and reptiles, ingest MPs, resulting in physiological stress, reproductive impairment, and broader ecosystem disruption. For human populations, the detection of MPs in drinking water and seafood raises concerns regarding chronic exposure. MPs can adsorb toxic substances and heavy metals, thereby facilitating their transfer through food webs and increasing the likelihood of accumulation within human tissues.

India has introduced policy measures to combat plastic pollution, most notably the nationwide ban on single-use plastics implemented in 2022 (Bansal et al., 2024). Nevertheless, enforcement challenges remain, particularly in rural regions and within informal waste management sectors. Effective mitigation of MP contamination requires strengthened waste management frameworks, investment in advanced wastewater treatment technologies capable of MP removal and sustained public awareness campaigns promoting reduced plastic consumption and safer alternatives Overall, the continued presence of MPs within India's water cycle reflects the combined pressures of rapid urban expansion, industrial development, and population growth. Addressing the growing challenge will require coordinated efforts across policy, technological innovation, and societal behavior to protect ecosystems and public health (Majeed and Sharma, 2024).

8.15.7 Implications for environmental protection and management

The outcomes of this review identify several practical and policy-relevant pathways for reducing MP contamination in India's freshwater ecosystems. A primary requirement is the rigorous implementation of existing regulatory instruments, particularly the Plastic Watse Management (PWM) Rules, 2023, which incorporate the principle of Extended Producer Responsibility (EPR). Strengthening adherence to EPR producer accountability for the collection, recycling, and environmentally responsible disposal of plastic waste, thereby minimizing plastic leakage into rivers, lakes, and other freshwater bodies.

Urban wastewater represents one of the most significant routes through which MPs enter freshwater systems. Modernizing sewage treatment facilities with advanced filtration technologies capable of capturing MPs, enforcing mandatory pre-treatment standards for textile industry effluents, and encouraging the use of biodegradable material alternatives can collectively reduce MP inputs. Additional source -reduction measures include the installation of microfiber filters in household and industrial washing machines, as well as improvements in urban stormwater management infrastructure to limit the transport of plastic debris into surface waters.

Inadequate solid waste management practices also substantially contribute to freshwater MP contamination. Expanding door-to-door waste in regions collection services, discouraging reliance on single-use plastic products, and promoting decentralized and community-based recycling systems can significantly curb plastic leakage into rivers and lakes. Active community engagement is equally critical. Localized clean-up programs and targeted awareness campaigns should be emphasized, particularly in regions with high tourist influxes or frequent religious and cultural gatherings near freshwater ecosystems.

Technological advancement and focused research play a central role in achieving long-term mitigation objectives. The development of portable MP detection tools, artificial intelligence-based monitoring systems, and open-access data repositories cam enhance environmental surveillance and early warning capabilities Importantly, integrated research frameworks that connect environmental science, public health, and policy analysis are required to improve understanding of long-term risks and to support evidence-driven decision-making processes.

8.15.8 Knowledge gaps

Despite increasing recognition of the issue, effective management of MP pollution in India's freshwater environments is hindered by several critical knowledge deficiencies. Comprehensive data on MP abundance, polymer composition, and spatial distribution across rivers, lakes, and wetlands remain scarce, particularly in rural and semi-urban areas. Although marine systems have received relatively greater scientific attention. Freshwater environments-which often serve as primary sources of drinking water and food-continue to be inadequately investigated.

The ecological consequences of MPs for freshwater biodiversity are not well characterized, especially in ecosystems that support endemic or ecologically sensitive species. In addition, limited information is available regarding MP accumulation and biomagnification within freshwater food webs, particularly in rural communities that depend heavily on locally harvested fish and aquatic vegetation.

A further critical gap concerns wastewater treatment infrastructure. The majority of existing wastewater treatment plants in India are not specifically designed to remove MPs, allowing these particles to pass through treatment processes and enter natural water bodies. Comprehensive assessments of current treatment technologies, along with the development of cost-effective and scalable filtration solutions, ae urgently needed. Moreover, public awareness of MP pollution—especially in rural settings-remains low. Gaining insights into local waste handling practices and consumption patterns is essential for designing effective, context-sensitive awareness programs and behavior-change interventions.

Addressing these knowledge gaps will require coordinated collaboration among governmental agencies, academic and research institutions, and environmental organizations to generate reliable datasets and implement targeted, evidence-based management strategies.

8.15.9 Key research directions

Enhancing scientific knowledge of MP pollution in Indian freshwater ecosystem is fundamental to sustainable water resource management and development of effective mitigation measures. Given the ecological, economic, and social importance of India's extensive network of rivers, lakes, and wetlands, future research should adopt an integrated, interdisciplinary, and policy-relevant perspective.

Priority areas for future investigation include:

- Establishing nationwide baseline database on MP concentrations, polymer types, and source contributions across freshwater environments.
- Examining the fate, transport, and transformation of MPs under diverse hydrological regimes and climatic conditions.
- Evaluating the ecotoxicological impacts of MPs on freshwater organisms and food web including their combined effects with co-occurring chemical pollutants.
- Assessing the effectiveness and practical feasibility of advanced wastewater treatment technologies for MP removal under Indian conditions.
- Investigating human exposure pathways associated with drinking water supplies and freshwater-derived food resources.
- Incorporating social science approaches to analyze public perceptions, behavioral determinants, and the effectiveness of policy interventions.

Collectively, these research initiatives can generate the scientific evidence require to strengthen regulatory frameworks, support technological innovations, and safeguard India's freshwater ecosystem and public health.

8.16 Overall summary and conclusion

8.16.1 Microplastic morphology and polymer type

Fibers were the most abundant MP type in both the water column and sediment samples, consistent with observations from multiple marine monitoring programs in the Midwest Pacific Ocean, the

Northwestern Mediterranean Sea, and the Northeast Pacific including coastal British Columbia. Sources of these fibers are diverse, encompassing wastewater effluents, textile shredding, cigarette filters, and personal care items like wipes, textile fibers are particularly pervasive due to continuous release throughout production, usage, laundering and disposal. However, further research is required to clarify the primary emission pathways and the mechanisms driving their transport in aquatic systems. In the water column, polyethylene (PE) and polypropylene (PP) were the most frequently detected polymers. This aligns with global production patterns, as PE and PP make up approximately 74% of worldwide plastic output and dominate marine MP inventories. Their widespread use in single-use products-such as packaging, bags and bottle caps-explain their high environmental presence. Their buoyancy further facilitates accumulation at the sea surface, enhancing detectability during surface sampling.

Although cellulose acetate is anthropogenic, other cellulose-type particles may originate naturally. In this study, such cellulose particles represented less than 1% of all samples. Previous studies similarly report PE and PP as the dominant polymers, although regional differences have been noted, with PET and polyamide (PA) being more abundant in some areas. These variations reflect local plastic usage patterns, riverine inputs and methodological differences. The fiber morphology complicates polymer identification because long, thin fibers often produce weak or incomplete FTIR spectra. Full-field scanning approaches may improve spectral accuracy and reduce identification bias. Vertical distribution patterns of MPs are influenced not only by polymer density but also by wave mixing, particle size, and shape. In sediments, polymer identity does not always correlate with negative buoyancy, suggesting that biofouling, turbulence, and aggregation affect settling. Limiting sampling to the sea surface risks underestimating higher-density polymers and other particles influenced by seasonal hydrodynamics or rainfall-runoff events.

8.16.2 Microplastic size distribution

MP sizes exhibited seasonal variability, with larger particles observed during the dry season. This contrasts with studies from Taiwan, where larger particles were more prevalent in the wet season fur to increased land-based inputs during rainfall. Seasonal changes in riverine inputs and winter hydrodynamic forcing may influence particle weathering rates, while ocean mixing contributes to fragmentation and long-distance dispersal. Most MPs in this study were <2.5 mm and classified as secondary MPs, indicating they originated from the beginning of larger debris. Further research is needed to understand how source strength, weathering and transport dynamics determine size distributions in coastal waters. Smaller, buoyant particles are more prone to downward transport due to low rising velocities. MP size generally decreases with depth because smaller particles have a higher surface-area-to-volume ration, accelerating density changes via biofouling, aggregation, and adsorption. These processes increase sinking rates regardless of polymer type, contributing to vertical stratification of particle sizes.

8.16.3 Limitations of microplastic sampling in near-coastal environments

Sampling MPs in dynamic costal zones in inherently challenging. Constrains including limited time, logistical difficulties, instrumentation, and financial resources restrict sampling frequency, spatial coverage, and depth resolution. High variability of MP concentrations across horizontal and vertical gradients can mask temporal trends or spatial patterns. Net-based methods introduce biases: mesh size limits the smallest particles captured, and clogging may restrict filtered volume. Fibers are particularly underestimated using nets. Sampling inconsistencies, human error, particle size fraction analyzed and differences in analytical methods can obscure true concentration-size relationships.

Bulk water sampling with Niskin bottles in the Northern Gulf of Mexico captured twice the MP concentration compared to manta nets, due to higher collection of small particles. These finding underscore the importance of considering methodological context when comparing studies. Diverse shapes, sizes, colors and polymers complicate generalizations about environmental behavior, and differences in ocean circulation, water column structure, sampling methods and reporting units further increase uncertainty. This chapter emphasizes the necessity of robust spatially comprehensive sampling strategies for monitoring MPs in near-coastal marine environments. MP concentrations and characteristics vary with proximity to land-based sources, seasonal hydrodynamics, and vertical position in the water column. Detection of MPs through the wate column and sediments indicates that's surface-only monitoring substantially underestimates total MP loads.

Larger MP predominate at the surface, while smaller particles are found at depth, highlighting the need for depth-integrated sampling approaches. To accurately assess ecological and environmental risks of MP pollution, monitoring programs:

- Geographic variability
- Proximity to urban and riverine inputs
- Seasonal changes in oceanographic conditions
- Particle transport and weathering processes

A comprehensive, multi-layered sampling framework is essential for reliable characterizing MP contamination in coastal marine systems.

References

Accinelli, C., Abbas, H.K., Shier, W.T., Vicari, A., Little, N.S., Aloise, M.R., Giacomini, S., 2019. Degradation of microplastics seed film-coating fragments in soil. Chemosphere 226, 645–650. https://doi.org/10.1016/j.chemosphere.2019.03.161.

Adeel, M., Mina, T., Rizzo, L., Fatta-Kassinos, D., 2024. The Impact of Microplastics on the efficacy of urban wastewater treatment processes. J. Environ. Chem. Eng. 12, 113625.

Ahmed, M.F., Mokhtar, M.B., Alam, L., 2020. Factors influencing people's willingness to participate in sustainable water resources management in Malaysia. J. Hydrol. Region. Stud. 31, 100737. https://doi.org/10.1016/j.ejrh.2020.100737.

Akyildiz, S.H., Fiore, S., Bruno, M., Sezgin, H., Yalcin-Enis, I., Yalcin, B., Bellopede, R., 2024. Release of microplastic fibers from synthetic textiles during household washing. Environ. Pollut. 357, 124455.

Alava, J.J., 2019. Ocean pollution and warming oceans: toward ocean solutions and natural marine bioremediation. In: Predicting Future Oceans. Elsevier, pp. 495–518.

Alava, J.J., Kazmiruk, T.N., Douglas, T., Schuerholz, G., Heath, B., Flemming, S.A., Bendell, L., Drever, M.C., 2021. Occurrence and size distribution of microplastics in mudflat sediments of the Cowichan-Koksilah estuary, Canada: a baseline for plastic particles contamination in an anthropogenic-influenced estuary. Mar. Pollut. Bull. 173, 113033.

All India Plastics Manufacturers Association (AIPMA, 2019), 2019. Plastic Waste in India. https://www.aipma.net/new/Past-Events/Presentation/02July21-1.pdf.

AMAP, 2021. Arctic Climate Change Update 2021: Key Trends and Impacts (Tromso, Norway).

Amrutha, K., Warrier, A.K., 2020. The first report on the source-to-sink characterization of microplastic pollution from a riverine environment in tropical India. Sci. Total Environ. 739, 140377.

Amrutha, K., Warrier, A.K., Rangel-Buitrago, N., 2023. Did the COVID-19 pandemic play a role in the spatial and temporal variations of microplastics? Evidence from a tropical river in southern India. Mar. Pollut. Bull. 192, 115088.

An, L., Liu, Q., Deng, Y., Wu, W., Gao, Y., Ling, W., 2020. Microplastics in terrestrial environments. In: He, D., Luo, Y. (Eds.), Handbook of Environmental Chemistry, vol 95. Springer, Cham, pp. 143–159. https://doi.org/10.1007/698_2020_449.

Andrady, A.L., 2011. Microplastics in the marine environment. Mar. Pollut. Bull. 62 (8), 1596–1605.

Baensch-Baltruschat, B., Kocher, B., Stock, F., Reifferscheid, G., 2020. Tyre and road wear particles (TRWP) A review of generation, properties, emissions, human health risk, ecotoxicity, and fate in the environment. Sci. Total Environ. 733, 137823. https://doi.org/10.1016/j.scitotenv.2020.137823.

Bansal, S., Mittal, S., Agarwal, R., 2024. Evaluating drivers of single-use plastic ban in India: an empirical study (January-April 2023). Int. J. Green Econ. 18 (4), 358–373.

Bao, M.R., Huang, Q.H., Lu, Z.B., et al., 2022. Investigation of microplastic pollution in Arctic fjord water: a case study of Rijpfjorden, Northern Svalbard. Environ. Sci. Pollut. Res. 29 (37), 56525–56534. https://doi.org/10.1007/s11356-022-19826-3.

Bergmann, M., Muetzel, S., Primpke, S., et al., 2019. White and wonderful? Microplastics prevail in snow from the alps to the Arctic. Sci. Adv. 5 (8). https://doi.org/10.1126/sciadv.aax1157.

Bergmann, M., Collard, F., Fabres, J., et al., 2022. Plastic pollution in the arctic. Nat. Rev. Earth Environ. 3 (5), 323–337. https://doi.org/10.1038/s43017-022-00279-8.

Bergmann, M., Allen, S., Krumpen, T., et al., 2023. High levels of microplastics in the Arctic sea ice Alga Melosira Arctica, a vector to ice-associated and benthic food webs. Environ. Sci. Technol. 57 (17), 6799–6807. https://doi.org/10.1021/acs.est.2c08010.

Borah, A., Lekshmy, J., Sajna, P., Devipriya, S.P., 2024. Assessment of microplastic and heavy metal pollution in agricultural soils of Ernakulam District, Kerala, India. Environ. Monit. Assess. 196 (11), 1090.

Cervantes-Audelo, I., 2013. Analisis de circulacion y dispersion en la Bahía de Todos Santos, Baja California. MSc Thesis. Universidad del Mar. Oaxaca, Mexico. pp. 61. http://rdurazo.ens.uabc.mx/estudiantes/tesis/Tesis%20Irving%20Audelo.pdf.

Capparelli, M.V., et al., 2022. Ecological traits influence the bioaccumulation of microplastics in commercially important estuarine crabs from the southeastern Gulf of Mexico. Mar. Pollut. Bull. 183, 114088.

Cesa, F.S., Turra, A., Baruque-Ramos, J., 2017. Synthetic fibers as microplastics in the marine environment: a review from textile perspective with a focus on domestic washings. Sci. Total Environ. 598, 1116–1129.

Chauhan, S.S., Trivedi, M.K., 2023. Assessment of surface water quality index of Gwalior Chambal region using Fuzzy-based approach. J. Inst. Eng.: Series A 104 (1), 1–17.

Chubarenko, I., Bagaev, A., Zobkov, M., Esiukova, E., 2016. On some physical and dynamical properties of microplastic particles in marine environment. Mar. Pollut. Bull. 108, 105–112. https://doi.org/10.1016/j.marpolbul.2016.04.048.

Chubarenko, I., et al., 2018. Behavior of microplastics in coastal zones. In: Zeng, E.Y. (Ed.), Microplastic Contamination in Aquatic Environments. Elsevier, pp. 175–223. https://doi.org/10.1016/B978-0-12-813747-5.00006-0.

Claessens, M., De Meester, S., Van Landuyt, L., De Clerck, K., Janssen, C.R., 2011. Occurrence and distribution of microplastics in marine sediments along the belgian coast. Mar. Pollut. Bull. 62 (10), 2199–2204.

Covernton, G.A., Cox, K.D., Fleming, W.L., Buirs, B.M., Davies, H.L., Juanes, F., Dudas, S.E., Dower, J.F., 2022. Large size (> 100-μm) microplastics are not biomagnifying in coastal marine food webs of British Columbia. Canada. Ecol. Appl. 32 (7), e2654.

Crew, A., Gregory-Eaves, I., Ricciardi, A., 2020. Distribution, abundance, and diversity of microplastics in the upper St. Lawrence River. Environ. Pollut. 260, 113994. https://doi.org/10.1016/j.envpol.2020.113994.

D'Angelo, A., Trenholm, N., Loose, B., et al., 2023. Microplastics distribution within western Arctic seawater and sea ice. Toxics 11 (9), 792. https://doi.org/10.3390/toxics11090792.

Das, S., Sarkar, S., Kanungo, D.P., 2023. A Critical Review on Landslide Susceptibility zonation: recent trends, techniques, and practices in Indian Himalaya. Nat. Hazards. 115 (1), 23–72.

Ding, J., Sun, Y., He, C., Li, J., Li, F., 2022. Towards risk assessments of microplastics in bivalve mollusks globally. J. Mar. Sci. Eng. 10 (2), 288.

Doyle, M.J., Watson, W., Bowlin, N.M., Sheavly, S.B., 2011. Plastic particles in coastal pelagic ecosystems of the Northeast Pacific Ocean. Mar. Environ. Res. 71, 41–52.

Duis, K., Coors, A., 2016. Microplastics in the aquatic and terrestrial environment: sources (with a specific focus on personal care products), fate and effects. Environ. Sci. Eur. 28 (1), 2.

D'Costa, A.H., 2022. Microplastics in decapod crustaceans: accumulation, toxicity and impacts, a review. Sci. Total Environ. 832, 154963. https://doi.org/10.1016/j.scitotenv.2022.154963.

Eduardo, A.L.H., et al., 2021. Microplastic concentrations in cultured oysters in two seasons from two bays of Baja California, Mexico. Environ. Pollut. 290, 118031.

Elkhatib, D., Oyanedel-Craver, V., 2020. A critical review of extraction and identification methods of microplastics in wastewater and drinking water. Environ. Sci. Technol. 54, 7037–7049. https://doi.org/10.1021/acs.est.9b06672.

Eriksen, M., Mason, S., Wilson, S., Box, C., Zellers, A., Edwards, W., Farley, H., Amato, S., 2013. Microplastic pollution in the surface waters of the laurentian great lakes. Mar. Pollut. Bull. 77, 177.

Eriksen, M., Lebreton, L.C.M., Carson, H.S., Thiel, M., Moore, C.J., Borerro, J.C., Galgani, F., Ryan, P.G., Reisser, J., 2014. Plastic pollution in the world's oceans: more than 5 trillion plastic pieces weighing over 250,000 tons afloat at sea. PLoS One 9, e111913.

Eriksen, M., Borgogno, F., Villarrubia-Gomez, P., et al., 2020. Mitigation strategies to reverse the rising trend of plastics in Polar regions. Environ. Int. 139, 105704. https://doi.org/10.1016/j.envint.2020.105704.

Evans, M.S., 2000. The large lake ecosystems of Northern Canada. Aquat. Ecosyst. Heal. Manag. 3, 65–79. https://doi.org/10.1080/14634980008656992.

Farooq, M., Nisa, F.U., Manzoor, Z., Tripathi, S., Thulasiraman, A.V., Khan, M.I., Khan, M.Y.A., Gani, K.M., 2023. Abundance and characteristics of microplastics in a freshwater river in northwestern Himalayas, India-Scenario of riverbank solid waste disposal sites. Sci. Total Environ. 886, 164027.

Fauziah, S.H., Rizman-Idid, M., Cheah, W., Loh, K.H., Sharma, S., NoorMaiza, M.R., Bordt, M., Praphotjanaporn, T., Samah, A.A., Sabaruddin, J.S., George, M., 2021. Marine debris in Malaysia: a review on the pollution intensity and mitigating measures. Mar. Pollut. Bull. 167, 112258. https://doi.org/10.1016/j.marpolbul.2021.112258.

Fu, D., Kurniawan, T.A., Wang, Y., Zhou, Z., Wei, Q., Hu, Y., Othman, M.H.D., Goh, H.H., Chew, K.W., Gui, H., 2023. Applicability of magnetic biochar derived from Fe-enriched sewage sludge for chromate removal from aqueous solution. Chem. Eng. Sci. 281, 119145. https://doi.org/10.1016/j.ces.2023.119145.

Galindo Montero, A.A., Costa-Redondo, L.C., Vasco-Echeverri, O., Arana, V.A., 2023. Microplastic pollution in coastal areas of Colombia: review. Mar. Environ. Res. 190, 106027. https://doi.org/10.1016/j.marenvres.2023.106027.

Gatidou, G., Arvaniti, O.S., Stasinakis, A.S., 2019. Review on the occurrence and fate of microplastics in sewage treatment plants. J. Hazard. Mater. 367, 504–512. https://doi.org/10.1016/j.jhazmat.2018.12.081.

GESAMP, 2015. Sources, fate and effects of microplastics in the marine environment: a global assessment. In: Kershaw, P.J. (Ed.), Rep. Stud. GESAMP No. 90, 96. IMO, Exeter, p. 1.

GESAMP, 2016. Sources, fate and effects of microplastics in the marine environment: part two of a global assessment. In: Kershaw, P.J., Rochman, C.M. (Eds.), IMO/FAO/UNESCOIOC/UNIDO/WMO/IAEA/UN/UNEP/UNDP Joint Group of Experts on the Scientific Aspects of Marine Environmental Protection.

Gibson, J.J., Prowse, T.D., Peters, D.L., 2006. Hydroclimatic controls on water balance and water level variability in Great Slave Lake. Hydrol. Process. 20, 4155–4172. https://doi.org/10.1002/hyp.6424.

Goh, H.H., Zong, L., Zhang, D., Liu, H., Dai, W., Lim, C.S., Kurniawan, T.A., Teo, K.T.K., Goh, K.C., 2022. Mid- and long-term strategy based on electric vehicle charging unpredictability and ownership estimation. Int. J. Electric. Power Energy Syst. 142, 108240. https://doi.org/10.1016/j.ijepes.2022.108240.

Gong, T., Xu, G., Chen, L., et al., 2023. Current state of research on microplastics in the marine atmosphere environment of the arctic region. Adv. Polar Sci. 34 (4), 251–261. https://doi.org/10.12429/j.advps.2023.0005.

Gopinath, K., Seshachalam, S., Neelavannan, K., Anburaj, V., Rachel, M., Ravi, S., Bharath, M., Achyuthan, H., 2020. Quantification of microplastic in red hills lake of Chennai city, Tamil Nadu, India. Environ. Sci. Pollut. Control Ser. 27, 33297–33306.

Guo, W., Xi, B., Huang, C., Li, J., Tang, Z., Li, W., Ma, C., Wu, W., 2021. Solid waste management in China: policy and driving factors in 2004–2019. Resourc., Conserv., Recycling 173, 105727. https://doi.org/10.1016/j.resconrec.2021.105727.

Gupta, D.K., Choudhary, D., Vishwakarma, A., Mudgal, M., Srivastava, A.K., Singh, A., 2023. Microplastics in freshwater environment: occurrence, analysis, impact, control measures and challenges. Int. J. Environ. Sci. Technol. 20 (6), 6865–6896.

Gupta, P., Saha, M., Naik, A., Kumar, M.M., Rathore, C., Vashishth, S., et al., 2024. A comprehensive assessment of macro and microplastics from Rivers Ganga and Yamuna: unveiling the seasonal, spatial and risk factors. J. Hazard Mater. 469, 133926.

Heller, M.C., Mazor, M.H., Keoleian, G.A., 2020. Plastics in the US: toward a material flow characterization of production, markets and end of life. Environ. Res. Lettrs. 15, 094034. https://doi.org/10.1088/1748-9326/ab9e1e.

Hickey, B., 1992. Circulation over the Santa monica-san Pedro basin and shelf. Prog. Oceanogr. 30, 37–115.

Hidalgo-Ruz, V., Gutow, L., Thompson, R.C., Thiel, M., 2012. Microplastics in the marine environment: a review of the methods used for identification and quantification. Environ. Sci. Technol. 46 (6), 3060–3075. https://doi.org/10.1021/es2031505.

Hopwood, M.J., Carroll, D., Dunse, T., et al., 2020. Review article: how does glacier discharge affect marine biogeochemistry and primary production in the Arctic? Cryosphere 14 (4), 1347–1383. https://doi.org/10.5194/tc-14-1347-2020.

Hurley, R., Braaten, H.F.V., Nizzetto, L., Steindal, E.H., Lin, Y., Clayer, F., van Emmerik, T., Buenaventura, N.T., Eidsvoll, D.P., Økelsrud, A., Norling, M., 2023. Measuring riverine macroplastic: methods, harmonisation, and quality control. Water Res. 235, 119902. https://doi.org/10.1016/j.watres.2023.119902.

Ikenoue, T., Nakajima, R., Fujiwara, A., et al., 2023. Horizontal distribution of surface microplastic concentrations and water-column microplastic inventories in the Chukchi Sea, western Arctic Ocean. Sci. Total Environ. 855, 159564. https://doi.org/10.1016/j.scitotenv.2022.159564.

INEGI. 1997. Anuario estadístico del estado de Campeche. Instituto Nacional de Estadística, Geografía e Informatica.

INEGI. 2017. Anuario estadístico y geografico de Baja California 2017. p. 425p. http://www.datatur.sectur.gob.mx/ITxEF_Docs/BCN_ANUARIO_PDF.pdf.

IPCC, 2021. In: Masson-Delmotte, V., Zhai, P., Pirani, A., et al. (Eds.), Climate Change 2021: The Physical Science Basis. Contribution of Working Group I to the Sixth Assessment Report of the Intergovernmental Panel on Climate Change.

Jones, J., Eicken, H., Mahoney, A., et al., 2016. Landfast sea ice breakouts: stabilizing ice features, oceanic and atmospheric forcing at Barrow, Alaska. Continent. Shelf Res 126, 50–63. https://doi.org/10.1016/j.csr.2016.07.015.

Kabir, A.H.M.E., Michon, E., Mingelbier, M., Robert, D., Soubaneh, Y.D., Xie, H., Lu, Z., 2025. Microplastics in benthic fish from the Canadian St. Lawrence River and Estuary: occurrence, spatial distribution, and ecological risk assessment. Mar. Pollut. Bull. 212, 117509.

Kamaruddin, H., Patittingi, F., Assidiq, H., Bachril, S.N., Al Mukarramah, N.H., 2022. Legal aspect of plastic waste management in Indonesia and Malaysia: addressing marine plastic debris. Sustainability 14 (12), 6985. https://doi.org/10.3390/su14126985.

Kanhai, L.K., Gardfeldt, K., Krumpen, T., et al., 2020. Microplastics in sea ice and seawater beneath ice floes from the Arctic Ocean. Sci. Rep. 10 (1), 5004. https://doi.org/10.1038/s41598-020-61948-6.

Kauffman, E., 2003. Climate and topography. www.coastal.ca.gov/coastalvoices/resources/Biodiversity_Atlas_Climate_and_Topography.pdf,10.31.23.

Kazmiruk, T.N., Kazmiruk, V.D., Bendell, L., 2018. Abundance and distribution of microplastics within surface sediments of a key shellfish growing region of Canada. PLoS One 13 (5), e0196005.

Kazour, M., Terki, S., Rabhi, K., Jemaa, S., Khalaf, G., Amara, R., 2019. Sources of microplastics pollution in the marine environment: importance of wastewater treatment plant and coastal landfill. Marine Pollut. Bull. 146, 608–618. https://doi.org/10.1016/j.marpolbul.2019.06.066.

Keerthika, K., Padmavathy, P., Rani, V., et al., 2023. Microplastics accumulation in pelagic and benthic species along the thoothukudi coast, South Tamil Nadu, India. Mar. Pollut. Bull. 189, 114735. https://doi.org/10.1016/j.marpolbul.2023.114735.

Kessouri, F., McWilliams, J.C., Bianchi, D., Sutula, M., Renault, L., Deutsch, C., Feely, R.A., McLaughlin, K., Ho, M., Howard, E.M., Bednaršek, N., Damien, P., Molemaker, J., Weisberg, S.B., 2021. Coastal eutrophication drives acidification, oxygen loss, and ecosystem change in a major oceanic upwelling system. Proc. Natl. Acad. Sci. U. S. A.

Kim, S.K., Lee, H.J., Kim, J.S., et al., 2021. Importance of seasonal sea ice in the western Arctic ocean to the Arctic and global microplastic budgets. J. Hazard Mater. 418, 125971. https://doi.org/10.1016/j.jhazmat.2021.125971.

Koongolla, J.B., Lin, L., Pan, Y.-F., Yang, C.-P., Sun, D.-R., Liu, S., et al., 2020. Occurrence of microplastics in gastrointestinal tracts and gills of fish from Beibu Gulf, South China Sea. Environ. Pollut. 258, 113734. https://doi.org/10.1016/j.envpol.2019.113734.

Kumar, G.,.A., Anjana, K., Hinduja, M., Sujitha, K., Dharani, G., 2020. Review on plastic wastes in marine environment – biodegradation and biotechnological solutions. Marine Pollut. Bull. 150, 110733. https://doi.org/10.1016/j.marpolbul.2019.110733.

Kurniawan, T.A., Haider, A., Ahmad, H.M., Mohyuddin, A., Aslam, H.M.U., Nadeem, S., Javed, M., Othman, M. H.D., Goh, H.H., Chew, K.W.J.C., 2023a. Source, occurrence, distribution, fate, and implications of microplastic pollutants in freshwater on environment: a critical review and way forward. Chemosphere 325, 138367. https://doi.org/10.1016/j.chemosphere.2023.138367.

Kurniawan, T.A., Lo, W., Othman, M.H.D., Goh, H.H., Chong, K.K., 2023b. Influence of Fe2O3 and bacterial biofilms on Cu(II) distribution in a simulated aquatic environment: a feasibility study to sediments in the Pearl River estuary (PR China). J. Environ. Manage. 329, 117047. https://doi.org/10.1016/j.jenvman.2022.117047.

Laju, R.L., Jayanthi, M., Jeyasanta, K.I., Patterson, J., Asir, N.G.G., Sathish, M.N., Edward, J.K.P., 2022. Spatial and vertical distribution of microplastics and their ecological risk in an Indian freshwater lake ecosystem. Sci. Total Environ. 820, 153337.

Lamberti, F.M., Roman-Ramírez, L.A., Wood, J., 2020. Recycling of bioplastics: routes and benefits. J. Polym. Environ. 28, 2551–2571. https://doi.org/10.1007/s10924-020-01795-8.

Lattin, G.L., Moore, C.J., Zellers, A.F., Moore, S.L., Weisberg, S.B., 2004. A comparison of neustonic plastic and zooplankton at different depths near the southern California shore. Mar. Pollut. Bull. 49, 291–294.

Law, K.L., Morét-Ferguson, S.E., Goodwin, D.S., Zettler, E.R., Deforce, E., Kukulka, T., Proskurowski, G., 2014. Distribution of surface plastic debris in the eastern Pacific Ocean from an 11-year data set. Environ. Sci. Technol. 48, 4732–4738.

Lebreton, L.C.M., Van Der Zwet, J., Damsteeg, J.-W., Slat, B., Andrady, A., Reisser, J., 2017. River plastic emissions to the world's oceans. Nat. Commun. 8 (1), 15611.

Lenaker, P.L., Corsi, S.R., Mason, S.A., 2021. Spatial distribution of microplastics in surficial benthic sediment of Lake Michigan and Lake Erie. Environ. Sci. Technol. 55, 373–384. https://doi.org/10.1021/acs.est.0c06087.

Li, C., Busquets, R., Campos, L.C.J., 2020. Assessment of microplastics in freshwater systems: a review. Sci. Total Environ. 707, 135578. https://doi.org/10.1016/j.scitotenv.2019.135578.

Lim, X., 2021. Microplastics are everywhere but are they harmful. Nature 593 (7857), 22–25.

Lin, X., Gowen, A.A., Pu, H., Xu, J.L., 2023. Microplastic contamination in fish: critical review and assessment of data quality. Food Control 153, 109939. https://doi.org/10.1016/J.FOODCONT.2023.109939.

Lusher, A., Hurley, R., Vogelsang, C., 2019. Microplastics in sewage sludge: captured but released. In: Karapanagioti, H.K., Kalavrouziotis, I.K. (Eds.), Microplastics in Water Wastewater, pp. 85–100. https://doi.org/10.2166/9781789060034_0085.

Ma, Z.F., Ibrahim, Y.S., Lee, Y.Y., 2020. Microplastic pollution and health and relevance to the Malaysia's roadmap to zero single-use plastics 2018–2030. Malaysian J. Med. Sci. 27 (3), 1–6. https://doi.org/10.21315/mjms2020.27.3.1.

Mackas, D., Thomson, R.E., Galbraith, M., 2001. Changes in the zooplankton community of the British Columbia continental margin, 1985-1999, and their covariation with oceanographic conditions. Can. J. Fish. Aquat. Sci. 58 (4), 685–702.

Madelaine, P.T.B., Jennifer, F.P., Jessica, H., Noah, J., Jesse, C.V., 2024. Microplastics and anthropogenic microparticles in surface waters from Yellowknife bay, great Slave Lake, northwest territories, Canada. J. Great Lake. Res. 50, 102348.

Mahara, N., Alava, J., Kowal, M., Grant, E., Boldt, J., Kwong, L., Hunt, B., 2022. Assessing size-based exposure to microplastic particles and ingestion pathways in zooplankton and herring in a coastal pelagic ecosystem of British Columbia, Canada. Mar. Ecol. Prog. Ser. 683, 139–155.

Maheswaran, B., Karmegam, N., Al-Ansari, M., Subbaiya, R., Al-Humaid, L., Raj, J.S., Maitre-Ekern, E., 2021. Re-thinking producer responsibility for a sustainable circular economy from extended producer responsibility to pre-market producer responsibility. J. Clean. Prod. 286, 125454. https://doi.org/10.1016/j.jclepro.2020.125454.

Majeed, L.R., Sharma, D., 2024. Unraveling geomorphometric anomalies in the Lolab basin of Kashmir Himalaya and its role in shaping sustainable land use policies. Discover Water 4 (1), 110.

Masura, J., Baker, J.E., Foster, G.D., Arthur, C., Herring, C., 2015. Laboratory Methods for the Analysis of Microplastics in the Marine Environment. in: Recommendations for Quantifying Synthetic Particles in Waters and Sediments. Siver Spring, MD. NOAA Marine Debris Division. https://doi.org/10.25607/OBP-604, 31pp.

Mitrano, D.M., Wohlleben, W., 2020. Microplastics regulation should be more precise to incentivize both innovation and environmental safety. Nat. Commun. 11, 5324. https://doi.org/10.1038/s41467-020-19069-1.

Moore, C.J., Moore, S.L., Leecaster, M.K., Weisberg, S.B., 2001. A comparison of plastic and plankton in the north Pacific central gyre. Mar. Pollut. Bull. 42, 1297–1300.

Morici, E., Cammilleri, G., Scirè, S., Bonomo, F.P., Tranchina, L., Terracina, F., Galluzzo, P., Ferrantelli, V., Monteverde, V.P., Galluzzo, F.G., Davì, S., Licciardi, M., Dara, S., 2025. Survey on the presence of floating microplastics, trace metals and metalloids in seawater from Southern Italy to the United States of America. Ecotoxicol. Environ. Saf. 290 (2025), 117507.

Mu, J.L., Qu, L., Jin, F., et al., 2019. Abundance and distribution of microplastics in the surface sediments from the northern Bering and Chukchi Seas. Environ. Pollut. 245, 122–130. https://doi.org/10.1016/j.envpol.2018.10.097.

Naher, H., Sarker, N.J., 2014. Food and feeding habits of white-throated kingfisher (*Halcyon smyrnensis*) in Bangladesh. Bangladesh J. Zool. 42 (2), 237–249. https://doi.org/10.3329/bjz.v42i2.23366.

Naji, A., Nuri, M., Amiri, P., Niyogi, S., 2019. Small microplastic particles (S-MPPs) in sediments of mangrove ecosystem on the northern coast of the Persian Gulf. Mar. Pollut. Bull. 146, 305–311. https://doi.org/10.1016/j.marpolbul.2019.06.033.

Nazir, A., Hussain, S.M., Riyaz, M., Zargar, M.A., Kere, Z., 2024. Microplastic and heavyMetal contamination in sediments of the high-altitude nundkol lake of Northwestern Himalayas-Kashmir. Environmental Pollution and Management 1, 167–178.

Neelavannan, K., Sen, I.S., Lone, A.M., Gopinath, K., 2022. Microplastics in the highaltitude Himalayas: assessment of microplastic contamination in freshwater lake sediments, Northwest Himalaya (India). Chemosphere 290, 133354.

Noel, M., et al., 2022. Microplastics distribution in sediment and mussels along the British Columbia Coast, Canada. Mar. Pollut. Bull. 185, 114273.

Novikov, M.A., Gorbacheva, E.A., Prokhorova, T.A., et al., 2021. Composition and distribution of marine anthropogenic litter in the Barents Sea. Oceanology 61 (1), 48–57. https://doi.org/10.1134/s0001437021010148.

Ory, N.C., Sobral, P., Ferreira, J.L., Thiel, M., 2017. Amberstripe scad Decapterus muroadsi (Carangidae) fish ingest blue microplastics resembling their copepod prey along the coast of Rapa Nui (Easter Island) in the South Pacific subtropical gyre. Sci. Total Environ. 568, 430–437. https://doi.org/10.1016/j.scitotenv.2017.01.175.

Pan, Z., Liu, Q., Sun, X., Li, W., Zou, Q., Cai, S., Lin, H., 2022. Widespread occurrence of microplastic pollution in open sea surface waters: evidence from the mid-North Pacific Ocean. Gondwana Res. 108, 31–40.

Pandey, N., Verma, R., Patnaik, S., Anbumani, S., 2023. Abundance, characteristics, and risk assessment of microplastics in indigenous freshwater fishes of India. Environ. Res. 218, 115011.

Parashar, N., Hait, S., 2022. Occurrence and removal of microplastics in a hybrid growth sewage treatment plant from Bihar, India: a preliminary study. J. Clean. Prod. 376, 134295.

Parashar, N., Hait, S., 2023. Abundance, characterization, and removal of microplastics in different technology-based sewage treatment plants discharging into the middle Stretch of the ganga river, India. Sci. Total Environ. 905, 167099.

Pathak, G., Nichter, M., Hardon, A., Moyer, E., Latkar, A., Simbaya, J., Pakasi, D., Taqueban, E., Love, J., 2023. Plastic pollution and the open burning of plastic wastes. Glob. Environ. Change 80, 102648. https://doi.org/10.1016/j.gloenvcha.2023.102648.

Patil, S., Kamdi, P., Chakraborty, S., Das, S., Bafana, A., Krishnamurthi, K., Sivanesan, S., 2023. Characterization and removal of microplastics in a sewage treatment plant from urban Nagpur, India. Environ. Monit. Assess. 195 (1), 47.

Peeken, I., Primpke, S., Beyer, B., et al., 2018. Arctic sea ice is an important temporal sink and means of transport for microplastic. Nat. Commun. 9, 1505. https://doi.org/10.1038/s41467-018-03825-5.

Pelletier, N., Chételat, J., Cousens, B., Zhang, S., Stepner, D., Muir, D.C.G., Vermaire, J.C., 2020. Lead contamination from gold mining in Yellowknife Bay (Northwest Territories), reconstructed using stable lead isotopes. Environ. Pollut. 259. https://doi.org/10.1016/j.envpol.2019.113888.

PlasticsEurope, 2023. Plastics –the Facts 2023. https://plasticseurope.org/.

Provencher, J.F., Aliani, S., Bergmann, M., et al., 2023. Future monitoring of litter and microplastics in the Arctic–challenges, opportunities, and strategies. Arct. Sci. 9 (1), 209–226. https://doi.org/10.1139/as-2022-0011.

Qaiser, N., et al., 2023. Microplastics abundance in abiotic and biotic components along aquatic food chain in two freshwater ecosystems of Pakistan. Chemosphere 313, 137177.

Robin, R.S., Karthik, R., Purvaja, R., Ganguly, D., Anandavelu, I., Mugilarasan, M., Ramesh, R., 2020. Holistic assessment of microplastics in various coastal environmental matrices, southwest coast of India. Sci. Total Environ. 703, 134947. https://doi.org/10.1016/j.scitotenv.2019.134947.

Roch, S., Ros, A.F.H., Friedrich, C., Brinker, A., 2021. Microplastic evacuation in fish is particle size-dependent. Freshwater Biol. 66 (5), 926–935. https://doi.org/10.1111/fwb.13687.

Rochman, C.M., Tahir, A., Williams, S.L., Baxa, D.V., Lam, R., Miller, J.T., Teh, F.C., Werorilangi, S., Teh, S.J., 2015. Anthropogenic debris in seafood: plastic debris and fibers from textiles in fish and bivalves sold for human consumption. Sci. Rep. 5. https://doi.org/10.1038/srep14340.

Ross, P.S., Morales-Caselles, C., 2015. Out of sight, but no longer out of mind: microplastics as a global pollutant. Integr. Environ. Assess.Manag. 11 (4), 721–722. https://doi.org/10.1002/ieam.1681.

Rowenczyk, L., Cai, H., Nguyen, B., Sirois, M., Cote-Laurin, M.C., Toupoint, N., Ismail, A., Tufenkji, N., 2022. From freshwaters to bivalves: microplastic distribution along the Saint-Lawrence river-to-sea continuum. J. Hazard. Mater. 435, 1–10. https://doi.org/10.1016/j.jhazmat.2022.128977.

RSIS, 2006. Area de Proteccion de Flora y Fauna Laguna de Terminos. https://rsis.ramsar.org/ris/1356. (Accessed 21 June 2022).

Rymbai, P.N., Jha, L.K., 2012. Estimation of Sediment Production Rate of the Umbaniun Micro-watershed, Meghalaya, India. J. Geogr. Reg. Plann.. 5 (11), 293–297.

Ryu, S.K., 2019. Variation in access to safe drinking water across different countries: an explanatory framework. Soc. Sci. 8 (2). https://doi.org/10.3390/socsci8020068.

Sarkar, D.J., Sarkar, S.D., Manna, R.K., Samanta, S., Das, B.K., 2020. Microplastics pollution: an emerging threat to freshwater aquatic ecosystem of India. J. Inland Fish Soc. India 52 (1), 5–15.

Sathish, K., Saraswat, S., Anusha, B.S., Kandpal, H., Kokila, A., 2024. Assessing microplastic pollution in Indian lake ecosystems: a mini review. Water, Air, Soil Pollut. 235 (4). https://doi.org/10.1007/s11270-024-07014-7.

Satya, A., Harimawan, A., Haryani, G.S., Johir, M.A.H., Vigneswaran, S., Kurniawan, T.A., Setiadi, T., 2021. Integrated treatment of submerged membrane and adsorption using dried Aphanothece sp for removing cadmium from synthetic wastewater. J. Water Proc. Eng. 41. https://doi.org/10.1016/j.jwpe.2021.102022.

Schuerholz, T., Meyer, M.C., Friedrich, L., Przemeck, M., Sümpelmann, R., Marx, G., 2006. Reliability of continuous cardiac output determination by pulse-contour analysis in porcine septic shock. Acta Anaesthesiol. Scand. 50 (4), 407–413. https://doi.org/10.1111/j.1399-6576.2006.00982.x.

Schuur, E.A.G., Abbott, B.W., Commane, R., Ernakovich, J., Euskirchen, E., Hugelius, G., Grosse, G., Jones, M., Koven, C., Leshyk, V., Lawrence, D., Loranty, M.M., Mauritz, M., Olefeldt, D., Natali, S., Rodenhizer, H., Salmon, V., Schädel, C., Strauss, J., Treat, C., Turetsky, M., 2022. Permafrost and climate change: carbon cycle feedbacks from the warming arctic. Ann. Rev. Environ.Resour. 47, 343–371. https://doi.org/10.1146/annurev-environ-012220-011847.

Selamat, M.H., Tamam, A.H., Soon, F.K., 2024. From risk to resilience: analyzing key success factors in Malaysian water risk management. Earth Syst. Environ. 8 (3), 723–745. https://doi.org/10.1007/s41748-024-00413-0.

Sharma, S., Basu, S., Shetti, N.P., Nadagouda, M.N., Aminabhavi, T.M., 2021. Microplastics in the environment: occurrence, perils, and eradication. Chem. Eng. J. 408. https://doi.org/10.1016/j.cej.2020.127317.

Shaw, D.B., Li, Q., Nunes, J.K., Deike, L., 2023. Ocean emission of microplastic. PNAS nexus 2 (10). https://doi.org/10.1093/pnasnexus/pgad296.

Siddique, M.A.M., Shazada, N.E., Ritu, J.A., Turjo, K.E.Z., Das, K., 2024. Does the mouth size influence microplastic ingestion in fishes? Mar. Pollut. Bull. 198. https://doi.org/10.1016/j.marpolbul.2023.115861.

Singh, S., Gray, A.B., Murphy-Hagan, C., Hapich, H., Cowger, W., Perna, J., Le, T., Nogi, H., Badwal, B., McLaughlin, K., Kessouri, F., Moore, C., Lattin, G., Thornton Hampton, L.M., Wong, C.S., Sutula, M., 2025. Microplastic pollution in the water column and benthic sediment of the San Pedro Bay, California, USA. Environ. Res. https://doi.org/10.1016/j.envres.2025.120866.

Spano, C., Muccifora, S., Castiglione, M.R., Bellani, L., Bottega, S., Giorgetti, L., 2022. Polystyrene nanoplastics affect seed germination, cell biology and physiology of rice seedlings in-short term treatments: evidence of their internalization and translocation. Plant Physiol. Biochem. 172, 158–166. https://doi.org/10.1016/j.plaphy.2022.01.012.

Srinivasalu, S., Natesan, U., Ayyamperumal, R., Kalam, N., Anbalagan, S., Sujatha, K., Alagarasan, C., 2021. Microplastics as an emerging threat to the freshwater ecosystems of Veeranam lake in south India: a multi-dimensional approach. Chemosphere 264, 128502.

Stabeno, P., Kachel, N., Ladd, C., Woodgate, R., 2018. Flow patterns in the eastern Chukchi Sea: 2010–2015. J. Geophys. Res., Oceans 123 (2), 1177–1195. https://doi.org/10.1002/2017jc013135.

Stenekes, S., Parlee, B., Seixas, C., 2020. Culturally driven monitoring: the importance of traditional ecological knowledge indicators in understanding aquatic ecosystem change in the northwest territories' Dehcho region. Sustainability 12 (19), 7923. https://doi.org/10.3390/su12197923.

Sutton, R., Mason, S.A., Stanek, S.K., Willis-Norton, E., Wren, I.F., Box, C., 2016. Microplastic contamination in the San Francisco bay, California, USA. Mar. Pollut. Bull. 109 (1), 230–235. https://doi.org/10.1016/j.marpolbul.2016.05.077.

Sutula, M., Ho, M., Sengupta, A., Kessouri, F., McLaughlin, K., Kenny, M.C., Bianchi, D., 2021. A baseline of terrestrial freshwater and nitrogen fluxes to the Southern California Bight, USA. Mar. Pollut. Bull. 170, 112669. https://doi.org/10.1016/j.marpolbul.2021.112669.

Tahir, R., Abdullah, Z., 2023. Revolutionising delivery in TVET for industry 4.0: strategies from the field–NGT and ISM approach. J. Curr. Sci. Res. Rev. 1 (05), 139–157.

Thacharodi, A., Meenatchi, R., Hassan, S., Hussain, N., Bhat, M.A., Arockiaraj, J., Ngo, H.H., Le, Q.H., Pugazhendhi, A., 2024. Microplastics in the environment: a critical overview on its fate, toxicity, implications, management, and bioremediation strategies. J. Environ. Manag. 349, 119433.

Tsering, T., Sillanpää, M., Sillanpää, M., Viitala, M., Reinikainen, S.P., 2021. Microplastics pollution in the Brahmaputra River and the Indus river of the Indian Himalaya. Sci. Total Environ. 789. https://doi.org/10.1016/j.scitotenv.2021.147968.

Tsering, T., Sillanpää, M., Viitala, M., Reinikainen, S.P., 2022. Variation of microplastics in the shore sediment of high-altitude lakes of the Indian Himalaya using different pretreatment methods. Sci. Total Environ. 849. https://doi.org/10.1016/j.scitotenv.2022.157870.

Usman, S., Abdull Razis, A.F., Shaari, K., Azmai, M.N.A., Saad, M.Z., Mat Isa, N., Nazarudin, M.F., 2022. The burden of microplastics pollution and contending policies and regulations. Int. J. Environ. Res. Publ. Health 19 (11). https://doi.org/10.3390/ijerph19116773.

Vage, K., Papritz, L., Havik, L., Spall, M.A., Moore, G.W.K., 2018. Ocean convection linked to the recent ice edge retreat along east Greenland. Nat. Commun. 9, 1287. https://doi.org/10.1038/s41467-018-03468-6.

Veerasingam, S., Ranjani, M., Venkatachalapathy, R., Bagaev, A., Mukhanov, V., Litvinyuk, D., Verzhevskaia, L., Guganathan, L., Vethamony, P., 2020. Microplastics in different environmental compartments in India: analytical methods, distribution, associated contaminants and research needs. TrAC, Trends Anal. Chem. 133, 116071. https://doi.org/10.1016/j.trac.2020.116071.

Viehman, S., Vander Pluym, J.L., Schellinger, J., 2011. Characterization of marine debris in North Carolina salt marshes. Mar. Pollut. Bull. 62 (12), 2771–2779. https://doi.org/10.1016/j.marpolbul.2011.09.010.

Walters, C., English, K., Korman, J., Hilborn, R., 2019. The managed decline of British Columbia's commercial salmon fishery. Mar. Pol. 101, 25–32. https://doi.org/10.1016/j.marpol.2018.12.014.

Wang, Y., Karasik, R., 2022. Plastic Pollution Policy Country Profile: Indonesia. NI PB 22-05. Duke University, Durham, NC.

Wang, T., Li, B., Shi, H., Ding, Y., Chen, H., Yuan, F., Liu, R., Zou, X., 2024. The processes and transport fluxes of land-based macroplastics and microplastics entering the ocean via rivers. J. Hazard. Mater. 133623. https://doi.org/10.1016/j.jhazmat.2024.133623.

Wen, Z., Xie, Y., Chen, M., Dinga, C.D., 2021. China's plastic import ban increases prospects of environmental impact mitigation of plastic waste trade flow worldwide. Nature Commun. 12 (1). https://doi.org/10.1038/s41467-020-20741-9.

Wesch, C., Elert, A.M., Wörner, M., Braun, U., Klein, R., Paulus, M., 2017. Assuring quality in microplastic monitoring: about the value of clean-air devices as essentials for verified data. Sci. Rep. 7 (1). https://doi.org/10.1038/s41598-017-05838-4.

Yang, Y., Liu, W., Zhang, Z., Grossart, H.P., Gadd, G.M., 2020. Microplastics provide new microbial niches in aquatic environments. Appl. Microbiol. Biotechnol. 104 (15), 6501–6511. https://doi.org/10.1007/s00253-020-10704-x.

Yu, X., Ladewig, S., Bao, S., Toline, C.A., Whitmire, S., Chow, A.T., 2018. Occurrence and distribution of microplastics at selected coastal sites along the southeastern United States. Sci. Total Environ. 613–614, 298–305. https://doi.org/10.1016/j.scitotenv.2017.09.100.

Yuan, B., Gan, W., Sun, J., Lin, B., Chen, Z., 2023. Depth profiles of microplastics in sediments from inland water to coast and their influential factors. Sci. Total Environ. 903, 166151. https://doi.org/10.1016/j.scitotenv.2023.166151.

YuLan, Z., ShiChang, K., Xi, L., ZhaoQing, W., ShiWei, S., Li, C.J., 2025. TingFeng Dou. Res. Cold Arid Reg. 17, 65–72.

Zhang, F., Xu, J., Wang, X., Jabeen, K., Li, D., 2021. Microplastic contamination of fish gills and the assessment of both quality assurance and quality control during laboratory analyses. Mar. Pollut. Bull. 173, 113051. https://doi.org/10.1016/j.marpolbul.2021.113051.

Zhang, Y.L., Gao, T.G., Kang, S.C., et al., 2023. Cryosphere as a temporal sink and source of microplastics in the Arctic region. Geosci. Front. 14 (4), 101566. https://doi.org/10.1016/j.gsf.2023.101566.

Zheng, S., Zhao, Y., Liangwei, W., Liang, J., Liu, T., Zhu, M., Li, Q., Sun, X., 2020. Characteristics of microplastics ingested by zooplankton from the Bohai Sea, China. Sci. Total Environ. 713, 136357. https://doi.org/10.1016/j.scitotenv.2019.136357.

Zhou, Y., Liu, X., Wang, J., 2019. Characterization of Microplastics and the Association of heavy metals with microplastics in suburban soil of central China. Sci. Total Environ. 694, 133798.

Zoveidadianpour, Z., Alava, J.J., Drever, M.C., Schuerholz, G., Pierzchalski, C., Douglas, T., Heath, W.A., Juurlink, B., Bendell, L., 2025. Microplastic distribution and composition in mudflat sediments and varnish clams (Nuttallia obscurata) at two estuaries of British Columbia, Canada: an assessment of potential anthropogenic sources. Mar. Pollut. Bull. 211, 117367. https://doi.org/10.1016/j.marpolbul.2024.117367.

CHAPTER 9

Regulatory frameworks and global perspectives on microplastic pollution

Melvin S. Samuel
Department of Civil, Construction, and Environmental Engineering, Marquette University, Milwaukee, WI, United States

Highlights

- MP regulation remains largely indirect and precautionary at the global level
- National policies vary widely, with few enforceable standards currently in place
- Prevention-focused strategies dominate governance approaches
- International efforts primarily focus on restricting internationally added MPs and reducing MP precursors

9.1 Introduction

Microplastic (MP) pollution has evolved from an emerging environmental concern into a globally recognized regulatory challenge affecting water quality, ecosystem integrity, and public health. The pervasive presence of MPs in surface waters, sediments, drinking water, food systems, and even atmospheric compartments has exposed significant limitations in the existing environmental governance structure. Unlike conventional chemical pollutants, MPs are heterogeneous in size, shape, polymer composition, and origin, making them difficult to regulate using traditional contaminant-based frameworks.

International organizations and national governments have increasingly acknowledged the need for coordinated action. However, regulatory responses remain fragmented, with most policies indirectly addressing MPs through broader plastic waste reduction strategies rather than direct numerical standards or enforceable thresholds. Reports from global agencies highlight that while MPs are widely detected across environmental matrices, standardized monitoring protocols, exposure limits, and risk-based criteria are still largely absent. Recent policy discussions emphasize prevention over remediation, focusing on source control, product bans, extended producer responsibility, and improved waste management systems. At the same time, growing scientific evidence on MP interactions with biota, toxic additives, and cocontaminants has intensified calls for more explicit regulatory recognition. This section synthesizes current international and national policy approaches, identifies governance gaps, and outlines future directions for developing effective, science-informed regulatory frameworks for MP pollution.

At the international level, governance of MP pollution is largely embedded within broader plastic and chemical management initiatives rather than addressed as an independent regulatory category. Global efforts, including those led by the United Nations, prioritize upstream plastic reduction through

Microplastics in Aquatic Environments. https://doi.org/10.1016/B978-0-443-45619-0.00016-9

resolutions aimed at curbing plastic pollution at its source. Within this landscape, the European Union has taken a comparatively proactive stance by introducing restrictions on intentionally added MPs under chemical regulatory frameworks, particularly for consumer and industrial products. However, most intentional measures remain focused on source control and prevention, with limited attention given to environmental monitoring requirements, exposure thresholds, or enforceable concentration limits, reflecting the early and evolving stage of global MP regulation. At the national level, regulatory responses vary widely across regions. Some countries have introduced mandatory monitoring and reporting requirements for MPs in drinking water and aquatic environments, while others emphasize bans on microbeads, single-use plastics, and the implementation of extended producer responsibility schemes. Industry-led standards increasingly promote improved plastic handling practices, recycling targets, and pellet loss prevention measures. Despite these initiatives, MPs are generally regulated indirectly, as secondary pollutants generated through the degradation of larger plastic materials, rather than as contaminants warranting standalone regulatory frameworks.

Significant challenges persist in the governance of MP pollution. The lack of harmonized definitions, standardized analytical methods, and universally accepted numerical thresholds limits regulatory enforcement across regional comparability. Most jurisdictions do not yet define legally enforceable limits for MPs in water, sediment, air, or food, complicating risk assessment and compliance. Furthermore, ongoing scientific uncertainty regarding toxicity, exposure pathways, and long-term human and ecological health effects continues to constrain policy development. These gaps highlight the urgent need for coordinated international frameworks that integrate scientific advances with adaptive regulatory mechanisms. This chapter is unique in that it consolidates fragmented regulatory, policy, and governance perspectives across global and national scales into a coherent science–policy narrative. By translating complex legal and regulatory structures into an accessible framework, it bridges the divide between environmental science and policymaking. The importance of this chapter lies in its relevance to current environmental and public health challenges, as MPs are increasingly detected in water, food systems, and human tissues. Clear regulatory guidance is essential to support evidence-based decision-making and effective mitigation strategies. The chapter further underscores the novelty and necessity of harmonized governance in a world of rapidly increasing plastic production, where aligning scientific understanding with regulatory action is critical for sustainable environmental management. This chapter also consolidates regulatory developments across international, national, and sectoral levels into a coherent governance narrative. Unlike many studies that focus solely on scientific occurrence or toxicology, this section bridges environmental science with policy analysis, highlighting regulatory gaps and future needs. Its novelty lies in emphasizing governance as a dynamic process shaped by scientific uncertainty. The chapter provides actionable insights for regulators, researchers, and decision makers navigating the evolving MP policy landscape.

9.2 Existing international regulations on microplastics

9.2.1 Microplastic pollution in aquatic ecosystems

Emerging major environmental issues are due to MP pollution as they are affecting both marine and terrestrial ecosystems. Almost every part of the environment, including rivers, lakes, oceans, soils, sediments, and living organisms are found to be with these tiny plastic particles as they are produced intentionally or formed when larger plastics break down (Dauvergne, 2018). Because of human activities such as industrial production, agriculture, textile manufacturing, and the disposal of everyday plastic products, these plastics are found widespread in the environment. Estimates made globally show that hundreds of thousands of tons of plastic debris are already found circulating in the world's oceans, and this number rises in number each year. This pollution originates on land and is carried in large amounts to the marine environment by mismanaged waste, littering, and uncontrolled discharge into rivers and stormwater systems. Studies prove that more than 80% of the total plastic pollution, with millions of tons entering the ocean annually, is from land-based sources. It poses long-lasting ecological risks, as they are released in only smaller fractions that float on the surface, and most of the plastic sink to the seafloor or accumulate along coastlines. Governments, scientific communities, nongovernmental organizations, and the public are affected by this growing plastic pollution. The factors affecting marine ecosystems are habitat destruction, declining biodiversity, and reduced water quality. The impacts caused by this pollution weaken the ocean system's flexibility, and to recover from disturbances is difficult.

The most important and urgent threat faced by the environment is plastics, as they contribute to more than 85% of marine pollution (Lincoln et al., 2022). As MPs have been detected throughout the seafood chain, concerns about human health have increased in recent years. As MPs are frequently ingested by fish, shellfish, and other marine organisms, these particles can accumulate in tissues consumed by humans. Due to the continuous exposure and widespread availability, these MPs are found to be in human lungs, blood, and placental tissue. The findings from the above studies suggest that strong policies and regulations are to be developed to reduce the source of plastic pollution, improve waste management, and protect both environmental and public health. Elevated levels of risks like cancer, gastrointestinal disorders, endocrine disruption, breast disease, and respiratory illnesses are associated with airborne MPs contaminants, which are found by cross-referencing toxicological databases. To develop strategies for reducing airborne MP pollution, it is necessary to understand inhalation exposure pathways in rapidly urbanizing regions.

9.2.2 Plastic management, global policies, and the need for stronger regulations

By 2030, to achieve the United Nations Sustainable Development Goals (SDGs), it is essential to develop effective plastic management for protecting ecosystems (UN, 2020). Studies show that not only biodiversity but also the global economy is affected by plastic pollution in marine environments. Because of the declining water quality, damaged habitats, and increased operational costs, the coastal communities and industries face losses in tourism, fishing, transportation, and recreation (Stoll et al., 2020). The annual losses ranging from USD 3300 to USD 33,000 per ton of plastic are caused by the plastic pollution in the ocean, which causes considerable impact on the economy (UN, 2012, 2014). Highlighting these growing risks, the United Nations urges countries to implement effective measures by 2025 to limit marine waste. To encourage the reduction in plastic inputs into ocean systems by governments, the private sectors, and communities, the United Nations Environment Program (UNEP) established the Global Partnership on Marine Litter (GPML) to support this global effort. Action plans focusing on marine plastic litter have been approved by the United Nations Environment Assembly (UNEA), which highlights the urgent need and coordinated efforts to reduce environmental and public-health risks (Henderson and Green, 2020; World Health Organization (WHO), 2022). Even though regulations have been implemented to limit plastic pollution in many countries, to address the growing problems, particularly the MPs, the current policies are not sufficient. More than 80% of marine waste is caused by MPs, but the existing measurements focus on general plastic pollution rather than the challenges caused by MPs. Short-term approaches to prevent environmental contamination, like investment in research, recycling systems, waste treatment, and biodegradation technologies, have increased for over a decade, but the result is not been as fruitful. Comprehensive and long-term solutions are required for addressing MP pollution. Measures such as enhancing recycling technologies, reducing discharges from wastewater treatment plants, improving plastic life-cycle management, and increasing public awareness about plastic use and disposal are adopted. This collaboration must be made globally because plastic pollution crosses national boundaries and accumulates in international waters.

9.2.3 Human rights and microplastic pollution

Worldwide recognized fundamental human rights is the right to health. In the 1946 Constitution, the World Health Organization (WHO) included this principle, and later in 1948 it was reinforced by the Universal Declaration of Human Rights. Today with the International Covenant on Economic, Social and Cultural Rights (ICESCR) many international and regional treaties protect this right carrying the highest legal authority (WHO, 2022). In ICESCR, Article 12 states that everyone in the constitution has the right to enjoy "the highest attainable standard of physical and mental health." Moreover, health is recognized as a basic human right formally by more than 115 national constitutions (Human Rights Council, R.O., 2021). Challenges come from MP pollution in multiple ways. To support human well-being with essential clean air, safe water, fertile soil, and sustainable food production, a healthy environment is required. When these natural resources are contaminated by MPs, they undermine food safety, degrade ecosystems, and break the internationally recognized human rights protections. For over hundreds of years, these MPs have remained in the environment, causing destruction in marine and terrestrial ecosystems. Further, it reduces the important natural process of oceans' ability to act as carbon sinks, which helps in regulating climate change. So, the protection of the environment from MPs is not only an ecological duty but also necessary to safeguard human dignity, health, and future generations (Human Rights Council, 2020).

It is widely recognized that there is a connection between human rights and environmental protection. To highlight this, the principle is echoed in the 1972 Stockholm Declaration and the 1992 Rio Declaration on Environment and Development by ICESCR on environmental quality to maintain good health. The United Nations strengthened this relationship recently by approving the 2021 Joint Declaration on the Right to a Clean, Healthy, and Sustainable Environment. The UN Human Rights Council passed Resolution 48/13 in 2021, which states that it is right for every person to live in an environment that supports well-being and sustainable development. The three pillars of sustainability, environmental, economic, and social, are highlighted in this resolution. And recognizes the rights to live, clean water, adequate food, housing, and good physical and mental health, as healthy ecosystems are essential for achieving human rights. These resolutions gained widespread global support, even though they are "soft laws," because they are not legally compulsory. Through national laws, constitutional provisions, regional agreements, and judicial decisions, more than 150 UN member states now recognize the right to a healthy environment. When addressing climate change in the 2015 Paris Agreement, it insists that countries consider human rights and further incorporate environmental protection into global governance frameworks.

Throughout the world, MPs have become a global threat to human health. The toxic chemicals carried by these MPs get accumulated in the human body, leading to respiratory, reproductive, metabolic, and inflammatory diseases. To address these risks, the UN Human Rights Council recognizes the right to a healthy environment as an important point and takes serious steps. The resolution boots policies to reduce plastic pollution, improve waste management, protect vulnerable populations, and promote long-term sustainability, as human rights are linked with environmental protection. A powerful foundation for global action is created by recognizing MP pollution as a human rights issue. It highlights that protecting the environment is not optional; to ensure good health, well-being, and rights of present and future generations is absolutely a necessary condition.

9.2.4 International agreements and treaties for microplastic prevention and management

Because of the activities driven by humans and inadequate waste management systems, MP pollution has now emerged as a major global challenge. The release of MPs into rivers, lakes, and oceans is contributed to by poor handling of solid waste, stormwater runoff, flooding, and landslides. The plastics can easily break down and fragment into tiny particles very slowly. These tiny particles can be transported by wind and ocean currents easily and accumulate far from their original source. This provides the reason for a global address, as MP pollution has a transboundary issue. A coordinated global action and long-term international cooperation are required in this issue. In recent years, to address MP pollution, either directly or indirectly, several international treaties and agreements have been passed. The agreements made in these issues are voluntary, and many of them are known as "soft law." These laws play a major role in shaping global norms and guiding national policies. These soft-law instruments may advance into binding commitments over a period (Stoll et al., 2020).

The important efforts taken internationally include the United Nations Sustainable Development Goals (SDGs), particularly SDG 14.1, which suggests reducing marine litter by 2025 in a considerable amount (Lohr et al., 2017). During the United Nations Ocean Conferences held in 2017 and 2018, support for global action was encouraged, and it insists the countries to adopt plastic management with stronger policies. The legal framework used to restrict the flow of plastic waste into the environment was provided by other major treaties such as the United Nations Convention on the Law of the Sea (UNEA, 2014), the London Convention and Protocol, the Stockholm Convention on Persistent Organic Pollutants, and the Basel Convention. A major advancement in international plastic governance is represented by the recent amendment to the Basel Convention, which regulates the global plastic waste trade. Despite various efforts and various treaties, there is a significant gap in management. The larger plastic debris or secondary MPs that form from the breakdown of larger items are mostly focused on these existing agreements. The primary MPs, such as microbeads or industrial pellets, which are directly entering the environment, are addressed by only a few. Stronger international regulations are needed to target MPs across their entire life cycle.

The political momentum to challenge MP pollution is increasing around the world. To reduce marine plastics and MPs, the UNEA has passed several resolutions. Guidelines were introduced by the Food and Agriculture Organization (FAO) for marking fishing gear to reduce marine litter, and the International Maritime Organization (IMO) has framed rules regarding the actions to address waste from ships (Lohr et al., 2017). Marine plastics endanger ecosystems, economies, and public health, as clearly reflected in these steps. Several directives that directly and indirectly address MP contamination are adopted within Europe by the European Union (EU). The Marine Strategy Framework Directive, the Waste Directive, the Single-Use Plastics Directive, and the Port Reception Facilities Directive are some. The intentionally added MPs are targeted with advanced regulations by the EU and include measures to reduce plastic waste by adopting a comprehensive Circular Economy Package, and encourage sustainable material usage. The progress across regions varies in Baltic countries. A legal ban on "rinse-off" cosmetic products containing MPs is only adopted in Sweden. Countries like Denmark, Finland, and Poland are conducting studies and drafting policies to inform potential bans (UNEP, 2018).

9.2.5 Global policies for microplastic management and prevention

During the late twentieth century, international efforts to address marine litter, including MPs, began, and since then scope and visibility have expanded. At the 2015 G7 Summit, the global concern about MPs has grown steadily, and even ocean health was highlighted as a priority issue. A vital policy to regulate marine pollution was formed by early international agreements, such as MARPOL Annex V and the UNCLOS (Axelsson and Sebille, 2017). There are limitations in these conventions. Because of difficulties in monitoring vast ocean areas, exemptions for certain vessels, and the lack of adequate port facilities in many regions, these implementations are not possible. Because of these limitations, the source of land pollution is always considered. Over the past 3 decades, global governance has evolved to address these challenges. A wide range of policies to reduce plastic use has been introduced by countries, especially on single-use plastic bags. In many regions Levies, bans, and fees have proven effective measures. For example, a ban on thin plastic bags in 2002 by Bangladesh has motivated the introduction of similar measures in more than 60 countries (Khan et al., 2022). The consumers' behavior changes by the pricing mechanisms for example, strict penalties for producing or using banned plastic bags in Kenya (Conti et al., 2021), 90% of plastic bag usage is reduced in Ireland by a levy, and about 74% of plastic consumption is reduced in Portugal by a 2015 tax (Martinho et al., 2017). Reduced plastic bag distribution and promoted reusable alternatives resulted from voluntary campaigns in Australia and parts of Europe.

A combination of legal restrictions, economic incentives, and public awareness campaigns is followed by many countries to reduce plastic waste, which reflects the broader global trends. Depending on strong enforcement, accessible alternatives, and public acceptance, the effectiveness of these measures is accounted. In South Africa, the levy on thicker bags was very low, so these policies became less successful (Bruch et al., 2016). On the other hand, when retailers and consumers cooperate, there can be significant reductions without formal bans, such as voluntary supermarket initiatives.

The central components of plastic mitigation strategies are recycling and improved waste management. With high recycling rates supported by legislation, producer responsibility schemes, and coordinated national efforts, Europe has made substantial progress. To reduce landfill volumes and extend the material's life cycle, reprocessing plastics can be adopted, and plastic waste can be converted into fuel by technologies such as pyrolysis. However, there are risks during energy recovery and the release of toxic substances during combustion, which need attention. Inadequate recycling infrastructure and systems prevail in many regions, which cannot keep pace with rising plastic consumption. Countries such as China have reshaped the recycling patterns in global plastic trade by reprocessing most imported plastic waste. Domestic recycling solutions must be adopted by many nations, as there are many restrictions on international waste shipments. This highlights the need for reduced reliance on landfills and incineration.

9.2.6 Global governance of microbeads

In recent years, the regulation of microbeads, tiny plastic particles widely used in personal care products, has reached global efforts, but progress has not been made well. Even though many countries aim to limit the production or sale of microbead-containing products by introducing voluntary initiatives, only some countries have proposed legally binding bans (Eriksen et al., 2020). National legislation prohibiting the manufacture, distribution, and/or sale of microbeads in personal care products has been adopted by only eight countries: Canada, New Zealand, South Korea, France, Sweden, the United Kingdom, the United States, and Italy. The only country with a comprehensive ban covering all rinse-off cosmetic products containing plastic particles smaller than 5 mm is Sweden. An important milestone is represented by these actions. But this also highlights the continued use and environmental release of microbeads worldwide by broader international cooperation.

9.2.7 Regional policies and regulations for microplastic management

MPs are often addressed within broader marine litter and waste management frameworks. Regional efforts to manage and prevent MP pollution have expanded through targeted regulations in recent years. Even though there is a continuous growth regarding awareness and research on MP contamination, comprehensive and MP-specific policies still lack. The most active regional body to reduce MP pollution by developing coordinated strategies is the EU. The EU has introduced measures based on the polluter pays principle, linking producer responsibility to environmental impacts through its "European Strategy for Plastics in a Circular Economy". Single-Use Plastics Directive, revisions to the Waste Framework Directive, and the Eco-design Directive are the key regulations to be included. These aims to restrict the use of intentionally added MPs, restrict the use of intentionally added MPs, and limit pollution from fishing gear and ship-generated litter. By improving product design, enhancing monitoring methods, and supporting innovation and research, unintentional MP release can be reduced proposed by the EU. A substantial progress is represented by these initiatives; there is a lack of a unified, long-term roadmap for MP prevention.

EU policy is further complemented by several regional seas conventions. To address marine litter, including MPs, provisions are included in the Barcelona Convention, Bucharest Convention, HELCOM Convention, and the OSPAR Convention (Vince and Hardesty, 2018). In 2015, a Regional Action Plan on Marine Litter was adopted by the Baltic Sea countries. Under the Barcelona Convention in 2013, the Regional Plan for Marine Litter Management was introduced by Mediterranean nations. Although implementation varies by country and region, frameworks are highlighting growing regional cooperation. Targeted initiatives are launched by other regions beyond Europe. To emphasize global cooperation, producer responsibility, and innovation in product design, the 2017 Call to Action on microbeads was adopted by the Secretariat of the Pacific Regional Environment Programme (SPREP). A major political commitment was represented by the ASEAN Bangkok Declaration on Combating Marine Debris (2019), in Southeast Asia, even though it follows a voluntary "soft law" approach without binding enforcement (Gunningham et al., 1998). Shared norms are established by these initiatives, and their fruitful results depend on national governments.

9.3 National and international level policies and industry standards

9.3.1 Policy implications and governance requirements for addressing microplastic pollution

The expanding scientific literature documenting MP contamination underscores an urgent need for coordinated and coherent policy action at both national and international scales. A central priority is the development of harmonized definitions, standardized sampling and measured protocols, and consistent analytical methodologies for MPs. In the absence of these core elements, comparisons among regulatory frameworks across jurisdictions remain challenging, and the establishment of enforceable limits or systematic monitoring obligations is significantly constrained. Existing policy initiatives have largely concentrated on intentionally added MPs, most notably microbeads used in personal care products. Although these measures represent a meaningful initial step, they address only a minor proportion of total MPs' inputs into the environment. By contrast, microfibers released from synthetic textiles and particles generated through tire wear represent far larger and more spatially diffuse sources, yet policy instruments targeting these pathways are still relatively limited. These sectors therefore present substantial opportunities for future regulatory advancement, including the introduction of product design requirements, extended producer responsibility frameworks, and infrastructure-based mitigation strategies. A further shortcoming in current governance approaches is the limited systematic assessment of policy performance. Only a small number of studies have critically examined whether existing regulations lead to measurable reductions in MP emissions or environmental concentrations. Measures such as bans on cosmetic microbeads or restrictions on plastic bags have been widely adopted, yet their environmental effectiveness, economic implications, and potential unintended outcomes remain insufficiently quantified. Rigorous policy evaluation is thus necessary to ensure that mitigation efforts deliver meaningful environmental benefits while avoiding unnecessary social or economic burdens.

9.3.2 Implications for policy development, policy context, and objectives

MPs are now widely distributed across environmental compartments and are increasingly recognized as potential threats to ecosystem integrity and human health. In response, governments at local, national, and international levels have launched research initiatives and regulatory actions aimed at better understanding and reducing MP pollution. Among these efforts, California has positioned itself as a leading jurisdiction, adopting some of the most comprehensive MP-oriented policies worldwide/ Comparable initiatives, however, are also emerging across other national and international contexts (Burton, 2015). The purpose of this chapter section is to analyze the current spectrum of governmental actions addressing MPs and to assess their effects on MP production, usage patterns, environmental distribution, and mitigation efforts. Through a synthesis of policy developments across multiple jurisdictions, this discussion seeks to support evidence-based policymaking and to inform future regulatory approaches within California and beyond (Pouikli, 2020).

This section is organized into five components:

- An overview of policy actions related to MPs in California.
- Government mandates supporting research on environmental and human health impacts.
- Regulatory interventions addressing major MP sources, including microbeads, textiles, and tires.

- A high-level assessment of multinational agreements and international initiatives; and
- Lessons derived from MP policy evaluations that may guide future MP governance.

9.3.3 Policy data sources and methodological approach

The policy information presented in this section is primarily derived from the Plastics Policy Inventory developed by the Nicholas Institute for Energy, Environment & Sustainability at Duke University. This resource compiles legislation, governmental actions, and multinational agreements addressing plastic pollution, including policies specific to MPs. To supplement this database, narrative analyses from *20 Years of Government Responses to Global Plastic Pollution* were used to situate contemporary policies within a broader historical context (Karasik et al., 2017, 2020).

Additional legislative information was obtained from LegiScan and BillTrack50, which offer comprehensive, publicly accessible records of enacted legislation across all U.S states (WHO, 2019; 2020). Searches were limited to laws that explicitly reference MPs to ensure relevance and completeness. Policies addressing textile- and tire-derived MPs were further informed by the Organization for Economic Cooperation and Development (OECD) (2020a,b) report on waterborne MPs, as well as targeted reviews of governmental and nongovernmental organization websites to capture recent policy developments.

9.3.3.1 Spotlight: Microplastics legislation in California

California has implemented several landmark policies aimed at mitigating MP pollution. Initial legislative efforts focused in preventing the release of preproduction plastic pellets from industrial facilities into aquatic environments. Subsequent regulations banned the sale of personal care products containing plastic microbeads, reflecting a precautionary approach to the intentional use of MPs. More recently, the state has transitioned toward more comprehensive monitoring and long-term strategic planning. Current legislation requires the adoption of standardized definitions and analytical methods for detecting MPs in drinking water, alongside routine statewide monitoring programs. In parallel, California has developed a statewide MPs strategy designed to address pollution through source reduction, pathway-based interventions such as wastewater and stormwater treatment, and public outreach and education efforts. The strategy also proposes forward-looking measures targeting secondary MPs, including fibers from textiles and particles from torn wear. Proposed actions include assessments of appliance-based filtration technologies, establishment of fiber-shredding performance standards, encouragement of alternative material designs, and transportation policies intended to reduce tire abrasion (California Legislature, 2007, 2015, 2018, 2022). Complementary packaging regulations further contribute to MP mitigation by curbing MP production, highlighting the interconnected nature of plastic governance across scales.

9.3.4 Summary and conclusion

Overall, current policy responses to MP pollution reflect increasing recognition of the issue but remain uneven in both scope and effectiveness. While notable progress has been achieved in regulating intentionally added MPs and strengthening monitoring frameworks, substantial gaps persist for dominant sources such as synthetic textiles and tires. The continued absence of harmonized definitions, limited evaluation of policy outcomes, and fragmented governance structures pose significant barriers to effective mitigation.

Looking ahead, effective governance of MPs will depend on coordinated international cooperation, scientifically grounded regulatory frameworks, and systematic assessment of policy performance. A comprehensive approach that integrates prevention, monitoring, and accountability across the entire plastic life cycle will be critical to reducing environmental contamination and safeguarding ecosystems and public health.

9.4 Microplastic policies: Research, regulation, and mitigation pathways

Policy responses to MPs pollution beyond California can be broadly categorized into three main approaches:

1. Initiatives that advance research on the occurrence, fate, and impacts of MPs.
2. Regulatory instruments intended to reduce or prevent releases of MPs; and
3. International and multinational agreements are designed to foster coordinated global responses.

Collectively, these approaches reflect increasing acknowledgment that scientific uncertainty—particularly concerning measurement techniques, exposure pathways, and health implications—has limited effective governance. Consequently, many governments have emphasized research and monitoring as essential precursors to evidence-based regulatory action.

9.4.1 Government support for microplastic research

Governments increasingly recognize that the lack of standardized definitions and analytical methodologies constrains informed policymaking. In response, numerous jurisdictions have enacted legislation or dedicated funding to support research on MP sources, environmental distribution, toxicological impacts, and mitigation options. Most such initiatives have emerged in the United States and Europe and tend to be broad in scope rather than focused on specific resources. At the international scale, several countries have assumed leadership roles. Norway played a pivotal part in elevating MPs on the global policy agenda by introducing marine litter and MPs into UNEA discussions (UNEA, 2016). Sustained funding commitments have supported research, monitoring, and capacity-building activities, particularly in Arctic regions and developing nations. Comparable efforts in the Netherlands, Sweden, and Malaysia prioritize standardized monitoring, evaluation of human health risks, and the development of technologies to limit environmental releases of MPs.

Multinational research collaborations have also developed. Regional initiatives, including those involving Pacific Island nations, have aimed to harmonize data collection and monitoring systems, although implementation has been delayed in certain cases. In the United States, federal legislation has required studies on MP contamination in food and drinking water has promoted funding mechanisms to support mitigation projects at the state and local levels.

9.4.1.1 Subnational research and policy initiatives

In addition to national governments, several subnational authorities have undertaken targeted research efforts. U.S states such as Illinois, New Jersey, Minnesota, and Vermont have passed legislation mandating assessments of MP prevalence, environmental transport pathways, and potential health effects. These initiatives frequently include recommendations for standardized testing protocols, identification of key data gaps, although some reports and advisory outcomes remain pending or unpublished, these actions indicate growing engagement with MP issues at regional scales.

9.4.2 National and multinational microplastics research initiatives

Governments worldwide have increasingly acknowledged that scientific uncertainty-especially related to detection methods, exposure routes, and health outcomes-impedes effective governance of MPs. As a result, many countries have launched dedicated research programs, often in partnership with international organizations, to strengthen the scientific foundation for policy development.

9.4.2.1 Multinational and international initiatives

9.4.2.1.1 Norway

- Played a central role in elevating MPs on the global policy agenda by proposing the inclusion of marine litter and MPs in UNEA deliberations in 2014.
- Has actively supported numerous UNEA resolutions addressing plastic pollution through its environmental authorities.
- Committed substantial long-term funding to national and international research initiatives on marine litter and MPs, including capacity-building support for developing countries.
- Arctic-focused research programs emphasize long-term monitoring, source identification, plastic life-cycle assessments, and the development of tools to evaluate ecological and human health risks.

9.4.2.1.2 The Netherlands

- Incorporated MPs' research into its national marine policy framework.
- Mandated investigations into potential human health effects of MPs, particularly those arising from tie wear and river transport.
- Called for harmonized monitoring protocols and standardized assessment methodologies across the EU.
- Advocated restrictions on MPs in selected consumer and industrial applications.

9.4.2.1.3 Sweden

- Adopted legislation allocating public funds to environmental agencies for research on plastic pollution.
- Research priorities include analysis of environmental transport pathways, assessment of mitigation strategies, and evaluation of regulatory effectiveness.

9.4.2.1.4 Malaysia

- Implementation of a long-term national marine litter policy with a strong emphasis on MPs' research.
- Focus areas include standardization of detection techniques, refinement of definitions for MPs and nanoplastics, and development of technologies to reduce environmental emissions.
- Encourages regional cooperation through the Association of Southeast Asian Nations (ASEAN) and alignment with UNEA commitments.

9.4.2.1.5 Pacific island nations Secretariat of the Pacific Regional Environment Programme (SPREP) initiative

- Established regional collaboration among Pacific Island countries to develop shared metrics and standardized systems for monitoring MPs.
- This is a great deal of implementation of the project that requires regional data comparability and collective environmental management, although implementation has encountered logistical challenges.

9.4.2.1.6 United States (federal level)

- Federal legislation mandates comprehensive assessments of MP occurrence in environmental media, food systems, and drinking water.
- Requires the development of science-based definitions and recommendations for standardized monitoring methodologies.
- Authorizes federal agencies to provide funding for state and local initiatives aimed at reducing MP pollution.

9.4.2.2 Subnational Microplastics research initiatives (United States)

9.4.2.2.1 Illinois

- Required a state-affiliated research institution to review scientific literature and regulatory frameworks related to MPs.
- Mandated the formulation of recommendations for legislative and regulatory measures to protect environmental quality and public health.

9.4.2.2.2 New Jersey

- Established a Plastics Advisory Council to evaluate the environmental and health impacts of plastics and MPs.
- Intended to inform future legislative actions through expert analysis and reporting.

9.4.2.2.3 Minnesota

- Allocated funding to examine plastic use within agricultural supply chains.
- Research scope includes MP contamination of terrestrial and aquatic systems, as well as interactions with other emerging contaminants.

9.4.2.2.4 Vermont

- Required a comprehensive assessment of MPs and associated chemicals in food waste and packaging materials.
- Mandated the development of standardized testing protocols, identification of data gaps, and recommendations for health-based standards.

9.4.2.3 Overall insight

Taken together, these initiatives indicate a global transition toward science-driven governance of MPs, with increasing emphasis on standardized methodologies, coordinated monitoring, and evidence-based policy formulation. Although progress differs across regions, national and multinational research efforts provide the essential foundation for future regulatory frameworks and effective mitigation strategies.

9.4.3 Microbead prevention and mitigation policies

Microbeads represent one of the earliest and most clearly regulated sources of MPs. Scientific evidence linking cosmetic microbeads to aquatic pollution—particularly in freshwater environments such as the Great Lakes—prompted swift policy action beginning in the mid-2010s. Subnational initiatives in the United States and Canada laid the groundwork for national bans that restrict the manufacture, sale, or distribution of rinse-off products containing plastic materials.

National prohibitions in the United States, Canada, and the United Kingdom were subsequently followed by comparable legislation across numerous countries in Europe, Asia, and Oceania. While specific requirements differ among jurisdictions, most bans were implemented between 2018 and 2022 and focus on personal care and cosmetic products. In parallel, several regions adopted voluntary industry agreements promoting microbead phase-outs, supported by multinational efforts such as UNEA resolutions and G7 commitments (G7, 2018). These agreements, however, generally lack binding enforcement provisions.

9.4.3.1 Implications for policy and governance

Despite significant advances in regulating microbeads, policies addressing other major MP sources, particularly textile fibers and tire water, remain comparatively limited. These sources contribute substantially greater quantities of MPs to the environment but present regulatory challenges due to their diffuse release pathways. This imbalance underscores a critical gap in current governance and highlights an important priority area for future policy development. Another key limitation is the absence of systematic assessments of policy effectiveness. Only a limited number of studies have examined whether bans, phase-outs of voluntary agreements, lead to measurable declines in environmental outcomes, and economic implications are therefore necessary to ensure that mitigation strategies maximize benefits while minimizing unintended effects.

9.4.3.2 Conclusion

Over the past decade, global and national policies addressing MP have expanded rapidly in response to mounting scientific evidence and public concern. While investments in research and the implementation of microbead bans represent important achievements, governance frameworks remain fragmented and incomplete. Advancing MP policy will require harmonized measurement techniques, increased attention to dominant secondary sources, and robust evaluation of regulatory outcomes. Coordinated action at national and international levels will be essential to move from precautionary measures toward comprehensive, science-based regulation.

9.4.4 Global trends in plastic production, use, and postuse management

9.4.4.1 Trends in global plastic production

Over recent decades, global plastic production has grown at an unprecedented pace. Evaluations conducted by the OECD show that worldwide plastic production rose from slightly more than 230 million metric tons (Mt) in the early 2000s to roughly 460 Mt by 2019 (OECD, 2022a,b). This accelerated expansion is driven by rising demand across multiple sectors, including packaging, construction, transportation, and consumer goods. Forecasts indicate that plastic production will continue to increase substantially through the middle of the century. Growth is expected to be dominated by primary plastics produced from virgin fossil-based or bio-based feedstocks, with annual output potentially surpassing 1000 Mt by 2060. While production of secondary plastics sourced from recycled materials is also anticipated to rise, its overall share remains relatively limited, highlighting continued reliance on virgin plastic inputs (OECD, 2022a,b).

9.4.4.2 Plastic production in the United States

The United States continues to be among the largest global producers of plastic materials. Industry statistics indicate that domestic production of plastic resins has increased in recent years, with output in 2024 recording year-over-year growth relative to 2023 (American Chemistry Council, 2024). Economic assessments further show that plastic material and resin manufacturing constitutes a major industrial sector, generating more than $100 billion in annual output, when plastic product manufacturing is also considered, the total economic value of the U.S. plastics industry exceeds $200 billion, underscoring both its substantial economic role and associated environmental responsibilities (Bureau of Economic Analysis, 2023; U.S. Environmental Protection Agency (EPA), 2023). Accordingly, to the BEA data summarized in Table 9.1, gross output from plastic product manufacturing in 2023 was approximately $248.3 billion.

Table 9.1 Gross output, industrial manufacturing of plastic products valued in USD billion (U.S. Environmental Protection Agency (EPA), 2023).

Naics—code	Industry	Gross production volume (USD)
32,611	Plastic packaging materials and unlaminated film manufacturing	47.6
32,613	Laminated plastic plate, sheet, and shape manufacturing	4.4
32,615	Urethane and other foam products manufacturing	14.0
32,616	Plastic bottle manufacturing	15.6
32,619	Other plastic products manufacturing	128.6

9.4.4.3 Global patterns of plastic use

Worldwide plastic consumption has shown a consistent upward trajectory, rising from roughly 130 Mt in 1990 to about 460 My by 2019 (OECD, 2022) (Table 9.2). In earlier decades, roughly, plastic use was largely concentrated in industrialized nations; however, accelerated economic development across Asia has substantially altered global demand dynamics. By 2019, China emerged as the leading consumer of plastics, followed by the United States and OECD countries in Europe. When assessed on a per-capita basis, plastic consumption remains highest in North America, particularly in the United States and Canada. Projections suggest that, if current trends continue, global plastic use could exceed 1.2 billion Mt per year by 2060.

9.4.4.4 Postuse plastic waste and disposal pathways

Although plastic production and consumption have increased markedly, the expansion of waste management systems has not kept pace. Estimates of plastic waste generation and treatment differ across studies because of variations in data availability and methodological frameworks. In the United States, the Environmental Protection Agency (EPA) indicates that most plastic waste is sent to landfills, while only limited fractions are recycled or utilized for energy recovery (EPA, 2020, 2021, 2022, 2023). In contrast, OECD evaluations apply a life-cycle approach that also considers plastics that are mismanaged or lost to the environment. Approximately 32 Mt of plastic waste were generated in the United States in 2018, of which around 8% was recycled, 16% was combusted with energy recovery, and 76% was disposed of in landfills (Table 9.3).

Discrepancies between EPA and OECD estimates arise from differences in system boundaries and underscore ongoing difficulties in accurately accounting for plastics that bypass formal waste management pathways. Taken together, these observations suggest that existing disposal practices remain inadequate to manage the growing volumes of plastic waste, thereby increasing the potential for environmental leakage and the formation of MPs.

9.4.4.4.1 Implications

The ongoing growth in plastic production, coupled with persistently low recycling rates and heavy reliance on landfilling, poses substantial long-term environmental risks. In the absence of major advances in material design, waste management infrastructure, and policy measures, continued increase in plastic use are likely to exacerbate plastic pollution and the accumulation of MPs across both aquatic and terrestrial ecosystem.

Organization for economic Cooperation and development (OECD) (2022). Global Plastic Outlook: Economic Drivers, Environmental Impacts and Policy Options.

9.5 Regulation of plastic pollution under the Clean Water Act

The Clean Water Act (CWA) constitutes the principal federal statute for safeguarding surface waters in the United States from pollution. Operating under a cooperative federalism approach, the Act is implemented jointly by the U.S EPA and state authorities. The CWA governs pollutant discharges into waters of the United States through an integrated set of regulatory instruments, permitting programs, and watershed-based management strategies that are applicable to plastic pollution, encompassing both macroplastics and MPs.

9.5.1 Clean Water Act (CWA) tools relevant to plastic pollution

9.5.1.1 Several provisions within the CWA offer mechanisms for addressing plastic contamination

- **Water Quality Standards (WQA):** States are responsible for establishing standards to protect designated uses of water bodies. In certain jurisdictions, trash—including plastic debris—has been explicitly identified as a pollutant that impairs water quality.
- **National Pollutant Discharge Elimination System (NPDES):** This permitting framework controls point source discharges from wastewater treatment facilities and stormwater systems, enabling the inclusion of limits or management requirements for plastic-related releases (U.S. Environmental Protection Agency (EPA), 2020, 2021).
- **Nonpoint Source Management Programs:** Diffuse sources of pollution, such as urban runoff, are addressed through state-led programs supported by federal funding under Section 319 of the CWA.
- **Place-Based Restoration Programs:** The CWA also authorizes ecosystem-focused initiatives, including the National Estuary Program, the Great Lakes Restoration Initiative, and the Chesapeake Bay Program, within which plastic pollution reduction can be integrated into broader water quality objectives.

9.5.2 Regulation of trash and macroplastics

Existing CWA authorities allow states and the EPA to address trash and large plastic debris in aquatic environments. Several states have designated trash as a cause of water quality impairment and, in accordance with CWA requirements, have developed total maximum daily loads (TMDLs), which establish pollutant load limits for impaired waters. A small number of states have adopted TMDLs specifically targeting trash to mitigate plastic debris in affected water bodies.

Stormwater permits issued under the NPDES program may also incorporate requirements aimed at reducing trash discharges associated with urban runoff. Although such provisions are not compulsory, their use has become more common in recent years. To facilitate implementation, the EPA issued guidance in 2021 outlining best practices for integrating trash control measures into stormwater permits, drawing on experiences from multiple jurisdictions.

9.5.3 Addressing MPs under the CWA

The CWA potentially provides authority to regulate MPs, although practical application remains limited (Clean Water Act, 1972). The Act mandates that the EPA develop Effluent Limitation Guidelines (ELGs) for industrial discharges and establish water quality criteria to inform state standards. In principle, the regulatory tools could be applied to MPs, and states may have permitted specific limits even in the absence of nationally established criteria. To date, however, the EPA has not promulgated ELGs or water quality criteria that specifically address MPs. A key constraint is the absence of standardized and reproducible analytical techniques for detecting and quantifying MPs and nanoplastics. This lack of methodological consistency limits reliable exposure assessment, risk analysis, and informed regulatory action. In response to the limitations, the EPA has designated MPs as a priority area for research. Expert workshops and subsequent research agendas have highlighted the need to advance analytical methodologies, improve source identification, enhance fate and transport modeling, and strengthen ecological and human health risk assessments. MPs have since been explicitly incorporated into the EPA's Safe and Sustainable Water Resources research priorities (EPA, 2020, 2021, 2022, 2023). The EPA has also implemented targeted measures to control preproduction plastics pellets. Under industrial stormwater permitting requirements, facilities that handle these materials must apply best management practices to prevent pellet loss and discharge, representing one of the most direct regulatory actions addressing MPs under the CWA.

9.5.3.1 Key policy implications

Although the CWA provides several avenues for addressing plastic pollution, its application to MPs remains largely indirect. Progress toward effective regulation will depend on advances in scientific understanding, the development of standardized measurement techniques, and clearer policy guidance to translate existing statutory authority into enforceable regulatory measures.

9.5.4 Federal authorities addressing marine debris and plastic pollution

9.5.4.1 Marine debris act: Federal framework for marine plastic management

The principal federal law governing marine debris in the United States is the Marine Debris Research, Prevention, and Reduction Act, which was first enacted by Congress in 2006 and has since undergone multiple amendments (Marine Debris Research, Prevention, and Reduction Act, 2006). Through these revisions, the statute—commonly referred to as the Marine Debris Act has been broadened to enhance federal coordination, scientific research, and response activities related to marine debris, including plastic pollution. Implementation is largely carried out through National Oceanic and Atmospheric Administration (NOAA's) Marine Debris Program, which is structured around five core pillars: debris removal, prevention efforts, scientific research, regional coordination, and emergency response. Collectively, these elements support field-based cleanup operations, sustained monitoring programs, development of innovative debris reduction approaches, and timely responses to acute debris incidents. To strengthen coordination across federal agencies, the Act also established the interagency Marine Debris Coordinating Committee (IMDCC). This body convenes senior representatives from multiple federal agencies to align research agendas, facilitate data sharing, and promote collaboration with state, local, and nongovernmental partners.

9.5.5 Save Our Seas 2.0 act: Expanding federal action on plastic pollution

In 2020, Congress passed the Save Our Seas 2.0 Act, which built upon earlier marine debris legislation and substantially expanded federal engagement in addressing plastic pollution (Save Our Seas 2.0 Act, 2020). This statute amended the Marine Debris Act and introduced new requirements affecting a range of federal agencies. Key elements of the Act include the creation of the Marine Debris Foundation, the establishment of innovation incentives such as the Genius Prize for Save Our Seas Innovations, and mandates directing federal agencies to prepare reports on plastic waste, recycling, reuse, and materials management. The legislation also formalized U.S policy in support of international collaboration to combat marine debris and authorized new grant programs administered by the U.S Environmental Protection Agency (EPA). These grants are intended to enhance postconsumer materials management, reduce plastics in drinking water and wastewater systems, and support community-based "trash-free waters" initiatives.

9.5.6 Regulation of plastic-associated chemicals under the Toxic Substances Control Act

Beyond legislation focused specifically on debris, the Toxic Substances Control Act (TSCA) grants the EPA authority to regulate chemicals associated with the manufacture and use of plastics. Under Title One of TSCA, the EPA may collect data on industrial chemicals and assess whether they pose unreasonable risks to human health or the environment throughout their lifecycle. Although polymers such as polyethylene terephthalate (PET) and polyvinyl chloride (PVC) comprise a large share of plastic materials, they are generally regarded as lower risk because of their high molecular weight and limited biological reactivity. As a result, EPA oversight has historically focused on monomers and additives used in plastic production, including plasticizers, flame retardants, stabilizers, and pigments, which may present greater toxicological risks.

9.5.6.1 Policy implications

Collectively, the Marine Debris Act, the Save Our Seas 2.0 Act, and TSCA establish a multitiered federal framework for addressing plastic pollution, spanning chemical manufacture and product design, debris prevention and removal, and international cooperation. Although these statutes provide substantial regulatory authority, their effectiveness relies on sustained scientific research, effective regulatory authority, and alignment with state, local, and global efforts.

9.5.7 Regulating Microplastics: The need for targeted policies and strategic action

Recent scientific studies have identified an expanding group of substances that may pose a risk to ecosystem and human health and are now categorized as emerging pollutants (EPs) (Kalita and Devi, 2023). Conventional environmental risk assessments have traditionally emphasized contaminants such as heavy metals, persistent organic pollutants, pesticides, and pharmaceuticals. However, improvements in analytical methodologies have demonstrated that contemporary consumer and industrial products-including personal care products, veterinary drugs, nanomaterials, coatings, industrial additives, antibiotics, and pharmaceuticals-are increasingly detected in environmental media. These substances are regarded as "emerging" not because of recent introduction, but because their extensive use, environmental release, and long-term consequences were previously underestimated or insufficiently recognized. Wastewater treatment systems are now acknowledged as major conduits for the transport of EPs, including MPs, into aquatic environments. Although substantial knowledge gaps persist, accumulating evidence indicates that EPs represent a significant environmental challenge (Gkika et al., 2022; Roy et al., 2022). Some EPs also originate from natural sources, yet their toxicity can vary according to concentration, environmental conditions, and interactions with cooccurring contaminants. Consequently, scientific interest has intensified, resulting in expanded monitoring efforts, discharge guidelines, regulatory instruments, and case studies assessing ecological and human health effects (On and Kara, 2019). Despite these advances, EPs—including MPs—remain largely excluded from routine monitoring programs, and their cumulative or source-specific risks are still poorly characterized.

Regulatory bodies worldwide establish drinking water standards to protect public health, taking into account EPs identified by health authorities. However, regulatory guidelines vary across regions due to differences in socioeconomic conditions, cultural context, and technological capacity. The WHO periodically issues updated, evidence-based guidelines intended to support global application. Historically, public health protection has focused primarily on acute waterborne diseases associated with conventional contaminants. In contrast, emerging and often unrecognized health outcomes potentially linked to EP exposure are not yet adequately reflected in health statistics, largely because of limited epidemiological data (Abdelbasir et al., 2020). This underscores the need for regular evaluations by national and regional health authorities to identify EPs in drinking and environmental waters, with public health protection as the primary objective. Effective regulation depends on coordinated information exchange among agencies and their implementation of consistent water safety measures. Public health institutions must assume a leading role in incorporating EP considerations including MPs and nanomaterials into water quality management frameworks.

9.5.8 World Health Organization initiatives addressing emerging pollutants

EPs present a multifaceted challenge for both drinking water and wastewater management, with implications for human health, ecosystem integrity, and sustainable development. Monitoring wastewater discharges, treatment performance, and water reuse practices has been identified as a crucial strategy for reducing risk associated with Eps. Through initiatives such as the UNESCO International Hydrological Program (IHP) International Water Quality Initiative, the WHO and partner organizations collaborate with UN member states to strengthen institutional capacity and policy frameworks aimed at managing health and environmental risks linked to EPs (Khan et al., 2020). Future challenges are expected to intensify as new contaminants emerge from evolving waste management practices, including composting, landfill operations, demographic change, land-use transitions, and climate change. Climate-induced modifications to hydrological processes and chemical behavior are likely to affect the distribution, persistence, and toxicity of EPs, potentially exacerbating risks associated with MPs and nanomaterials.

The WHO has identified thousands of synthetic and naturally occurring substances of emerging concern. However, existing water quality guidelines largely focus on traditional contaminants. This situation highlights the need to reevaluate regulatory standards governing the production, use, and disposal of EPs and to develop new monitoring systems capable of assessing their occurrence, accumulation, and long-term impacts. National monitoring programs, supported by reliable and standardized analytical methods, are essential for robust risk assessment and informed regulatory decision-making (Abdelbasir et al., 2020).

9.5.9 Global agreements and international frameworks on emerging pollutants

International collaboration has been central to efforts aimed at addressing transboundary water pollution. A significant milestone was the adoption of the convention on the protection and use of transboundary watercourses and international lakes (Water Convention) under the United Nations Economic Commission for Europe (UNECE). Adopted in 1992 and entering into force in 1996, the convention was subsequently amended to permit accession by all UN member states, thereby becoming a global legal framework for transboundary water governance in 2016. Under the convention, participating countries are required to prevent, control, and reduce transboundary environmental impacts, including those affecting human health and socioeconomic conditions. Signatories must implement sustainable water management practices, monitor ecological effects, reduce pollution as its source, treat wastewater discharges, and apply best available technologies across relevant sectors (Van der Ploge et al., 2012). Joint monitoring initiatives, shared data exchange, coordinated assessments, and collaborative pollution reduction strategies are based on the foundation of this cooperative approach.

Access to adequate and safe water is increasingly recognized as one of the defining challenges of the 21st century, driven by population growth, pollution pressure, resource limitations, and climate change. International organizations emphasize cooperation to achieve global sustainability objectives, particularly those articulated in the United Nations SDGs. SDG six specifically focuses on universal access to safe drinking water and sanitation, while other goals address ecosystem conservation, public health, and climate resilience (Durel et al., 2015). Achieving these targets requires the identification of priority Eps, comprehensive risk evaluations, and alignment of national polices with international standards. Integrating a regulatory framework with effective management strategies is essential to reduce pollutant loads, safeguard biodiversity, and ensure long-term water security (National Academies of Sciences, Engineering, and Medicine, 2022).

9.5.10 The need for more rigorous regulatory controls

Despite the availability of a wide range of wastewater treatment technologies, regulatory effectiveness and implementation remain inconsistent due to competing commercial priorities, weak legal enforcement, and limited institutional capacity. In many regions, stringent treatment requirements are undermined by inadequate regulatory backing, allowing obsolete or inefficient technologies to continue in use. Furthermore, heavy dependence on bottled drinking water often shifts the attention away from the critical need to upgrade wastewater treatment infrastructure. Because treated wastewater is frequently perceived as having low economic value relative to the cost of infrastructure development, investments in wastewater management are commonly deprioritized. Overcoming these barriers requires coordinated global action to position wastewater treatment as a core element of sustainable water management. Regulatory authorities must proactively engage stakeholders to strengthen political commitment and public support for improved implementation and enforcement. Industrial wastewater, in particular, remains underregulated in many parts of the world, despite its recognized importance for achieving SDGs. Although international organizations emphasize the significance of industrial wastewater control, regulatory agendas often remain fragmented and focused primarily on the national level.

Effective governance relies on the availability of reliable data, sustained political commitment, sufficient financial resources, and strong enforcement capacity. Monitoring frameworks must be both scientifically robust and adaptable, ensuring an appropriate balance between technical rigor, affordability, and practical feasibility. Fully integrating wastewater treatment into water-cycle assessment is essential for advancing sustainability objectives. While point sources such as surface runoff continue to contribute substantially to overall pollution loads. Industrial activities account for roughly one-fifth of global freshwater withdrawals, yet regulatory effectiveness varies considerably among regions. Countries with well-established legal systems and long-standing enforcement mechanisms tend to exhibit higher compliance, whereas many developing regions lack comprehensive regulatory structures. Consequently, a substantial proportion of wastewater is discharged without adequate treatment, intensifying environmental contamination and MP release. Although innovative treatment technologies have been proposed, robust regulatory frameworks are necessary to promote their widespread adoption. Treatment processes should be selected based on contaminant profiles and supported by cost-benefit evaluations and infrastructure planning. Sustainable wastewater management ultimately depends on strong regulatory oversight, technological advancements, cross-sectoral collaborations, and effective monitoring tools. Cooperation among researchers, engineers, environmental professionals, and policymakers is critical for identifying priority pollutants and addressing key knowledge gaps related to occurrence, exposure pathways, ecotoxicity, combined effects, food-chain transfer, and mitigation options. Comprehensive evaluations of future risks to agriculture and food security are also urgently required.

9.5.11 Conclusion and final perspectives

This review provides a comprehensive and integrative evaluation of MP pollution, examining the issue through environmental, social, economic, political, and ethical lenses. A multidimensional framework is advanced that incorporates principles of the circular economy, green chemistry, environmental justice, and public participation to address the escalating MP challenges. By synthesizing existing knowledge and identifying critical research and policy gaps, this work delivers practical guidance for mitigating MP contamination and supports informed decision-making by policymakers, researchers, practitioners, and stakeholders. EPs comprise a diverse group of largely unregulated substances, including pesticides, pharmaceuticals, personal care products, microbeads, plastic additives, nanomaterials, and micro- and nanoparticles. Although awareness of their potential impacts has grown, effective removal during wastewater treatment remains difficult. Conventional treatment systems are often inadequate, although early-stage screening and targeted control measures can reduce environmental release. Advanced treatment technologies—such as membrane filtration, adsorption, and advanced oxidation processes—demonstrate potential for specific contaminants, yet overall removal performance remains inconsistent.

Treatment efficiency is strongly dependent on the stage of processing, with primary treatment achieving limited reductions, secondary treatment providing moderate removal, and tertiary systems delivering the highest efficiencies. Nevertheless, complete elimination of Eps, including MPs, remains challenging. Continued scientific investigation and technological innovation are therefore essential to enhance treatment effectiveness. Risk assessment research underscores the close linkage between EP source, exposure pathways, and associated environmental and health risks. Emissions of EPs are driven by industrial expansion, consumer practices, and contemporary lifestyles. Although risk assessment frameworks and treatment guidelines are available globally, their application remains uneven. Advancing effective management requires an improved understanding of contaminant behavior, analytical limitations, wastewater occurrence, and ecological and human health consequences. Existing risk prioritization approaches require further refinement and broader implementation. MP pollution represents one of the most complex and pressing environmental challenges of the modern era. MPs are now ubiquitous in aquatic and terrestrial environments, where their small size facilitates entry into food webs and diverse ecological compartments. Their variability in shape, composition, persistence, potential for bioaccumulation, and biological interactions poses serious concerns for ecosystem integrity and human health.

Several overarching conclusions emerge from this synthesis. First, MPs are highly heterogeneous contaminants that challenge traditional monitoring and risk assessment methods, underscoring the need for standardized sampling and analytical protocols. Second, substantial uncertainties regarding MP fate, behavior, toxicity, and dose-response relationships under real-world conditions remain insufficiently assessed. Integrated treatment approaches may improve performance, but practical implementation continues to face constraints. Addressing MP pollution requires a coordinated and systemic response that integrates regulatory measures, economical instruments, infrastructure investments, education, and innovation. Policy actions should be prioritizing key plastic products, industrial sectors, and waste streams, while promoting a transition away from linear, petrochemical-dependent economic models. Achieving this transition requires not only technological solutions but also behavioral change, social innovation, and strengthened producer responsibility.

Future research priorities include evaluating MP removal technologies using life-cycle assessment and cost-benefit analysis; enhancing public engagement and education initiatives; incentivizing industrial innovation through regulatory and economic mechanisms; clarifying adverse outcome pathways linking MP exposure to biological effects; improving understanding of environmental fate and transport processes; assessing MP accumulation in soils; and fostering collaborative efforts to advance detection methods, sustainable materials, and circular business models.

Ultimately, effective mitigation of MP pollution depends on prevention-oriented strategies, sustained investment in research, and collective action. Decisions made at present will shape future environmental trajectories. Through coordinated policy development, scientific programs, and societal engagement, it is possible to confront the MP challenge and protect both ecosystem and human health.

9.6 Climate change, free radicals, and the environmental fate of microplastics

9.6.1 Climate change driven changes in free radicals in soil and water

Studies show that climate change plays a major role in determining the formation of reactive molecules, especially free radicals, and their behavior in soil and water systems. The MPs break down, and movement through the environment is influenced by these changes. The shift in rainfall patterns contributes to the major factor. The moisture of the soil increases due to strong and frequent rainstorms, which dissolve more oxygen into rivers and lakes. This action speeds up the natural processes of microbial activity and sunlight-driven reactions. More reactive oxygen species, particularly hydroxyl radicals, are formed in these conditions. Oxygen-dependent species are suppressed by flooding as they change the balance in soil microbes, which favor those flourishes without oxygen and produce superoxide radicals. These radicals quickly transform them by altering the soil's overall chemical balance when floodwaters retreat, and oxygen returns. Pollutants like organic chemicals and heavy metals from soil are washed out into the nearby waters by heavy rainfall. The production of reactive radicals is further boosted by these pollutants, which makes the environment even more chemically active. MPs' weather, fragment and their interaction with other pollutants in the environment are affected by these climate-driven changes.

9.6.2 Drought, salinization, and heatwave effects on free radical formation

In soil and water environments, the production of free radicals is strongly influenced by the rising soil salinity and during the drought periods. Microbial activity slows when the moisture level drops during the drought as the soil dries, which reduces the natural breakdown of organic matter and the microbial pathways that normally generate radicals. The concentration of pollutants such as heavy metals and persistent organic chemicals is increased in drought as there is less water to dilute them. When exposed to strong sunlight, these accumulated contaminants can promote free radical formation. Photochemical reactions in soil are increased due to high solar radiation during drought. Reactive elements such as hydroxyl and superoxide radicals are produced under ultraviolet light by minerals such as titanium dioxide and zinc oxide, which act as natural photocatalysts. The radical production is supported by clay minerals, including kaolinite and montmorillonite by reactions on the soil surface.

Another layer of complexity is added by heatwaves. Chemical reactions are accelerated by a rise in temperature that encourages the formation of reactive molecules. Additional radicals are created when the organic matter decomposes rapidly due to the increase in heat, which stimulates the production of hydrogen peroxide in water bodies. Soil microbial communities are shifted due to thermal stress, which reduces heat-sensitive species and increases the microbes that live in higher temperatures. This affects radical formation based on their metabolic pathway. The environmental factors such as moisture, oxygen content, decomposition rates, and pollutant movement are altered by various climate-driven changes such as more frequent heatwaves, extended drought, and shifts in rainfall patterns. These changes affect the production of free radicals and their transformation in soil and water systems. The breakdown of MPs is greatly influenced by the free radicals. The abundance changes in number, and the behavior influences the qualities of MPs, like how quickly they degrade, how long they will be present in the environment, and the impacts they may have on ecosystems.

9.6.3 Impact of free radical types and concentrations on microplastic degradation

The types and amounts of free radicals present in soil and water are greatly influenced by the climate-driven shifts in environmental conditions, and these radicals are responsible for breaking down MPs. When ultraviolet (UV) light is passed through plastic particles, the polymer surface absorbs the energy, which splits the chemical bonds and triggers photooxidation. The reactive species—such as hydroxyl radicals (•OH), superoxide anions ($O_2\bullet^-$), and singlet oxygen ($^1O_2$) are produced from this process that attack the plastic surface and accelerate its degradation. The MP endorsement rate depends on the radical numbers and their survival time in the environment. The radical formation and their effective interaction with MP surfaces are influenced by factors such as changes in sunlight intensity, wavelength, temperature, moisture, and the presence of natural oxidants or dissolved ions. Photooxidative reactions are increased by the stronger UV light, the formation and diffusion of free radicals are promoted by high temperature. The degradation process is further influenced by moisture fluctuations as they affect the oxygen availability and the formation of dissolved reactive species. Furthermore, the radical is disturbed by the chemical makeup of the plastic itself. The aromatic structures in polystyrene are highly sensitive to photooxidation. In contrast, the functional groups on the plastic surface can either increase the plastic's reactivity or shift the degradation pathway, such as carbonyl or hydroxyl groups. The study of these results shows that climate change directly influences the various factors in the environment and how they are responsible for the breakdown of MPs into the ecosystem.

9.6.4 Direct radical attack and secondary reactive oxygen species (ROS) cycles

By directly attacking carbon–carbon and carbon–hydrogen bonds on their surfaces, free radicals break down the MP polymers. The plastic structure is weakened by this process, causing chain scission and small fragments, oxidized molecules, and short-chain oligomers are formed. These breakdown products involve additional cycles of reactive oxygen species (ROS) formation, which further accelerates MP degradation by creating a feedback process. These reactions are intensified by natural catalysts such as biochar and metal oxides, which contain iron or manganese speeds up the polymer breakdown by activating oxidants, such as persulfate, producing sulfate radicals and more hydroxyl radicals. MPs can generate environmentally persistent free radicals (EPFRs) under strong ultraviolet (UV) radiation, which remain active for extended periods and continue driving oxidative degradation.

9.6.4.1 Synergistic interactions with coexisting pollutants

These persistent radicals, in addition to MPs interact with pollutants commonly found in soil and water. Polycyclic aromatic hydrocarbons (PAHs) are oxidized by EPFR, which increases their mobility and toxicity. The surface chemistry of MPs, such as charge, roughness, and functional groups, is altered by UV-driven radical reactions. These changes determine how MPs clump together, disperse, and react with their surroundings. Studies made in laboratories show that radicals formed by polystyrene and polyethylene can activate peroxymonosulfate, a strong oxidant. This enhances the MP degradation by the rapid production of sulfate and hydroxyl radicals. Another layer of complexity is added by climate change, which alters the number of heavy metals and organic pollutants in the environment, resulting in the enhancement or inhibition of radical-mediated breakdown.

9.6.4.2 Role of extreme weather and advanced oxidation processes (AOPs)

The physical and chemical weathering of MPs is altered by extreme climate events such as heatwaves and periods of intense UV sunlight. The polymer characteristics, including surface texture, color, crystallinity, particle size, and chemical functional groups, are changed by elevated temperatures and stronger sunlight. The high level of ROS and EPFRs is promoted by advanced oxidation processes (AOPs) due to the climate-related changes, which further speed up degradation. MP properties such as polarity, hydrophobicity, and surface composition are altered by these environmental changes, which affect the interaction with metals, organic contaminants, and microbial communities. The surface roughness and surface area are increased by enhanced oxidation, which pays the path for pollutants to attach easily and radicals to fragment the plastic.

9.6.5 Climate change effects on the ecological distribution of Microplastics

Climate change reshapes the ecological distribution and mobility of MPs in addition to free radical chemistry. MP fragmentation, transport, and bioavailability are influenced by temperature fluctuations, shifting precipitation patterns, extreme weather events, and changes in biodiversity. The ecological footprint of smaller MPs and nanoplastics is increased as they exhibit higher mobility and oxidative susceptibility.

Influence of precipitation patterns and extreme events

- MP transport is strongly affected by changes in rainfall patterns.

- MPs move from terrestrial environments into rivers, lakes, and oceans due to increased precipitation and runoff.
- Large volumes of MPs are transferred across land–water interfaces by extreme rainfall and flooding.
- MP accumulation in surface soils happens because of reduced precipitation, where enhanced contact with soil particles and radicals increases weathering.
- Dispersal of MP into remote ecosystems is further influenced by glacial melt, thawing permafrost, and altered atmospheric circulation. Deposition of MPs in terrestrial and aquatic habitats is contributed to by high wind events, which redistribute airborne MPs across long distances.

9.6.6 Biological drivers of Microplastic distribution under climate change

MP transport is influenced by climate-driven biodiversity shifts. Because of warming temperatures or changing hydrological conditions, the feeding behavior may be altered, or the organism may migrate, and the interactions with MPs might change accordingly.

- Through ingestion and excretion, MPs might be introduced into new regions by migration of aquatic organisms.
- Altered metabolic rates may increase or decrease organismal uptake and depuration of MPs.
- Redistribution of MPs through trophic interactions by climate-modified food webs.

9.6.7 Marine systems: Circulation driven long distance transport

Marine environments are particularly sensitive to climate-driven changes as they act as the largest sink for MPs. The hotspots for radical production and MP degradation are mainly in coastal interfaces and the water–air boundary area. Long-distance MP transport is due to the variability in ocean currents, driven by changes in temperature and salinity.

These current shifts may cause:

- Alteration in MPs accumulation area (e.g., subtropical gyres),
- MP degradation rates are influenced by modified radical exposure,
- Dynamics and vertical transport change aggregation,
- Affect coastal ecosystems MP concentrations, which are vital for fisheries.

Altogether, the influence on MP degradation, mobility, and ecological distribution is majorly affected by climate change. The environmental fate of MPs is affected by these effects and also their interactions with pollutants, organisms, and biogeochemical cycles.

9.7 Summary and conclusion

Current regulatory frameworks addressing MP pollution are fragmented and predominantly preventive, relying on broader plastic waste reduction strategies. While international and national initiatives demonstrate growing recognition of the issue, enforceable standards and harmonized monitoring remain limited.

Table 9.2 Regional plastic use—valued in metric tons (Mt).

		Plastic usage (1990)	Plastics usage (2019)	Plastic usage per capita (kg/ Population) - 2019
OECD Europe	OECD EU	32.8	63.7	152.9
	OECD non-EU	8.1	21.4	124.3
OECD America	United States	42.0	84.3	255.2
	Canada	3.7	7.5	202.2
	Other OECD America	4.8	13.1	65.4
Other America	Latin America	3.7	22.9	50.9
OECD Pacific	OECD Asia	5.9	18.2	102.4
	OECD Oceania	0.6	4.4	143.9
Other Asia	China	7.8	94.0	69.0
	India	1.9	29.3	22.1
Middle East & Africa	Middle East & North Africa	5.6	21.2	47.1
	Other Africa	2.8	17.9	15.9

Effective governance of MP pollution will require a transition from indirect, precautionary measures to comprehensive, science-informed regulatory frameworks. Standardized methodologies, improved risk assessment, and coordinated international action are essential to address the diffuse and persistent nature of MPs. Strengthening regulatory alignment across jurisdictions will be critical for protecting aquatic ecosystems, safeguarding public health, and achieving long-term sustainability goals (Tables 9.2 and 9.3).

Table 9.3 Plastic in municipal solid waste - in million metric tons (Mt).

Year	Plastics generated	Plastics recycled
1960	0.4	–
1970	2.6	–
1980	6.2	0.01
1990	15.5	0.3
2000	23.2	1.3
2005	26.7	1.6
2015	28.5	2.3
2017	31.3	2.8
2018	32.1	2.7

References

Abdelbasir, S.M., et al., 2020. Emerging pollutants in water systems: challenges and future perspectives. J. Environ. Chem. Eng. 8, 103770.

American Chemistry Council, 2024. U.S. Plastic Resin Production Statistics. American Chemistry Council, Washington, DC, USA.

Axelsson, C., Sebille, E.V., 2017. Prevention through policy: urban macroplastic leakages to the marine environment during extreme rainfall events. Mar. Pollut. Bull. 124 (1), 211–227. https://doi.org/10.1016/j.marpolbul.2017.07.024.

Bruch, C., Mengerink, K., Harrison, E., Flanagan, D., Carey, I., Casey, T., Davis, M., Hessami, E., Lombardi, J., Michelen, N., Parts, C., 2016. Marine Litter Legislation: A Toolkit for Policymakers, vol 109. Environmental Law Institute (ELI) United Nations Environment Programme (UNEP).

Bureau of Economic Analysis (BEA), 2023. Industry Economic Accounts: Plastics and Rubber Manufacturing. U. S. Department of Commerce, Bureau of Economic Analysis, Washington, DC, USA.

Burton, G.A., 2015. Losing sight of science in the regulatory push to ban microbeads from consumer products and industrial use. Integrated Environ. Assess. Manag. 11 (3), 346–347. https://doi.org/10.1002/ieam.1645.

California Legislature, 2007. Assembly and Senate Bills Addressing Microplastics and Plastic Pollution. California State Legislature, Sacramento, CA, USA.

Clean Water Act, 1972. 33 U.S.C. §§ 1251–1387. United States Congress, Washington, DC, USA.

Conti, I., Simioni, C., Varano, G., Brenna, C., Costanzi, E., Neri, L.M., 2021. Legislation to limit the environmental plastic and microplastic pollution and their influence on human exposure. Environ. Pollut. 288, 117708. https://doi.org/10.1016/j.envpol.2021.117708.

Dauvergne, P., 2018. The power of environmental norms: marine plastic pollution and the politics of microbeads. Environ. Polit. 27 (4), 579–597.

Durel, M., et al., 2015. International cooperation for water sustainability under the sustainable development goals. Water Int. 40 (5–6), 915–930.

Environmental Protection Agency (EPA), 2020. Advancing Sustainable Materials Management: Facts and Figures. U.S. Environmental Protection Agency, Washington, DC, USA.

EPA, 2021. National Recycling Strategy: Part One of a Series on Building a Circular Economy for All. https://www.epa.gov/circulareconomy/national-recycling-strategy.

EPA, 2023. Biden-Harris Administration Announces Latest Steps to Reduce Plastic Pollution Nationwide. https://www.epa.gov/newsreleases/biden-harris-administration-announces-latest steps-reduce-plastic-pollution-nationwide. (Accessed 21 April 2023).

Eriksen, M., Mason, S., Wilson, S., Box, C., Zellers, A., Edwards, W., Farley, H., Amato, S., 2020. 20 Years of Government Responses to Global Plastic Pollution. Nicholas Institute for Energy, Environment & Sustainability. Duke University, Durham, NC, USA.

G7. 2018. Ocean Plastics Charter. Government of Canada. Charlevoix, Quebec, Canada.

Gkika, E., et al., 2022. Emerging pollutants in wastewater and aquatic environments. Environ. Sci. Pollut. Control Ser. 29, 1–15.

Gunningham, N., Grabosky, P., Sinclair, D., 1998. Smart Regulation: Designing Environmental Policy. Clarendon Press, Oxford.

Henderson, L., Green, C., 2020. Making sense of microplastics? Public understandings of plastic pollution. Mar. Pollut. Bull. 152, 110908. https://doi.org/10.1016/j.marpolbul.2020.110908.

Human Rights Council. 2020. Right to a healthy environment: good practices, Report of the Special Rapporteur on the issue of human rights obligations relating to the enjoyment of a safe, clean, healthy, and sustainable environment. https://documents-dds-ny.un.org/doc/UNDOC/GEN/G19/355/14/PDF/G1935514.pdf?OpenElement.

Human Rights Council, R.O. 2021. The human right to a clean, healthy and sustainable environment Preamble. https://documents-dds-ny.un.org/doc/UNDOC/GEN/G21/289/50/PDF/G2128950.pdf?OpenElement.

Kalita, S., Devi, A., 2023. Emerging pollutants: sources, impacts, and regulatory challenges. Environ. Chall. 11, 100705.

Karasik, R., Vegh, T., Diana, Z., Bering, J., Caldas, J., Pickle, A., Rittschof, D., Virdin, J., 2017. Microplastic pollution in the great lakes. J. Great Lake. Res. 43 (5), 1–12.

Karasik, R., et al., 2020. 20 Years of Government Responses to Global Plastic Pollution. Nicholas Institute for Energy, Environment & Sustainability. Duke University, Durham, NC, USA.

Khan, S., et al., 2020. UNESCO-IHP initiatives on water quality and emerging contaminants. Water Pol. 22 (S1), 85–98.

Khan, N.A., Khan, A.H., López-Maldonado, E.A., Alam, S.S., López López, J.R., Méndez Herrera, P.F., Mohamed, B.A., Mahmoud, A.E.D., Abutaleb, A., Singh, L., 2022. Microplastics: occurrences, treatment methods, regulations and foreseen environmental impacts. Environ. Res. 215. https://doi.org/10.1016/j.envres.2022.114224.

Lincoln, S., Andrews, B., Birchenough, S.N.R., Chowdhury, P., Engelhard, G.H., Harrod, O., Pinnegar, J.K., Townhill, B.L., 2022. Marine litter and climate change: inextricably connected threats to the world's oceans. Sci. Total Environ. 837, 155709. https://doi.org/10.1016/j.scitotenv.2022.155709.

Lohr, A., Savelli, H., Beunen, R., Kalz, M., Ragas, A., Belleghem, F.V., 2017. Solutions for global marine litter pollution. Curr. Opin. Environ. Sustain. 28, 90–99. https://doi.org/10.1016/j.cosust.2017.08.009.

Marine Debris Research, Prevention, and Reduction Act. 2006. Public Law 109–449. United States Congress. Washington, DC, USA.

Martinho, G., Balaia, N., Pires, A., 2017. The Portuguese plastic carrier bag tax: the effects on consumers' behavior. Waste Manag. 61, 3–12. https://doi.org/10.1016/j.wasman.2017.01.023.

NASEM, 2022. Reckoning With the U.S. Role in Global Ocean Plastic Waste. National Academies Press, Washington, DC, p. 32. https://nap.nationalacademies.org/catalog/26132/reckoning-with-the-us-role-in-global-ocean-plastic-waste.

National Academies of Sciences, Engineering, and Medicine, 2022. Assessment of Microplastics in the Environment. National Academies Press, Washington, DC, USA.

OECD, 2023. Climate Change and Plastics Pollution: Synergies Between Two Crucial Environmental Challenges. https://www.oecd.org/content/dam/oecd/en/publications/reports/2023/05/climate-change-and-plastics_d3364145/5e0bfe87-en.pdf.

On, J., Kara, D., 2019. Environmental occurrence and risk assessment of emerging contaminants. Environ. Monit. Assess. 191, 612.

Organisation for Economic Co-operation and Development (OECD), 2022a. Policies to Reduce Microplastics in Water: Focus on Textile and Tyre Sources. OECD Publishing, Paris, France.

Organisation for Economic Co-operation and Development (OECD), 2022b. Global Plastics Outlook: Policy Scenarios to 2060. OECD Publishing, Paris, France.

OECD, 2022c. Global Plastics Outlook: Economic Drivers, Environmental Impacts, and Policy Options. https://doi.org/10.1787/de747aef-en.

Pouikli, K., 2020. Concretising the role of extended producer responsibility in European Union waste law and policy through the lens of the circular economy. ERA Forum 20 (4), 491–508. https://doi.org/10.1007/s12027-020-00596-9.

Roy, S., et al., 2022. Pathways and impacts of emerging pollutants in aquatic ecosystems. Sci. Total Environ. 806, 150673.

Save Our Seas 2.0 Act, 2020. Public Law 116–224. United States Congress, Washington, DC, USA.

Stoll, T., Stoett, P., Vince, J., Hardesty, B.D., 2020. Governance and Measures for the Prevention of Marine Debris. Handbook of Microplastics in the Environment. Springer, Cham, pp. 1–23. https://doi.org/10.1007/978-3-030-10618-8_26-1.

UN, 2012. Report of the United Nations Conference on Sustainable Development 20–22 June 2012.
UN, 2014. United Nations Resolutions and Decisions Adopted by the United Nations Environment Assembly of the United Nations Environment Programme at its First Session on 27 June 2014: Resolution 1/6 Marine Plastic Debris and Microplastics United Nations.
UN, 2020. Sustainable Development Goals – Knowledge Platform. United Nations. https://sustainabledevelopment.un.org/?menu¼1300.
UNEP, 2018. Legal Limits on Single-Use Plastics and Microplastics: A Global Review of National Laws and Regulations. United Nations Environment Programme (UNEP), Nairobi, 2018 Jan 15.
United Nations Environment Assembly (UNEA), 2016; 2018. Resolutions on Marine Litter and Microplastics. United Nations Environment Programme, Nairobi, Kenya.
United Nations Environment Assembly (UNEA), 2014. Resolution 1/6: Marine Plastic Debris and Microplastics. Retrieved from. https://www.ecolex.org/details/decision/marine-plastic-debris-and-microplastics-55d61f84-1850-42b7-ba62-7fc07d1829ec/. (Accessed 14 June 2022).
U.S. Environmental Protection Agency (EPA), 2020. Toxic Substances Control Act (TSCA) Overview. U.S. Environmental Protection Agency, Washington, DC, USA.
U.S. Environmental Protection Agency (EPA), 2021. NPDES Stormwater Compendium for Trash-Related Provisions. U.S. Environmental Protection Agency, Washington, DC, USA.
U.S. Environmental Protection Agency (EPA), 2023. Strategic Research Action Plan (StRAP): Safe and Sustainable Water Resources. U.S. Environmental Protection Agency, Washington, DC, USA.
U.S. Environmental Protection Agency (EPA), 2022. Plastics: Material-Specific Data. https://www.epa.gov/facts-and-figures-about-materials-waste-and-recycling/plastics-material-specific-data. (Accessed 12 November 2022).
Van der Ploge, S., et al., 2012. Transboundary water governance and pollution control. Int. J. Water Resour. Dev. 28 (3), 483–498.
Vince, J., Hardesty, B.D., 2018. Governance solutions to the tragedy of the commons that marine plastics have become. Front. Med. S.A., Austr. Front. Mar. Sci. 5. https://doi.org/10.3389/fmars.2018.00214.
World Health Organization, 2019. Microplastics in Drinking-Water. World Health Organization, Geneva, Switzerland.
World Health Organization (WHO, 2022. Human Rights. www.who.int/news-room/fact-sheets/detail/human-rights-and-health.

CHAPTER 10

Technological solutions for microplastic remediation: Innovations in detection, removal, and prevention

Melvin S. Samuel

Department of Civil, Construction, and Environmental Engineering, Marquette University, Milwaukee, WI, United States

Highlights

- Advanced filtration and membrane technologies enable efficient removal of Microplastics from water systems.
- Green materials and biodegradable alternatives address Microplastic pollution at its source.
- Integrated cleanup and waste management strategies reduce secondary formation
- Technological innovation bridges the gap between monitoring and actionable remediation.
- Sustainable solution require alignment between engineering advances and environmental policy.

10.1 Introduction

The rapid proliferation of MPs in aquatic environments has necessitated the development of technological solutions that move beyond conventional pollution control strategies. While early research primarily focused on identifying sources and documenting environmental occurrence, recent efforts have shifted toward active remediation, prevention at source, and system-level intervention. Technological innovation now plays a central role in addressing MP pollution across freshwater, estuarine, and marine systems, particularly where natural attenuation processes are insufficient. Filtration-based technologies have emerged as some of the most effective tools for MP removal, especially within drinking water and wastewater infrastructures. Advances in membrane science including microfiltration, ultrafiltration, and hybrid membrane systems have demonstrated high removal efficiencies for a wide range of particle sizes. However, challenges such as membrane fouling, energy demand, and secondary waste generation underscore the need for continued optimization and integration with upstream control measures (Rafiq and Xu, 2023).

Parallel to removal technologies, the development of biodegradable alternatives and green materials represents a preventative strategy aimed at reducing MP generation at the source. Bio-based polymers, compostable packaging, and environmentally benign additives are increasingly being explored to replace conventional plastics, thereby limiting long-term environmental persistence. The

Microplastics in Aquatic Environments. https://doi.org/10.1016/B978-0-443-45619-0.00015-7

innovation aligns with circular economy principles, emphasizing material redesign, reuse, and end-of-life sustainability. Advances in cleanup strategies and waste management further complement technological remediation. Improved solid waste segregation, enhanced recycling infrastructure, and targeted recovery of plastic debris before fragmentation can significantly reduce MP formation (Seo et al., 2024). Together, these technological approaches form a multilayered mitigation framework that integrates detection, removal, and prevention, offering a pathway toward sustainable management of MP pollution in aquatic systems.

This book chapter is significant because it consolidates emerging remediation technologies within a unified sustainability framework, rather than treating detection, removal, and prevention as isolated efforts. Unlike many published works that focus on single technologies of laboratory-scale study, this chapter emphasizes system integration, scalability and long-term environmental relevance. Its novelty lies in combining filtration advances, green material innovation, and waste management strategies into a cohesive roadmap that aligns engineering solution with circular economy principles and policy needs. This integrative perspective is essential for translating research advances into real-world implementation.

10.2 Filtration technologies for microplastic removal

Landfills serve as the ultimate repository for a large fraction of plastic waste generated worldwide. Over extended periods, plastics deposited in these environments experience progressive, fragmentation driven by mechanical stress, ultraviolet radiation, oxidative processes, and biological interactions. Rather than undergoing complete mineralization, these materials disintegrate into progressively smaller fragments, ultimately generating MPs, defined as plastic particles with dimensions below 5 mm. Once formed, these particles are redistributed into adjacent environmental compartments, including aquatic systems, terrestrial soils and biota (Mitrano et al., 2021).

Within these compartments the pedosphere the soil layer formed at the interface of the hydrosphere, biosphere, and the atmosphere functions as a major sink for MPs. Estimates suggest that approximately one-third of global plastic waste accumulates in soils, positioning terrestrial environments as significant and long-term reservoirs for MPs. Given their persistence over timescales ranging from decades to centuries, the continuous buildup of MPs raises serious concerns regarding soil health and ecosystem performance. MPs have been shown to modify numerous soils characterizes, such as structural stability, moisture retention, nutrient dynamics, enzymatic processes and microbial community, composition. The magnitude and nature of these effects are strongly influence by the particle size, morphology, chemical composition, ad concentration (Huang et al., 2022). Such changes can affect plant development and in the long term, influence agricultural yields and food security. Furthermore, MPs present in soil can be transported into freshwater bodies via surface runoff and leaching process, thereby amplifying broader environmental contamination.

10.2.1 Need for reliable microplastic monitoring

A comprehensive understanding of the extent and implications of MP pollution in the soil relies on the availability of robust monitoring and analytical techniques. Precise identification and quantification are critical for determining contamination levels evaluating ecological risks and guiding policy development and remediation planning. Early MP research primarily targeted marine systems, where relatively simple matrices such as seawater facilitated extraction procedures. In contrast, investigations focusing on soils emerged almost a decade later due to the considerably higher complexity of terrestrial matrices. Soils are characteristics by heterogenous textures, elevated organic matter content, and strong aggregate formation, all of which complicate MP isolation. Unlike seawater, soils can entrap MPs within aggregates or bind them to organic components, significantly hindering separation. Organic matter represents a major analytical challenge, as plant residues and humic substances may adhere to MP surfaces or resemble synthetic polymers, leading to potential misidentification (Lehmann et al., 2021). Consequently, an essential step in soil sample preparation involves the removal of organic matter, commonly achieved through chemical digestion or enzymatic treatments to enable accurate isolation and analysis of MPs.

10.2.2 Current extraction challenges

The success of MPs extraction from soils is governed by multiple interacting factors:

- Soil type and texture, which affect aggregations behavior and particle binding
- Organic matter content, which increases separation complexity
- Polymer density, influencing recovery during density-based separation
- Experimental procedures, including sample handling protocols and processing time
- Analytical instrumentation, such as filtration systems, membranes, centrifuges, and ultrasonic equipment (Steinmetz et al., 2020).

As a result of these variables, reported extraction efficiencies differ substantially among studies. Low density polymers may be readily recovered by floatation methods, whereas higher density polymers may remain unrecovered. Centrifugation techniques can enhance separation efficiency but tend to lose effectiveness at high soil loadings (Grause et al., 2022). Ultrasonication is often employed to disrupt aggregates and liberate embedded MPs; however, excessive application can alter or damage particle integrity. These methodological differences hinder direct comparison of recovery rates across studies, underscoring the need for methodological harmonization (Büks et al., 2021). The adoption of standardized protocols, uniform equipment specifications, and soil-type adapted procedures would substantially enhance inter-laboratory reproducibility.

10.2.3 Toward standardized approaches

Advancing research in this field requires that extraction methods to be assessed not only for recovery efficiency but also for reproducibility, feasibility and applicability across diverse soil types. Key measures to improve soil-based MP extraction include:

- Disrupting soil aggregates while preserving MP integrity
- Effectively eliminating organic matter prior to separation
- Optimizing sample mass to reduce matrix interference
- Selecting density separation media appropriate for targeted polymer classes
- Implementing size-selective techniques to recover small MPs and nanoplastics

The established of standardized and validated methodological will facilitate more reliable cross-study comparisons and contribute to a more accurate assessment of MP contamination in terrestrial environments.

10.2.4 Microplastic separation techniques

10.2.4.1 Density separation

Density separation (DS) is among the most frequently employed methods for isolating MPs from soil and sediment matrices. This technique is based on a simple physical principle: when soil is suspended in a liquid with a known density, denser mineral particles settle, while lighter plastic particles rise to the surface. Most plastics are hydrophobic and exhibit minimal water absorption, allowing them to remain buoyant in solutions with sufficiently high density. To prepare the separation medium, the salts are dissolved in water to adjust its specific gravity. Sodium chloride (NaCl) and zinc chloride ($ZnCl_2$) are the salts most used for this purpose. NaCl is low-cost, readily available, and relatively safe for routine laboratory applications; however, its density is adequate only for separating low-density polymers such as polyethylene (PE) and polypropylene (PP). In contrast, $ZnCl_2$ produces a substantially denser solution enabling the recovery of higher density polymers and enhancing the separation of weathered MPs. To balance performance and cost, mixed salt solutions, typically combining NaCl and $ZnCl_2$ are also employed to increase solution density while limiting expenses.

It should be recognized that the environmental density of MPs often differs from their original manufactured density. Many commercial plastics contain additives, fillers or pigments that after their physical properties including density. Furthermore, MPs that have interacted with soil minerals or organic matter may acquire surface coatings, which effectively increase their apparent density. For this reason, high-density floatation solutions are frequently favored, as they enhance overall recovery efficiency and assist in removing adhered impurities from particles surfaces. Solutions based on $ZnCl_2$, sodium iodide (NaI), and sodium bromide (NaBr) have proven particularly effective or separating MPs from mineral rich soils. DS performs optimally in sandy soils, where the density contrast between mineral grains and plastics is more pronounced, with reported recovery rates reaching up to 98.6%. Conversely, the method is less effective in clay-rich soils where fine particles and high surface area impede efficient separation. Even under optimized conditions, maximum recovery on clay soils may not exceed 87.5% for particles larger than 250 μm (Zhang and Liu, 2018). Importantly, recovery efficiency is influenced by both soil type and MP concentration and does not always exhibit a linear relationship. Centrifugation is sometimes applied to accelerate the separation process; however, excessive centrifugal force can fragment MPs, generating smaller particles and biasing results. Therefore, careful optimization of centrifugation speed and duration is essential to ensure effective separation without compromising particle integrity.

In samples characterized by high organic matter content, combining density separation with digestion of organic material (DO) can markedly enhance MP recovery. Nevertheless, conventional filtration steps that often follow digestion may result in MPs becoming trapped within filter pores or being lost during handling. To reduce such losses, filtration should be minimized wherever feasible. In cases, where filtration is unavoidable, or when precise quantification of small MPs is required, analytical techniques that eliminate the need for filtration such as pyrolysis GC–MS (Py–GC–MS) or Raman spectroscopy can provide effective alternatives. When digestion is required, procedures that do not necessitate subsequent filtration are preferred, as they reduce handling-related errors and improve overall extraction accuracy.

10.2.4.2 Oil extraction

Oil extraction represents another widely applied approach for isolating MPs from soils and sediments. This method exploits the lipophilic characteristics of plastic surfaces. When soil particles and MPs are mixed with oil and water, distinct phases form, and plastics preferentially associate with the oil layer and float. While this behavior allows MPs to be separated from mineral components, isolating the oil phase and subsequently cleaning the recovered particles can be technically demanding. In controlled sediment systems, oil extraction has demonstrated recovery efficiencies exceeding 90% (Scopetani et al., 2020). However, its performance declines substantially in field soils with elevated organic matter content. Organic materials tend to bind to MPs, reducing their affinity for oil and diminishing separation efficiency. Consequently, the inclusion of an organic matter digestion step is recommended when applying oil extraction to complex soil matrices (Radford et al., 2021).

10.2.4.3 Other extraction approaches

In addition to gravimetric methods, several alternative techniques have been developed for MP isolation.

Electrostatic Separation: This approach separates particles according to differences in electrical conductivity between plastic materials and mineral constituents. Under controlled conditions, electrostatic separation can achieve recovery of 90%–100% for MPs ranging from 63 μm to 5 mm. However, its applicability is limited in highly aggregated soils, where tightly bound particles or MPs attached to organic matter cannot be efficient separated.

Magnetic Nanoparticle Tagging: An emerging strategy involves coating MPs with iron-based nanoparticles, enabling their retrieval through the application of electromagnets. Although this technique shows considerable potential, it requires specialized reagents and instrumentation. Despite ongoing technological advancements, density separation remains the most used method due to its cost-effectiveness, broad accessibility, and compatibility with standard laboratory workflows.

Removal of Organic Matter: In soils with substantial organic matter content, density separation alone is often inadequate. Organic components frequently exhibit densities like those of many polymers, resulting in poor discrimination between mineral particles and MPs. As a result, digestion of organic matter (DO) is a crucial step for improving extraction performance Approximately 80% of published studies incorporate an organic matter digestion step either before or after density separation. Commonly reported workflow sequences include:

DS - > DO
DO - > DS
DO - > DS - > DO

However, meaningful comparison among these workflows is challenging because studies differ in salt composition, soil texture, polymer type, and extraction conditions.

Polymer Stability During Digestion: A critical issue associated with digestion of organic matter (DO) procedures is the potential for MP degradation or changes in particle mass. An effective digestion protocol should effectively eliminate organic constituents while preserving the chemical and physical integrity of MPs. Published studies employ Fenton's reagent or hydrogen peroxide (H_2O_2), as these oxidative treatments successfully decompose organic matter while causing minimal damage to commonly encountered polymers. In contrast, acidic or alkaline digestion approaches are used much less frequently because of their higher potential to degrade plastics. Certain polymers, including polyamide, polyethylene terephthalate (PET), and polystyrene (PS), are particularly susceptible to partial or complete degradation when exposed to strong acids or bases (Pfeiffer and Fischer, 2020). Only a single study has reported the use of highly concentrated acidic digestion (98% H_2SO_4), and in that case, the extent of polymer alteration was not comprehensively assessed (Li et al., 2020).

Fenton's Reagent: Fenton's reagent operates through the catalytic action of ferrous ions, which promote H_2O_2-driven oxidation of organic matter. While this method is highly efficient, it may leave minor residues on MP surfaces, resulting in slight increase in particle mass. To prevent damage to aged or weathered MPs, digestion temperatures are generally maintained below 50°C, although this constraint necessities longer reaction times (Savino et al., 2022). Standard protocols typically apply 30–50 mL of reagent per 10 g of soil (Radford et al., 2021). Irrespective of the selected workflow sequence, filtration is commonly required after digestion or density separation. However, filtration introduces additional uncertainty, as MPs may become trapped within membrane pores or lost during handling.

Enzymatic Digestion: Recent investigations have explored multi-stage enzymatic digestion protocols in combination with Fenton's reagent. These approaches enhance digestion efficiency, particularly for MPs in the 100–400 μm size range. Despite their effectiveness and relatively gentle impact on polymer integrity, enzymatic methods are labor-intensive. They require continuous monitoring and adjustment of pH and temperature and involve extended processing times ranging from 7 to 22 days (Mbachu et al., 2021). Consequently, although enzymatic digestion minimizes polymer damage, its prolonged duration limits its suitability for rapid analyses or high-throughput MP extraction.

Evaluating the Recovery Rate of Extraction Methods: Spike-recovery experiments constitute akey quality-control measure for validating MP extraction methodologies. In such experiments, a known quantity and type of MPs are intentionally introduced into soil samples initially confirmed to be MP-free. Following application if the extraction protocol, the mass or number of recovered MPs is compared with the amount originally added. This comparison enables quantification of extraction efficiency and identification of potential methodological biases. Given the wide variability of soil properties, polymer characteristics, and separation techniques, spike-recovery tests are essential for demonstrating the reliability and reproducibility of extraction procedures (Hurley et al., 2018). Nevertheless, only approximately 26% of studies that combine density separation with organic matter digestion have conducted these validation experiments, indicating a substantial lack of methodological standardization in soil-based MP research.

Influence of Density Separation Solutions: Reported recovery efficiencies show considerable variability even identical salt solutions are used. For example:

- Polystyrene (PS) recoveries using $ZnCl_2$ range from 100% to a slow as 29%.
- Polyethylene (PE) recoveries using $CaCl_2$ range between 94% and 42%.

This wide range indicates that recovery efficiency is governed by factors beyond solution density alone. In most studies, density separation is coupled with an organic matter digestion step, making it difficult to isolate the independent effect of the density solution. Only a limited number of comparative investigations have directly assessed different solutions under identical conditions. One such study compared low-density NaCl with high-density NaCI and $ZnCl_2$ and found that higher-density solutions consistently yielded superior recovery. This outcome is likely attributable to interactions between MPs and soil density. From a purely technical standpoint, high-density solutions are therefore recommended, even for low-density polymers such as PE, PP, and PS. However, these advantages must be weighed against considerations of cost, toxicity and environmental impact (Kedzierski et al., 2017).

A pragmatic strategy is to select density solutions by balancing extraction efficiency with economic and environmental sustainability.

Effects of Particle Size: Particle size exerts a strong influence on MP extraction efficiency, with larger particles generally showing higher recovery rates. Larger polyethylene (PE) particles exhibit improved recovery when density separation is combined with organic matter digestion. Several mechanisms explain the reduced recovery of smaller MPs:

Aggregation with fine soil particles

- Small MPs can adhere to cohesive clay or slit particles, increasing their effective density and causing them to sink during density separation (Fang et, al., 2024).
- Binding to organic matter
- Organic compounds can strongly attach to small MPs, decreasing buoyancy and complicating separation processes (Zhang et al., 2022).

Detection Limitations: MPs smaller than approximately 10 μm are difficult to detect using conventional microscopic or spectroscopic techniques. This limitation often results in underestimation of MP abundance and artificially low recovery values. Because of these constraints, a universal extraction protocol suitable for all particle sizes is not achievable. For MPs below 50 μm, specialized and advanced methodologies-such as alternative separation techniques, thermal analytical methods, or high-resolution spectroscopic tools-are required to ensure accurate recovery and quantification.

10.2.4.4 Soil sample preprocessing for microplastics extraction

Successful extraction of MPs begins with appropriate soil preprocessing. Samples are commonly dried and sieved to remove coarse debris and to determine particle-size distribution. Sieving is rapid and effective for MPs larger than 100 μm; however, electrostatic interactions can cause MPs to adhere to sieve surfaces, and smaller particles may be trapped within the soil aggregates. To enhance size characterization, advanced techniques such as laser light diffraction (e.g., Mastersizer analysis) maybe applied after extraction. Although these methods show promise, their direct application for MP detection in soils remains limited.

For MPs smaller than 100 μm, high-resolution analytical tools are required (Table 10.1), including:

Table 10.1 The comparison of quantification methods of microplastics.

Identification methods	Techniques	Tested MPs	Advantages	References
FTIR	–	PP, PE, PS, PVC, PVA, PU, PTFE, PET, Others	A quick analysis with precision	Choi et al. (2021)
μ FTIR	IR-compatible filters	PE, PET, PAN	Quick bulb analysis, automated particle analysis of sizes and polymer types	Liu et al. (2019), Pérez-Reverón et al. (2022)
Microscopy	Heating or hot needle	PE, PP	Low cost	Zhang and Liu (2018)
Raman spectroscopy	Pretreatment by H_2O_2 or alcohol	PS, PE, PMMA, PA, PP	Quick analysis	Li et al. (2022a), Zada et al. (2018)

- Micro-FTIR (μFTIR)
- Raman spectroscopy
- Pyrolysis–GC–MS (Py–GC–MS) (Huang et al., 2021)

These techniques enable precise identification and quantification of particles that cannot be reliably detected by visual methods. Additionally, knowledge of local plastic usage patterns and the collection of larger plastic debris from sampling sites can aid in identifying potential sources of MPs.

10.2.5 Challenges in extracting MPs from soil

The most challenging aspect of soil-based MP extraction is the removal of organic matter and adhered impurities. Digestion reagents must be sufficiently aggressive to decompose organic material while remaining mild enough to prevent polymer degradation. Similarly, procedures used to disrupt soil aggregates may unintentionally fragment plastics into smaller particles. Implementing a predigestion step can assist in breaking aggregates and removing organic coatings prior to the main extraction. Oxidative digestion using hydrogen peroxide (H_2O_2)is generally favored over acidic or alkaline treatments, as strong acids and bases can damage certain polymers. Although sulfuric acid (H_2SO_4) is highly effective at organic matter removal, it poses a risk to the structure integrity of some plastics. Soil texture and organic matter content strongly influence digestion performance (Han et al., 2022). Testing digestion reagents on a larger MP fragment collected from the same samples can help assess polymer stability. In some cases, repeated or extended digestion cycles maybe necessary to achieve effective organic matter removal without compromising MP integration.

10.2.5.1 Density separation of microplastic extraction

High-density salt solution, including $ZnCl_2$, Nal, and NaBr, are extensively employed to separate MPs from soil mineral fractions. The application of centrifugation can shorten separation time; however, excessive rotational speeds may increase the likelihood of MP fragmentation and should therefore be applied with caution. In soils containing elevated levels of organic matter, coupling organic matter digestion (DO) with density separation (DS) substantially enhances extraction performance. Nevertheless, dependence on filtration following digestion or DS remain problematic, as MPs are frequently retained within the membrane pores, leading to analytical uncertainty and reduce recovery. To limit the need for filtration, researchers may implement direct identification and mass-based quantification techniques such as:

- Py–GC–MS
- Raman spectroscopy
- Chemical fingerprinting approaches

When digestion is unavoidable, protocols that do not require subsequent filtration are preferred to minimize particle loss and analytical bias.

10.2.5.2 Identification and quantification of microplastics

For visual-based identification, stereomicroscopy combined with FTIR spectroscopy remains the most widely used approach (Table 10.1). These methods yield reliable results for relatively larger MPs but show reduced accuracy for particles smaller than approximately 50 μm.

For the analysis of smaller MPs:

- μFTIR is effective for particles down to ~50 μm
- Raman spectroscopy can identify particles as small as ~20 μm

While microscopy and spectroscopic methods typically report MPs as particle counts per unit soil mass, greater comparability across studies can be achieved by expressing MP concentrations as mass per soil mass (Guo et al., 2022). Mass-based reporting provides a more standardized and quantitative basis for assessing contamination intensity.

10.2.5.3 Emerging methods for mass-based quantification

Mass-based quantification of MPs is an area of rapid methodological development. Advanced analytical techniques currently in use include:

- Py–GC–MS, which enables polymer-specific mass determination
- Accelerated Solvent Extraction (ASE) which isolates polymer mass prior to FTIR analysis
- Pressurized Liquid Extraction (PLE), which supports polymer-specific detection via combustion ion chromatography

Integrated approaches, such as PLE combined with Py–GC–MS or ASE coupled with Py–GC–MS, offer improved analytical accuracy. However, these methods remain sensitive to interference from soil organic matter, indicating that further refinement is required to improve their applicability to complex soil matrices.

10.2.5.4 Uncertainty introduced by environmental factors

Environmental variability introduces substantial uncertainty in MP extraction and analytical outcomes:

- Temperature fluctuations may induce thermal degradation of MPs, modifying their chemical signatures and complicating identification by FTIR or Py–GC–MS.
- Rainfall events and flooding can redistribute MPs with vertically and horizontally within soil profiles, increasing sampling complexity.
- Soil texture strongly influences MP adhesion and aggregation, with clay-rich soils generally exhibiting stronger binding than sandy soils.

These variables directly affect extraction efficiency and quantification reliability. Consequently, environmental conditions and soil properties must be carefully accounted for during experimental design to ensure accurate and reproducible MP assessments. The techniques currently applied for extracting, identifying and quantifying MPs in soil systems demonstrated that successful MP analysis depends heavily on the thorough removal of soil organic matter, as residual organics can interfere with both extraction efficiency and polymer identification. Density separation remains a fundamental method, with markedly improved performance when high-density solutions are employed, particularly for smaller particle sizes. However, given the wide variability in soil characteristics and polymer properties, the selection of both digestion strategies and density solutions must be tailored to specific sample conditions.

For identification and quantification, stereomicroscopy and FTIR spectroscopy continue to be the most used tools. Nonetheless, prevailing reporting units-primarily particle counts, or particle counts per soil mass-are inadequate for comprehensive environmental risk assessment or modeling applications. The advancement of mass-based quantification approaches is therefore essential for accurately determining contamination levels, assessing removal efficiencies, and evaluating MP transport and fate within soils.

Although methodological inconsistencies currently hinder direct comparison among studies, the importance of spike-recovery experiments for validating extraction protocols is clear. Similarly, digestion procedures must be systematically tested to determine whether they alter or degrade MPs. The integration of robust extraction techniques with reliable mass-based analytical methods will be central to the development of standardized protocols for soil MP assessment. Establishing validated, reproducible, and integrated methodologies will enhance environmental monitoring efforts and support improved management and mitigation of MP pollution in terrestrial ecosystems.

10.3 Microplastic mitigation techniques

10.3.1 Source and impacts

Global plastic production reached 400 million metric ton in 2022, reflecting a sustained upward trend compared with previous years (Statista, 2023). As plastic use continues to expand, the generation of MPs has increased correspondingly, and these particles have now been identified in air, water, soil, and wide range of organisms. Their pervasive distribution has raised serious concerns regarding potential risks to wildlife and human health through ingestion, inhalation, and dermal exposure.

The European Food Safety Authority (EFSA) defines MPs as a synthetic polymer particle smaller than 5 mm, whereas nanoparticles are characterized as particles with diameters above 0.1 μm. Based on their source, MPs are broadly categorized into two main groups:

- **Primary MPs:** deliberately manufactured particles used in cosmetic, personal-care products, cleaning formulations, and as feedstock in synthetic textile production.
- **Secondary MPs:** particles formed through the fragmentation of larger plastic items as a result of physical abrasions, chemical weathering, and biological degradation processes.

The rate and mechanism of plastic degradation are strongly controlled by the environmental conditions. Variables such as solar radiation intensity, temperature, oxygen availability, and microbial activity determine the pace at which plastics fragments. For instance:

- Photo-oxidative processes driven by ultraviolet radiation accelerate plastic breakdown in surface environments
- Cold, low-oxygen settings, such as deep-sea sediments, markedly slow degradation, allowing plastic to persist and accumulate over extended periods.

Because degradation typically leads to fragmentation rather than complete mineralization, MPs continue to disperse within the ecosystem and biological systems, posing long-term risks that are not yet fully understood.

10.3.2 Microplastic mitigation techniques

Strategies aimed at mitigating MP pollution incorporate physical, biological, and chemical approaches (Table 10.2); each associated with specific benefits and limitations. Given the wide variability in MP shape, size, polymer composition, and environmental behavior, no single remediation technique is universally effective. Consequently, effective mitigation often depends on integrated, multi-stage treatment strategies.

***Physical mitigation approaches*:** Physical methods-including filtration, sorption, sedimentation, and mechanical removal-are among the most applied techniques for MP control. Filtration systems, particularly those integrated into wastewater treatment facilities, play a vital role in intercepting MPs before they are discharged into natural waterbodies. Treatment processes such a coagulation, flocculation, and settling can remove substantial proportions of MPs from influent wastewater. Although these physical approaches are generally effective, their performance is highly dependent on the MP size, shape, and density. Larger particles are more readily captured, whereas smaller MPs and nano plastics often bypass filtration systems, resulting in incomplete removal. This limitation highlights the need for advanced membrane materials and improved mechanical separation technologies.

***Biological mitigation approaches*:** Biological remediation method employs living organisms-including microorganisms, fungi, and plants-to degrade, absorb, or immobilize MPs. Certain

Table 10.2 Summary of physical and biological microscopic remediation techniques.

Technique	Description	Weakness	References
Membrane filtration	Utilizes membranes with specific pore sizes to physically filter out microplastics from wastewater	High energy consumption, membrane fouling, and frequent maintenance	Adewuyi and Li. (2024)
Electro-coagulation	Electrodes are used to generate coagulants in situ, which aggregate microplastics for easier removal	High operational costs and potential release of metals from electrodes	Kim and Park (2021)
Magnetic carbon nanotubes	Magnetic carbon nanotubes are used to attract and remove microplastics from water via magnetic separation	Potential ecological risks due to the introduction of nanomaterials into the environment	Tang et al. (2021)
Biodegradation by microorganisms	Certain bacteria and fungi can break down microplastics under specific environmental conditions	Slow degradation rates and uncertainty efficiency at large scales	Osman et al. (2024)
Chemical oxidation	Oxidative agents such as ozone are used to degrade microplastics chemically	It can generate harmful by-products, incomplete degradation, and is energy intensive	Padervand et al. (2020)

microbial strains produce enzymes capable of degrading specific plastic polymers, while some plant species can take up or stabilize MPs within their tissue or rhizosphere (Namasivayam and Avinash, 2024).

Despite their potential, biological approaches face several constraints:

- Degradation rates are typically slow compared with physical or chemical methods
- Effectiveness depends on favorable environmental conditions, such as appropriate moisture, temperature, and pH
- Limited capacity to degrade a broad range of polymer types
- Potential ecological risks associated with introducing nonnative species for remediation

As a result, although biological strategies are environmentally benign, their scalability and long-term efficiency remain areas requiring further investigation.

***Chemical mitigation approaches*:** Chemical remediation techniques seek to transform plastic into smaller, potentially less harmful compounds through processes such as:

- Photocatalytic degradation
- Electrochemical oxidation
- Advanced oxidation process
- Hydrolytic and chemical dissolution reactions

An emerging chemical strategy is ferrate peroxidation, which has demonstrated the ability to degrade MPs while simultaneously breaking down associated additives, plasticizers and monomeric compounds (Wang et al., 2021). This dual functionality positions ferrate as a promising option for the concurrent removal of MPs and cooccurring chemical contaminants.

However, chemical mitigation methods present notable challenges:

- Many reactions require tightly controlled pH, temperature, or light conditions
- Degradation byproducts may be toxic, persistent, or environmentally harmful
- Chemical reagents and catalysts can be costly and may pose environmental risk if improperly managed.

Accordingly, thorough assessment of reaction pathways, byproduct toxicity, operational demands, and ecological impacts is essential prior to large-scale application.

***Integrated mitigation strategies*:** Effective MP remediation increasingly relies on the integration of physical, chemical, and biological methods. Representative combined approaches include:

- Filtration → biological degradation
- Chemical oxidation → sedimentation
- Photocatalysis → membrane separation

Table 10.2 provides a comparative overview of major MP remediation strategies-encompassing photolytic, electrochemical, mechanical, biological, chemical and thermal methods-highlighting the chemical bonds targeted by each approach, along with their respective advantages and limitations. The development of integrated and optimized mitigation pathways will be critical for future environmental management, particularly as MP pollution continues to intensify across both terrestrial and aquatic ecosystems.

10.4 Biodegradable alternatives and green materials

10.4.1 Geopolymer in microplastic remediation

Research into the use of geopolymers for MP removal is still in its early phase. The first experimental evidence by (Bhuyan et al., 2024) demonstrated that a superhydrophobic geopolymer foam, modified with a saline-coupling agent, can effectively separate MPs from water. This pioneering work underscores the potential of geopolymers as emerging materials for MP capture, drawing on extensive literature on geopolymer adsorption-particularly their interactions with metals, dyes, organic pollutants, and emerging contaminants-this section examines the anticipated mechanism by which geopolymers may remove MPs. Critical properties, including surface chemistry, pore architecture, and functional groups, are expected to influence adsorption behavior and retention efficiency, these insights provide a conceptual foundation to guide future experimental research and the development of geopolymer-based remediation strategies.

10.4.2 Anticipated mechanisms

MPs removal using geopolymers can be broadly categorized into:

- Physical adsorption mechanism
- Chemical adsorption mechanism

Both approaches exploit the intrinsic structure and tuneable surface chemistry of geopolymers.

10.4.2.1 Physical adsorption

Geopolymer possess a highly interconnected aluminosilicate network, forming a multi-scale porous structure. This architecture, combined with surface interactions, enable the effective physical capture of MPs. Major physical adsorption mechanism include:

- Physical entrapment within pores
- Electrostatic interactions
- Van der Waals forces

Understanding these interactions are crucial for optimizing geopolymer performance in environmental applications.

***Physical entrapment*:** The three-dimensional framework of cross-linked aluminosilicate chains in geopolymers generates micro-, meso-, and macropores, offering a substantial internal surface area (Burduhos Nergris et al., 2020). These pores act as natural "traps," immobilizing particles across a broad size range, including MPs of several milometers in diameter.

- Micropores (<2 nm): contribute to high surface area
- Mesopores (2–50 nm): facilitated particle movement with the network
- Macropores (>50 nm): accommodated larger MP fragments

As MPs transverse geopolymer matrices, they encounter numerous pore openings and surface sites, enhancing entrapment probability. This passive mechanical capture requires minimal energy, making it well-suited for scalable, ecofriendly water treatment systems.

Electrostatic and Van der Waals forces: Beyond mechanical entrapments, geopolymers can absorb microparticles via electrostatic interactions and van der waal's forces. MPs often carry charges due to additives, oxidation, or environmental exposure. When the geopolymer surface possesses an opposite charge, electrostatic attraction facilitates MP binding and retention (Wimalasinghe et al., 2018). Surface charge modulation-through pH adjustment or functionalization-can further increase affinity for specific polymer types. Van der Waals forces additionally stabilize trapped MPs, particularly at short distances within confined pores. Although weaker than ionic or covalent bonds, these forces reinforced attachments, reducing the likelihood of MP release. Combined, electrostatic and van der Waals interactions create a robust mechanism capable of capturing MPs with diverse surface properties.

10.4.2.2 Chemical and hydrogen bonding

Geopolymers can also engage in chemical interactions with MPs. Their aluminosilicate frameworks and modifiable surfaces allow formation of hydrogen bonds, covalent bonds, and ion-exchange interactions (Arokiasamy et al., 2023). These chemical pathways often provide stronger and more selective adsorption than conventional adsorbents such as activated carbon. Functionalization of geopolymers with silanes, amines, thiols, epoxies, or other reactive groups enhance geopolymer-plastic interactions. Environmental aging processes (UV exposure, oxidation) introduce functional groups such as–OH and –COOH onto MP surfaces (Atugoda et al., 2021). These groups can form hydrogen bonds with complementary functionalities on modified geopolymers, including amines, carboxyl's, or thiols. Hydrogen bonding provides:

- Strong, reversible interactions
- Enhanced adhesion in flowing water
- Compatibility with oxidized and weathered MPs

Tailoring geopolymer surfaces with hydrogen-bond-forming groups enables the capture of a broad spectrum of MPs, including environmentally aged particles (Liu et al., 2019)

10.4.2.3 Covalent bonding and ion exchange

Covalent bonds form permanent, stable linkage between geopolymers and MPs (Della Rocca et al., 2021). Functionalization with silanol (-SiOH) or epoxy groups allow reaction with MP surface groups (e.g., hydroxyls, carboxyls, amines) to generate:

- Ester bonds
- Amide bonds
- Siloxane linkages

  Silanol groups can react with carboxylated MP surfaces, producing durable covalent attachments (Lu et al., 2021). These interactions are irreversible and maintain strong retention under variable pH, salinity, or flow conditions, making them valuable for long-term environmental stabilization.

Geopolymers naturally contain exchangeable cation (e.g., Na^+, K^+, Ca^{2+}) within their aluminosilicate networks (Tang et al., 2024). MPs carrying ionic additives or residues can engage in ion-exchange processes:

- Geopolymer cations replace ions on MP surface

- Electrostatic attraction is enhanced
- Repulsive forces are minimized
- Adsorption strength increases

This mechanism is particularly effective in saline or coastal environments, where MPs frequently carry ions such as Na^+ or Cl^-. Ion exchange allows geopolymers to compete for these ions, improving MP capture and retention efficiency.

10.4.3 Microplastic control using geopolymer membrane

10.4.3.1 Microplastic control with hydrophilic geopolymer membranes

MPs due to their hydrophobic surface properties, tend to strongly interact with hydrophobic materials, including materials conventional polymeric filtration membranes. These interactions are primarily driven by noncovalent forces, causing MPs to adhere to similarly nonpolar surfaces. Consequently, traditional hydrophobic membranes often experience MP accumulation, a phenomenon, lowers water flux, and diminishes overall filtration efficiency and operational lifespan (Thomas et al., 2015). Research by Hadidi and Zydney (2014) indicates that hydrophobic membranes demonstrate notably lower flux recovery, especially under neutral to basic pH conditions. This suggests that while hydrophobic membranes capture MPs via size exclusion, hydrophobic-hydrophobic interactions further exacerbate fouling. In contrast, geopolymer membranes provide a viable solution. By introducing hydrophilic functional groups onto their surfaces, these membranes can markedly reduce MP adhesion. Hydrophilic surfaces attract water molecules, forming hydration layers that act as protective barriers, preventing direct MP attachment.

Key advantages of hydrophilic geopolymer membranes include:

- Lower incidence of MP induced fouling
- Enhance water flux and filtration performance
- Extended operational lifespan
- Reduced cleaning and maintenance demands

BY minimizing hydrophobic containment buildup, including MPs, hydrophilic geopolymer membranes address major filtration challenges and support the development of sustainable wastewater treatment challenges.

10.4.3.2 Microplastic biofilm management with geopolymer membranes

Interactions between MPs and microorganisms significantly influence fouling within wastewater systems. MPs frequently serve as substrates for microbial colonization forming biofilms composed of bacteria, fungi and other microorganisms (Nurlina et al., 2024). MP-associated biofilms:

- Affect the buoyancy and mobility of MPs
- Complicate their separation and removal
- Accelerate membrane fouling
- Reduce treatment efficiency

  Geopolymer membranes can be engineered to either suppress or facilitate biofilm formation depending on treatment objectives.

Biofilm Suppression: In cases, where biofilms are detrimental, membranes can be modified with materials like zirconium dioxide (ZrO_2), which exhibits strong anti-adhesive properties. Incorporation of ZrO_2 reduces microbial adhesion to the membrane surface (Kaali et al., 2011), thereby minimizing:

- Biofilm development
- Clogging
- Operational fouling

Controlled Biofilm Support: In scenarios where biofilm formation is advantageous—such as promoting microbial degradation of pollutants- geopolymer membranes can be tailored support beneficial microbial communities. Embedding antimicrobial metals like copper or silver nanoparticles enables membranes to:

- Inhibit harmful bacterial growth
- Promote functional microbial colonization
- Facilitate pollutant biodegradation
- Enhance containment removal efficiency

Thus, geopolymer membranes offer a versatile platform to manage biofilm-MP interactions within wastewater treatment systems.

10.4.4 Microplastic electrostatic filtration using geopolymer membranes

Electrostatic interactions constitute another critical mechanism affecting MP removal during membrane filtration. Both MPs and membrane surfaces can possess electrical charges, and their interactions are governed by principles of electrostatic attraction and repulsion (Della Rocca et al., 2021).

10.4.4.1 Electrostatic repulsion for anti-fouling

When both the membrane and MPs carry similar electrical charges, electrostatic repulsion occurs. This phenomenon prevents MPs from adhering to the membrane surface, thereby lowering fouling and improving filtration efficiency. Under alkaline conditions, many MPs acquire a negative charge (Joo et al., 2021) and membranes exhibiting similar negative zeta potentials strongly repel these particles. A high membrane zeta potential in such conditions:

- Enhances the rejection of pollutants.
- Preserves membrane cleanliness
- Minimizes the accumulation of fouling over time

***Electrostatic Attraction and Fouling*:** In contrast, if MPs and membrane surfaces carry opposite charges, attraction occurs, increasing the risk of fouling. This underscores the necessity of designing geopolymermembranes with surface charges tailored to the specific water chemistry and the MP composition.

***Charge-Modified Polymers*:** Geopolymer membranes can also integrate anionic polymers or other additives to modify surface charge (Elgarahy et al., 2023). By aligning the membrane's surface charge with that of the MPs present, electrostatic repulsion is maximized, resulting in more effective MP removal.

10.4.4.2 Integrated mechanism of microplastic removal

MP capture by geopolymer membranes extends beyond simple size exclusion. Instead, it relies on a combination of hydrophobic interactions, biofilm dynamics, electrostatic forces, and engineered surface characteristics. When membranes are properly designed, they can:

- Repel hydrophobic MPs via hydrophilic surfaces
- Mitigate microbial fouling or support beneficial biofilms
- Exploit electrostatic repulsion for enhanced removal
- Use engineered porosity to physically trap particles

Advancing geopolymer membrane technology requires precise surface engineering, including control of charge, hydrophilicity, functional groups, and pore architectures. These improvements enhance filtration efficiency, positioning geopolymer membranes as powerful tools for MP mitigation in wastewater treatment.

10.4.5 A case study on the application of geopolymers for microplastic mitigation

Bhuyan et al. (2024) presented a novel approach for MP removal from water using a super hydrophobic geopolymer foam modified with silane-coupling agents. This study directly addresses the pressing environmental concern of MP contamination in aquatic systems and demonstrates the potential of advanced geopolymer-based materials in wastewater treatment.

10.4.5.1 Foam preparation and surface modification

In the study by Bhuyan et al. (2024), the geopolymer foam was formulated with metakaolin, sodium silicate, sodium hydroxide, hydrogen peroxide, and the surfactant Triton X-100. This combination enhanced both the structural integrity and functional properties of the foam. To further optimize performance, the foam surface was modified using a silane-coupling agent, imparting superhydrophobic properties. The modified material exhibited a water contact angle of 152 degrees, confirming strong hydrophobicity. This characteristic is particularly advantageous for capturing MPs, as many, especially polyolefins, are inherently hydrophobic. The surface modification thus facilities stronger interactions with MPs, improving their capture and separation from water.

10.4.5.2 Performance in microplastic separation

In laboratory tests, the superhydrophobic geopolymer foam demonstrated outstanding performance. Using polyethylene microspheres (53–63 μm), the foam achieved approximately 99% removal efficiency, even after processing nearly 200 bed volumes of water (Bhuyan et al., 2024). This result indicates the material's durability and sustainability for repeated or continuous operation. By contrast, unmodified geopolymer foams exhibited significantly lower performance, maintaining function for only about five bed volumes before structural deterioration. This comparison highlights the critical role of surface modification in improving adsorption capacity and operational longevity. The foam was further tested with real wastewater, specifically laundry effluent containing MPs ranging from 2 to 2000 μm Even in these complex matrices, the foam maintained around 8% separation efficiency over 60 bed volumes, demonstrating promising potential for real world applications despite the heterogeneity of household wastewater.

10.4.5.3 Dominant removal mechanism

Bhuyan et al. (2024) emphasize that the foam's high removal efficiency is primarily driven by chemical interaction linked to surface hydrophobic rather than by simple physical filtration or size-exclusion. This distinction is critical, suggesting that engineered surface chemistry, rather than pore size alone, can significantly enhance MP capture. This mechanism sets the geopolymer foam apart from conventional filtration materials that rely mainly on mechanical sieving and often struggle with smaller MPs.

***Environmental and Practical Implications*:** The outcomes of this study highlight the considerable potential of superhydrophobic geopolymer foams as efficient. Economical, and scalable materials for mitigating MP pollution. Their notable durability, high removal performance, and demonstrated capability to treat real wastewater suggest strong feasibility for incorporation into both centralized and decentralized water treatment infrastructures.

Beyond the introduction of a novel material platform, this work identifies multiple directions for future research, including:

- Scaling up manufacturing processes for application in industrial wastewater treatment
- Incorporation of the foam into established filtration and treatment configurations
- Investigations of surface modifications to enable selective adsorption of diverse MP categories
- Evaluation of long-term performance and stability under variable environmental conditions

Overall, this case study offers a critical proof of concept, illustrating that engineered geopolymer materials represent a promising and sustainable approach for MP remediation. By leveraging low-cost precursor materials alongside tuneable surface functionalities, geopolymer-based solutions have the potential to contribute substantially to global water quality improvement and broader environmental protection efforts.

Overall, geopolymers demonstrate strong potential as transformative materials for addressing MP contamination, offering as sustainable, versatile, and potentially scalable solution to contemporary water treatment challenges. Their adjustable chemistry, structural robustness, and compatibility with a wide range of functional enhancements make them well suited for capturing diverse MP types through physical, chemical and electrostatic interactions. Nevertheless, realizing this potential will require continued research to address existing limitations, optimize surface engineering strategies, enhance long-term durability and confirm performance under real-world conditions. The establishment of standardized testing methodologies and the exploration of large-scale production pathways will also be critical for advancing geopolymer technologies component in global efforts to combat plastic pollution, supporting cleaner water systems, protecting ecological health and fostering a more sustainable future for coming generations.

10.5 Microalgae-based approaches for microplastic mitigation

In comparison with conventional remediation strategies, the use of microalgae for MP removal presents several distinct advantages, including the ability to perform repeated removal cycles, minimal environmental disturbance and lower energy demands. Microalgae photosynthetic, unicellular organisms ranging from a few to several hundred micrometers serve as fundamental components of aquatic ecosystems by acting as primary producers of carbohydrates, proteins, and lipids. Their capacity to form biofilms and secrete extracellular polymeric substances (EPS) facilitates close interactions with MPs, promoting aggregation and potential extraction from aquatic environments (Ugya et al., 2022; Borges et al., 2021). In recent years, microalgae have gained attention as promising biological agents for MP degradation. A growing body of research has reported interactions between aquatic microalgae and MPs. These organisms can proliferate in nutrient-rich waste streams, accumulate substantial lipid reserves and importantly, do no compete with agricultural food resources, making them an attractive biomass source for bioplastic production (Chia et al., 2020). Their rapid growth rates, straightforward cultivation requirements, and heightened sensitivity to pollutants further position microalgae as valuable tools for ecotoxicological studies (Li et al., 2022a,b). EPS produced by microalgae supports the formation of hetero aggregates with MPs, enabling a repeatable and readily scalable removal mechanism. While earlier studies primarily emphasized MP toxicity and microalgal colonization behavior, more recent investigations have begun to examine the role of microalgae in promoting MP aggregation and facilitating separation from water bodies (Hossain et al., 2023). Despite increasing research interest, comprehensive reviews addressing both microalgae-assisted degradation and MP-derived bioplastic production remain scarce. Accordingly, the central objective of this review is to assess the potential of microalgae as biological agents for MP degradation in aquatic environments. This includes an examination of:

- Microbial enzymatic activities
- Metabolic pathways associated with polymer breakdown
- Interactions between microalgae and MPs
- The practical feasibility of microalgal systems for real world MP mitigation

Given the rapidly intensifying global challenge posed by MP pollution, elucidating and leveraging the role of microalgae in remediation strategies is of critical importance.

10.5.1 Microplastics in aquatic ecosystems

MPs, commonly defined as plastic particles smaller than 5 mm, have emerged as pervasive global pollutants. It is estimated that oceans alone contain approximately 4.85 trillion MP particles (A.K (Priya et al., 2022). Despite extensive research spanning several decades, effective strategies for large-scale MP removal remain largely insufficient.

MPs originate from:

- Primary sources, such as intentionally manufactured particles incorporated into cosmetics, detergents and various industrial products
- Secondary sources, resulting from the degradation and fragmentation of larger plastic debris through ultraviolet radiation, mechanical abrasion, chemical weathering, or biological activity.

Once released into aquatic environments, MPs disrupt ecological processes by interfering with nutrient cycling, impairing feeding behaviors, and altering physiological functions in organisms (Vanapalli et al., 2020). Their small size facilitates ingestion by a broad range of species, including plankton, fish and shellfish, leading to physical obstructions, oxidative stress, and the accumulation of hazardous chemicals. Of particular concern, MPs have been detected in human stool samples and within the bloodstream, underscoring their extensive penetration into food webs and the human body (Zhang et al., 2017).

10.5.2 Polymer composition of microplastics

Primary MPs comprise a broad spectrum of commonly utilized polymer types including Polyethylene (PE), Polypropylene (PP), Polyvinyl chloride (PVC), Polystyrene (PS), Polyester (PES), Polyamide (PA), Polyacrylonitrile (PAN), Polyethylene terephthalate (PET), Polyphthalamide (PPA), Polycarbonate (PC), and Polymethyl methacrylate (PMMA) (Ncube et al., 2021). In addition to these primary materials, secondary MPs are generated through the degradation of larger plastic products in response to environmental stressors such as ultraviolet radiation, oxidative processes, and mechanical forces from wave action. Although the mechanisms governing the transition from primary to secondary MPs are not yet fully elucidated, MPs are frequently associated with chemical additives, including phthalates, PBDEs, PVC stabilizers, and TBBPA. These compounds are incorporated to improve polymer performance but present significant taxological risks to the environment (Debroy et al., 2022).

***Forms and Environmental* Presence:** MPs occur in a wide range of physical forms, including pellets, fragments, fibers, films, filaments and ropes, microbeads, foams and sponges, and rubber particles. Among these, fibrous MPs are particularly prevalent in aquatic environments and represent a major exposure pathway for human through food consumption, drinking water, and inhalation. Comprehensive characterization of MPs encompassing polymer composition, particle size, color, and morphology is critical for assessing ecological impacts, potential human health risks, and environmental transport and fate (Menendez-Manjon et al., 2022).

10.5.3 The potential of microalgae for microplastic degradation in aquatic ecosystems

Microalgae have attracted substantial scientific interest as potential agents for the degradation and bioremediation of MPs (Fig. 10.1), offering an environmentally sustainable and low-impact approach to addressing plastic contamination (Madadi et al., 2021). As the accumulation of MPs in aquatic environments continues to escalate, research efforts have increasingly focused on elucidating the effects of MPs on ecosystem functioning, nutrient dynamics, and energy transfer within aquatic food webs (Borges et al., 2021). A diverse array of MP forms-including microfilms, microfoams, microbeads, microfibers, and microfragments-enter environments from numerous anthropogenic sources. For instance, microfilms originating from disposable tissues and medical dressings progressively fragment over time; microfoams derived from food packaging degrade under thermal stress and UV exposure; microbeads from personal care products are discharged into wastewater streams; and microfibers are released from synthetic textiles during washing processes. In addition, fragmentation of larger plastic debris, such as bottles and contains, contributes significantly to MP contamination in rivers, lakes, and marine environments.

10.5.4 Microalgae-microplastic interactions

The capacity of microalgae to contribute to MP transformation is closely associated with their physiological and metabolic attributes. Microalgae harness solar energy and available carbon sources to generate biomass, and MPs can interact with these organisms through several pathways:

- **Surface attachment:** MP particles can adhere to algal cell surfaces.

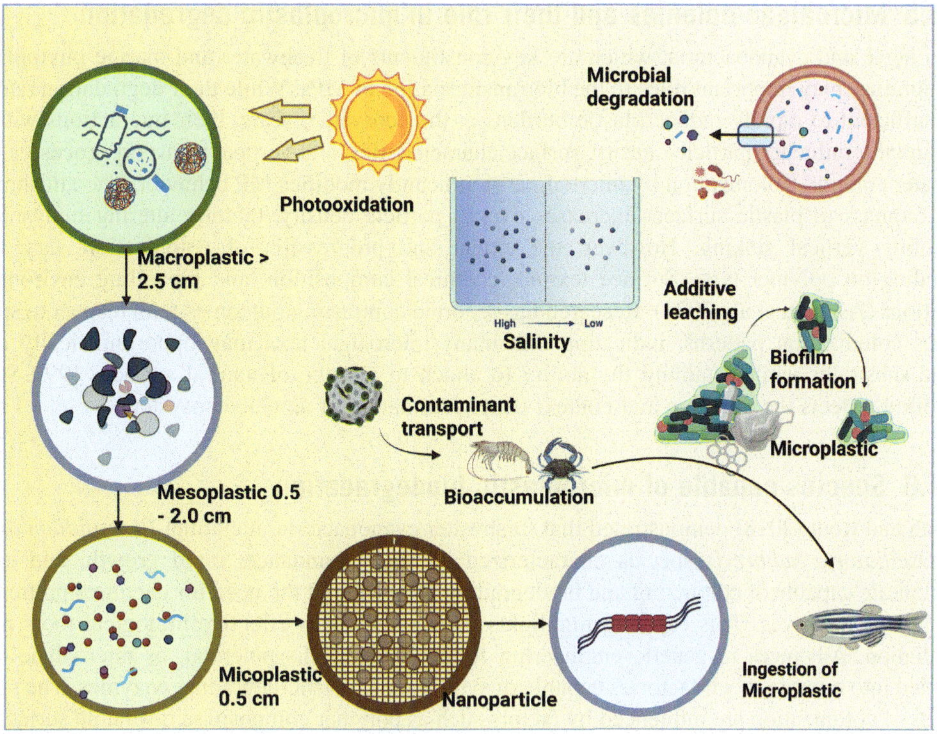

FIGURE 10.1

Conceptual illustration of plastic degradation pathways in aquatic environments, showing the progressive fragmentation of macroplastics into mesoplastics, microplastics, and nanoplastics driven by ultraviolet radiation, photo-oxidation, mechanical wave action, salinity, thermal-oxidative processes, and microbial activity. The figure also highlights additive leaching, biofilm development, contaminant transport, bioaccumulation across aquatic organisms, and the eventual formation of low-molecular-weight oligomers and monomers, emphasizing the multifaceted environmental impacts of plastic transformation in water systems.

- **Aggregate formation:** Extracellularpolymeric substances (EPS) secreted by microalgae can entrap MPs within hetero aggregates.
- **Biofilm development:** Microalgae may colonize MP surfaces, forming biofilms that promote subsequent microbial succession.
- **Limited carbon utilization:** Although direct assimilation of MPs as carbon sources is rare, certain microalgae may utilize polymer additives or partially oxidized fragments depending on MP size and chemical composition.

Microbial colonization of both natural and synthetic substrates enhances MP degradation by promoting physical abrasion, enzymatic breakdown, and the formation of complex microbial consortia.

10.5.5 Microalgae colonies and their role in microplastic degradation

Green algae and cyanobacteria, which are key constituents of freshwater and marine phytoplankton communities, play important roles in the biotransformation of MPs. While their degradation rates may be insufficient to rapidly reduce plastic burdens at the ecosystem scale, their interactions with MPs substantially influence particle density, surface characteristics, and vertical transport processes within the water column. Colonization by microalgae significantly modifies MP behavior. The attachment of algal biomass to plastic surfaces increases overall particle density, thereby altering buoyancy and promoting vertical sinking. However, the extent and progression of colonization vary widely depending on polymer type. Surface texture, chemical composition, and prevailing environmental conditions (Asiandu et al., 2022). Research has reported contrasting outcomes with respect to species-species colonization patterns, indicating that many microalgal taxa may opportunistically exploit plastic substrates while retaining the ability to attach to a range of natural surfaces [99]. Such as variability reflects the dynamic and context-dependent nature of aquatic ecosystems.

10.5.6 Species capable of microplastic biodegradation

Sarmah and Rout (2018) demonstrated that freshwater cyanobacteria, including *Phormidium lucidum* and *Oscillatoria subbrevis* species characterized by high abundance, rapid growth, and ease of isolation-are capable of colonizing and biodegrading MPs without the need for chemical pretreatment or oxidative additives. This capacity highlights their potential as effective biological tools for MP remediation. Advances in genetic engineering further expand this potential, as microalgae can be modified into microbial cell factories capable of synthesizing plastic-degrading enzymes. The rate and success of colonization are influenced by factors such as polymer composition, available surface area, surface energy, and surface roughness (Kaiser et al., 2017). Under specific conditions, certain colony-forming microalgae may partially utilize MP components as carbon sources, although this behavior is not commonly observed.

Microalgae Succession on Plastic Surfaces: The initial stages of microalgal colonization on MP surfaces resemble early biofouling processes, wherein generalist species rapidly occupy previously uncolonized substrates. This early phase is largely stochastic and is governed by factors such as growth rate, dispersal capacity and surface availability (Stoodley et al., 2004). As colonization progresses, intensified competition for space and nutrients leads to the establishment of niche-specialist communities. These organisms produce:

- Extracellular polymeric substances (EPS) to provide structural stability
- Secondary metabolites that facilitate intercellular communication through quorum sensing
- Antimicrobial compounds that inhibit competing microorganisms (Dang and Lovell., 2016).

Such shifts in community composition reflect a dynamic ecological equilibrium among microalgae, bacteria, and other microorganisms inhabiting MP surfaces-microhabitats commonly referred to as the “plastisphere”.

10.5.7 Influence of polymer type on algal colonization

The characteristics of the plastic substrate exert a decisive influence on algal colonization dynamics. A study conducted by Nava et al. (2021), across 45 lentic ecosystems demonstrated that microalgae are capable of colonization and community composition were strongly dependent on polymer properties as well as prevailing environmental conditions.

10.5.7.1 Biotechnological advances

A notable breakthrough in this field was reported by Moog et al. (2019) who genetically engineered the marine microalga *Phaeodactylum tricornutum* to synthesize and secrete PETase-an enzyme capable of degrading PET and PETG plastics under relatively mild temperature conditions ($\approx$21°C). This achievement highlights the practicality of employing engineered microalgae for plastic degradation, even within saline aquatic environments. Collectively, these findings emphasize the complex and evolving nature of microalgae-MP interactions. From initial surface attachment to enzymatic breakdown, microalgae play a pivotal role in influencing the transport, transformation, and ultimately fate of MPs in aquatic systems, thereby offering a biologically mediated strategy for mitigating plastic pollution.

10.5.7.2 Microplastic microalgae interactions

Microalgae are fundamental constituents of aquatic ecosystems, functioning as primary producers and forming the base of aquatic food webs. Their use in MPs (MP) removal is increasingly recognized as a sustainable alternative to conventional remediation approaches. Microalgae based systems are characterized by low energy requirements, reduced environmental impacts, and the capacity for reuse across multiple treatment cycles (Abdelfattah et al., 2023). Nevertheless, the escalating concentration of MPs in aquatic environments poses significant risks to microalgae, triggering stress responses that modify their physiology and compromise their ecological roles (Geremia et al., 2023). Upon exposure to MPs, microalgae activate a range of defense mechanisms that enable tolerance to MP-induced stress while simultaneously supporting biodegradation processes. MPs can disrupt essential cellular pathways, including chlorophyll biosynthesis and the photosynthetic electron transport chain, resulting in electron accumulation and oxidative damage (Podbielska and Szpyrka, 2023). These effects reflect three primary mechanisms through which MPs influence microalgae:

- Adsorption and internalization of MP particles
- Declines in photosynthetic efficiency

A clear understanding of these processes is critical for evaluating the dual role of microalgae as both sensitive organisms and active participants in MP transformation.

10.5.7.3 Adsorption and internalization

Microalgae facilitate MP removal through surface adsorption, whereby MPs adhere to algal cells, and through internalization, in which particles are taken up into the cell interior. Attachment of MPs to microalgae occurs not only via physical interactions but also through biochemical processes. Many microalgal species secrete extracellular polymeric substances (EPS), adhesive matrices composed primarily of polysaccharides and proteins, which markedly enhance MP binding capacity. In certain species, MPs may also be internalized. During endocytic processes, the algal cell membrane envelope small MP particles, enclosing them with intracellular compartments. This mechanism contributes to the physical elimination of MPs from surrounding waters and represents an adaptive cellular response to particulate stress.

Tunali et al. (2020), reported that *Chlorella vulgaris* is capable of both adsorbing and internalizing 0.5 μm polystyrene MPs, as confirmed by scanning electron microscopy. Adsorption increased with MP concentration, while growth inhibition rose from 15.7%at 50 mg/L to 28.9% at 1000 mg/L. Although *C. vulgaris* demonstrates substantial potential for MP removal, this capacity is accompanied by diminished algal growth, underscoring the trade-off between remediation efficiency and organismal health. Subsequent investigations employing atomic force microscopy (AFM) and FluidFM showed that MP exposure stimulates increased EPS production in microalgae (Demir-Yilmaz et al., 2022). The resulting EPS-enriched conditions enhance hetero aggregation and significantly improve MP removal efficiency. For instance, microalgae preconditioned with MP exposure for 6 days achieved removal efficiencies of up to 98% for PMMA microparticles, indicating that acclimated microalgae can function as highly effective biological “filters”.

10.5.7.4 Reduction of photosynthetic efficiency

MPs can disrupt the photosynthetic apparatus of microalgae, indirectly influencing their capacity to contribute to MP degradation. MPs suspended in their water column or accumulating at the surface can attenuate light penetration, thereby reducing the energy available for photosynthesis (Atugoda et al., 2022). Consequently, chlorophyll-a concentrations decline, carbon fixation rates decrease, and overall microalgal growth is impaired (Villalobos et al., 2019; Liu et al., 2020) observed that Scenedesmus obliquus exposed to MPs exhibited pronounced reductions in photosynthesis performance, effects that subsequently propagated to zooplankton reliant on microalgae as a food source. These findings illustrate the cascading ecological impacts of MP contamination that extend beyond microalgal physiology. Despite these limitations, microalgae remain promising candidates for wastewater treatment applications, including MP removal during tertiary treatment stages. Their high biomass productivity, capacity for pollutant uptake, and ability to convert contaminants into value-added products support their integration into bioremediation systems (Abdelfattah et al., 2023). However, MP exposure can negatively affect microalgal colony viability, potentially due to nutrient adsorption onto MP surfaces (Abdelfattah et al., 2023).

10.5.7.5 Oxidative stresses

MP exposure also includes oxidative stress in microalgae. MPs and their associated chemical additives can stimulate the generation of reactive oxygen species (ROS) within algal cells, overwhelming antioxidant defense systems and causing damage to lipids, proteins and nucleic acid (Pikula et al., 2019). In response, microalgae upregulate antioxidant enzymes, including superoxide dismutase, and catalase (Natarajan et al., 2022). demonstrated that *Dunaliella salina* exposed to MPs produced elevated ROS levels and countered this stress through enhance antioxidant activity. These protective responses not only facilitate tolerance to MP exposure but also contribute to degradation processes. ROS generated through photosynthesis or stress-related pathways can oxidize MP surfaces, initiate polymer breakdown when combine with UV radiation or other environmental factors (Sharma et al., 2021). Furthermore, oxidative stress often promotes increased EPS secretion, encouraging biofilm formation on MP surfaces. Such biofilms accelerate biodegradation by creating microenvironments conducive to the development of microbial consortia capable of further polymer decomposition.

10.5.7.6 Physiological transformations of microplastics

In addition to experiencing MP-induced stress, microalgae actively participate in MP transformation through physiological processes. Within aquatic environments, MPs undergo fragmentation, surface erosion, and chemical modification because of interacting environmental and biological factors (Lu et al., 2023). Microalgae can markedly accelerate MP aging, leading to observable changes such as altered surface morphology, decreased particle size, and increased surface roughness following exposure to algal metabolism. For example, polyethylene (PE) and PVC MPs subjected to photo-Fenton or biologically mediated degradation exhibit reductions in size and surface integrity. MP uptake by microalgae is strongly size-dependent. Nanoparticles may traverse cell wall pore channels, whereas microparticles larger than 1–2 μm were previously considered too large for cellular entry. However, recent studies challenge this assumption: after 72 h, 24% of *P. helgolandica* cells were found to contain 1 μm MP, and 43% of *S. quadricauda* cells internalized particles of the same size (Chen et al., 2022a). Additionally, *Heterocapsa triquetra* has been showed to exhibit phagotrophic behavior, internalizing 2 μm MP (Long et al., 2017). These interactions underscore the ability of microalgae to facilitate MP degradation by producing enzymes and metabolites that weaken polymer chains (Jain et al., 2023). Over time, MPs become smoother, less angular, and more rounded morphological changes indicative of progressive degradation. While smaller particles may increase ingestion risks for aquatic organisms, their modified surfaces are generally more susceptible to continued biological breakdown.

10.5.7.7 Chemical interactions between microalgae and microplastics

Microalgae driven chemical degradation of MPs proceeds through three principal pathways:

- Enzymatic activity
- Physiochemical transformations
- Biological transformations

Together, these mechanisms destabilize polymer structures and accelerate the breakdown of MPs.

10.5.7.8 Enzymatic activity

Microalgae secrete a range of enzymes that initiate and facilitate the chemical degradation of plastic materials. Esterases are particularly important, as they hydrolyze ester bonds commonly present in polymers such as PET. Falah et al., (2020) demonstrated that *Chorella vulgaris* produces esterases capable of fragmenting PET MPs into smaller units, thereby increasing their susceptibility to biodegradation. In addition, oxidative enzymes such as laccases and peroxidases generate ROS that chemically modify plastic surfaces, further enhancing their vulnerability to degradation. These enzymatic processes represent the initial stage of a multi-step biodegradation pathway.

10.5.7.9 Physiochemical transformations

In addition to enzymatic mechanisms, microalgae contribute to MP degradation through nonenzymatic physical and chemical processes. Physical abrasions can occur as microalgae move across or graze on MP surfaces. Wu et al., (2022) reported that *Nannochloropsis oculate* induced surface abrasion on PVC MPs, removing superficial layers and exposing underlying polymer surfaces. Biofilm formation is a critical process: MPs are embedded in algal biofilms, which create microenvironments enriched with metabolites and enzymes that regulate degradation (Barboza et al., 2020).

10.5.8 Biological transformations

Microalgae also indirectly facilitate MP degradation by promoting microbial colonization. Bacteria, fungi, and other microorganisms associated with algal biofilms secrete complimentary enzymes that further contribute to polymer breakdown (Prata et al., 2019; Xiao et al., 2023) demonstrated that polyethylene MP degradation was significantly enhanced when microalgae functioned synergistically with associated bacterial communities. In addition, microalgae can internalize MPS, enabling -partial intercellular degradation. Gowthami et al., (2023) observed that marine microalgae exposed to polyethylene MPs ingested and partially oxidized these particles, resulting in the formation of smaller fragments and metabolic byproducts. Collectively, these biological interactions emphasize the central role of microalgae as catalysts for both direct and indirect MP degradation in aquatic environments.

10.5.9 Microplastic impacts on microalgae

Research on freshwater and marine algal species has consistently shown that plastic particle toxicity increases as particle size decreases. Nano and micro-sized plastics are of particular concern due to their small dimensions and high surface area, which enable them to adsorb toxic substances such as organic pollutants and heavy metals. As a result, these microorganisms and small invertebrates (Law, 2017).

10.5.9.1 Effects on growth and physiology

MPs can directly disrupt algal growth and metabolic processes. Investigations involving mixotrophic algae such as *Cryptomonas* sp. Indicate that direct exposure to polyethylene (PE) MPs or their leachates can elicit measurable toxic responses (S.J (Taipale et al., 2019). In general, increasing MP concentrations are associated with reduced algal growth rates (Khoironi et al., 2019). At elevated exposure levels, microalgae commonly display pronounced declines in photosynthetic performance, followed by partial recovery over time, suggesting some capacity for physiological acclimation under prolonged MP stress (Zhang et al., 2017). Experimental studies further indicate that charged nano-particles (~0.02 μm) can adhere to algal cell surfaces and suppress photosynthesis, likely through interference with light harvesting or electron transport pathways (Bhattacharya et al., 2010). Across multiple investigations, smaller particles were consistently more harmful than larger ones, particularly when the ration of algal cell size to MP size fell within a critical logarithmic range (0.75–3.07) (Lagarde et al., 2016). Conversely, in certain freshwater systems, exposure to polystyrene beads measuring 63–75 μm at concentrations of 130 mg/L over 5 days was linked to enhanced algal growth (Chae et al., 2019). These observations indicate that MP effects are not universally negative and are strongly influenced by particle size, exposure concentration, and polymer chemistry.

10.5.9.2 Light shading and cell damage

MPs may also affect microalgae indirectly by modifying light availability. Larger MPs suspended in the water column or floating at the surface can scatter or block incoming light, reducing the intensity available for photosynthesis and thereby decreasing photosynthetic efficiency (Yurtsever et al., 2017). In such scenarios, light shading represents the dominant stressor, constraining energy acquisition and carbon fixation. Smaller MPs, in contrast, are more likely to induce physical and mechanical damage. These particles can attach to cell wall, abrade cell surfaces, or compromise membrane integrity, resulting in physiological stress and inhibited growth. For example, small polystyrene particles (0.05 mm) markedly suppressed the growth of *Dunaliella tertiolecta,* whereas larger particles measuring 6 mm produced negligible effects (Sjollema et al., 2016). PE MPs have similarly been shown to impair algal growth and photosynthesis, although some species demonstrate the ability to acclimate to their presence over time.

10.5.9.3 Role of additives and leachates

The effects of MPs on microalgae are influenced not only by particle size and morphology but also by chemical composition, including additives such as stabilizers, phthalates, and endocrine-disrupting compounds. These substances can leach into surrounding waters and modulate algal physiological responses. In certain cases, MPs containing low concentrations of leachable additives may even stimulate algal growth. For instance, *Dunaliella salina* exposed to larger MPs that did not alter cell morphology enhanced photosynthetic activity (Chae et al., 2019). It has been proposed that trace amounts of released compounds may function as supplemental nutrients or growth promoters under specific conditions. However, such responses are highly context-dependent, and their long-term ecological consequences remain poorly understood. Overall, the net impact of MPs on microalgae is governed by an interplay of multiple factors, including:

- Particle size, shape and concentration
- Polymer composition and associated chemical additives
- Duration of exposure and environmental conditions
- Algal species-species tolerance and adaptive capacity

Further investigation is required to comprehensively elucidate how MPs and their additives interact with diverse taxa across environmentally relevant gradients.

10.5.10 Conclusion

MP contamination in aquatic environments has rapidly emerged as one of the most widely reported global environmental challenges. Concurrently, microalgae are increasingly recognized as valuable renewable resources, not only for sustaining ecosystem function but also as sustainable sources of environmentally friendly polymers and bioproducts for applications in health, packaging and agriculture. A thorough understanding of the complex interactions between MPs and microalgae is therefore essential. Existing research reveals substantial variability and inconsistency in reported outcomes, reflecting differences in experimental design, MP characteristics, algal species, and environmental contexts. This review integrates current knowledge on:

- Microalgal responses to MP exposure, including growth inhibition, photosynthetic disruption, and oxidative stress
- Microalgae—MP interactions involving adsorption, internalization, biofilm formation, and disruption, and oxidative stress
- The dual role of these interactions in both impairing algal communities and facilitating MP transformation and removal

Collectively, available evidence supports the feasibility of microalgae-based bioremediation strategies for MP mitigation. Microalgae can colonize plastic surfaces, form biofilms, and influence nutrient dynamics, thereby shaping both MP fate and aquatic ecosystem productivity. Strategically harnessing these interactions under controlled conditions offers a promising pathway toward more sustainable management of MP pollution. Nevertheless, significant knowledge gaps persist. Future research priorities should include:

- Standardization of experimental protocols to enhance comparability
- Assessment of long-term and community-levels effects under realistic environmental conditions
- Improved understanding of the roles of additives, leachates and MP aging
- Scaling and optimization of microalgae-based systems for application in wastewater treatment and natural waters

10.6 Advancing in clean-up strategies, waste management and remediation of microplastics

10.6.1 Source control of microplastic pollution

Source Control represents the most effective and sustainable approach for lowering the environmental load of MPs. Preventing MPs from entering natural systems reduces reliance on costly downstream treatment processes and significantly limits ecological and human health impacts. A combination of individual actions, industrial accountability, and policy-based interventions is therefore essential for minimizing MP generation and environmental release.

10.6.1.1 Reduction of plastic use

One of the most fundamental strategies involves decreasing overall plastic consumption, with particular emphasis on single-use plastics. The adoption of reusable bags, bottles, containers, and packaging materials can subsequently reduce plastic waste generation (Cunha et al., 2019). In parallel, increased use of biodegradable and compostable plastics provides a more environmentally benign alternative, as these materials tend to degrade more rapidly and safely under natural conditions (Xi et al., 2022). Encouraging industries to reduce excessive packaging and transition toward sustainable material choices further reinforces these reduction efforts (Liu et al., 2023a,b).

10.6.1.2 Avoidance of microbead containing products

Consumer goods that contain microbeads-such as certain toothpastes exfoliating scrubs, and cosmetic cleansers-constitute a direct and preventable source of MP pollution. Choosing microbead-free alternatives, in combination with regulatory bans, can markedly decrease the quantity of MPs discharged into wastewater systems (Xi et al., 2022).

10.6.1.3 Improving water filtration and household exposure reduction

Household-level filtration of tap water is an effective means of reducing MP ingestion particularly given the widespread detection of MPs in municipal water supplies worldwide (Gopalakrishnan and Kashian, 2023). At larger scales, upgrading wastewater treatment plants (WWTPs) with advanced filtration technologies-including membrane bioreactors, enhanced coagulation-flocculation processes, and tertiary filtration units-can significantly improve MP capture prior to discharge into rivers, lakes and coastal waters.

10.6.1.4 Industrial responsibility and cleaner manufacturing

Industrial activities are major contributors to MP pollution through resin pellet loss, fiber-shredding from synthetic textiles, and degradation of industrial polymers. To address these sources, industries should:

- Design products that do not contain MPs or microbead additives
- Implement closed-loop recycling and material recovery systems
- Transition toward sustainable alternative materials
- Minimize emissions through cleaner and more efficient production technologies

  Collaboration among industrial sectors, environmental agencies, and regulatory authorities can accelerate the adoption of best practices and substantially reduce industrial MP emissions.

10.6.1.5 Policy and regulatory measures

Robust regulatory frameworks are critical for achieving large-scale reductions in MP pollution. Policies that restrict microbeads, limit single-use plastics, establish thresholds for industrial MP emissions, and mandate pollution control measures can deliver significant environmental benefits (Cvetkovska et al., 2022). Public awareness and stakeholder engagement play an important role in influencing decision-makers and expediting the implementation of such legislation.

10.6.2 Mechanisms of treatment technologies in microplastic removal

Technologies designed for MP removal operate through a range of fundamental mechanisms that enable efficient separation of MPs from water and wastewater streams. Core processes-including filtration, coagulation-flocculation, settling, flotation, electrocoagulation, sedimentation, adsorption, and biological degradation-form the basis of modern treatment systems (Table 10.3). Each mechanism targets MPs according to their physical, chemical, or electrostatic characteristics.

Table 10.3 Microplastics degradation methods: Mechanisms and limitation.

Degradation method	Description	Bonds involved	References
Biodegradation	Microbes degrade MPs by breaking down various bonds through enzymatic activity	C–C, C–H, and polymer chain bonds	Agarwal (2020)
Chemical degradation	MP destruction through chemical reactions with specific substances	Varies based on the chemical used	Zhou et al. (2021)
Mechanical degradation	Physical forces like grinding and abrasion lead to particle size reduction	Polymer chains, causing bond breaking	Song et al. (2017)
Thermal degradation	MPs degrade when subjected to high temperatures	Polymer chain, C–C, and C–H bonds	Magnusson and Noren (2014)
Photodegradation	Breakdown of MPs under exposure to UV light	Mainly C–C and C–H bonds	Tofa et al. (2019a,b)

***Filtration*:** Filtration is among the most straightforward and widely applied mechanisms for MP removal. It employs physical barriers or porous media to retain MPs based on size exclusion principles (Abdelfattah et al., 2023). As water passes through filter materials with pore sizes smaller than the particles, MPs are retained in a manner analogous to sieving. Filtration performance is strongly influenced by pore size distribution, membrane composition, surface hydrophobicity, and operating pressure.

***Coagulation and Flocculation*:** Coagulation and Flocculation processes involve the addition of coagulants (e.g., alum, ferric salts) and flocculants (e.g., polyacrylamide) to neutralize the surface charges of MPs (Geremia et al., 2023). Charge neutralization destabilizes dispersed particles, promoting their aggregation into larger, denser flocs. These flocs can then be efficiently removed through settling, floatation or filtration. This approach is particularly effective for small or neutrally buoyant MPs that are difficult to capture using filtration alone.

***Settling (Gravity Separation)*:** Settling processes exploit differences in density between MPs and water. Over sufficient retention times, gravity causes denser polymer particles to accumulate at the bottom of the treatment tanks or clarified (Podbielska and Szpyrka, 2023). This method is most effective in low-flow systems and is frequently combined with coagulation-flocculation to improve the removal of smaller or low-density MPs.

***Flotation*:** Flotation techniques employ air bubbles or buoyant agents that attach to MP particles and transport them to the water surface, where they can be mechanically removed (Martínez et al., 2023). This mechanism is particularly effective for low-density polymers such as polyethylene and polypropylene, which naturally exhibit buoyancy. Dissolved air floatation (DAF) and induced air floatation (IAF) are widely used in wastewater treatment facilities for MP separation.

***Electrocoagulation*:** Electrocoagulation utilizes an applied electrical current to induce the dissolution of sacrificial metal electrodes, typically iron or aluminum, releasing ions that neutralize MPs surface charges (Babiak and Krzeminska, 2021). This process facilitates the formation of microflocs and macroflocs, which may either settle or float depending on system design. Additionally, hydrogen gas generated during electrolysis can enhance floatation efficiency.

***Sedimentation*:** Sedimentation is closely related to settling but involves deliberate control of hydraulic conditions, such as reduced flow velocities, within sedimentation basins or clarifiers. These conditions promote the separation of MPs from the water column, particularly aggregated particles formed following coagulation-flocculation process.

***Adsorption*:** Some treatment systems employ adsorbent materials-including activated carbon, biochar, geopolymers, and engineered clays-that exhibit strong affinities for MPs (Tunali et al., 2020). Removal occurs through surface interactions such as hydrophobic bonding, electrostatic attraction, and van der Waals forces. Adsorption is especially effective for capturing small MPs or aged particles with functionalized surfaces.

***Biological Degradation*:** Although biological degradation is not a primary removal mechanism in conventional treatment systems, certain approaches rely on microorganisms—such as microalgae, bacteria, and fungi—to gradually biodegrade MPs into smaller or less harmful compounds (Demir-Yilmaz et al., 2022). This process involves enzymatic oxidation, biofilm formation, extracellular polymeric substances (EPS), and microbial metabolic pathways. While comparatively slow, biological degradation represents a promising long-term strategy for reducing MP persistence in aquatic environments.

***Density Separation Method*:** Density Separation remains one of the most frequently employed techniques for isolating MPs from soils and sediments due to differences in density between plastic particles and mineral matter. Typical plastic densities range from 0.9 to 2.3 g/cm^3, whereas common soil materials generally fall between 2.6 and 2.7 g/cm^3. By preparing solutions with tailored densities, MPs can be selectively floated or separated from denser soil particles using sedimentation or centrifugation techniques (Paul et al., 2024).

***Choice of Density Media*:** Sodium chloride (NaCl) is commonly utilized owing to its low cost, nontoxicity, and ease of preparation, although its maximum density (~1.2 g/cm^3) restricts recovery of higher-density plastics. Magnesium Chloride ($MgCl_2$), which reaches ~1.31 g/cm^3; improves recovery efficiency for denser polymers (Atugoda et al., 2022). Other brine solutions, including sodium iodide (NaI) and zinc bromide ($ZnBr_2$), have demonstrated superior recovery rates, particularly for MPs smaller than 1 mm. Quinn et al., (2017), highlighted $ZnBr_2$ (density 1.7 g/cm$^{3)}$ as highly effective, representative a significant advancement in density-based separation from sediment matrices.

***Centrifugation-Assisted Separation*:** Centrifugation enhances phase separation by accelerating sedimentation. Speeds of 3700 min^{-1} for 10 minutes have been shown to yield high recovery rates (Grause et al., 2022). In soils spiked with mixed MPs, overall extraction efficiencies consistently exceed 95%. Polyethylene terephthalate (PET), the densest polymer tested, displayed slighter lower recovery, likely due to its hydrophilic surface reducing air bubble adhesion and buoyancy.

***High-Density Solutions for Robust Extraction*:** Konechnaya et al. (2020), demonstrated that zinc chloride ($ZnCl_2$ density 1.63 g/cm^3, pH 3) could recover MPs from sea sand with nearly 100% efficiency across multiple polymer types and grain sizes. When applied to additional sediment types, including sandy–silty and fine-grained sediments, recovery rates ranged from 94% to 106%, illustrating the reliability and versatility of $ZnCl_2$ based density separation.

***Chemical Digestion Method*:** Chemical Digestion is widely applied to remove organic matter from soils and sediments while leaving MPs intact for downstream separation, typically via filtration (Munno et al., 2018). String oxidizing or alkaline agents—such as hydrochloric acid (HCl), sulfuric acid (H_2SO_4), and nitric acid (HNO_3)—effectively degrade organic matrices but may compromise polymer integrity (Mai et al., 2018).

***Effect of Reagents on Polymer Integrity*:** Although acids efficiently remove organics, some polymers are vulnerable to degradation. Polyamide (PA) and polyurethane (PUR) are especially susceptible, whereas polymers like acrylonitrile butadiene styrene (ABS), polymethyl methacrylate (PMMA), and polyvinyl chloride (PVC) show moderately but notable changes (Enders et al., 2017). Alkaline digestion using KOH or KOH: NaCIO mixtures is increasingly preferred to preserve polymer structures (Munno et al., 2018). compared alkaline digestion with vet peroxide oxidation (WPO) for MP extraction from organic-rich matrices. WPO generated temperatures exceeding 70°C, leading to compete degradation of certain plastics. Elevated temperature was identified as the primary cause of polymer loss across multiple MP types.

***Temperature and Reagent Effects*:** Pfeiffer and Fischer (2020); evaluated protocols for removing biogenic organic matter in water and sediment samples at 20–70°C over 1–7 days. They reported that:

- Carbonate sheets remained largely unreactive to oxidizing agents (H_2O_2, NaClO, Fenton's reagent) and alkaline solutions (NaOH, KOH).
- Carbonates dissolved rapidly in strong acids such as HNO_3 and HCL.
- Sodium hypochlorite (NaClO) effectively digested both soft and hard tissues at 40–50°C

However, higher reagent concentrations and temperatures accelerated polymer degradation. PA and PET were most affected, with 15%–100% mass loss depending on the treatment. PA degraded fastest in acidic conditions, while PET was most sensitive to alkaline digestion.

***Risks of Polymer Alteration During Digestion*:** Acidic (HNO_3), alkaline (KOH, NaOH) and oxidative (H_2O_2) treatments can alter MP size, shape, color and surface morphology (Cole et al., 2014). These changes can lead to:

- Misidentification under microscopy or spectroscopy
- Over-or underestimation of particle counts
- Incorrect polymer classification

  Selecting digestion methods that balance organic matter removal with MP preservation is therefore critical.

10.6.3 Microplastic separation from water

The presence of MPs in treated tap and bottled water has become a global concern due to potential health risks (WHO, 2019). MPs may contain hazardous additives or carry other contaminants, highlighting the need for improved removal technologies and understanding of MP behavior during water treatment.

10.6.3.1 Drinking water treatment plants

Drinking water treatment plants (DWTP) are central to public health protection, but MP removal efficiency is often incomplete (Wang et al., 2020). Conventional processes may capture some MPs, yet smaller fractions frequently remain. Wang et al., (2020), analyzed MPs in an advanced DWTP effluent, finding that coagulation at pH 7.70–7.84 followed by sedimentation and granular activated carbon (GAC) filtration removed 40.5%–54.5% and 56.8%–60.9% of MPs, respectively. Ozonation (1.5–2.5 mg/L) decreased removal efficiency for 1–5 μm MP, likely due to oxidative fragmentation. MP removal efficiency was strongly dependent on particle characteristics. Pivokonský et al., (2020), compared two DWTPs along the same river. A plant with a more complex, multi-step train removed 88% of MPs $\geq$1 μm, whereas a simpler plant achieved 40% removal. MPs were primarily <10 μm fragments and fibers, with polymers including cellulose acetate, PET, PVC, PE and PP.

Sarkar et al. (2021) evaluated a pulse-clarification DWTP using Ganga River water. Raw water contained 17.88 MP/L; pulse clarification removed 63%, and sand filtration removed 85%, with sand filters capturing more particles due to physical screening. These studies indicate conventional DWTPs often fail to eliminate all MPs, particularly small or low-density particles, emphasizing the need for advanced technologies like membrane filtration. Barbier et al., (2022) compared three DWTPs in Paris. Nanofiltration achieved >99%% MP removal, outperforming conventional treatment, though recontamination during distribution was noted (Dronjak et al., 2022). reported influent concentrations of 4.23 $\pm$ 1.26 MP/L near Barcelona, reduced to 0.075 $\pm$ 0.019 MP/L after advanced oxidation, density separation, and multi-stage treatment. Jung et al., (2022) assessed DWTPs along Korea's Nakdong

River. MP > 20 μm were detected, with higher concentrations in river water than reservoirs. PE and PP dominated. Preozonization and sedimentation removed 70%–80% of MPs, while PE/PP removal reached 81%–88%. PET and PMMA were effectively captured during filtration. MP levels correlated with water quality indices, demonstrating environmental factors influence MP fate.

10.6.3.2 Coagulation

Coagulation represents a fundamental approach for MP removal in drinking water treatment systems. This process involves destabilizing suspended MPs encouraging their aggregation into larger flocs, which can then be separated through sedimentation, filtration, or floatation (Rajala et al., 2020). Coagulants function by neutralizing the negative charges on MP surfaces, thereby reducing electrostatic repulsion and promoting particle collisions and aggregation. Ma et al., (2019a,b), assessed the removal efficiency of polyethylene (PE) MPs using aluminum- and iron-based coagulants. Experiments were conducted at pH values between six and 8, beginning with rapid mixing at 300 rpm, followed by slow mixing at 100 rpm. Aluminum salts demonstrated superior performance compared with iron salts particularly for smaller PE particles. Nevertheless, even high doses of aluminum coagulants offered only limited improvements, indicating that water chemistry had minor influence, whereas pH and polyacrylamide addition markedly affected removal efficiency. The cost between various methods for MP removal are shown in Table 10.4.

Liu et al., (2019), further investigated PE behavior under iron-based coagulation and reported lower removal rates for larger PE particles. The addition of polyacrylamide, especially anionic variants, significantly enhanced removal, illustrating that polymer bridging can compensate for the limitations of conventional charge-neutralization coagulation (Skaf et al., 2020). examined the elimination of model plastic microspheres (1–5 μm) using standard coagulation. Alum dosages of 5–10 mg/L achieved turbidity reduction even in the presence of surfactants. Rapid mixing at 100 rpm was followed by 30 minutes of flocculation at 30 rpm, with sweep flocculation identified as the predominant mechanism. PE microfibers were also effectively removed, through surfactants influenced the removal dynamics. Overall, coagulation shows considerable potential for MP mitigation in drinking water, but its efficiency depends on MP size, polymer type, surface properties, pH, and polymeric coagulant aids. Optimization, particularly with tailored flocculants, may further improve removal performance in full-scale applications.

10.6.3.3 Sand filtration

Sand filtration is widely regarded as an effective barrier for MP removal in drinking water treatment, often serving as a critical polishing step following coagulation and sedimentation (Muthulakshmi et al., 2023). Prior studies emphasize its importance in conventional treatment trains. Na et al., (2021) reported that MP removal efficiency increases with particle size during both coagulation and subsequent sand filtration. Their protocol included rapid mixing at 160 rpm for 2 minutes, followed by slow mixing at 60 rpm for 30 minutes and a 1-hour settling period. Quartz sand (0.3–0.45 mm) was used as the filter medium, with a 250 μm mesh screen downstream to capture breakthrough particles. MP retention in the sand column was governed primarily by two mechanisms: attachment, which depends on particle-collector interactions as predicted by single-collector collision theory, and straining, which occurs when particle diameter approaches a critical ratio relative to pore size. Straining was particularly effective for particles ≥67.5 μm.

Table 10.4 Comparison between the various methods for MP removal.

Criteria	MPs separation from soil and sediments				
	Density separation	**Chemical digestion**	**Filtration**	**Oil-based extraction**	**Floatation**
Cost	Moderate cost	High cost due to use of chemical	Cost depends on type of filter	High cost due to use of solvents	High cost due to use of chemicals
Efficiency	High efficiency to remove small-sized MPs	High efficiency for degrading MPs, but may be selective	High efficiency	Effective for hydrophobic MPs but less for hydrophilic types	High efficiency
Applicability	Suitable for removing MPs from soil	Suitable for treating diverse types of MPs	Flexible for different MP sizes	Used for isolating MPs from water	Used for removing floating MPs
Secondary pollution	No secondary pollution unless improperly disposed	Possible secondary pollutants	Possible secondary pollutants	Possible secondary pollutants	Potential for residual chemicals or effluent contamination
Sustainability	Environmentally sustainable	Not very sustainable due to usage of chemical reagents	Sustainable	Sustainability depends on the type of oils and waste	Sustainability depends on the type of floatation agents used
Scalability	Scale for large-scale applications	Limited scalability due to cost and complexity	Scale in wastewater treatment facilities	Scalable in industrial processes	Scalable in water treatment facilities

Sembiring et al. (2021) evaluated rapid sand filters for the removal of artificially introduced plastic and tire flakes. Tire flakes showed high removal efficiencies (90.72%–93.03%), while plastic flakes ranged from 47.78%–97.7%, with larger particles (>200 μm) exhibiting improved performance. Filtration rates of 4–10 m/h were tested, reflecting hydraulic loading comparable to real-world systems. Cherniak et al., (2022) demonstrated that MP removal is strongly influenced by particle size and morphology:

- Sedimentation primarily removed filters
- Sand filtration captured fibers and fragments >50 μm, and
- Granular activated carbon (GAC) filtration removed smaller fragments (1–10 μm)

These findings confirm that sand filtration effectively targets medium- and large-sized MPs through a combination of attachment and straining, though smaller MPs may require complementary treatments such as GAC or membrane filtration.

10.6.3.4 Magnetic purification

Magnetic purification has emerged as an innovative technique for MP removal, offering distinct advantages over traditional methods. Unlike filtration, it reduces adsorbent loss and simplifies handling. In contrast to chemical coagulation or flocculation, it avoids residual chemicals in treated water because magnetic adsorbents are recoverable (Table 10.5).

Bakhteeva et al. (2023), introduced a "green" magnetic sedimentation approach using Fe–C–NH_2 nanoparticles (4–8 nm) to remove PE (10–200 μm) and PET (5–30 μm) MPs from model aqueous suspensions (>0.1 mg/L). MPs were magnetized by forming heteroaggregates with the nanoparticles, which were then separated using a gradient magnetic field from permanent magnets. PET removal required lower nanoparticle dosages (0.005 g/L) than PE, attributed to PET's higher hydrophilicity and stronger affinity for nanoparticle attachment. Martin et al., (2022) investigated hydrophobic-coated iron oxide nanoparticles for magnetizing MPs in both freshwater and saline systems. Nanoparticles (92.65–107.02 nm) were coated with polydimethylsiloxane-based hydrophobic layers to enhance binding with hydrophobic MPs. Using a 2-inch NdFrB permanent magnet, the system achieved:

- 100% removal of large MPs (2–5 mm), and
- Nearly 90% removal of nanoparticles (100–1000 nm).

These results underscore magnetic purification as a rapid, chemical-free, and effective strategy for MP removal, adaptable to a range of particle sizes and water matrices.

10.6.4 Removal of microplastics in wastewater treatment plants

Wastewater treatment plants (WWTPs) demonstrate variable efficiency in removing MPs across different treatment stages. Nonetheless, the persistence of MPs in treated effluents and their significant accumulation in sewage sludge remain pressing issues. These challenges emphasize the necessity for treatment technologies specifically designed to capture MPs and prevent their discharge into aquatic and terrestrial ecosystems.

10.6.4.1 Preliminary and primary treatment

In conventional WWTPs, wastewater progresses through sequential stages: pretreatment, primary treatment, secondary treatment, and tertiary treatment. The pretreatment stage relies solely on physical processes to remove large debris such as sand, oils, grease and solid wastes that could interfere with downstream operations. Typical equipment includes solid wastes that could interfere with downstream operations. Typical equipment includes solid wastes that could interfere with downstream operations. Typical equipment includes screening units, grit chambers, and grease traps. MPs are separated during grit and grease removal through mechanisms such as sand settling and grease flotation, which operate via buoyancy- and density-driven sinking or floating. Masi et al., (2020; Lusher et al., (2017) reported that approximately 62% of MPs removed in WWTPs are captured during this initial stage, with microbeads being particularly effectively retained. Combined, pretreatment, usually involving primary clarifiers, targets suspended solids via sedimentation. The efficiency of sedimentation depends on factors including solid composition, concentration, hydraulic retention time, and clarified design. Sheng et al., (2008); Bayo et al., (2020) documented a ~90% reduction in microliter (90.1%) and MPs

Table 10.5 Comparison between the various methods for MP removal.

Criteria	MPs separation from drinking water (coagulation, filtration)	MPs separation from marine water (sieving/ filtration/ Trawl/Net)	Magnetic separation	Biodegradation	Advanced oxidation processes
Cost	High cost due to chemical consumption	Moderate to high cost depending on the scale	Low cost	Low cost	High cost depending on the type of oxidizing agent
Efficiency	Effective in removing large MPs	High efficiency for removing floating MPs	High efficiency for removing magnetic MPs	Efficiency depends on microbial and environmental conditions	High efficiency for organic polymers and low efficiency for others
Applicability	Suitable for small MPs in municipal drinking water treatment systems	Suitable for marine water or coastal water treatment	Produce secondary MPs	Limited to biodegradable plastics	Applicable for a broad range of MPs
Secondary pollution	Minimal secondary pollution	Possible secondary pollutants from trawling or sieving	Produce secondary MPs	Possible secondary pollutants	Possible secondary MPs
Sustainability	Sustainable if chemical use is minimal	Sustainability depends on scale operation	Environmentally sustainable	Environmentally sustainable	Sustainability depends on the type of oxidant used and the energy required
Scalability	Scalable for large water treatment facilities	Scalable for large water bodies	Scalability depends on amount of magnetic MPs present	Scalable in natural environments	Suitable for industrial scale operations

(90.3%) in the primary effluent of an urban WWTP, noting than 67% of microliter was removed during primary clarification alone. MP fibers, which really attach to organic solids, are efficiently removed at this stage (Lares et al., 2018). Combined, pretreatment and primary treatment stages can eliminate 78%–98% of MPs (Prata, 2018), largely because MPs tend to adhere to suspended solids and settle with primary sludge.

10.6.4.2 Secondary treatment

Following primary classification, wastewater enters secondary treatment, which relies on aerobic biological processes to degrade dissolved and particulate organic matter. Aeration promotes microbial activity but may also facilitate aerosolization of MPs, allowing some particles to disperse into the atmosphere (Chen et al., 2020a). Secondary treatment typically reduces MP concentrations by 96%–98% (Lares et al., 2018). However, removal efficiency varies among studies due to differences in sampling protocols, extraction procedures, analytical techniques, and particle size detection limits. Despite high removal rates, WWTPs remain substantial sources of environmental MPs because:

- Large volumes of treated wastewater are discharged daily, allowing smaller MPs and nanoplastics to escape into receiving waters.
- Sewage sludge, which concentrates MPs, is commonly applied to agricultural soils, leading to persistent terrestrial accumulation.

Smaller MPs, which are more difficult to filter or settle, are particularly concerning due to their mobility and ecological impact.

10.6.4.3 Tertiary treatment

Some WWTPs include tertiary treatment stages, driven by regulatory standards, water reuse requirements, or the need for advanced effluent polishing. Common tertiary processes include coagulation-flocculation, chlorination, UV irradiation, ozonation, and membrane filtration. In many cases, conventional tertiary treatment contributes minimally to additionally MP removal beyond secondary treatment. However, some studies indicate tertiary processes can achieve up to 4% additional MP removal after secondary treatment (Blair et al., 2019). Chlorination, widely applied for disinfection, only marginally reduces MP (~7%) (Prata, 2018). In contrast, ozonation achieves significantly higher removal (~90%) due to its ability to react with polymers and generate radical oxidants that fragment plastic chains. Membrane-based tertiary treatments, particularly membrane bioreactors (MBRs), are highly effective in retaining MPs. Studies indicate MBRs can capture ~99% of microfibers, leading to an overall reduction of 96.2% in MP concentrations from primary influent to final effluent (Talvitie et al., 2017). The enclosed design also produces MPs in relatively clean form, as most organic and inorganic solids are removed upstream, which is beneficial for downstream recycling or forensic analysis (Sol et al., 2020). Innovative tertiary methods have been explored as well:

Rapid sand filtration (Sembiring et al., 2021): reported removal efficiencies of 47.78%–93.03%, especially for particles $\geq$200 μm. Silica sand of two mesh sizes (20–40 and 40–70) was used with artificial MPs such as tire fragments and plastic flakes.

Visible light photocatalysis (Uheida et al., 2021): immobilized zinc oxide nanorods on glass fiber to capture and degrade suspended polypropylene MPs under visible light, achieving a 65% reduction in particle volume after 2 weeks.

Sol–gel methods: Organosilane precursors induce MP agglomeration into larger flocs through pH-induced reactions, facilitating filtration (Adegoke et al., 2023). Particle sizes increased due to van del Waals and hydrophobic interactions, and the chemistry can be tailored to target specific polymer types (Herbort et al., 2018). However, residual silicon and competing contaminants may reduce efficiency or introduce secondary contamination.

10.6.4.4 Retention of microplastics in sewage sludge

Despite high MP removal efficiencies, most MPs accumulate in sewage sludge rather than being fully degraded (Huang et al., 2023). MP concentration in sludge very widely, ranging from 1000 to 301,400 particles/kg od dry sludge (Nguyen et al., 2022). It is estimated that 10^6–10^{14}. MPs are released globally each year through sludge disposal (Hatinoglu and Sanin, 2021). A study of 28 WWTPs across 11 Chinese provinces found an average MP concentration of 22,700 particles/kg, predominately while fibers (59.6%) composed mainly of polyolefins, acrylic fibers, polyethylene (PE), and polyamides (Li et al., 2018). Larger MPs accumulate preferentially in sludge, as they are more effectively removed during preliminary and primary treatment (Sun et al., 2019). Comparisons across studies are challenging due to the lack of standardization sampling, digestion, extraction, and polymer identification protocols in sludge matrices. Nonetheless, sludge consistently contains MP concentrations several orders of magnitude higher than untreated influent. The use of sludge in agricultural represents a major pathway for terrestrial MP accumulation. Reports indicate:

- Each sludge application increases soil MP concentration by ~280 particles/kg for light-density MPs and ~430 particles/kg for heavy-density MPs (van den Berg et al., 2020).
- Repeated sludge applications raise MP abundance from 0.6 to 10.4 particles/g soil, depending on the number of amendments (Corradini et al., 2019).

These observations underscore the importance of developing sludge specific MP remediation strategies, especially given sludge's role as a widely used fertilizer.

10.6.5 Integrated perspective

Effective remediation of MPs necessitates a comprehensive approach combining source control optimized treatment processes, and environmental management strategies. In soils and sediments, MPs are addressed through methods such as density separation, floatation, chemical digestion, filtration, and oil-assisted extraction. Water treatment strategies depend on context: drinking water systems employ coagulation, sand filtration, and magnetic purification, whereas marine environments utilize sieving, trawling, and density-based separation. Wastewater treatment plants (WWTPs) remain critical for MP mitigation, although they can also serve as pathways for technologies, alongside standardized analytical methodologies, is vital for advancing MP remediation and reducing long-term ecological risks.

10.6.5.1 Novel microplastic remediation approaches

Traditional MP remediation approaches-such as filtration, coagulation, sedimentation, and density separation-face inherent technical limitations. Their efficiency varies according to environmental parameters, polymer types, particle sizes, and wastewater characteristics. Most research relies on synthetic laboratory samples, with few studies assessing real-world performance in natural waters or actual wastewater. To studies assessing real-world performance in natural waters or actual wastewater. To overcome these limitations, researchers are developing novel, high-efficiency, and scalable techniques. The following section summarizes these emerging approaches.

10.6.5.2 Advanced oxidations processes (AOPs) for microplastic degradation

Advanced oxidation processes (AOPs) are chemical treatments that generate highly reactive hydroxyl radicals (·OH) among the most potent oxidants, to degrade organic and inorganic contaminants in water and air. Due to their ability to cleave stable C–C and C–H bonds, AOPs have gained interest for MP degradation (Kallawar et al., 2021). Techniques such as photocatalysis, ozonation, sulfate radical foxidation, and Fenton-like processes have shown promise in accelerating polymer chain fragmentation and surface oxidation (Sacco et al., 2023).

Fenton Reaction-Based AOPs: The Fenton reaction produces hydroxyl radicals by reacting ferrous ions (Fe^{2+}) with hydrogen peroxide (H_2O_2), while Fenton-like reactions use alternatives transition metals to activate peroxide species. Hydroxyl radicals attack MP polymer chains, causing surface oxidation, new functional group formation, reduced molecular weight, and eventual fragmentation. These processes also enhance compatibility of MPs with downstream separation or biodegradation techniques. Fenton reagents can further aid in isolating MPs from organic-rich matrices by oxidizing interfering natural organic matter. Tagg et al., (2017); Li et al., (2022a,b) investigated the oxidative degradation of nylon-6 and polystyrene MPs using Fe^{2+} -activated peroxymonosulfate and Fenton reagents. In their study:

- 0.5g of MPs were treated in 19.3 mL of ultrapure water.
- pH was adjusted to three using sulfuric acid to prevent iron hydroxide precipitation
- Reactions were initiated with 0.3 mL of 30% H_2O_2 and 0.3 mL of 0.2 M Fe^{2+} solution
- An alternative system used 0.913 g peroxymonosulfate instead of H_2O_2
- Stirring was maintained at 150 rpm for 5 h

After four oxidation cycles, MPs experienced mass reductions of 25.6% (nylon-6) and 22.1% (PS). FTIR and other analyses revealed oxygenated groups such as C–O, C=O, and O–C=O, with degradation products including alkanes, alcohols, and carboxylic acids identified via SERS and GC–MS.

Ortiz et al. (2022) assessed MPs from five common polymers (PET, PE, PVC, PP, expanded PS) under severe Fenton conditions:

- MPs exhibited modest mass losses (~10%), with smaller particles degrading more efficiently
- Oxidation introduced hydrophilic oxygenated groups, reducing particle size and increasing acidity
- Aromatic polymers (expanding PS, PET) degraded more readily that aliphatic polymers (PE, PP, PVC)

Fenton processes not only alter solid MP particles but also rapidly oxidize MP leachates, demonstrating their ability to mineralize dissolved organic contaminants.

H_2O_2 Reaction-Based AOPs: Hydrogen Peroxide (H_2O_2) is commonly applied in AOPs because it produces highly reactive hydroxyl radicals (·OH) when activated by UV light or catalysts, effectively oxidizing and fragmenting MPs (Easton et al., 2023). studied H_2O_2 with UVC irradiation, achieving 52.7% mass loss using 500 mg/L H_2O_2 and 4mW/cm_2 UVC over 48 h. The process induced substantial surface oxidation, reflected by a 10.6% in the carbonyl index, along with structural changes and enhanced removal of soluble organics from the MP surfaces (Li et al., 2023) examined PLA-derived MPs from commercial straws and food bags under UV/H_2O_2 exposure at 25°C and 30% RH. Commercial PLA MPs aged slower than pure PLA MPs likely due to additives. Functional group transformations differed between MP types, with adsorption capacity increasing 88% for pure PLA and 64%–56% for MPs from straws and bags. These findings highlight that oxidation extent and degradation depend on polymer composition and additive content.

Heterogenous Photocatalysis: Heterogenous photocatalysis has emerged as a promising MP degradation approach, using semiconductor photocatalysis-typically TiO_2 or ZnO—activated by light, often UV. Light excitation generates electron-hole pairs, with valence-band holes reacting with water or adsorbed oxygen to produce highly reactive hydroxyl radicals (·OH) that cleave polymer chains, accelerating naturally slow degradation processes (Li et al., 2023). UV or solar photons also excite MPs, promoting chain scission, branching, cross-linking, and surface oxidation, although catalysts are necessary for practical degradation rates.

Vital-Grappin et al. (2021) described radical-driven MP degradation, where ·OH radicals and superoxide ($O_2^{\cdot-}$) initiate C–H bond cleavage, producing alkyl radicals. Photo-Kolbe reactions with photogenerated holes (h^+) generate additional radicals. Alkyl radicals react with oxygen to form peroxyl radicals, abstracting hydrogen from neighboring chains to produce hydroperoxides, decomposing into oxy radicals and ·OH, reducing polymer molecular weight. Termination occurs via radical recombination. While MP adsorption onto C, N-doped TiO_2 catalysts may be limited, degradation proceeds through $O_2^{\cdot-}$ and ·OH- mediated oxidation, demonstrating photogenerated electrons' critical role in HDPE MP breakdown. Degradation occurs under both oxygen-rich and inert conditions, with pathways varying accordingly (Fig. 10.1).(Ariza-Tarazona et al., 2019) first applied heterogeneous photocatalysis to degrade HDPE MPs from cosmetic scrubs using N–TiO_2 photocatalysts synthesized via a green route from mussel extrapallial fluid, achieving effective degradation in solid and aqueous media. Tofa et al., (2019a,b) used ZnO nanorods under visible light (~60–70 klux) to degrade LDPE MPs over 175 h, achieving a 30% increase in carbonyl index, with notable surface cracking, wrinkling, and cavity formation. Degradation correlated with catalyst surface area.

Cao et al. (2022) combined photocatalytic H_2 evolution with MP oxidation using MXene/$Zn_xCd_{1-x}S$ photocatalysts irradiated by a 300 W xenon lamp (>420 nm), achieving enhanced charge separation, optimized band alignment, and improved oxidative capacity at 5°C (Feng et al., 2022). developed a NH_2-MIL-88B (Fe)/MoS_2 heterostructure, achieving 96.44% photodegradation of tetracycline, 72.36% salinity removal, and degradation of MP fragments and HDPE films under visible light (400–760 nm). The MoS_2 interface enhanced light adsorption, charge carrier lifetime, recombination resistance, and surface area, promoting efficient pollutant transformation. LC-MS confirmed degradation intermediates toxicity assessments indicated reduced toxicity, and field emission SEM showed progressive HDPE morphological degradation. DFT calculations elucidated charge transfer pathways and photoreactive species formation, emphasizing engineered heterojunction photocatalysts' potential for MP remediation.

10.6.5.3 Biodegradation of microplastics

Biodegradation of MPs involves the microbial-mediated breakdown of plastic particles into simpler, environmentally compatible compounds. Various bacteria and fungi possess enzymatic machinery capable of attacking polymer surfaces, with hydrolytic enzymes-such as lipases and esterases-playing a crucial role in cleaving ester, amide and other functional bonds in plastic polymers. These enzymatic activities initiate surface alteration, particle fragmentation, and gradual reduction in particle size. Under aerobic conditions, microorganisms metabolize CH2 and C–C structures, producing biomass components containing C–C, C–O, and C–O–C linkages (Li et al., 2023). For instance (Paço et al.,

2017), demonstrated that the marine *fungus Zalerion maritimum* could utilize polyethylene (PE) pellets at 25°C, reducing both mass and particle dimensions. As illustrated in Fig. 10.2, enzymatic biodegradation converts MPs into intermediate compounds and ultimately CO_2, achieving complete mineralization while generating microbial biomass and energy. Scaling such enzymatic systems for industrial applications requires careful consideration of chemical, physical and biological constraints, optimization of catalytic steps, and enzyme engineering tailored to specific applications (Zurier and Goddard, 2021).

Biodegradation typically produces MPs with narrower size distributions and smoother, glossier surfaces due to selective erosion of polymer chains. Combining biological and chemical treatments can enhance efficiency by disrupting protective surface layers that impede microbial activity. Chen et al., (2020a,b), reported that hyperthermophilic composting at 40–70°C over 8 weeks removed 43.7% of MPs nearly nine times higher than conventional composting-attributed to synergistic effects of elevated temperature and intensified bacterial activity. Shabbir et al., (2020), employed periphytic biofilms to degrade polypropylene (PP), polyethylene (PE), and polyethylene terephthalate (PET). Batch reactor experiments at 30°C and pH seven showed that glucose supplementation enhanced biodegradation, with 60-day efficiencies of 9.52%–18.02% for PP, 5.95%–14.02% for PE, and 13.24%–19.72% for PET, surpassing rates achieved by unamended natural biofilms. As MP remediation strategies evolve, biodegradation complemets advanced oxidation processes-including Fenton chemistry, H_2O_2 based AOPs and heterogenous phtocatalysis-providing a sustainable, biologically driven solution. These approaches, combined with upstream interventions such as improved waste management and microbially assisted recycling pathways, are critical to reducing MP release and persistence in the environment.

10.6.6 Treatment of plastic waste

Despite their societal and economic advantages, improper plastic waste management has accelerated global plastic pollution. Accumulation debris threatens wildlife through ingestion, entanglement, and suffocation and poses risks to human health, food security, and livelihoods, while also facilitating zoonotic disease transmission. Unmanaged plastics impose economic burdens via losses in fisheries, tourism, and ecosystem services. Due to impose economic burdens via losses in fisheries, tourism, and ecosystem services. Due to the high resistance to synthetic polymers to biodegradation, plastics persist for centuries, fragmentating slowly into MPs and nanoplastics (NPs) that contaminate terrestrial, aquatic, and atmospheric systems. Controlling plastic leakage requires robust policy frameworks and enforcement. While many countries have regulations for solid waste and wastewater management, coverage gaps and poor implementation often limit their effectiveness. Government interventions are essential to ensure plastic collection, storage, transportation, recycling and disposal systems are technically feasible, economically viable, socially acceptable, and environmentally sound (Nikiema and Asiedu, 2022).

Natural degradation of conventional plastics can take millennia, whereas some biopolymers degrade within months under suitable conditions. Nonetheless, both bioplastics and conventional plastics require controlled waste management to prevent MP and NP formation. Biodegradation processes depend on specific environmental conditions, and bioplastic streams must be processed separately to avoid contamination and efficiency loss. Emerging technologies targeting plastic conversion or breakdown into benign or valuable secondary products are increasingly important for circular plastic strategies (Chen et al., 2022a,b). Plastic recycling and recovery are classified into four categories: primary (re-extrusion), secondary (mechanical recycling), tertiary (chemical or feedstock recycling), and quaternary (energy recovery). Primary recycling converts clean, uncontaminated plastics into products with properties similar to the original material but is milted to homogenous waste streams. Secondary recycling involves sorting, washing, shredding, melting, and reforming plastics; it is economical but often produces lower-quality materials due to polymer degradation.

Tertiary recycling encompasses chemical depolymerization-pyrolysis, gasification, solvolysis-to produce monomers, fuels, or chemical feedstocks. It can handle heterogenous or contaminated waste and regenerate raw materials for new plastics. Biological tertiary recycling has emerged as an eco-friendly alternative using microbial or enzymatic depolymerization under mild conditions. Quaternary recycling converts plastic waste into energy through combustion, incineration with energy recovery, cofiring in industrial kilns, liquefaction, or high-temperature pyrolysis, suitable for mixed or contaminated waste streams (Hopewell et al., 2009). Effective plastic waste management requires integrated strategies combining policy, technological innovation, and improved infrastructure. Enhancing recycling efficiency, expanding chemical and biological recycling, and reducing source-level plastic leakage are key to mitigating MP and NP generation and protecting ecosystems and human health.

10.6.7 Prevention of microplastic generation and release

Preventing MP generation and environmental release is essential for reducing ecological and human health impacts. MPs enter ecosystems via mismanaged plastic waste, synthetic textile abrasion, industrial emissions, and effluent discharges. Reducing plastic consumption-through reusable bags, bottles, and durable containers-is a key preventive strategy. Reformulating plastics to minimize additives such as biocides and stabilizers improves biodegradability and lowers secondary MP formation. Regulatory interventions, including single-use plastic restrictions or bans, have been effective, though events like the COVID-19 pandemic temporarily reversed such measures. Improving waste collection, recycling, and disposal infrastructure prevents plastic leakage. Effective waste management and public participation reduce debris accumulation, while offer sustainable alternatives. Waste-to-energy methods, such as plasma gasification and pyrolysis, convert nonrecyclable plastics into energy or fuels, minimizing uncontrolled degradation (Puncochar et al., 2012). These strategies combine behavioral and technological measures, emphasizing coordinated action among consumers, industries, and policymakers.

10.6.8 Effective microplastic treatment processes

Various MP treatment methods exist, but effectiveness depends on polymer type, particle morphology, and environmental conditions. Membrane-based systems offer high removal efficiency but face fouling challenges, increasing operational costs (Table 10.6). No universally effective method currently exists; improving performance or integrating complementary approaches can generate synergistic effects. Nanotechnology has emerged as a valuable tool, offering high surface area,

Table 10.6 Removal of MPs using membrane filtration versus other primary treatment methods.

Membrane filtration versus other methods	Treatment performance		Removal rate (%)
	Pros	Cons	
Activated sludge + sedimentation	a. Popular technology in municipal WWTPs b. Uncomplicated treatment (by microorganisms)	a. Treatment efficacy depends on retention time and nutrient level b. Leach out or adsorption of toxic chemicals (in long retention time) c. Low removal efficiency of MPs	66.7
Sedimentation	a. Straightforward method of treatment through gravitational settling b. Removal of high density of MPs	a. Residual pollutants with MPs b. Removal efficacy depends on the density of MPs c. issues with MPs present in landfill leachate and storm water runoff returning to WWTPs	91.7
Flotation	a. Simple treatment method (floating by air bubble) b. Removal of low density of MPs	a. Potential residual pollutants of MPs b. Removal efficacy depends on the density of MPs	95.0
RO	a. Removal through semi-permeable membrane b. Ease of treatment	a. Lower removal efficiency of MPs than that by the MBR process b. Even after RO, plastic debris still remains in WWTPs	90.4
MBR	a. High removal efficiency of MPs as one of commonly applied membrane technologies b. Good exclusion of MPs due to small pore size	a. Difficulty in controlling multiple parameters influencing treatment efficiency b. Costlier and energy consuming than sedimentation	99.9

tuneable properties, and strong physiochemical interceptions to enhance MP capture. Nanoparticles are applied in photocatalysis membrane separation, advanced oxidation, and adsorption, accelerating pollutant breakdown and targeting diverse MPs. Bethi et al., (2016); Pasanen et al., (2023) demonstrated >98% removal of polystyrene microspheres in 5 minutes using a water-synthesized zeolitic imidazolate framework magnetic nanocomposite. Engineered enzymes and microbial biotechnologies also enable rapid polymer degradation without additional cofactors (Na et al., 2021; Liu et al., 2021) created a *Pseudomonas aeruginosa* biofilm with a "capture-and-release" mechanism, aggregating low-density MPs via enhanced EPS production for subsequent recovery. These hybrid microbial-nanomaterial systems show significant potential for scalable MP remediation.

10.6.9 Life cycle assessment as a tool for evaluating microplastic environmental impact

Life Cycle Assessment (LCA) provides a holistic framework for quantifying environmental impacts of MPs across production, use, release, treatment, and disposal (Fig. 10.2). LCA helps assess ecotoxicity, resource depletion, and pollution, guiding product design and waste management decisions. Salieri et al., (2021) applied worst-case assumptions to evaluate MP emissions from consumer products. Case studies with a polyester T-shirt and MP-containing shower gel indicated that advanced wastewater treatment significantly reduced freshwater ecotoxicity contributions. Sensitivity analysis showed that removal efficiency, environmental fate, and polymer fragmentation affect overall outcomes. Mahabeer et al., (2023) investigated methylene blue adsorption onto MPs, confirming environmental viability via LCA. Saling et al., (2020) developed models incorporating MP fragmentation and degradation kinetics, improving impact assessment accuracy. Remaining gaps include quantifying MPs in natural

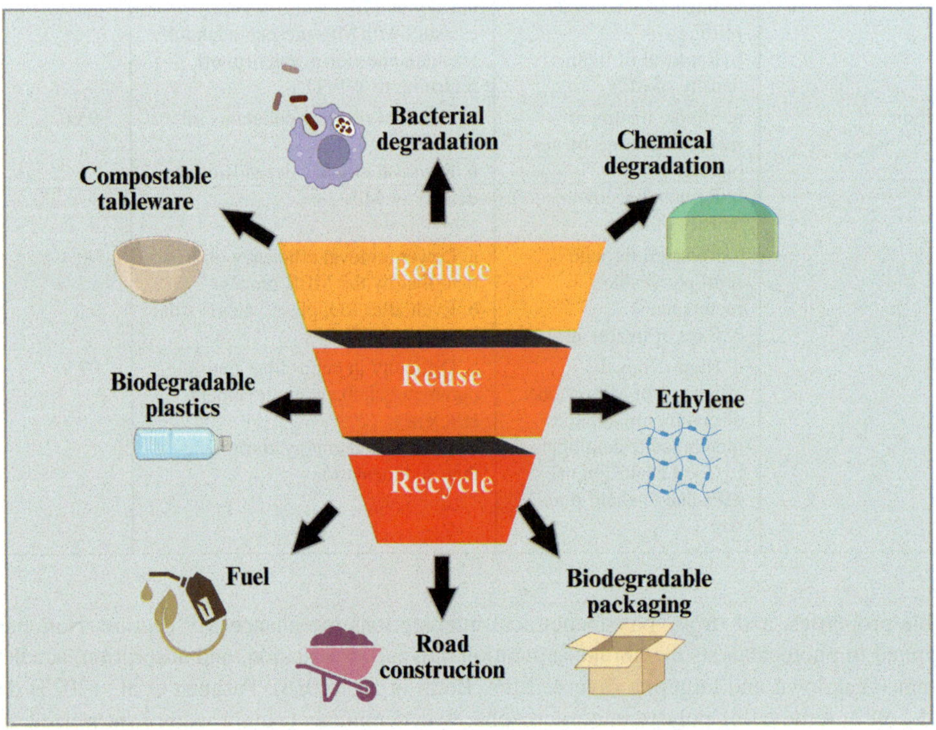

FIGURE 10.2

Schematic representation of integrated reduce–reuse–recycle strategies for mitigating plastic consumption and environmental pollution. The framework illustrates material flow pathways from conventional plastics toward sustainable alternatives, including biodegradable polymers, compostable products, fungal-based packaging, and bio-derived fuels, alongside chemical and microbial degradation routes that convert polymers into monomers and value-added products. This circular approach highlights waste minimization, resource recovery, and the transition toward a sustainable plastics economy.

systems, addressing nanoscale effects, and considering interactions with cooccurring pollutants. Refinement of LCA methodologies is critical to support sustainable plastic design, regulatory decisions, and environmental policies.

10.7 Summary and conclusion

MP pollution is a pressing global environmental challenge due to persistence, ubiquity, and potential ecosystem and human health risks. This review provided an overview of MP sources, characteristics, pathways and impacts, highlighting widespread presence in water, soil, and air, and transport through food webs. MPs originate from industrial activities, consumer products, agriculture, and wastewater treatment, contributing to pervasive distribution. Remediation techniques-including filtration, separation, advanced oxidation, photocatalysis, and biodegradation-show both progress and limitations. Challenges remain in detecting MPs in complex matrices understanding long-term ecological effects, and developing scalable, cost-effective treatments. Rising global plastic pollution will likely increase MP generation, intensifying environmental and health impacts. Future research should focus on eco-friendly degradation strategies, improved analytical methods, and synergistic technologies leveraging biotechnology, advanced materials, and crucial to reducing MP environmental impacts. Coordinated efforts among governments, industries researchers, and environmental agencies are essential to mitigate MP pollution and safeguard ecosystems and human health. Technological solutions are indispensable for mitigating MP contamination in aquatic environments. Filtration technologies offer immediate removal potential, biodegradable alternatives reduce further pollution loads and improved clean up and waste management strategies limit secondary MP generation. Together, these approaches address both symptoms and root cause of MP pollution.

The effective management of MP pollution requires a transition from reactive clean-up to proactive, technology-driven prevention and remediation. Advances in filtration systems, green material design, and waste management practices demonstrate that technological innovation can substantially reduce MP burdens in aquatic ecosystems. However, long-term success depends on integrating these solutions within broader sustainability frameworks, supported by regulatory action and societal engagement, by highlighting practical scalable technologies, this chapter contribute a forward-looking foundation for reducing MP contamination and safeguarding aquatic ecosystem health.

References

Abdelfattah, A., Ali, S.S., Ramadan, H., Aswar, E.I., Eltawab, R., Ho, S.H., Elsamahy, T., Li, S., Shhekh, M.M., Schagerl, M., Kornros, M., Sun, J., 2023. Microalgae based wastewater treatment: mechanisms, challenges, recent advances, and future prospects. Environ. Sci. Ecotechnol. 13, 100205. https://doi.org/10.1016/j.ese.2022.100205.

Adegoke, K.A., Adu, F.A., Oyebamiji, A.K., Bamisaye, A., Adigun, R.A., Olasoji, S.O., Ogunjinmi, O.E., 2023. Microplastics toxicity, detection, and removal from water/wastewater. Mar. Pollut. Bull. 187. https://doi.org/10.1016/j.marpolbul.2022.114546.

Adewuyi, A., Li, Q., 2024. Emergence of microplastics in African environmental drinking water sources: a review on sources, analysis and treatment strategies. J. Hazard. Mater. Adv. 100465.

Agarwal, S., 2020. Biodegradable polymers: present opportunities and challenges in providing a microplastic-free environment. Macromol. Chem. Phys. 221 (6), 2000017.

Ariza-Tarazona, M.C., Villarreal-Chiu, J.F., Barbieri, V., Siligardi, C., Cedillo-González, E.I., 2019. New strategy for microplastic degradation: green photocatalysis using a protein-based porous N-TiO 2 semiconductor. Ceram. Int. 45 (7), 9618–9624. https://doi.org/10.1016/j.ceramint.2018.10.208.

Arokiasamy, P., Abdullah, M.M.Al B., Rahim, S.Z.A., Sadique, M., Ming, L.Y., Salleh, M.A.A.M., Zainol, M.R. R.M.A., Ghazali, C.M.R., 2023. Diverse material based geopolymer towards heavy metals removal: a review. J. Mater. Res. Technol. 22, 126–156. https://doi.org/10.1016/j.jmrt.2022.11.100.

Asiandu, A.P., Wahyudi, A., Sari, S.W., 2022. Aquatic plastics waste biodegradation using plastic degrading microbes . Slovak university of agriculture. J. Microbiol. Biotechnol. Food Sci. 11 (5). https://doi.org/10.55251/jmbfs.3724.

Atugoda, T., et al., 2021. Interactions between microplastics, pharmaceuticals and personal care products: implications for vector transport. Environ. Int. 149, 106367.

Atugoda, T., Piyumali, H., Liyanage, S., Mahatantila, K., Vithanage, M., 2022. Fate and behavior of microplastics in freshwater systems. In: Rocha-Santos, T., Costa, M.F., Mouneyrac, C. (Eds.), Handbook of Microplastics in the Environment. Springer, Cham. https://doi.org/10.1007/978-3-030-39041-9_42.

Babiak, W., Krzeminska, I., 2021. Extracellular polymeric substances (EPS) as microalgal bioproducts: a review of factors affecting EPS synthesis and application in flocculation processes. Energies 14 (13), 4007. https://doi.org/10.3390/en14134007.

Bakhteeva, I.A., Medvedeva, I.V., Filinkova, M.S., Byzov, I.V., Minin, A.S., Zhakov, S.V., Uimin, M.A., Patrakov, E.I., Novikov, S.I., Suntsov, A.Y., Demin, A.M., 2023. Removal of microplastics from water by using magnetic sedimentation. Int. J. Environ. Sci. Technol. 20 (11), 11837–11850. https://doi.org/10.1007/s13762-023-04776-1.

Barbier, J.S., Dris, R., Lecarpentier, C., Raymond, V., Delabre, K., Thibert, S., Tassin, B., Gasperi, J., 2022. Microplastic occurrence after conventional and nanofiltration processes at drinking water treatment plants: preliminary results. Front. Water 4. https://doi.org/10.3389/frwa.2022.886703.

Barboza, L.G.A., Lopes, C., Oliveira, P., Bessa, F., Otero, V., Henriques, B., Raimundo, J., Caetano, M., Vale, C., Guilhermino, L., 2020. Microplastics in wild fish from North East Atlantic Ocean and its potential for causing neurotoxic effects, lipid oxidative damage, and human health risks associated with ingestion exposure. Sci. Total Environ. 717, 134625. https://doi.org/10.1016/j.scitotenv.2019.134625.

Bayo, J., Olmos, S., López-Castellanos, J., 2020. Microplastics in an urban wastewater treatment plant: the influence of physicochemical parameters and environmental factors. Chemosphere 238, 124593. https://doi.org/10.1016/j.chemosphere.2019.124593.

Bethi, B., Sonawane, S.H., Bhanvase, B.A., Gumfekar, S.P., 2016. Nanomaterials-based advanced oxidation processes for wastewater treatment: a review. Chem. Eng. Proc.: Proc. Intensif. 109, 178–189. https://doi.org/10.1016/j.cep.2016.08.016.

Bhattacharya, P., Lin, S., Turner, J.P., Ke, P.C., 2010. Physical adsorption of charged plastic nanoparticles affects algal photosynthesis. J. Phys. Chem. C 114 (39), 16556–16561. https://doi.org/10.1021/jp1054759.

Bhuyan, M.A.H., Busquets, R., Campos, L.C., Luukkonen, T., 2024. Separation of microplastics from water using superhydrophobic silane-coupling-agent-modified geopolymer foam. Separ. Purif. Technol. 339, 126709. https://doi.org/10.1016/j.seppur.2024.126709.

Blair, R.M., Waldron, S., Gauchotte-Lindsay, C., 2019. Average daily flow of microplastics through a tertiary wastewater treatment plant over a ten-month period. Water Res. 163, 114909. https://doi.org/10.1016/j.watres.2019.114909.

Borges, B.R., Carlos, V.M., Pompeo, M., 2021. Microplastics and freshwater microalgae: what do we know so far? Aquat. Ecol. 55, 363–377. https://doi.org/10.1007/s10452-021-09834-9.

Büks, F., Kayser, G., Zieger, A., Lang, F., Kaupenjohann, M., 2021. Particles under stress: ultrasonication causes size and recovery rate artifacts with soil-derived POM but not with microplastics. Biogeosciences 18 (1), 159–167. https://doi.org/10.5194/bg-18-159-2021.

Burduhos Nergis, D.D., et al., 2020. Revealing the influence of microparticles on geopolymers' synthesis and porosity. Materials 13 (14), 3211.

Cao, B., Wan, S., Wang, Y., Guo, H., Ou, M., Zhong, Q., 2022. Highly-efficient visible-light-driven photocatalytic H2 evolution integrated with microplastic degradation over MXene/ZnxCd1-xS photocatalyst. J. Colloid Interf. Sci. 605, 311–319. https://doi.org/10.1016/j.jcis.2021.07.113.

Chae, Y., Kim, D., An, Y.-J., 2019. Effects of micro-sized polyethylene spheres on the marine microalga Dunaliella salina: focusing on the algal cell to plastic particle size ratio. Aquat. Toxicol. 216, 105296. https://doi.org/10.1016/j.aquatox.2019.105296.

Chen, G., Feng, Q., Wang, J., 2020a. Mini-review of microplastics in the atmosphere and their risks to humans. Sci. Total Environ. 703, 135504. https://doi.org/10.1016/j.scitotenv.2019.135504.

Chen, Z., Zhao, W., Xing, R., Xie, S., Yang, X., Cui, P., Lu, J., Liao, H., Yu, Z., Wang, S., Zhou, S., 2020b. Enhanced in situ biodegradation of microplastics in sewage sludge using hyperthermophilic composting technology. J. Hazard. Mater. 384, 121271. https://doi.org/10.1016/j.jhazmat.2019.121271.

Chen, C., Fan, D., Ling, H., Huang, X., Yang, G., Cai, D., Zhao, J., Bi, Y., 2022a. Microwave catalytic co-pyrolysis of Chlorella vulgaris and high density polyethylene over activated carbon supported monometallic: characteristics and bio-oil analysis. Bioresour. Technol. 263, 127881. https://doi.org/10.1016/j.biortech.2022.127881.

Chen, J., Wu, J., Sherrell, P.C., Chen, J., Wang, H., Zhang, W.x., Yang, J., 2022b. How to build a microplastics-free environment: strategies for microplastics degradation and plastics recycling. Adv. Sci. 9 (6). https://doi.org/10.1002/advs.202103764.

Cherniak, S.L., Almuhtaram, H., McKie, M.J., Hermabessiere, L., Yuan, C., Rochman, C.M., Andrews, R.C., 2022. Conventional and biological treatment for the removal of microplastics from drinking water. Chemosphere 288. https://doi.org/10.1016/j.chemosphere.2021.132587.

Chia, W.Y., Tang, D.Y.Y., Khoo, K.S., Lup, A.N.G.K., Chew, K.W., 2020. Nature's fight against plastic pollution: algae for plastic biodegradation and bioplastics production. Environ. Sci. Ecotechnol. 4, 100065. https://doi.org/10.1016/j.ese.2020.100065.

Choi, Y.R., Kim, Y.N., Yoon, J.H., Dickinson, N., Kim, K.H., 2021. Plastic contamination of forest, urban, and agricultural soils: a case study of Yeoju City in the Republic of Korea. J. Soils Sediment. 21 (5), 1962–1973. https://doi.org/10.1007/s11368-020-02759-0.

Cole, M., Webb, H., Lindeque, P.K., Fileman, E.S., Halsband, C., Galloway, T.S., 2014. Isolation of microplastics in biota-rich seawater samples and marine organisms. Sci. Rep. 4. https://doi.org/10.1038/srep04528.

Corradini, F., Meza, P., Eguiluz, R., Casado, F., Huerta-Lwanga, E., Geissen, V., 2019. Evidence of microplastic accumulation in agricultural soils from sewage sludge disposal. Sci. Total Environ. 671, 411–420. https://doi.org/10.1016/j.scitotenv.2019.03.368.

Cunha, C., Faria, M., Nogueira, N., Ferreira, A., Cordeiro, N., 2019. Marine vs freshwater microalgae exopolymers as biosolutions to microplastics pollution. Environ. Pollut. 249, 372–380. https://doi.org/10.1016/j.envpol.2019.03.046.

Cvetkovska, M., Vakulenko, G., Smith, D.R., Zhang, X., Hüner, N.P.A., 2022. Temperature stress in psychrophilic green microalgae: minireview. Physiol. Plantarum 174 (6). https://doi.org/10.1111/ppl.13811.

Dang, H., Lovell, C.R., 2016. Microbial surface colonization and biofilm development in marine environments. Microbiol. Mol. Biol. Rev. 80 (1), 91–138.

Debroy, A., George, N., Mukherjee, G., 2022. Role of biofilms in the degradation of microplastics in aquatic environments. J. Chem. Technol. Biotechnol. 97 (12), 3271–3282. https://doi.org/10.1002/jctb.6978.

Della Rocca, D.G., Peralta, R.M., Peralta, R.A., Rodríguez-Castellón, E., de Fatima Peralta Muniz Moreira, R., 2021. Adding value to aluminosilicate solid wastes to produce adsorbents, catalysts and filtration membranes for water and wastewater treatment. J. Mater. Sci. 56 (2), 1039–1063. https://doi.org/10.1007/s10853-020-05276-0.

Demir-Yilmaz, I., Yakovenko, N., Roux, C., Guiraud, P., Collin, F., Coudret, C., Halle, A.T., Formosa-Dague, C., 2022. The role of microplastics in microalgae cells aggregation: a study at the molecular scale using atomic force microscopy. Sci. Total Environ. 832, 155036. https://doi.org/10.1016/j.scitotenv.2022.155036.

Dronjak, L., Exposito, N., Rovira, J., Florencio, K., Emiliano, P., Corzo, B., Schuhmacher, M., Valero, F., Sierra, J., 2022. Screening of microplastics in water and sludge lines of a drinking water treatment plant in Catalonia, Spain. Water Res. 225, 119185. https://doi.org/10.1016/j.watres.2022.119185.

Easton, T., Koutsos, V., Chatzisymeon, E., 2023. Removal of polyester fibre microplastics from wastewater using a UV/H2O2 oxidation process. J. Environ. Chem. Eng. 11 (1), 109057. https://doi.org/10.1016/j.jece.2022.109057.

Elgarahy, A.M., Ali, M., Eloffy, M.G., Zahran, M., Kharbish, S., Elwakeel, K.Z., Bhatnagar, A., 2023. Geopolymers as sustainable eco-friendly materials: classification, synthesis routes, and applications in wastewater treatment. Separ. Purif. Technol. 324, 124631. https://doi.org/10.1016/j.seppur.2023.124631.

Enders, K., Lenz, R., Beer, S., Stedmon, C.A., 2017. Extraction of microplastic from biota: recommended acidic digestion destroys common plastic polymers. ICES (Int. Counc. Explor. Sea) J. Mar. Sci. 74 (1), 326–331. https://doi.org/10.1093/icesjms/fsw173.

Falah, W., Chen, F.J., Zeb, B.S., Hayat, M.T., Mahmood, Q., Ebadi, A., Toughani, M., Li, E.Z., 2020. Polyethylene terephthalate degradation by Microalga Chlorella vulgaris along with pretreatment. Materiale Plastice 57 (3), 260–270. https://doi.org/10.37358/MP.20.3.5398.

Fang, Z., Sallach, J., Hodson, M.E., 2024. Size-and concentration-dependent effects of microplastics on soil aggregate formation and properties. J. Hazard Mater. 465, 133395. https://doi.org/10.1016/j.jhazmat.2023.133395.

Feng, X., Long, R., Liu, C., Liu, X., 2022. Visible-light-driven removal of tetracycline hydrochloride and microplastics (HDPE) by nano flower hybrid heterojunction NH2-MIL-88B(Fe)/MoS2 via enhanced electron-transfer. Separ. Purif. Technol. 302, 122138. https://doi.org/10.1016/j.seppur.2022.122138.

Geremia, E., Tomajoli, M.T.M., Murano, C., Petito, A., Fasciolo, G., 2023. The impact of Micro- and Nanoplastics on aquatic organisms: mechanisms of oxidative stress and implications for human health—a review. Environments 10 (9), 161. https://doi.org/10.3390/environments10090161.

Gopalakrishnan, K., Kashian, D.R., 2023. Complex interactions among temperature, microplastics and cyanobacteria may facilitate cyanobacteria proliferation and microplastic deposition. Ecotoxicol. Environ. Saf. 263, 115259. https://doi.org/10.1016/j.ecoenv.2023.115259.

Gowthami, A., Syed Marjuk, M., Raju, P., Nanthini Devi, K., Santhanam, P., Dinesh Kumar, S., Perumal, P., 2023. Biodegradation efficacy of selected marine microalgae against Low-Density Polyethylene (LDPE): an environment friendly green approach. Mar. Pollut. Bull. 190. https://doi.org/10.1016/j.marpolbul.2023.114889.

Grause, G., Kuniyasu, Y., Chien, M.F., Inoue, C., 2022. Separation of microplastic from soil by centrifugation and its application to agricultural soil. Chemosphere 288, 132654. https://doi.org/10.1016/j.chemosphere.2021.132654.

Guo, Z.Q., Li, P., Yang, X.M., Wang, Z.H., Lu, B.B., Chen, W.J., Wu, Y., Li, G.W., Zhao, Z.W., Liu, G.B., Ritsema, C., Geissen, V., Xue, S., 2022. Soil texture is an important factor determining how microplastics affect soil hydraulic characteristics. Environ. Int. 165, 107293. https://doi.org/10.1016/j.envint.2022.107293.

Hadidi, M., Zydney, A.L., 2014. Fouling behavior of zwitterionic membranes: impact of electrostatic and hydrophobic interactions. J. Membr. Sci. 452, 97–103. https://doi.org/10.1016/j.memsci.2013.09.062.

Han, N., Zhao, Q., Ao, H., Hu, H., Wu, C., 2022. Horizontal transport of macro-and microplastics on soil surface by rainfall induced surface runoff as affected by vegetations. Sci. Total Environ. 831, 154989.

Hatinoglu, M.D., Sanin, F.D., 2021. Sewage sludge as a source of microplastics in the environment: a review of occurrence and fate during sludge treatment. J. Environ. Manag. 295, 113028. https://doi.org/10.1016/j.jenvman.2021.113028.

Herbort, A.F., Sturm, M.T., Fiedler, S., Abkai, G., Schuhen, K., 2018. Alkoxy-silyl induced agglomeration: a new approach for the sustainable removal of microplastic from aquatic systems. J. Polym. Environ. 26 (11), 4258–4270. https://doi.org/10.1007/s10924-018-1287-3.

Hopewell, J., Dvorak, R., Kosior, E., 2009. Plastics recycling: challenges and opportunities. Phil. Trans. Biol. Sci. 364 (1526), 2115–2126. https://doi.org/10.1098/rstb.2008.0311.

Hossain, M.B., Banik, P., Nur, A.A., Choudhury, T.R., Liba, S.I., Albeshr, M.F., Yu, J., Arai, T., 2023. Microplastics in fish culture ponds: abundance, characterization, and contamination risk assessment. Front. Environ. Sci. 11. https://doi.org/10.3389/fenvs.2023.1251158.

Huang, D., Xu, Y., Lei, F., Yu, X., Ouyang, Z., Chen, Y., Jia, H., Guo, X., 2021. Degradation of polyethylene plastic in soil and effects on microbial community composition. J. Hazard Mater. 416, 126173.

Huang, D., Wang, X., Yin, L., Chen, S., Tao, J., Zhou, W., Chen, H., Zhang, G., Xiao, R., 2022. Research progress of microplastics in soil-plant system: ecological effects and potential risks. Sci. Total Environ. 812, 151487. https://doi.org/10.1016/j.scitotenv.2021.151487.

Huang, A., Zhi, D., Zhou, Y., 2023. Current progress of microplastics in sewage sludge . Springer science and business media Deutschland GmbH. In: Handbook of Environmental Chemistry, 114, pp. 167–187. https://doi.org/10.1007/698_2022_854.

Hurley, R.R., Lusher, A.L., Olsen, M., Nizzetto, L., 2018. Validation of a method for extracting microplastics from complex, organic-rich, environmental matrices. Environ. Sci. Technol. 52 (13), 7409–7417. https://doi.org/10.1021/acs.est.8b01517.

Jain, R., Gaur, A., Suravajhala, R., Chauhan, U., Pant, M., Tripathi, V., Pant, G., 2023. Microplastic pollution: understanding microbial degradation and strategies for pollutant reduction. Sci. Total Environ. 905, 167098. https://doi.org/10.1016/j.scitotenv.2023.167098.

Joo, S.H., Liang, Y., Kim, M., Byun, J., Choi, H., 2021. Microplastics with adsorbed contaminants: mechanisms and treatment. Environ. Chall. 3, 100042. https://doi.org/10.1016/j.envc.2021.100042.

Jung, J.W., Kim, S., Kim, Y.S., Jeong, S., Lee, J., 2022. Tracing microplastics from raw water to drinking water treatment plants in Busan, South Korea. Sci. Total Environ. 825. https://doi.org/10.1016/j.scitotenv.2022.154015.

Kaali, P., Perez-Madrigal, M.M., Stromberg, E., Aune, R.E., Czel, G., Karlsson, S., 2011. The influence of Ag+, Zn2+ and Cu2+ exchanged zeolite on antimicrobial and long term in vitro stability of medical grade polyether polyurethane. Exp. Polym. Lett. 5 (12), 1028–1040. https://doi.org/10.3144/expresspolymlett.2011.101.

Kaiser, D., Kowalski, N., Waniek, J.J., 2017. Effects of biofouling on the sinking behavior of microplastics. Environ. Res. Lett. 12 (12), 124003. https://doi.org/10.1088/1748-9326/aa8e8b.

Kallawar, G.A., Barai, D.P., Bhanvase, B.A., 2021. Bismuth titanate based photocatalysts for degradation of persistent organic compounds in wastewater: a comprehensive review on synthesis methods, performance as photocatalyst and challenges. J. Cleaner Prod. 318. https://doi.org/10.1016/j.jclepro.2021.128563.

Kedzierski, M., Le Tilly, V., César, G., Sire, O., Bruzaud, S., 2017. Efficient microplastics extraction from sand. A cost effective methodology based on sodium iodide recycling. Mar. Pollut. Bull. 115 (1–2), 120–129. https://doi.org/10.1016/j.marpolbul.2016.12.002.

Khoironi, A., Anggoro, S., Sudarno, 2019. Evaluation of the interaction among microalgae Spirulina sp, plastics Polyethylene terephthalate and Polypropylene in freshwater environment. J. Ecol. Eng. 20 (6), 161–173. https://doi.org/10.12911/22998993/108637.

Kim, K.T., Park, S., 2021. Enhancing microplastics removal from wastewater using electro-coagulation and granule-activated carbon with thermal regeneration. Processes 9 (4), 617.

Konechnaya, O., Lüchtrath, S., Dsikowitzky, L., Schwarzbauer, J., 2020. Optimized microplastic analysis based on size fractionation, density separation and μ-FTIR. Water Sci. Technol. 81, 834–844. https://doi.org/10.2166/wst.2020.173.

Lagarde, F., Olivier, O., Zanella, M., Daniel, P., Hiard, S., Caruso, A., 2016. Microplastic interactions with freshwater microalgae: hetero-aggregation and changes in plastic density appear strongly dependent on polymer type. Environ. Pollut. 215, 331–339. https://doi.org/10.1016/j.envpol.2016.05.006.

Lares, M., Ncibi, M.C., Sillanpää, M., Sillanpää, M., 2018. Occurrence, identification and removal of microplastic particles and fibers in conventional activated sludge process and advanced MBR technology. Water Res. 133, 236–246. https://doi.org/10.1016/j.watres.2018.01.049.

Law, K.L., 2017. Plastics in the marine environment. Ann. Rev. Marine Sci. 9 (1), 205–229. https://doi.org/10.1146/annurev-marine-010816-060409.

Lehmann, A., Leifheit, E.F., Gerdawischke, M., Rillig, M.C., 2021. Microplastics have shape- and polymer-dependent effects on soil aggregation and organic matter loss – an experimental and meta-analytical approach. Microplast. Nanoplast. 1 (1). https://doi.org/10.1186/s43591-021-00007-x.

Li, X., Chen, L., Mei, Q., Dong, B., Dai, X., Ding, G., Zeng, E.Y., 2018. Microplastics in sewage sludge from the wastewater treatment plants in China. Water Res. 142, 75–85. https://doi.org/10.1016/j.watres.2018.05.034.

Li, W., Wufuer, R., Duo, J., Wang, S., Luo, Y., Zhang, D., Pan, X., 2020. Microplastics in agricultural soils: extraction and characterization after different periods of polythene film mulching in an arid region. Sci. Total Environ. 749, 141420. https://doi.org/10.1016/j.scitotenv.2020.141420.

Li, D., Sheerin, E.D., Shi, Y., Xiao, L., Yang, L., Boland, J.J., Wang, J.J., 2022a. Alcohol pretreatment to eliminate the interference of micro additive particles in the identification of microplastics using Raman spectroscopy. Environ. Sci. Technol. 56 (17), 12158–12168. https://doi.org/10.1021/acs.est.2c01551.

Li, Z., Dong, S., Huang, F., Lin, L., Hu, Z., Zheng, Y., 2022b. Toxicological effects of microplastics and sulfadiazine on the microalgae Chlamydomonas reinhardtii. Front. Microbiol. 13. https://doi.org/10.3389/fmicb.2022.865768.

Li, W., Zhao, W., Zhu, H., Li, Z.J., Wang, W., 2023. State of the art in the photochemical degradation of (micro) plastics: from fundamental principles to catalysts and applications. J. Mater. Chem. A 11 (6), 2503–2527. https://doi.org/10.1039/d2ta09523h.

Liu, F., Vianello, A., Vollertsen, J., 2019. Retention of microplastics in sediments of urban and highway stormwater retention ponds. Environ. Pollut. 255, 113335. https://doi.org/10.1016/j.envpol.2019.113335.

Liu, G., Jiang, R., You, J., Muir, D.C.G., Zeng, E.Y., 2020. Microplastic impacts on microalgae growth: effects of size and humic acid. Environ. Sci. Technol. 54 (3), 1782–1789. https://doi.org/10.1021/acs.est.9b06187.

Liu, H., Jiao, Q., Pan, T., Liu, W., Li, S., Zhu, X., Zhang, T., 2023. Aging behavior of biodegradable polylactic acid microplastics accelerated by UV/H2O2 processes. Chemosphere 337, 139360. https://doi.org/10.1016/j.chemosphere.2023.139360.

Liu, H., Wen, X., Sun, Y., Zhang, H., Wang, H., Wei, J., 2023. Identification, quantification and biodegradation of microplastics from personal care products and detergents by microorganism. J. Water Process Eng. 53, 103754. https://doi.org/10.1016/j.jwpe.2023.103754.

Liu, S.Y., Leung, M.M.L., Fang, J.K.H., Chua, S.L., 2021. Engineering a microbial 'trap and release' mechanism for microplastics removal. Chem. Eng. J. 404. https://doi.org/10.1016/j.cej.2020.127079.

Long, M., Paul-Pont, I., Hégaret, H., Moriceau, B., Lambert, C., Huvet, A., Soudant, P., 2017. Interactions between polystyrene microplastics and marine phytoplankton lead to species-specific hetero-aggregation. Environ. Pollut. 228, 454–463. https://doi.org/10.1016/j.envpol.2017.05.047.

Lu, J., et al., 2021. Adsorption and desorption of steroid hormones by microplastics in seawater. Bull. Environ. Contam. Toxicol. 107, 730–735.

Lu, Q., Zhou, Y., Sui, Q., Zhou, Y., 2023. Mechanism and characterization of microplastic aging process: a review. Front. Environ. Sci. Eng. 17 (8). https://doi.org/10.1007/s11783-023-1700-6.

Lusher, A.L., Hurley, R.R., Vogelsang, C., Nizzetto, L., Olsen, M., 2017. Mapping Microplastics in Sludge.

Ma, B., Xue, W., Hu, C., Liu, H., Qu, J., Li, L., 2019a. Characteristics of microplastic removal via coagulation and ultrafiltration during drinking water treatment. Chem. Eng. J. 359, 159–167. https://doi.org/10.1016/j.cej.2018.11.155.

Ma, B., Xue, W., Ding, Y., Hu, C., Liu, H., Qu, J., 2019b. Removal characteristics of microplastics by Fe-based coagulants during drinking water treatment. J. Environ. Sci. 78, 267–275. https://doi.org/10.1016/j.jes.2018.10.006.

Madadi, R., Maljaee, H., Serafim, L.S., Ventura, S.P.M., 2021. Microalgae as contributors to produce biopolymers. MDPI, Iran Marine Drugs 19 (8). https://doi.org/10.3390/MD19080466.

Magnusson, K., Noren, F., 2014. Screening of Microplastic Particles in and Down-Stream a Wastewater Treatment Plant.

Mahabeer, J., Kumari, U., Lokhat, D., Carsky, M., Charan Meikap, B., 2023. Implementation of microplastics derived from waste plastic for uptake of MB dye: performance and LCA study. Desalination 546, 116214. https://doi.org/10.1016/j.desal.2022.116214.

Mai, L., Bao, L.J., Shi, L., Wong, C.S., Zeng, E.Y., 2018. A review of methods for measuring microplastics in aquatic environments. Environ. Sci. Pollut. Res. 25 (12), 11319–11332. https://doi.org/10.1007/s11356-018-1692-0.

Martin, L.M.A., Sheng, J., Zimba, P.V., Zhu, L., Fadare, O.O., Haley, C., Wang, M., Phillips, T.D., Conkle, J., Xu, W., 2022. Testing an iron oxide nanoparticle-based method for magnetic separation of nanoplastics and microplastics from water. Nanomaterials 12 (14). https://doi.org/10.3390/nano12142348.

Martínez, G.A., Carizzo, M.V., Devaud, L.Z., 2023. Microplastic Particles' effects on aquatic organisms and their role as transporters of organic pollutants. Water 15, 2915. https://doi.org/10.3390/w15162915.

Masi, P., Sol, D., Ardura, A., Laca, A., Borrell, Y.J., Dopico, E., Laca, A., Machado Schiaffino, G., Díaz, M., Garcia-Vazquez, E., 2020. Bioremediation as a promising strategy for microplastics removal in wastewater treatment plants. Mar. Pollut. Bull. 156, 111252. https://doi.org/10.1016/j.marpolbul.2020.111252.

Mbachu, O., Jenkins, G., Pratt, C., Prasad, K., 2021. Enzymatic purification of microplastics in soil. MethodsX 8, 101254. https://doi.org/10.1016/j.mex.2021.101254.

Menendez-Manjon, A., Martínez-Díez, R., Sol, D., Laca, A., Laca, A., Rancano, A., Díaz, M., 2022. Long-term occurrence and fate of microplastics in WWTPs: a case study in Southwest Europe. Appl. Sci. 12 (4), 2133. https://doi.org/10.3390/app12042133.

Mitrano, D.M., Wick, P., Nowack, B., 2021. Placing nanoplastics in the context of global plastic pollution. Nature Nanotechnol. 16 (5), 491–500. https://doi.org/10.1038/s41565-021-00888-2.

Moog, D., Schmitt, J., Senger, J., Zarzycki, J., Rexer, K.H., Linne, U., Erb, T., Maier, U.G., 2019. Using a marine microalga as a chassis for polyethylene terephthalate (PET) degradation. Microb. Cell Fact. 18 (1). https://doi.org/10.1186/s12934-019-1220-z.

Munno, K., Helm, P.A., Jackson, D.A., Rochman, C., Sims, A., 2018. Impacts of temperature and selected chemical digestion methods on microplastic particles. Environ. Toxicol. Chem. 37 (1), 91–98. https://doi.org/10.1002/etc.3935.

Muthulakshmi, L., Mohan, S., Tatarchuk, T., 2023. Microplastics in water: types, detection, and removal strategies. Environ. Sci. Pollut. Res. 30 (36), 84933–84948. https://doi.org/10.1007/s11356-023-28460-6.

Na, S.H., Kim, M.J., Kim, J.T., Jeong, S., Lee, S., Chung, J., Kim, E.J., 2021. Microplastic removal in conventional drinking water treatment processes: performance, mechanism, and potential risk. Water Res. 202. https://doi.org/10.1016/j.watres.2021.117417.

Namasivayam, S.K.R., Avinash, G., 2024. Review of green technologies for the removal of microplastics from diverse environmental sources. Environ. Qual. Manag. 33 (3), 449–465.

Natarajan, L., Soupam, D., Dey, S., Chandrasekaran, N., Kundu, R., Paul, S., Mukherjee, A., 2022. Toxicity of polystyrene microplastics in freshwater algae Scenedesmus obliquus: effects of particle size and surface charge. Toxicol. Rep. 9, 1953–1961. https://doi.org/10.1016/j.toxrep.2022.10.013.

Nava, V., Matiasa, M.G., Escriva, A.C., Messyasz, B., Leoni, B., 2021. Microalgae colonization of different microplastic polymers in experimental mesocosms across an environmental gradient. Glob. Change Biol. 8, 1402–1413. https://doi.org/10.1111/gcb.15989.

Ncube, L.K., Ude, A.U., Ogunmuyiwa, E.N., Zulkifli, R., Beas, I.N., 2021. An overview of plastic waste generation and management in food packaging industries. Recycling 6 (1), 12. https://doi.org/10.3390/recycling6010012.

Nguyen, M.K., Hadi, M., Lin, C., Nguyen, H.L., Thai, V.B., Hoang, H.G., Vo, D.V.N., Tran, H.T., 2022. Microplastics in sewage sludge: distribution, toxicity, identification methods, and engineered technologies. Chemosphere 308. https://doi.org/10.1016/j.chemosphere.2022.136455.

Nikiema, J., Asiedu, Z., 2022. A review of the cost and effectiveness of solutions to address plastic pollution. Environ. Sci. Pollut. Res. 29, 24547–24573. https://doi.org/10.1007/s11356-021-18038-5.

Nurlina, N., Pratama, J.H., Pambudi, A.B., Rahmawati, Z., Subaer, S., Abdullah, M.M.A.B., Gusrizal, G., Fansuri, H., 2024. A review of geopolymer membrane for water treatment. Appl. Clay Sci. 251. https://doi.org/10.1016/j.clay.2024.107301.

Ortiz, D., Munoz, M., Nieto-Sandoval, J., Romera-Castillo, C., de Pedro, Z.M., Casas, J.A., 2022. Insights into the degradation of microplastics by Fenton oxidation: from surface modification to mineralization. Chemosphere 309. https://doi.org/10.1016/j.chemosphere.2022.136809.

Osman, S.A., et al., 2024. Biodegradation of paraffin wax using Bacillus sp. SASH for quality improvement of crude oil properties. Petrol. Sci. Technol. 42 (10), 1193–1210.

Paço, A., Duarte, K., da Costa, J.P., Santos, P.S.M., Pereira, R., Pereira, M.E., Freitas, A.C., Duarte, A.C., Rocha-Santos, T.A.P., 2017. Biodegradation of polyethylene microplastics by the marine fungus Zalerion maritimum. Sci. Total Environ. 586, 10–15. https://doi.org/10.1016/j.scitotenv.2017.02.017.

Padervand, M., Lichtfouse, E., Robert, D., Wang, C., 2020. Removal of microplastics from the environment. A review. Environ. Chem. Lett. 18 (3), 807–828. https://doi.org/10.1007/s10311-020-00983-1.

Pasanen, F., Fuller, R.O., Maya, F., 2023. Fast and simultaneous removal of microplastics and plastic-derived endocrine disruptors using a magnetic ZIF-8 nanocomposite. Chem. Eng. J. 455, 140405. https://doi.org/10.1016/j.cej.2022.140405.

Paul, I., Biswas, R., Halder, G., 2024. Traversing the potential of phytoremediation and phycoremediation as pioneering technologies in microplastic mitigation – a critical review. Sci. Total Environ. 956, 177200. https://doi.org/10.1016/j.scitotenv.2024.177200.

Pérez-Reverón, R., González-Sálamo, J., Hernández-Sánchez, C., González-Pleiter, M., Hernández-Borges, J., Díaz-Peña, F.J., 2022. Recycled wastewater as a potential source of microplastics in irrigated soils from an arid-insular territory (Fuerteventura, Spain). Sci. Total Environ. 817, 152830. https://doi.org/10.1016/j.scitotenv.2021.152830.

Pfeiffer, F., Fischer, E.K., 2020. Various digestion protocols within microplastic sample processing—evaluating the resistance of different synthetic polymers and the efficiency of biogenic organic matter destruction. Front. Environ. Sci. 8, 572424. https://doi.org/10.3389/fenvs.2020.572424.

Pikula, K.S., Zakharenko, A.M., Aruoja, V., Golokhvast, K.S., Tsatsakis, A.M., 2019. Oxidative stress and its biomarkers in microalgal ecotoxicology. Current Opinion in Toxicology 13, 8–15. https://doi.org/10.1016/j.cotox.2018.12.006.

Pivokonský, M., Pivokonska, L., Novotna, K., Cermakova, L., Klimtova, M., 2020. Occurrence and fate of microplastics at two different drinking water treatment plants within a river catchment. Sci. Total Environ. 741, 140236. https://doi.org/10.1016/j.scitotenv.2020.140236.

Podbielska, M., Szpyrka, E., 2023. Microplastics – an emerging contaminants for algae. Critical review and perspectives. Sci. Total Environ. 885, 163842. https://doi.org/10.1016/j.scitotenv.2023.163842.

Prata, J.C., 2018. Microplastics in wastewater: state of the knowledge on sources, fate and solutions. Mar. Pollut. Bull. 129, 262–265. https://doi.org/10.1016/j.marpolbul.2018.02.046.

Prata, J.C., da Costa, J.P., Lopes, I., Duarte, A.C., Rocha-Santos, T., 2019. Effects of microplastics on microalgae populations: a critical review. Sci. Total Environ. 665, 400–405. https://doi.org/10.1016/j.scitotenv.2019.02.132.

Priya, A.K., Jalil, A.A., Dutta, K., Rajendran, S., Vasseghian, Y., Karimi-Maleh, H., Soto-Moscoso, M., 2022. Algal degradation of microplastic from the environment: mechanism, challenges, and future prospects. Algal Res. 67, 102848. https://doi.org/10.1016/j.algal.2022.102848.

Puncochar, M., Ruj, B., Chatterj, P.K., 2012. Development of process for disposal of plastic waste using plasma pyrolysis technology and option for energy recovery. Procedia Eng. 42, 420–430. https://doi.org/10.1016/j.proeng.2012.07.433.

Quinn, B., Murphy, F., Ewins, C., 2017. Validation of density separation for the rapid recovery of microplastics from sediment. Anal. Methods 9, 1491–1498. https://doi.org/10.1039/C6AY02542K.

Radford, F., Zapata-Restrepo, L.M., Horton, A.A., Hudson, M.D., Shaw, P.J., Williams, I.D., 2021. Developing a systematic method for extraction of microplastics in soils. Anal. Meth. 13 (14), 1695–1705. https://doi.org/10.1039/d0ay02086a.

Rafiq, A., Xu, J.L., 2023. Microplastics in waste management systems: a review of analytical methods, challenges and prospects. Waste Manag. 171, 54–70. https://doi.org/10.1016/j.wasman.2023.08.015.

Rajala, K., Grönfors, O., Hesampour, M., Mikola, A., 2020. Removal of microplastics from secondary wastewater treatment plant effluent by coagulation/flocculation with iron, aluminum and polyamine-based chemicals. Water Res. 183, 116045. https://doi.org/10.1016/j.watres.2020.116045.

Sacco, N.A., Zoppas, F.M., Devard, A., González Muñoz, M.d.P., García, G., Marchesini, F.A., 2023. Recent advances in microplastics removal from water with special attention given to photocatalytic degradation: review of scientific research. Microplastics 2 (3), 278–303. https://doi.org/10.3390/microplastics2030023.

Salieri, B., Stoudmann, N., Hischier, R., Som, C., Nowack, B., 2021. How relevant are direct emissions of microplastics into freshwater from an LCA perspective? Sustainability 13 (17), 9922. https://doi.org/10.3390/su13179922.

Saling, P., Gyuzeleva, L., Wittstock, K., Wessolowski, V., Griesshammer, R., 2020. Life cycle impact assessment of microplastics as one component of marine plastic debris. Int. J. Life Cycle Ass. 25 (10), 2008–2026. https://doi.org/10.1007/s11367-020-01802-z.

Sarkar, D.J., Das Sarkar, S., Das, B.K., Praharaj, J.K., Mahajan, D.K., Purokait, B., Mohanty, T.R., Mohanty, D., Gogoi, P., Kumar V, S., Behera, B.K., Manna, R.K., Samanta, S., 2021. Microplastics removal efficiency of drinking water treatment plant with pulse clarifier. J. Hazard Mater. 413. https://doi.org/10.1016/j.jhazmat.2021.125347.

Sarmah, P., Rout, J., 2018. Efficient biodegradation of low-density polyethylene by cyanobacteria isolated from submerged polyethylene surface in domestic sewage water. Environ. Sci. Pollut. Res. 25, 33508–33520. https://doi.org/10.1007/s11356-018-3079-7.

Savino, I., Campanale, C., Trotti, P., Massarelli, C., Corriero, G., Felice Uricchio, V., 2022. Effects and impacts of different oxidative digestion treatments on Virgin and aged microplastic particles. Polymers 14 (10), 1958. https://doi.org/10.3390/polym14101958.

Scopetani, C., Chelazzi, D., Mikola, J., Leiniö, V., Heikkinen, R., Cincinelli, A., Pellinen, J., 2020. Olive oil-based method for the extraction, quantification and identification of microplastics in soil and compost samples. Sci. Total Environ. 733, 139338. https://doi.org/10.1016/j.scitotenv.2020.139338.

Sembiring, E., Fajar, M., Handajani, M., 2021. Performance of rapid sand filter - single media to remove microplastics. Water Suppl. 21 (5), 2273–2284. https://doi.org/10.2166/ws.2021.060.

Seo, Y., Zhou, Z., Lai, Y., Chen, G., Pembleton, K., Wang, S., He, J.-zheng, Song, P., 2024. Micro- and nano-plastics in agricultural soils: assessing impacts and navigating mitigation. Sci. Total Environ. 931, 172951. https://doi.org/10.1016/j.scitotenv.2024.172951.

Shabbir, S., Faheem, M., Ali, N., Kerr, P.G., Wang, L.-F., Kuppusamy, S., Li, Y., 2020. Periphytic biofilm: an innovative approach for biodegradation of microplastics. Sci. Total Environ. 717, 137064. https://doi.org/10.1016/j.scitotenv.2020.137064.

Sharma, S., Sharma, P., Sharma, A., Srivastava, A., 2021. Microbial degradation of plastics: new insights and its emerging role in affecting the environmental sustainability. Environ. Technol. Innov. 24, 101836. https://doi.org/10.1016/j.eti.2021.101836.

Sheng, G.P., Yu, H.Q., Cui, H., 2008. Model-evaluation of the erosion behavior of activated sludge under shear conditions using a chemical-equilibrium-based model. Chem. Eng. J. 140 (1–3), 241–246. https://doi.org/10.1016/j.cej.2007.10.001.

Sjollema, S.B., Redondo-Hasselerharm, P., Leslie, H.A., Kraak, M.H.S., Vethaak, A.D., 2016. Do plastic particles affect microalgal photosynthesis and growth? Aquat. Toxicol. 170, 259–261. https://doi.org/10.1016/j.aquatox.2015.12.002.

Skaf, D.W., Punzi, V.L., Rolle, J.T., Kleinberg, K.A., 2020. Removal of micron-sized microplastic particles from simulated drinking water via alum coagulation. Chem. Eng. J. 386. https://doi.org/10.1016/j.cej.2019.123807.

Sol, D., Laca, A., Laca, A., Díaz, M., 2020. Approaching the environmental problem of microplastics: importance of WWTP treatments. Sci. Total Environ. 740, 140016. https://doi.org/10.1016/j.scitotenv.2020.140016.

Song, Y.K., et al., 2017. Combined effects of UV exposure duration and mechanical abrasion on microplastic fragmentation by polymer type. Environ. Sci. Technol. 51 (8), 4368–4376.

Statista. 2023. Global plastic production 1950-2022, 22nd April 2023. https://www.statista.com/statistics/282732/global-production-of-plastics-since-1950/.

Steinmetz, Z., Kintzi, A., Muñoz, K., Schaumann, G.E., 2020. A simple method for the selective quantification of polyethylene, polypropylene, and polystyrene plastic debris in soil by pyrolysis-gas chromatography/mass spectrometry. J. Analyt. Appl. Pyrol. 147. https://doi.org/10.1016/j.jaap.2020.104803.

Stoodley, H.L., Costerton, J.W., Stoodley, P., State, M., 2004. Bacterial biofilms: from the natural environment to infectious diseases. Nat. Rev. Microbiol. 2 (2), 95–108. https://doi.org/10.1038/nrmicro821.

Sun, J., Dai, X., Wang, Q., van Loosdrecht, M.C.M., Ni, B.J., 2019. Microplastics in wastewater treatment plants: detection, occurrence and removal. Water Res. 152, 21–37. https://doi.org/10.1016/j.watres.2018.12.050.

Tagg, A.S., Harrison, J.P., Ju-Nam, Y., Sapp, M., Bradley, E.L., Sinclair, C.J., Ojeda, J.J., 2017. Fenton's reagent for the rapid and efficient isolation of microplastics from wastewater. Chem. Commun. 53 (2), 372–375. https://doi.org/10.1039/c6cc08798a.

Taipale, S.J., Peltomaa, E., Kukkonen, J.V.K., Kainz, M.J., Kautonen, P., Tiirola, M., 2019. Tracing the fate of microplastic carbon in the aquatic food web by compound-specific isotope analysis. Sci. Rep. 9 (1). https://doi.org/10.1038/s41598-019-55990-2.

Talvitie, J., Mikola, A., Koistinen, A., Setälä, O., 2017. Solutions to microplastic pollution – removal of microplastics from wastewater effluent with advanced wastewater treatment technologies. Water Res. 123, 401–407. https://doi.org/10.1016/j.watres.2017.07.005.

Tang, Y., et al., 2021. Removal of microplastics from aqueous solutions by magnetic carbon nanotubes. Chem. Eng. J. 406, 126804.

Tang, Z.Q., et al., 2024. Enhancing geopolymer performance with silane-functionalized graphene oxide: efflorescence reduction, pore refinement, and improved high temperature properties. Constr. Build. Mater. 440, 137420.

Thomas, S., Shanks, R., Chandrasekharakurup, S., 2015. Design and Applications of Nanostructured Polymer Blends and Nanocomposite Systems Design and Applications of Nanostructured Polymer Blends and Nanocomposite Systems. Elsevier Inc., India, pp. 1–426. https://doi.org/10.1016/C2013-0-13999-2.

Tofa, T., et al., 2019a. Enhanced visible light photodegradation of microplastic fragments with plasmonic platinum/zinc oxide nanorod photocatalysts. Catalysts 9 (10), 819.

Tofa, T.S., Kunjali, K.L., Paul, S., Dutta, J., 2019b. Visible light photocatalytic degradation of microplastic residues with zinc oxide nanorods. Springer Verlag, Sweden Environmental Chemistry Letters 17 (3), 1341–1346. https://doi.org/10.1007/s10311-019-00859-z.

Tunali, M., Uzoefuna, E.N., Tunali, M.M., Yenigun, O., 2020. Effect of microplastics and microplastic-metal combinations on growth and chlorophyll a concentration of Chlorella vulgaris. Sci. Total Environ. 743, 140479. https://doi.org/10.1016/j.scitotenv.2020.140479.

Ugya, A.Y., Meguellati, K., Aliyu, A.D., Abba, A., Musa, M.A., 2022. Microplastic stress induce bioresource production and response in microalgae: a concise review. Environ. Pollut. Bioavail. 34 (1), 51–60. https://doi.org/10.1080/26395940.2022.2033138.

Uheida, A., Mejía, H.G., Abdel-Rehim, M., Hamd, W., Dutta, J., 2021. Visible light photocatalytic degradation of polypropylene microplastics in a continuous water flow system. J. Hazard Mater. 406, 124299. https://doi.org/10.1016/j.jhazmat.2020.124299.

van den Berg, P., Huerta-Lwanga, E., Corradini, F., Geissen, V., 2020. Sewage sludge application as a vehicle for microplastics in eastern Spanish agricultural soils. Environ. Pollut. 261, 114198. https://doi.org/10.1016/j.envpol.2020.114198.

Vanapalli, K.R., Dubey, B.K., Sarmah, A.K., Bhattacharya, J., 2020. Assessment of microplastic pollution in the aquatic ecosystems - an Indian perspective. Case Stud. Chem. Environ. Eng. 3, 100071. https://doi.org/10.1016/j.cscee.2020.100071.

Villalobos, M.B., Olmo, P.F., Guzman, J.L., Sevilla, J.M.F., Fernandez, F.G.A., 2019. Evaluation of photosynthetic light integration by microalgae in a pilot-scale raceway reactor. Bioresour. Technol. 280, 404–411. https://doi.org/10.1016/j.biortech.2019.02.032.

Vital-Grappin, A.D., Ariza-Tarazona, M.C., Luna-Hernández, V.M., Villarreal-Chiu, J.F., Hernández-López, J.M., Siligardi, C., Cedillo-González, E.I., 2021. The role of the reactive species involved in the photocatalytic degradation of hdpe microplastics using c,n-tio2 powders. Polymers 13 (7). https://doi.org/10.3390/polym13070999.

Wang, Z., Lin, T., Chen, W., 2020. Occurrence and removal of microplastics in an advanced drinking water treatment plant (ADWTP). Sci. Total Environ. 700, 134520. https://doi.org/10.1016/j.scitotenv.2019.134520.

Wang, X.S., Song, H., Liu, Y.L., Pan, X.R., Zhang, H.C., Gao, Z., Kong, D.Z., Wang, R., Wang, L., Ma, J., 2021. Quantitively analyzing the variation of micrometer-sized microplastic during water treatment with the flow cytometry- fluorescent beads method. ACS ES and T Eng. 1 (12), 1668–1677. https://doi.org/10.1021/acsestengg.1c00253.

WHO, 2019. Microplastics in Drinking-Water Key Messages. World Health Organization.

Wimalasinghe, R.M., Weatherly, C.A., Wahab, M.F., Thakur, N., Armstrong, D.W., 2018. Geopolymers as a new class of high pH stable supports with different chromatographic selectivity. Anal. Chem. 90 (13), 8139–8146. https://doi.org/10.1021/acs.analchem.8b01426.

Wu, D., Hou, Y., Cheng, J., Han, T., Hao, N., Zhang, B., Fan, X., Ji, X., Chen, F., Gong, D., Wang, L., McGinn, P., Zhao, L., Chen, S., 2022. Transcriptome analysis of lipid metabolism in response to cerium stress in the oleaginous microalga Nannochloropsis oculata. Sci. Total Environ. 838. https://doi.org/10.1016/j.scitotenv.2022.156420.

Xi, B., Wang, B., Chen, M., Lee, X., Zhang, X., Wang, S., Yu, Z., Wu, P., 2022. Environmental behaviors and degradation methods of microplastics in different environmental media. Chemosphere 299, 134354. https://doi.org/10.1016/j.chemosphere.2022.134354.

Xiao, X., Li, W., Li, S., Zuo, X., Liu, J., Guo, L., Lu, X., Zhang, L., 2023. The growth inhibition of polyethylene nanoplastics on the bait-microalgae isochrysis galbana based on the transcriptome analysis. Microorganisms 11 (5), 1108. https://doi.org/10.3390/microorganisms11051108.

Yurtsever, M., Kirkan, E.O., Sevindik, T.O., Tunca, H., 2017. The impact of PS microplastics on green algae Chlorella vulgaris growth. In: 15th International Conference on Environmental Science and Technology).

Zada, L., Leslie, H.A., Vethaak, A.D., Tinnevelt, G.H., Jansen, J.J., de Boer, J.F., Ariese, F., 2018. Fast microplastics identification with stimulated Raman scattering microscopy. J. Raman Spectrosc. 49 (7), 1136–1144. https://doi.org/10.1002/jrs.5367.

Zhang, G., Liu, Y., 2018. The distribution of microplastics in soil aggregate fractions in southwestern China. Sci. Total Environ. 642, 12–20. https://doi.org/10.1016/j.scitotenv.2018.06.004.

Zhang, C., Chen, X., Wang, J., Tan, L., 2017. Toxic effects of microplastic on marine microalgae Skeletonema costatum: interactions between microplastic and algae. Environ. Pollut. 220, 1282–1288. https://doi.org/10.1016/j.envpol.2016.11.005.

Zhang, J., Zhang, Q., Maa, J.P.-Y., Shen, X., Liang, J., Yu, L., Ge, L., Wang, G., 2022. Effects of organic matter on interaction forces between polystyrene microplastics: an experimental study. Sci. Total Environ. 844, 157186. https://doi.org/10.1016/j.scitotenv.2022.157186.

Zhou, D., Chen, J., Wu, J., Yang, J., Wang, H., 2021. Biodegradation and catalytic-chemical degradation strategies to mitigate microplastic pollution. Sustain. Mater. Technol. 28, e00251. https://doi.org/10.1016/j.susmat.2021.e00251.

Zurier, H.S., Goddard, J.M., 2021. Biodegradation of microplastics in food and agriculture. Curr. Opin. Food Sci. 37, 37–44. https://doi.org/10.1016/j.cofs.2020.09.001.

CHAPTER

11 Sustainable strategies for reducing microplastics contamination in aquatic systems

Melvin S. Samuel

Department of Civil, Construction, and Environmental Engineering, Marquette University, Milwaukee, WI, United States

Highlights

- Circular economy principles reduce microplastic pollution by targeting upstream plastic production and waste leakage.
- Public awareness and community participation strengthen the effectiveness of mitigation strategies
- Policy frameworks play a critical role in translating scientific evidence into enforceable action.
- Artificial intelligence and Machine Learning enhances microplastic detection, source tracking, and mitigation planning.
- Integrated socio-technical approaches offer scalable and long-term solutions for aquatic ecosystems.

11.1 Introduction

Microplastic (MP) contamination has emerged as a persistent and transboundary challenge in aquatic environments, affecting freshwater, coastal, and marine systems worldwide. Traditional mitigation approaches largely centered on end of pipe controls and localized cleanup efforts have proven insufficient in addressing the scale, complexity, and persistence of MP pollution. As plastics continue to circulate through production, consumption, degradation, and environmental redistribution, sustainable solutions must move beyond detection alone and integrate prevention, system redesign, and intelligent governance. Recent advances highlight the importance of circular economy frameworks, community engagement, policy innovation, and artificial intelligence (AI) as interconnected pillars for mitigating MP contamination. Circular economy strategies aim to reduce plastic leakage at its source by rethinking material design, reuse, and recovery pathways. Public participation and awareness are essential for translating policy and technology into behavioral changes. At the same time, AI-driven tools are transforming how MPs are detected, quantified, tracked, and managed, enabling faster, more accurate, and scalable interventions. This section synthesizes these complementary approaches to outline a sustainable and adaptive pathway for reducing MP contamination in aquatic systems, with particular emphasis on how AI can bridge scientific understanding, operational monitoring, and policy implementation.

Circular economy strategies aim to decouple economic growth from plastic waste generation by emphasizing reduction, redesign, reuse, and recycling. In the context of MPs, these approaches are particularly important because many particles originate from the degradation of everyday plastic products rather than deliberate disposal. Design innovations that minimize polymer shedding, promote

Microplastics in Aquatic Environments. https://doi.org/10.1016/B978-0-443-45619-0.00009-1

biodegradable alternatives, and extend product lifespan can significantly reduce MP generation before it enters aquatic systems. Equally important is the transition from linear waste management to closed-loop material flows. Improved recycling efficiency, extended producer responsibility, and material recovery systems reduce environmental plastic loads and limit secondary MP formation. When aligned with monitoring data, circular economy interventions can be targeted toward high-risk sectors such as textiles, packaging, tire wear and coatings.

Technological and policy solutions are most effective when supported by informed and engaged communities. Public awareness initiatives improve understanding of MP sources, pathways, and impacts, encouraging behavioral changes such as reduced plastic consumption, improved waste generation, and support for sustainable products. Community-based monitoring programs further enhance data collection while fostering environmental stewardship. Educational campaigns, and stakeholder participation create feedback loops between science and society. These mechanisms help ensure that mitigation strategies remain socially acceptable, culturally relevant, and adaptable to local conditions, particularly in regions where regulatory enforcement alone may be insufficient.

Policy instruments remain essential for scaling MP mitigation beyond voluntary action. Effective frameworks combine regulatory controls on plastic production and use with economic incentives that promote sustainable alternatives. Vans on selected single use plastics, standards for polymer labeling, and incentives for circular design are increasingly being adopted across regions. However, policy effectiveness depends on access to reliable, high-resolution data, integrating AI-supported monitoring systems into regulatory frameworks allows policymakers to identify pollution hotspots, assess intervention outcomes, and adapt regulations based on real-time evidence. Such adaptive governance models are particularly valuable for addressing diffuse and transboundary MP pollution.

Artificial Intelligence represents a transformative tool for both MP detection and mitigation planning. AI-based image analysis, spectral classification, and machine learning algorithm significantly reduce the time and subjectivity associated with traditional identification methods. These tools enable automated processing of large datasets from water, sediment, biota, and atmospheric samples, improving detection accuracy across diverse environmental matrices. Beyond detection, AL supports predictive modeling of MP transport, accumulation and source attribution. By integrating environmental data, hydrodynamics, and land use information. AI systems can forecast contamination trends and evaluate the effectiveness of mitigation strategies. This predictive capacity is critical for prioritizing interventions and aligning technological solutions with circular economy and policy goals.

This book chapter is novel and distinct from existing literature in three keyways. First, it moves beyond single discipline perspectives by integrating technology, policy, circular economy principles, and societal engagement into a unified mitigation framework. Second, rather than treating, prevention, and governance. Third, the chapter emphasizes actionable pathways, translating scientific advances into practical strategies relevant for regulators, industries, and communities. Unlike many published reviews that focus either on MP occurrence or detection methods, this chapter explicitly addresses how mitigation can be operationalized across scales. Thos integrative approach makes the chapter particularly valuable for interdisciplinary researchers, policymakers, and sustainability practitioners seeking solutions rather than isolated diagnostics.

11.2 Circular economy approaches to plastic waste reduction

11.2.1 Plastic and microplastic waste management

Managing plastic and micro plastic waste has become a global priority, and researchers agree that meaningful progress requires action at three different levels. First, society must greatly reduce its dependence on disposable plastic items such as shopping bags, straws, cutlery, takeaway containers, and bottled water. Many regions have already introduced restrictions or bans on these products, and early evidence shows that such policies can significantly decrease plastic litter in the environment. Second, countries need stronger and more reliable waste collection and recycling systems. A large portion of plastic pollution originates not from improper disposal from individuals, but from waste escaping during storage, transport, or landfilling. Improving recycling infrastructure, increasing public access to waste services, and adopting batter sorting technologies can prevent plastics from entering rivers, oceans, and soils. Traditional recycling methods where plastics are shredded, melted, and reshaped are useful but limited, highlighting the need for more efficient and widespread systems. Third, long-term solutions depend on scientific innovation. Researchers are exploring ways to break plastics down into their basic molecular units so they can be rebuilt into new materials, creating a true circular economy (Shen et al., 2019). Biodegradable plastics are also under development, but most currently degrade only under controlled industrial conditions with high temperatures. Recent studies have shown the widespread presence of MP in the environment and emphasized the need for stronger policies that prevent their release (Park and Kim, 2019). This chapter discusses a range of strategies and technologies that support the safe collection, treatment, and disposal of MP waste, helping reduce the burden of plastic pollution on ecosystems and human health.

11.2.2 Bio-circular green economy for microplastic control

The concept of a bio-circular green economy (BCGE) unites three sustainability frameworks Circular Economy (CE), Green Economy (GE), and Bio Economy (BE) to create an integrated pathway for environmental protection, economic resilience, and social well-being. Although each idea has developed independently, their combined application has gained significant momentum since early 2000s, especially withing academic research and global policy initiatives. In this integrated model, the Green Economy serves as an overarching framework that encompasses circular resource use, biological productivity, and nature-based solutions. GE emphasizes the value of ecological systems and recognizes the role of local communities through eco-tourism, environmental education, and socially inclusive development. The Bio Economy focuses on replacing fossil-based materials with renewable biological resources and using biotechnological innovations to transform biomass into chemicals, materials, fuels, and high-valued products. However, BE can only achieve true sustainability when its strategies incorporate circular design principles rather simply substituting biological feedstocks for conventional plastics (D' Amato et al., 2017).

Globally, research interest in CE, GE and BE has expanded rapidly. China has emerged as a leading contributor to CE literature, reflecting its national push for waste reduction and recycling policies implemented after 2008. Similarly, a significant rise in CE publications followed major international events such as the 2015 UN climate change conference and the European Union's circular economy action plan. Research on the green economy remains strongest in Europe but is gaining traction across the United States, Chine and developing nations, especially after UNEP's influential GE report in 2011. The Bioeconomy continuous to evolve, with growing attention to biosecurity, rural development, and the use with boarder GE priorities, including landscape restoration and nature-based climate solution, such as replacing invasive species with native varieties or applying bioremediation in agricultural systems (Pfau et al., 2014).

In the context of MP pollution, the biocircular economy (BCE) framework offers a promising roadmap for long-term prevention and mitigation. Addressing MPs requires coordinated actions across political, economic, and social sectors. Key strategies include redesigning plastic products for durability and recyclability, strengthening regulations on production and intentional MP additives, promoting biodegradable alternatives, and improving wastewater treatment to prevent MP release. Higher plants, algae, and cyanobacteria have the potential to function as biological factories for producing biodegradable plastics, while also storing carbon in biomaterial-based infrastructure an important contribution to climate-positive design. Although the BCE concept has existed for nearly 2 decades, its application to MP control is still emerging, and strict policy enforcement will be essential for meaningful progress (Gillespie and Philp, 2013).

11.2.3 A bio-circular green approach addresses microplastic pollution through several complementary routes

(a) Biodegradable plastics: Developing truly biodegradable polymers that mineralize into harmless products can reduce MP generation, provided they do not fragment into smaller particles during degradation.

(b) Advanced recycling technologies: New chemical and mechanical recycling processes are improving the recovery of plastics including MPs keeping them within the production loop and out of the environment.

(c) Consumer awareness: Public education plays a crucial role. By choosing products made from recycled materials, avoiding unnecessary plastics, and supporting circular design, consumers help reduce demand for virgin plastics and the associated MP burden.

Overall, the BCGE represents a holistic and forward-looking framework for MP reduction. By uniting renewable resource use, circular product design, sustainable policy development, technological innovation, and social engagement, this model provides a robust foundation for building a future in which MP pollution is significantly reduced and ecosystems are better protected.

11.2.4 Recycling of plastics and its role in reducing microplastic pollution

Recycling remains one of the most practical and widely promoted strategies for reducing plastic pollution and preventing the formation of MPs. In modern plastic recycling facilities, discarded plastics are first sorted by polymer type, then shredded, washed, and processed into small granules or pellets that can be used to manufacture new products. Although recycling reduces waste, the mechanical steps involved such as cutting, friction, and abrasion can unintentionally generate MPs. These tiny particles often end up in wash water, which may be discharged if not properly treated, contributing to environmental contamination. Despite these challenges, recycling remains essential given the enormous volume of plastic waste produced globally. Between 1950 and 2017, approximately 7000 million tons of plastic waste was generated; however, only about 700 million tons was recycled, while the majority was either incinerated or sent to landfills (Geyer, 2020).

Recycling offers several benefits, including conserving natural resources, reducing dependence on landfills, minimizing plastic leakage into the environment, and lowering greenhouse gas emissions. Most plastics used in consumer products are thermoplastics, which can be melted and reshaped, making mechanical recycling the most economical option. This method also has a lower carbon footprint compared to other treatment pathways. In closed-loop or primary recycling systems, materials are recycled back into products of similar quality PET beverage bottles are a well-known example, and PET bottle to bottle recycling is one of the few closed loop processes that has reached commercial scale success. In contrast, open loop or secondary recycling, often called downcycling, results in materials of lower quality with each cycle; PET bottles, for instance are commonly converted into textiles such as fleece fabrics rather than new bottles (Shen et al., 2010).

Chemical or feedstock recycling, also known as tertiary recycling, breaks polymers down into their fundamental building blocks, such as monomers or fuels, which can be used to create new plastics or chemical products. This method can handle mixed and heavily contaminated plastic waste that is unsuitable for mechanical recycling. Countries like Japan have adopted chemical recycling extensively, integrating it alongside mechanical methods as part of a broader national waste management strategy. Overall, while recycling alone cannot solve the global plastic crisis, improving recycling technologies, increasing recycling rates, and minimizing the release of MPs during processing are crucial steps toward creating a more sustainable and circular plastics economy.

11.2.5 Policy concerns and recommendations for addressing microplastic pollution

Despite growing global awareness of MP pollution, current policy responses remain fragmented, limited in scope, and often disconnected from the full life cycle of plastics. Much of the scientific literature highlights the severity of MP contamination, yet few studies translate these findings into actionable policy pathways. Effective long-term management of plastic waste requires well-designed regulatory frameworks that not only reduce plastic generation but also support innovation, strengthen waste systems, and minimize ecological and human health impacts. A major challenge is that existing policies often lack standardized tools for monitoring progress or evaluating environmental outcomes, especially regarding MPs in aquatic ecosystems, food systems, and human health. In many countries, regulations have historically focused on MPs, with MPs receiving attention only in recent years.

Several nations have begun implementing targeted policies. In 2015, the United States introduced the Microbead-Free Waters Act, banning the manufacture and sale of rinse-off cosmetics containing microbeads (Excell, 2018). Canada followed with extensive public consultations on plastic waste management and, by 2018, prohibited the production and import of personal-care products formulated with microbeads. That same year, the European Commission launched its first EU Plastics Strategy, which aims for all plastic packaging placed on the European market to be recyclable by 2030 and sets a target of recycling 55% of plastics waste. Currently, only about 30% of Europe's plastic waste is recycled, illustrating the scale of the challenge. Similar policy developments have emerged in Asia, Oceania, and South America, Highlighting growing recognition of the issue.

For policies to be genuinely effective, they must move beyond symbolic bans and become part of coherent national and international strategies. This requires the inclusion of measurable indicators to assess MP exposure, environmental fate, and removal efficiency. Policy frameworks must also ensure that waste collection, transport, storage, and final disposal or recycling is technically feasible, economically viable, environmentally sound, and socially acceptable. Importantly, governments should promote a transition toward low carbon- and circular economic models, integrating sustainable design, extended producer responsibility, improved recycling infrastructure, incentives for biodegradable alternatives, and stronger enforcement mechanisms. Such coordinated and science driven policy measures are essential for reducing MP pollution and safeguarding ecosystems and public.

11.2.6 Short-term policy standards for microplastic control

Short-term policy actions play a crucial role in immediately reducing MP inputs into the environment. One of the most important steps is strengthening waste management systems so that plastic waste is properly collected, transported, and disposed of before it can enter waterways. This includes routine improvements in municipal waste handling and stricter oversight of disposal practices. Another urgent action is the prohibition of microbeads and microfibers in personal care and hygiene products, including items such as dental floss and exfoliating cosmetics, which are direct contributors to MP pollution.

Industrial regulation is equally essential. Manufactures producing plastic pellets, micro-sized materials, or products prone to releasing MPs must follow strict protocols to prevent leaks during production, storage, and transport. Industries should be required to monitor emissions, adopt best practices to minimize loss of plastic material, and implement systems for recycling and reusing waste generated on-site. Scientific support is also needed in the short term. Funding should be made available for research on the ecological and health impacts of MPs, as well as for capacity-building programs aimed at informal waste pickers and recyclers. Strengthening their knowledge and operational safety can greatly enhance solid waste recovery and reduce leakage into the environment.

Transparent product labeling is another immediate step. Clearly marking items that contain MPs enables consumers to make informed, environmentally responsible choices. Additionally, governments and local bodies can support targeted cleanup efforts in highly contaminated areas especially coastal regions by using mechanical and chemical recycling technologies, installing sea bins and trash racks in waterways, improving stormwater management systems, and expanding wastewater treatment capacity. Public participation through voluntary cleanup campaigns also helps build awareness and community engagement (Fig. 11.1). Finally, short-term efforts must include widespread consumer education. Awareness campaigns can inform the public about the risks associated with MPs, encourage reduction in plastic use, and promote responsible disposal practices.

11.2.7 Long-term policy standards for microplastic control

Long-term strategies focus on systemic and sustainable changes needed to significantly reduce MP generation and prevent future pollution. Establishing continuous monitoring networks is a critical first step. These systems should track MP levels, evaluate policy effectiveness, and guide modifications to existing regulations. As noted by UNEP (2019a,b), over 67 countries have already adopted bans on single-use plastics, demonstrating the effectiveness of long-term regulatory measures. Extended Producer Responsibility (EPR) schemes must be enforced to hold manufacturers accountable for the full life cycle of their products from design and production to disposal and recycling. Strong EPR policies encourage companies to create more sustainable materials, develop take-back programs, and reduce plastic use overall. Educational integration is another essential long-term measure. Incorporating lessons on MPs, waste reduction, and environmental stewardship into school curricula can foster environmentally conscious behavior from an early age.

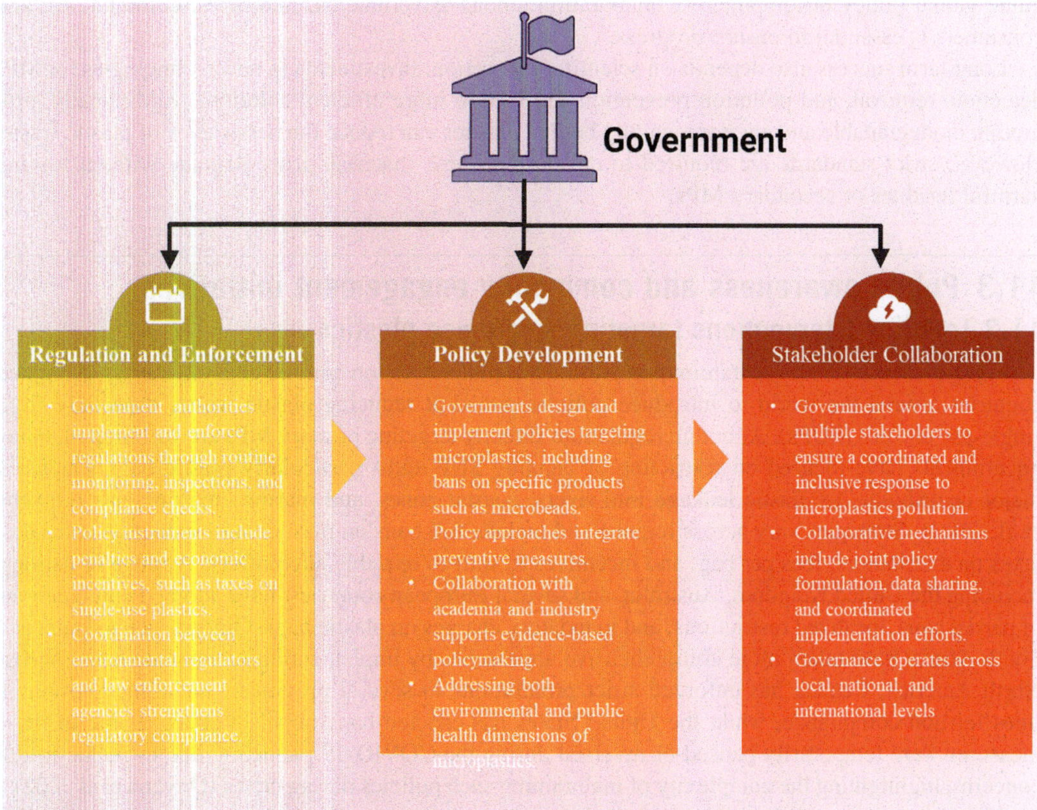

FIGURE 11.1

Schematic overview of governmental roles and interventions in microplastics management. The figure highlights regulatory formulation and enforcement, including monitoring, inspections, and economic instruments, alongside collaborative governance involving stakeholders across local, national, and international levels. It emphasizes integrated policy development, cross-sector collaboration with academia and industry, and coordinated actions addressing both environmental and human health impacts of microplastics.

Investing in advanced wastewater treatment systems is also crucial. Technologies capable of filtering MPs from sewage and industrial effluents such as membrane bioreactors, tertiary filtration, and adsorption systems can significantly reduce MPs entering aquatic environments. Changing consumer behavior remains a long-term priority. Encouraging sustained preference for reusable, recyclable, and plastic-free products helps reduce overall demand for virgin plastics. This transition must be supported by accessible alternatives and effective public outreach. International Cooperation is vital because MP pollution does not respect national boundaries. Countries must participate in global agreements to prevent unintentional releases of MPs into ocean and to regulate plastic waste trade. A zero-waste philosophy promoting reuse, repair, and recycling across all stages of a product life should guide global policy development. Collaboration among governments, industries, communities, and consumers is essential to ensure progress.

Long-term success also depends on scientific innovation. Investments in new technologies for MPs detection, removal, and pollution prevention can lead to more efficient solutions. Additionally, promoting biodegradable and environmentally benign plastics can reduce the persistence of plastic waste. However, strict standards are required to ensure that these materials truly degrade without leaving harmful residues or secondary MPs.

11.3 Public awareness and community engagement initiatives

11.3.1 Policy interventions targeting single-use plastics

Growing evidence of MP contamination and its harmful effects on aquatic ecosystems has prompted countries around the world to introduce policies aimed at reducing plastic waste, especially from single-use plastics (SUPs). Although regulations directly targeting primary MPs are still limited, many governments have focused on restricting or banning SUP items particularly plastic bags which are major contributors to plastic leakage into rivers, coastal zones, and marine environments (Onyena et al., 2021). Strategies vary across regions and include full bans on plastic bag production and sales, taxes on distributors, fees per bag, and incentives to shift to reusable alternatives. A range of nations, including the United Kingdom, Australia, and several European countries, have adopted combinations of partial bans, localized restrictions, and mandatory charges on plastic bags. Others, such as India and South Africa, have taken more comprehensive approaches by implementing nationwide bans. Across North America, Canada has enforced strict prohibitions in several provinces, including substantial fines for non-compliance, while the United States has enacted bans in select states. Some of these measures were temporarily paused or reversed during the COVID-19 pandemic due to public health concerns, highlighting the complexity of maintaining such policies during crises (Onyena et al., 2021).

Elsewhere, nations in Latin America and Asia have also pursued ambitious targets. Colombia, for example, set goals to reduce plastic bag consumption by 80% within a short timeframe and phase them out entirely soon after, while Buenos Aires Province in Argentina remains a regional leader with a complete ban in commercial markets (Poortinga et al., 2013). China's introduction of a plastic bag fee in 2008 drastically reduced usage in supermarkets, although illegal distribution continues to pose challenges. Similarly, Australian states have independently adopted their own bans and programs, such as South Australia's "Zero Waste" initiative, which helped cut hundreds of millions of bags annually (Block, 2013). Despite positive outcomes, the effectiveness of SUP policies varies significantly, with reductions in usage ranging widely from 33% to as high as 96%. In some countries including New Zealand policies have been introduced but their long-term impacts are still emerging. To broaden the scope beyond bags, several governments are also moving to restrict other commonly littered SUP items. A notable example is Canada's decision in 2021 to classify certain SUPs as toxic under national environmental law, enabling bans on hard-to-recycle food containers, checkout bags, cutlery, and other high-pollution products identified through coastal clean-ups (Onyena et al., 2021).

Although progress is visible, SUPs remain widely manufactured due to their convenience, low cost, and lack of readily available alternatives in certain sectors. For policy interventions to effectively curb MP pollution, governments must regularly review existing regulations, close enforcement gaps, and promote sustainable, accessible alternatives. Comprehensive, well-coordinated policies supported by public awareness campaigns and industrial innovation are essential to significantly reduce the flow of single-use plastics into the environment.

11.4 Regulatory frameworks implemented by international and regional organizations

Global concern over the rapid increase in plastic waste has prompted governments and international bodies to enact regulations aimed at limiting the production, use, and release of plastics into the environment. Over the last 3 decades, a wide range of legislative frameworks has been introduced, reflecting a growing awareness of the environmental and health risks posed by plastic pollution. Despite these efforts, many governance approaches still rely heavily on bans, taxes, and voluntary initiatives, while standardized protocols for managing MPs are largely absent. Studies emphasize that limited community participation, insufficient citizen science programs, and weak co-management strategies continue to hinder effective monitoring and remediation of MPs. Strengthening legislation through a coordinated, closed-loop approach one that reshapes consumer behavior, promotes eco-design, improves recycling, and limits landfill waste remains essential for advancing sustainable plastic management.

One major gap in existing policies is the limited consideration of MP contamination in agricultural systems. MPs enter farmlands through sewage sludge, plastics-coated fertilizers, and irrigation, yet regulations addressing these pathways remain minimal. The European Union has acknowledged sewage sludge as a major source of MP contamination in soils and has called for updates to earlier directives to incorporate MPs related risks and strengthen wastewater treatment standards. A revised version of Directive 86/278/EEC is expected to address safe sludge reuse, regulate potentially toxic elements associated with MPs, and ensure sustainable agricultural practices (Milojevic and Cydzik–Kwiatkowska, 2021).

At the global level, the United Nations (UN) has played a critical role in raising awareness of plastic pollution and its effects on marine ecosystems. According to the UN sustainable Developments Goals (SDGs) report (2021), over three billion people depend on oceans for their livelihoods, yet ocean health continues to decline due to plastic waste, warming, acidification, and overfishing. Marine dead zones have nearly doubled within a decade, and nearly half of key marine biodiversity areas lack adequate protection. MPs are linked to 13 of the 17 SDGs, with SDG 14 (Life Below Water) directly addressing marine pollution. However, progress has been slow, largely due to the absence of globally standardized indicators and limited investment only about 1.2% of national budgets, on average directed toward ocean research (UN, 2021).

Growing alarm over marine debris led the United Nations Environment Assembly (UNEA) to adopt several landmark resolutions during its fourth session in 2019. Resolution UNEP/EA.4/Res.6 urges member states to prevent and reduce plastic pollution through sustainable production, circular economy principles, and environmentally sound waste management. It emphasizes international cooperation and improved science-policy communication to better understand the ecological impacts of MPs. A companion resolution focusing on single-use plastics (UNEP/EA.4/Res.9) highlights the global failure to manage SUPs, noting that only 9% of all plastic ever produced has been recycled. Without intervention, an estimated 12 billion metric tons of plastic waste may accumulate in the environment by 2050. The resolution encourages nations to develop policies that restrict SUPs, promote alternatives, and improve public education. A third resolution (UNEP/EA.4/Res.1) links these efforts to broader sustainability goals, particularly SDG 12, promoting responsible production and consumption through circular economy thinking (Usman et al., 2022).

In parallel, organizations such as the International Union for Conservation of Nature (IUCN), the UN Environment Program (UNEP), and the Life Cycle Initiative have developed tools to help countries identify "plastic pollution hotspots." The National Guidance for Plastic Pollution Hot spotting and Shaping provides a standardized framework to assess plastic leakage, prioritize interventions, and design targeted policies across the plastic life cycle. The World Health Organization (WHO) has also emphasized the need to evaluate the health risks of MPs in water, highlighting the importance of improving drinking water and wastewater treatment systems. Although advanced treatment processes can remove up to 90% of MPs, many regions still lack access to adequate water and sanitation infrastructure, leaving populations vulnerable to both plastic pollution and associated pathogens (WHO, 2020).

Within Europe, Directive (EU, 2019/904) targets the environmental and health impacts of single-use plastics through bans, reduction measures, and penalties for non-compliance (The European Parliament, 2019). While it does not directly regulate MPs, it encourages manufacturers to decrease their use in products and adopt circular design principles. In Southeast Asia, the Association of Southeast Asian Nations (ASEAN) has introduced a regional action plan for 2021–25 to tackle marine debris. Supported by the World Bank's PROBLUE initiative, this plan focuses on reducing plastic waste generation, improving recycling systems, enhancing monitoring capacity, and building value chains for reuse. By harmonizing regional standards and strengthening investment and training, ASEAN aims to significantly reduce the flow of plastics and MPs into marine environments.

11.4.1 Critiques and potential ways forward

Policies aimed at reducing plastic pollution particularly measures targeting single-use plastic bags represent an important step toward strengthening environmental governance. However, despite these advances, many interventions suffer from weak statutory backing, insufficient enforcement, and poor compliance, especially in developing nations (Onyena et al., 2021). These shortcomings contribute to the continued and widespread release of plastics and MPs into coastal and marine ecosystems. Although national and regional initiatives such as bans, taxes, and international resolutions attempt to curb pollution, the absence of clear, standardized guidelines for regulating plastic discharge continues to hinder effective mitigation.

Numerous studies have documented the ecological and biological risks associated with MPs, reinforcing the urgent need for dedicated policies to address their proliferation (Nuamah et al., 2023). These findings make it evident that current policy approaches although helpful are insufficient to eliminate MPs from global aquatic systems. Persistent challenges include inadequate funding for ocean science, limited public participation, fragmented legal frameworks, and the lack of standardized indicators for monitoring plastic pollution. Even at the global level, major initiatives fall short; for example, although the United Nations recognizes the plastic crisis across 11 Sustainable Development Goals, only a single indicator directly measures marine plastic pollution. In addition, several influential directives, including those issued by the European Parliament, still overlook MP particles despite growing evidence of human exposure (Usman et al., 2022). Many of the World Health Organization's recommendations also remain unrealistic for low-income and developing countries.

Collectively, these issues demonstrate the need for substantial investment in research and the creation of practical, measurable tools for addressing the MPs crisis.

A major barrier to progress is the limited involvement of local communities in coastal regions, where public awareness of MPs remains low. Insufficient understanding of MP sources, impacts, and mitigation strategies often results in poorly executed bans and weak local compliance. Public education plays a central role in shifting behavior; studies reveal that awareness is extremely low in many regions for instance, only 12% of participants in Bangladesh and 26% in China had prior knowledge of MPs (Masud et al., 2024). Yet, once informed about the potential health risks, most individuals expressed strong concern and a willingness to adopt sustainable practices. Effective outreach through workshops, community forums, school programs, and media campaigns can encourage reductions in plastic use, greater participation in recycling efforts, and support for sustainable alternatives. Local initiatives such as coastal clean-ups, recycling drives, and community monitoring further reinforce environmental stewardship.

Government-led action remains a critical pillar of MP mitigation. Financial incentives such as deposit refund schemes for bottles and containers can significantly influence consumer behavior and promote reuse. Investments in waste handling, recycling systems, and modern treatment infrastructure are equally important. Across the world, numerous regions have demonstrated the power of strict legislation. For example, by 2017, 271 U S. municipalities had enacted plastic bag policies, most of which combined bans with fees (Wagner, 2017). India has taken decisive steps to phase out single-use plastics by 2022, while several European nations, including the Netherlands, the UK, and Germany, have enacted stringent restrictions (Onyena et al., 2021). In 2020, Jakarta implemented a comprehensive ban requiring retailers to offer environmentally friendly alternatives. China's efforts, including early bans and nationwide bag fees, resulted in dramatic reductions in plastic bag consumption. Across Africa, legislative momentum has been even stronger: 34 countries now enforce SUP restrictions, with Rwanda, Uganda, and Kenya among the earliest adopters. Ireland's levy on plastic carrier bags, supported by public education campaigns, cut consumption sharply and earned broad public support (Convery et al., 2007). Deposit-refund programs in the United States and northern Europe have similarly reduced plastic bottle waste (Cornago et al., 2021).

Collaborative approaches integrating governments, civil society, non-governmental organizations, and private industries are essential to advance sustainable materials, alternative packaging, and policy reforms. Research highlights the benefits of stakeholder engagement and co-design processes for ensuring effective implementation of single-use plastic bans (Adam et al., 2020). A forward-looking model is the contributory co-management framework, which emphasizes shared responsibility across stakeholders. As illustrated in Fig. 11.1, co-management assigns complementary roles to government agencies, industry actors, scientific experts, local communities, and consumer groups. This framework has been successfully applied to plastic waste regulation, fisheries management, coastal habitat restoration, and marine monitoring. For example, Rwanda's plastic bag ban benefited significantly from community-led initiatives and multi-stakeholder cooperation (Ogutu et al., 2023). The EU Marine Strategy Framework Directive similarly encourages cross-border cooperation to address marine litter (Maruf et al., 2024). Co-management efforts in Lofoten, Norway, show how the active participation of fishing communities can lead to successful resource governance (Jentoft and Kristoffersen, 1989). In contrast, limited coordination and inadequate government support, as seen in Sri Lanka's mangrove management programs, have hindered success (Nijamdeen et al., 2023). However, positive examples such as Greece's co-management structures for coastal development highlight the potential for integrating ecological conservation with local economic development through ecotourism and diversified livelihoods (Kyvelou et al., 2023).

The importance of capacity building is evident across these examples; practical training, education, and public–private partnerships such as those utilized in Seoul demonstrate how local initiatives can be strengthened to support nationwide waste management goals (Lee-Geiller and Kütting, 2021). Local governments play a central role in achieving Sustainable Development Goal 14 by reducing coastal pollution, improving climate resilience, supporting sustainable fisheries, and promoting blue economy initiatives. Clearly defining the responsibilities of each stakeholders fosters shared accountability and greatly enhances the impact of mitigation efforts (Fig. 11.2).

11.5 Artificial intelligence in microplastic detection and mitigation

11.5.1 AI-driven innovations in microplastic detection and pollution control

The rapid advancement of artificial intelligence (AI) has opened a transformative pathway for improving the efficiency, accuracy, and scalability of MP detection. As MP pollution intensifies worldwide, AI technologies are emerging as essential tools for monitoring contamination, accelerating analysis, and supporting global mitigation efforts. An overview of the current research landscape reveals significant growth: by July 31, 2024, over 1600 studies addressing MP detection (Bertoldi et al., 2021) and more than 100 focusing specifically on AI-based detection methods had been published (Web of Science search; keywords: MPs and detection, MPs and detection, machine learning or deep learning (Fig. 11.3). This rising interest reflects AI's growing role in reshaping environmental monitoring systems.

Traditional MP analysis often reliant on manual sorting, visual interpretation, and labor-intensive spectroscopy faces multiple challenges, including time constraints, analyst bias, and the difficulty of processing large sample volumes. AI integration helps overcome these limitations. For instance, computer-vision-assisted techniques can identify, segment, and quantify MPs far more efficiently than human observers. One study demonstrated the use of partial least squares with interval selection models to recognize MP images, thereby reducing analysis time and minimizing subjective interpretation (Bertoldi et al., 2021).

Deep learning approaches have shown particularly strong promise. Convolutional neural networks (CNNs), including architectures such as U-Net, can automatically detect and classify microbeads and various MP shapes in complex microscopic images (Fig. 11.3). These tools can process data in batches, significantly reducing manual input while maintaining high accuracy (Lee et al., 2022). More recently, advanced deep learning models such as VGG16, UNet, and MultiResUNet have demonstrated exceptional performance in identifying MPs in scanning electron microscopy (SEM) images, offering improved prediction capabilities and robust segmentation (Hu et al., 2024).

Beyond image recognition, machine learning (ML) greatly enhances spectral analysis. Environmental samples often generate massive Fourier-transform infrared (FTIR) and Raman spectral datasets that require careful interpretation. ML algorithms can rapidly process these data, extract key polymer signatures, and accurately classify plastic types present in environmental samples (Choi et al., 2024). In Raman spectroscopy, ML has been employed to enhance weak signals, streamline spectral interpretation, and accelerate the identification of MPs in complex real-world matrices (Sunil et al., 2024).

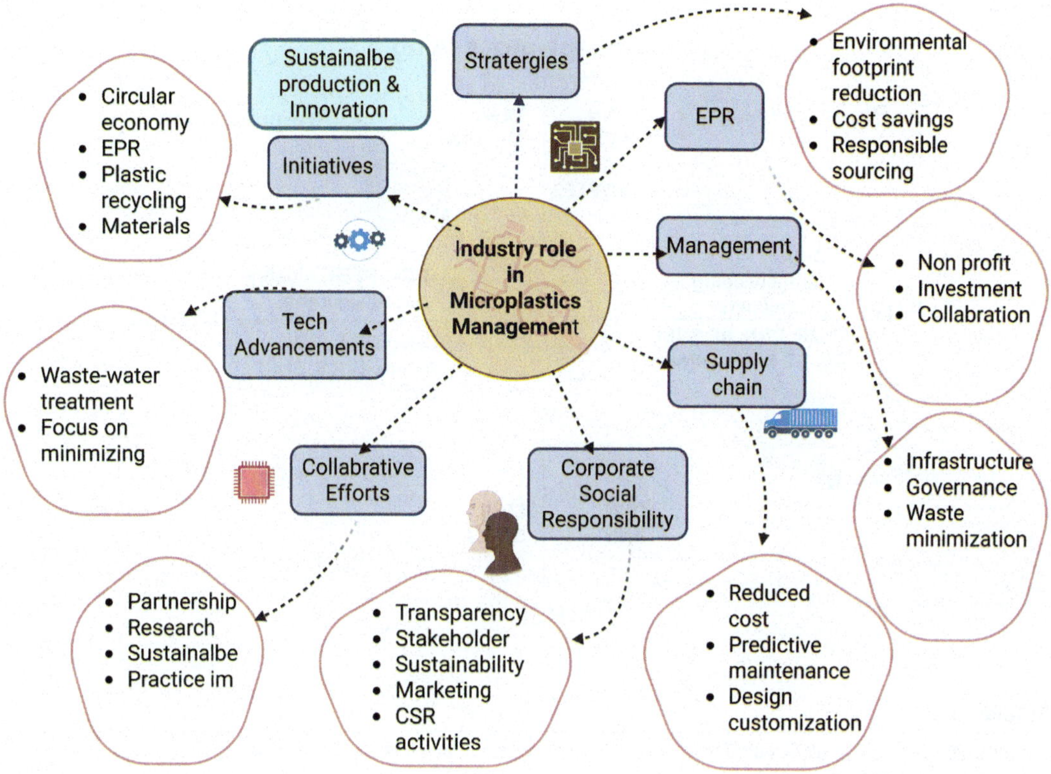

FIGURE 11.2

Conceptual illustration of a triple-helix framework highlighting the role of industry in microplastics management. The model integrates sustainable production and innovation, supply-chain management, and corporate social responsibility (CSR), emphasizing initiatives such as extended producer responsibility, circular economy practices, bio-based material adoption, and technological advancements. It also underscores collaboration with governments, non-profit organizations, and academia, alongside investments in infrastructure, transparency, and stakeholder engagement to reduce environmental footprints and microplastic generation.

As environmental monitoring generates increasingly large and complex datasets, ML provides powerful tools for automating classification and simplifying analytical workflows.

AI technologies are also being applied beyond detection to advance pollution control and MP source tracking. Predictive modeling offers a new dimension in environmental diagnostics: using historical and spatial data, AI can forecast pollution origin, transport pathways, and distribution patterns. This capability improves the reliability of MP traceability, addressing long-standing challenges associated with conventional tracking methods (Kida et al., 2024). By integrating hydrodynamic models, land-use data, and environmental variables, AI models can identify pollution hotspots and inform targeted mitigation strategies.

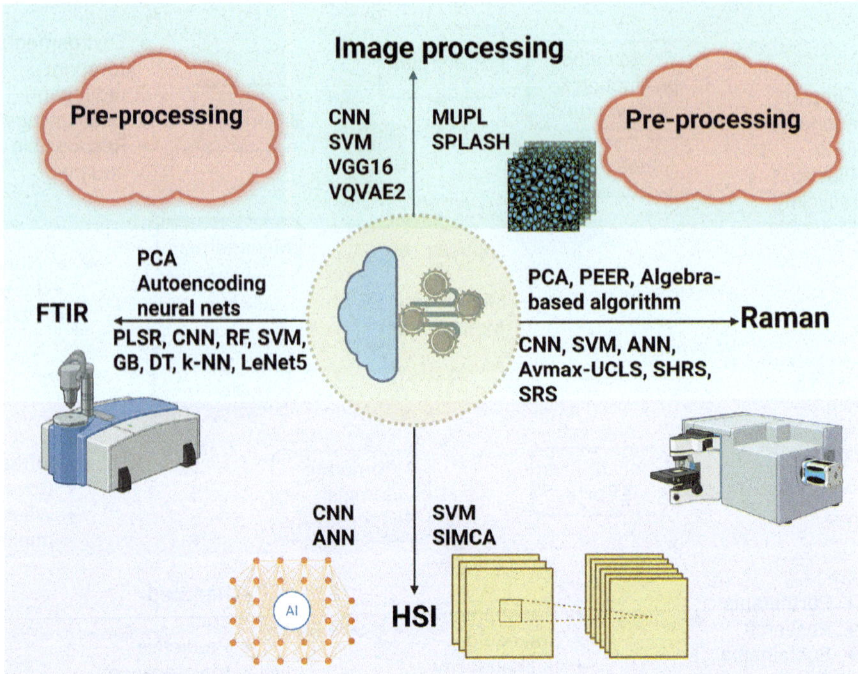

FIGURE 11.3

Schematic overview of artificial intelligence–assisted microplastic detection workflows, integrating spectroscopic (FTIR, Raman) and imaging-based approaches with hyperspectral imaging (HSI). The figure highlights key data pre-processing steps and machine learning algorithms used for feature extraction, classification, and identification of microplastics across different analytical platforms.

More broadly, embedding AI into environmental monitoring represents a critical step toward a more sustainable and responsive pollution-management system. AI-supported detection enables rapid, automated analysis of MPs in water, sediments, soil, and even atmospheric samples, helping scientists and policymakers better understand the scale and implications of MP contamination. The ability to aggregate, interpret, and visualize large datasets enhances our capacity to identify emerging trends, evaluate risks, and design informed policies.

This chapter provides a comprehensive exploration of these emerging frontiers, highlighting how machine learning and deep learning techniques contribute to MP detection, environmental assessment, and pollution mitigation. It also examines practical applications of AI in real-world monitoring systems and discusses how AI-driven data preprocessing can strengthen the accuracy and consistency of MP analyses. As the global community intensifies its response to plastic pollution, AI technologies will continue to shape the future of MP research, offering innovative and scalable solutions for safeguarding environmental and human health.

11.5.2 Artificial intelligence assisted image processing for microplastic detection

Artificial intelligence has become a cornerstone technology in the advancement of MP detection, particularly through its ability to analyze and interpret large volumes of MP image data. Machine learning (ML) and deep learning (DL), the two most widely applied branches of AI in this field, enable automated recognition of MP particles by learning from labeled image datasets. By training these models on thousands of images, researchers can accurately identify a wide range of MP types including fibers, fragments, films, and microbeads with high precision (Fig. 11.4).

One notable example is the deep learning model Mask R–CNN, applied by Han et al. (2023), which can simultaneously locate, segment, classify, and count MPs in both laboratory and field images. In their study, MP samples were collected using sieves of 4.76 and 0.85 mm, photographed under consistent indoor and outdoor conditions, and then processed into image tiles using Python. The trained Mask R–CNN achieved a classification accuracy of 93.3% under controlled laboratory

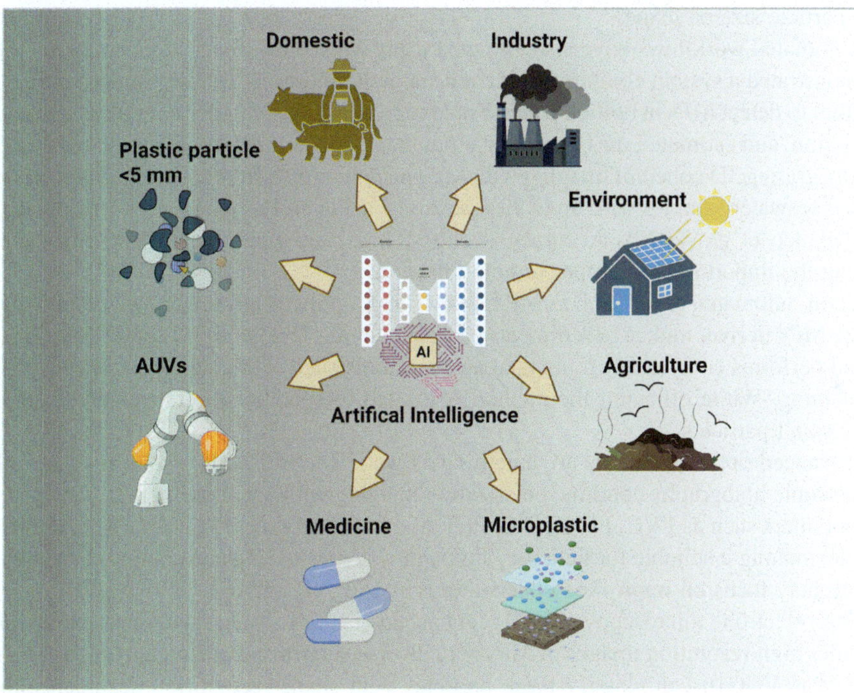

FIGURE 11.4

Conceptual illustration of the application of artificial intelligence (AI) in microplastics research and management. The schematic shows how AI-based tools support source identification of microplastics across domestic, industrial, environmental, medical, and agricultural sectors, as well as pollution assessment and real-time monitoring through AI-driven autonomous underwater vehicles (AUVs). This integrated approach highlights the role of intelligent systems in improving detection efficiency, source tracking, and continuous environmental surveillance of microplastic contamination.

lighting, although performance dropped slightly by about 15% when applied to natural outdoor backgrounds. Despite this reduction, the model could still reliably detect and categorize MPs in real-world environments, demonstrating strong potential for rapid field-based assessments.

A different AI-enhanced approach combines vector-quantized variational autoencoder (VQVAE2) feature learning with optical-resolution photoacoustic microscopy. In this technique, MPs are first separated and imaged using a photoacoustic microscope, producing high-resolution images that are then interpreted by a DL model (Huang et al., 2024). This system improves processing speed and classification accuracy, though it still relies on conventional preprocessing steps like traditional microscopy methods.

Other researchers have developed fully automated or semi-automated imaging systems to streamline MP detection. For instance (Zhu et al., 2021) introduced a lightweight DL network called CompNet, built upon the holographic classifier CNN architecture and capable of generating digital holograms using partially coherent light. Tested on both synthetic and real beach samples, CompNet successfully identified the number of MPs in each image and reduced model training time by approximately 72%. However, its performance is limited to small and unbalanced datasets and cannot determine particle size or shape.

Fully automated workflows have also emerged using ML-based image segmentation (Vitali et al., 2024). demonstrated a system combining Nile red staining, fluorescence microscopy, and a supervised ML algorithm to detect MPs in bottled water. Their approach automatically quantified particle number, size distribution, and estimated mass, offering a practical tool for high-throughput analysis. Similarly, a method integrating 3D coherent imaging with ML can distinguish MPs from algae and other particles in pretreated seawater samples with up to 99% accuracy using an H-support vector machine (H-SVM) classifier (Bianco et al., 2020). Accuracy declines, however, when larger impurities are present, underscoring the importance of proper sample filtration.

Other semi-automated tools, such as MP Lab (MUPL), apply traditional image processing to detect and analyze MPs in river and cave sediments (Giardino et al., 2023). MUPL processes images within seconds and performs comparably to manual analysis tools such as ImageJ, especially when aided by Nile red staining. While efficient, the method is limited to specific size ranges and cannot reliably detect very small particles.

More advanced optical methods are emerging as well. The SPLASH platform smart polarization and spectroscopic holography combine polarization imaging with ML classifiers to accurately identify common polymers such as PVC, PP, PS, PC, PET, and PMMA (Hu et al., 2024). SPLASH is compact and portable, making it suitable for field use, although water scattering can compromise image quality. Future work may focus on improving underwater scattering algorithms.

Scanning electron microscopy (SEM) enhanced by DL offers another powerful option. SEM provides high-resolution images of MPs with diverse shapes and chemical compositions, and DL models such as VGG16 can classify these particles with shape-recognition accuracies above 98% (Shi et al., 2022). Although highly accurate, SEM imaging remains slow and labor-intensive, and developing automated SEM workflows and larger MP datasets will be essential for wider adoption.

Across all these techniques, AI significantly improves the speed, accuracy, and scalability of image-based MP detection. Nevertheless, challenges remain model performance depends heavily on the quality and diversity of training datasets, preprocessing steps are often labor-intensive and distinguishing between visually similar particles such as synthetic fibers, natural fibers, and organic debris-remains difficult. Despite these limitations, the integration of AI with imaging technologies represents one of the most promising avenues for achieving rapid, high-throughput, and consistent monitoring of MP pollution at a global scale.

11.5.3 Artificial intelligence assisted FTIR for microplastic identification

Fourier transform infrared (FTIR) spectroscopy remains one of the most widely used analytical tools for identifying polymer types in MP samples. Its ability to generate characteristic spectral fingerprints makes it indispensable in environmental MP research. However, interpreting FTIR spectra is traditionally a labor-intensive task that requires expert knowledge, especially when dealing with environmental samples that contain spectral noise, weathered polymers, or mixed contaminants (Veerasingam et al., 2021). The integration of artificial intelligence (AI) has transformed this landscape by enabling rapid, automated, and highly accurate interpretation of large spectral datasets.

AI models can extract meaningful features from complex spectra, classify polymer types, and even quantify MP content far more efficiently than manual interpretation. For example, combining FTIR with partial least squares regression (PLSR) allows for pre-screening large sediment datasets for plastics such as LDPE and PET, significantly reducing the time required for manual review (Hahn et al., 2019). More advanced applications involve machine learning (ML) trained on extensive spectral libraries including open-source databases, commercially acquired polymers, and environmental samples (Liu et al., 2023). demonstrated this by evaluating a suite of ML algorithms, such as one-dimensional convolutional neural networks (1D-CNNs), random forests (RF), and decision trees (DT), achieving up to 97.44% accuracy using 1D-CNNs. RF models performed particularly well when spectral datasets were limited or contained noise, highlighting that different ML architectures excel under different conditions.

To enhance robustness, researchers have begun to adopt ensemble learning approaches. One such model integrates 7 ML algorithms and uses a neural network to determine the best-performing classifier, resulting in improved identification of imbalanced and contaminated MP datasets (Yan et al., 2022). These ensemble strategies help overcome challenges such as unknown polymer types and uncertainty in selecting a single optimal model.

AI is also improving both qualitative and quantitative MP analysis. Quaternion-based feature extraction, which fuses original FTIR spectra with their first and second derivatives, has been successfully paired with quaternion principal component analysis (QPCA) and support vector machines (SVMs) to enhance classification accuracy (Tan et al., 2023). Although powerful, this method is less suited for small sample sizes or cases involving a large variety of polymer types. For aged environmental MPs, which often exhibit altered surface chemistry, ML models paired with attenuated total reflection FTIR (ATR-FTIR) have proven highly effective. Studies using RF, gradient boosting, DT, and k-nearest neighbor (k-NN) have achieved up to 99% accuracy for PET MPs (Enyoh and Wang, 2024), while convolutional neural networks (CNNs) have shown even greater capability in identifying weathered MPs with minimal preprocessing (Zeng et al., 2024).

Preprocessing of FTIR spectra particularly noise reduction is another area where AI excels. Spectral artifacts arising from instrument drift, environmental noise, or sample heterogeneity can significantly hinder accurate detection. Integrating ML with preprocessing techniques such as baseline correction, principal component analysis (PCA), and oversampling helps reduce noise and improve classification performance. For instance, employing ML based denoising prior to classification enabled support vector classification (SVC) models to achieve accuracies of 95.99%, while deep learning models like LeNet5 reached nearly 97% (Thar et al., 2023). Autoencoder neural networks have also been successfully used to reconstruct distorted spectra and eliminate artifacts, dramatically improving data quality without manual intervention (Brandt et al., 2021).

Micro-FTIR (μFTIR) imaging adds another dimension by providing both chemical and spatial information about MPs. AI models trained on μFTIR images can rapidly identify polymer types and measure particle size and abundance across environmental samples. A random forest model developed by Hufnagl et al. (2019) detected five polymer types within minutes and was later expanded to identify over 20 types without increasing analysis time (Hufnagl et al., 2022). This demonstrates the scalability of AI-assisted μFTIR imaging for high-throughput environmental monitoring.

Overall, the integration of AI with FTIR spectroscopy has significantly enhanced the speed, accuracy, and automation of MP analysis. These techniques reduce dependency on expert interpretation, improve the handling of noisy or aged spectra, and enable rapid processing of large sample sets. Despite these advancements, challenges persist particularly when analyzing very small particles, highly weathered polymers, or complex environmental matrices. Future progress will depend on expanding spectral databases, improving preprocessing strategies, developing more adaptive ML models, and integrating multimodal analytical tools. Nonetheless, AI-assisted FTIR represents a major step toward standardized, scalable, and highly reliable MP detection across diverse environmental contexts.

11.5.4 Artificial intelligence assisted Raman spectroscopy for microplastic and nano plastic detection

Raman spectroscopy is an essential tool for identifying MPs and nano plastics (NPs), offering precise molecular information through vibrational fingerprints. However, its effectiveness is sometimes limited by challenges such as weak signal intensity, baseline drift, and resolution constraints imposed by laser diffraction (Yang et al., 2023). To overcome these limitations, artificial intelligence (AI) and advanced Raman based technologies have increasingly been integrated into analytical workflows, significantly improving detection accuracy, speed, and robustness.

One of the most promising developments is spatial heterodyne Raman spectroscopy (SHRS), which enhances signal-to-noise ratios and enables rapid identification of common polymers such as PP, PC, PS, and HDPE (Xue et al., 2022). SHRS relies on specialized spectral processing including interference correction, baseline removal, and spectral calibration and can be fully automated through optimized preprocessing algorithms. This makes SHRS a strong candidate for high-throughput environmental monitoring.

AI also plays a pivotal role in improving nano plastic detection, an area were traditional Raman imaging struggles due to particle size and environmental interference. Machine learning (ML) models trained on nanoscale Raman spectra-such as random forest (RF) have demonstrated accuracies above 98% in both laboratory datasets and environmental samples (Xie et al., 2023). Similarly, hyperspectral stimulated Raman scattering (SRS) imaging combined with data-driven spectral matching allows rapid single-particle analysis and successful classification of nanoscale polymer types (Qian et al., 2024). These advancements underscore the increasing value of AI-enhanced Raman technologies for resolving extremely small particles that previously eluded detection.

For complex environmental samples containing heterogeneous particles and interfering signals, advanced preprocessing methods are essential. The PEER (peak extraction and retention) algorithm effectively isolates relevant Raman features and improves the quality of low-intensity nanoparticle spectra. In addition, algorithms based on principal component analysis (PCA) and algebraic reconstruction enable improved deconvolution of Raman images and enhance the signal-to-noise ratio (Luo et al., 2023). By integrating Raman data with scanning electron microscopy (SEM), these approaches also enable parallel evaluation of particle morphology and chemical composition.

Deep learning (DL) has further transformed Raman spectral interpretation. Multi-layer 1D convolutional neural networks (1D-CNNs) can classify MP polymers with accuracy levels approaching 97%, outperforming traditional ML models while reducing the need for extensive preprocessing (Qin et al., 2024). These networks automatically extract key spectral features, streamline data processing, and simplify analysis workflows. Additional innovations include combining convolutional neural networks with average mapping spectrum (AMS) techniques, which help correct baseline drift and enable quantitative assessment of MPs in various aquatic environments (Luo et al., 2024). This advancement addresses a major limitation of Raman-based methods, which often struggle to provide accurate quantitative data.

Raman spectroscopy coupled with multivariate analysis also offers practical benefits for field applications. Using PCA-LDA for dimensionality reduction and SVM for classification, researchers have demonstrated rapid automatic identification of MPs in environmental matrices, though nanoscale detection remains challenging due to strong background interference (Jin et al., 2022). New imaging approaches, such as the AVmax-UCLS hyperspectral Raman method, provide fast identification of particle type and shape and show strong resilience to low similarity data typical of aged environmental MPs (Li *et al.*, 2024a). This technique requires minimal preprocessing and may hold significant potential for large-scale, real-world monitoring.

Beyond MPs, hyperspectral Raman imaging combined with artificial neural networks (ANNs) has shown great potential for rapid, high-accuracy detection. Although its performance decreases for particles smaller than 10 μm, future improvements in laser power and detector sensitivity could expand its applicability (Sim et al., 2024). Overall, AI-driven Raman techniques dramatically improve MP detection by enhancing spectral clarity, enabling faster classification, and reducing reliance on manual interpretation.

In summary, the combination of AI and Raman spectroscopy represents a transformative step forward in micro- and nano plastic analysis. Techniques such as SHRS, SRS imaging, 1D-CNN classification, and multivariate analysis have significantly improved the resolution, speed, and reliability of both qualitative and quantitative assessments. Despite ongoing challenges associated with complex sample matrices, spectral noise, and particle heterogeneity, continued development of AI algorithms, improved spectral libraries, and optimized imaging systems will further strengthen the role of Raman techniques in environmental monitoring. As research progresses, these technologies are expected to become integral tools for detecting a broader range of MP mixtures especially those involving smaller and more diverse particle sizes ultimately enabling more accurate assessments of plastic pollution in real-world environments.

11.5.5 Application of artificial intelligence in microplastic pollution control

Artificial intelligence (AI) is rapidly emerging as a transformative tool for understanding, monitoring, and mitigating MP pollution. As MP sources and pathways become increasingly complex, conventional tracing methods often struggle to provide accurate and timely information. AL overcomes many of these limitations by revealing hidden patterns, predicating pollutant behavior, and linking MP distributions to their origins. This capability is crucial for designing targeted mitigations strategy and informing evidence based environmental policies.

Traditional source-tracing methods such as cluster analysis (CA), principal component analysis (PCA), redundancy analysis (RDA), and absolute principal component score multiple linear regression (APCS-MLR) have long been used to estimate potential pollutant sources. While effective under controlled circumstances, these approaches require strict data assumptions and may generate substantial uncertainties in heterogeneous natural environments (Mohan and Lakshmanan, 2023). In contrast, machine learning (ML) can process complex, nonlinear datasets, allowing more reliable interpretation of the factors influencing MP abundance and distribution.

AI-driven models now play a significant role in identifying potential MP sources. For example, ML classifiers including decision trees (DT), random forests (RF), and support vector machines (SVM) have been used alongside PCA to distinguish between multiple pollution sources with high accuracy, offering new insights into MP origins (Wu et al., 2024). Artificial neural networks (ANNs) have similarly been used to simulate the migration and transformation of plastic particles. Early ANN models show strong correlation with experimental results, although continuous improvements are needed to incorporate broader environmental conditions beyond idealized laboratory settings (Kida et al., 2024).

AI also excels in predicting the spatial extent and intensity of MP pollution. Machine learning algorithms such as RF, SVM, long short-term memory (LSTM) networks, and least-squares SVM have achieved high predictive accuracy when applied to long-term environmental datasets (Tran et al., 2023). These models can classify pollution severity, evaluate ecological risk, and identify emerging hotspots. Additionally, near-infrared sensors integrated with ML particularly when combined with transfer learning algorithms such as manifold embedded distribution alignment (MEDA) provide a cost-effective option for mapping soil MP contamination in real time (Qiu et al., 2020).

Beyond source identification and prediction, AI contributes to real-time monitoring and pollution control. The integration of deep-ultraviolet Raman spectroscopy with AI enables continuous monitoring of water quality parameters including MPs, nitrates, and pharmaceuticals providing a more comprehensive understanding of environmental conditions (Post et al., 2021). Real-time ML analytics further help track MP movement, identify transport pathways, and determine pollutant fate, enabling tailored remediation strategies for specific ecosystems (Phan and Luscombe, 2023).

Advanced robotics equipped with AI including autonomous underwater vehicles (AUVs) and aerial drones are revolutionizing MP surveillance across marine, freshwater, and terrestrial systems (Watanabe et al., 2019). These platforms utilize cameras, optical sensors, and machine learning algorithms to map MP concentrations over large areas with minimal human intervention. Deep learning models such as YOLOv8, which performs rapid object detection under changing underwater lighting conditions, can identify MPs in real time with high precision (Shah et al., 2023). Although current models are trained on relatively small datasets, ongoing development and expanded training libraries will enable broader application in natural environments.

AI also enhances environmental decision-making by automating large-scale data processing and accelerating policy-relevant analyses. However, these tools depend heavily on the availability of comprehensive, high-quality datasets a persistent limitation in MP research. Challenges also arise from the need for robust algorithms that can adapt to diverse environmental matrices and account for noise, variability, and incomplete data. As a result, ongoing refinement of ML and deep learning (DL) algorithms remains essential.

Despite these challenges, AI has demonstrated immense potential to reshape MP pollution management. Integrated AI frameworks can analyze complex environmental datasets, detect subtle pollution trends, and automate time-consuming processes such as image classification and spectral analysis. When combined with remote sensing, Raman spectroscopy, and near-infrared imaging, AI provides unprecedented insight into MP behavior from source to sink.

Ultimately, the future of MP pollution control lies in multidisciplinary collaboration, where AI technologies complement traditional environmental science. Continued development of low-cost, scalable, and environmentally friendly AI systems will be essential for accurate monitoring, predictive modeling, and effective mitigation across diverse ecosystems. By addressing challenges related to data quality, model generalization, and real-world deployment, AI can become a cornerstone technology in global efforts to reduce MP pollution and safeguard ecological and human health.

11.5.6 Machine learning techniques for microplastic quantification

Machine learning (ML) has become an essential analytical tool for identifying, classifying, and quantifying MPs across diverse environmental samples. These techniques can process large, complex datasets generated from spectroscopy, microscopy, and chemical imaging, thereby improving both accuracy and efficiency compared with traditional manual interpretation. Depending on the task such as polymer classification, toxicity prediction, or exploratory analysis ML methods can be broadly grouped into supervised and unsupervised learning approaches.

11.5.6.1 Supervised learning methods

Supervised learning models, including support vector machines (SVM) and artificial neural networks (ANN), rely on labeled datasets where each spectrum or particle is assigned a known polymer type, morphology, or concentration. These models excel at classification and quantification tasks, often delivering high precision when sufficient high-quality annotated data is available. Because supervised algorithms learn directly from examples, they are extremely effective for:

- Polymer type identification
- Prediction of MP toxicity or behavior
- Quantification of MPs in complex mixtures

However, their performance is tightly linked to the quality, balance, and representativeness of the training dataset. Creating and curating these annotated datasets often requires significant labor, technical skill, and time.

11.5.6.2 Unsupervised learning methods

Unsupervised learning techniques, such as K-means clustering, hierarchical cluster analysis (HCA), and dimension-reduction methods like principal component analysis (PCA), operate on unlabeled datasets. These methods are valuable for:

- Detecting unknown or unexpected MP morphologies
- Grouping spectra with similar features
- Identifying patterns, trends, or anomalies without prior assumptions

Although unsupervised methods facilitate exploration of complex datasets and assist in revealing hidden structure, they typically offer lower interpretability and precision compared to supervised models, especially when dataset variability is high.

11.5.7 General considerations in machine learning model development

Machine learning algorithms learn by processing training data much like students absorb information through lessons. Their performance, however, cannot be predicted in advance. For this reason, researchers often train multiple algorithms simultaneously, comparing accuracy, precision, and robustness to determine the most suitable model. Most ML models require the optimization of hyperparameters training and architectural settings that govern model complexity, regularization, and performance. Techniques such as cross-validation, as recommended by (Back et al., 2022), are used to fine-tune hyperparameters and prevent overfitting.

ML also plays a key role in automated spectral enhancement, enabling the correction of noise, baseline drift, and instrumental artifacts. Autoencoder-based approaches, for example, can reconstruct clean spectra from distorted signals (Brandt et al., 2021).

11.5.7.1 Federated, local, and swarm learning

Emerging approaches such as federated learning (FL) and swarm learning (SL) offer new possibilities for secure, scalable MP detection:

- Local learning trains models on individual datasets but suffers from poor generalizability across regions or laboratories.
- Centralized learning aggregates all data to a single server, providing highly accurate models but raising concerns regarding privacy, data sharing, and regulatory compliance.
- Federated learning distributes the training process across multiple devices or institutions while keeping raw data local. Only model updates are shared, allowing data privacy and scalability.
- Swarm learning builds upon federated learning by incorporating blockchain technology, enhancing data integrity and trust among distributed collaborators, and offering a secure framework for large-scale MP identification.

11.5.7.2 Unsupervised machine learning techniques

Unsupervised learning is especially valuable when analyzing MP datasets without prior knowledge of polymer types or spectral patterns. The goal is to uncover natural groupings within the data based on shared spectral or morphological characteristics. This is particularly helpful for complex mixtures where MPs co-occur with biological matter, sediments, or chemical contaminants.

***K-means Clustering*:** K-means clustering is one of the most widely used unsupervised algorithms in environmental MP research. It partitions data into k clusters by minimizing the distance between each data point and its assigned centroid. The choice of k is critical:

- Larger k values generate finer, more specific clusters
- Smaller k values produce broader groups that may overlook subtle distinctions

The algorithm begins by selecting k initial centroids, assigning each spectrum or particle to the nearest centroid, recalculating cluster centers, and iterating until assignments stabilize.

A practical example can be seen in the work of (Wander et al., 2020), who applied K-means to FTIR spectra derived from atmospheric MP samples. Their results showed that the algorithm identified broad material categories such as natural cellulose fibers, synthetic polymers (polyamide, polyester, polyacrylic, polypropylene), and animal-derived fibers like wool or silk. However, K-means lacked the resolution to differentiate between closely related polyester variants (PET, PBT, PTT), demonstrating that while useful, clustering alone may not fully resolve fine spectral differences among chemically similar polymers.

11.5.8 Decision trees and random forests

Decision trees (DTs) are among the most intuitive and widely used machine learning algorithms for classification tasks, including MP identification. They operate by recursively splitting a dataset into smaller subsets based on feature thresholds that best separate the classes. Each internal "node" represents a decision rule, while the terminal "leaf" nodes represent classification outcomes. Although DTs are easy to interpret and computationally efficient, they tend to generate deep and overly complex trees when trained on large datasets. This often results in overfitting, where the model performs well on training data but poorly on new, unseen samples (Deepnarain et al., 2019).

To overcome these limitations, Random Decision Forests (RDFs)-commonly referred to as random forests combine the predictions of many individual decision trees. Each tree is trained on a random subset of the data and features, and the final classification is determined through majority voting. This ensemble approach significantly improves predictive accuracy, reduces overfitting, and increases robustness to noise. Random forests have shown exceptional performance in MP research, particularly when applied to FTIR spectral datasets. Reported sensitivities range from 94% to 100%, demonstrating their ability to distinguish subtle differences in spectroscopic signatures (Hufnagl et al., 2022).

RDFs have been applied successfully across a wide range of environmental samples-such as sediment, soil, water, sea salt, plankton, and compost highlighting their adaptability to complex and heterogeneous matrices (Weisser et al., 2021). One study achieved 95.45% classification accuracy when identifying 17 classes of synthetic polymers, natural fibers, and background materials, underscoring the model's capability to handle high variability and overlapping spectra.

Random forest approaches have also been tested using Raman spectroscopy. A recent study reported flawless classification within a small test set, indicating strong potential for Raman-based MP analysis (Vinay Kumar et al., 2021). However, broader validation with larger and more diverse datasets is needed before universal adoption. Overall, random forests offer a scalable, interpretable, and highly accurate solution for automating MP identification across spectroscopic platforms. Fig. 11.3 illustrates how DTs and RDFs integrate into the broader MP detection workflow from spectral preprocessing to feature extraction and final classification.

11.5.9 Artificial neural networks

Artificial Neural Networks (ANNs), inspired by the information-processing architecture of the human brain, have become increasingly influential in environmental science applications, including hydrological prediction, pollution monitoring, and wastewater modeling (Yang et al., 2019). ANNs consist of interconnected nodes (neurons) that transform input data by applying weighted sums and nonlinear activation functions. These networks can learn highly complex relationships but often require careful design, extensive training data, and substantial computational resources. Without proper regularization and validation, ANNs are prone to overfitting, like deep decision trees.

11.5.10 Convolutional neural networks (CNNs)

Convolutional Neural Networks represent a specialized class of ANNs designed for image-based analysis. By processing images through localized pixel filters, CNNs can automatically extract color, shape, texture, and spatial features. Their strength lies in reducing the need for manual feature engineering while achieving high accuracy. CNNs have outperformed traditional multilayer perceptron's in polymer identification tasks using Raman, SERS, and optical spectra. Several studies highlight their capabilities:

- MP detection in soil: A CNN achieved 78.5% accuracy when classifying plastics from visible/NIR soil spectra (Ng et al., 2020).
- Spectral enhancement: Autoencoder networks have successfully improved FTIR and Raman signal quality, helping downstream classification models perform more reliably (Brandt et al., 2021).
- FTIR classification: Although ANNs performed strongly in ATR-FTIR classification, some researchers favored SVMs due to ANN computational demands (Back et al., 2022).

CNNs are also promising candidates for transfer learning, enabling pre-trained models to adapt quickly to new polymer types using only small additional datasets a valuable feature for laboratories with limited annotated data.

11.5.11 Image segmentation and microplastic measurement

Segmentation algorithms are crucial for isolating MPs from their background in microscopic images and for measuring characteristics such as fiber length, diameter, and morphology. Approaches originally developed for medical imaging such as U-Net and Multi Res UNet have been adapted to MP research (Lee and Jhang, 2021). These architectures can reliably detect irregular shapes but may struggle when image noise, atypical particle structures, or overlapping objects are present.

Diffraction-based sensing methods, paired with neural networks, offer another promising direction. By analyzing scattering patterns, networks can identify particles from a single diffraction image (Grant-Jacob et al., 2018). developed a system where scattering patterns were cropped, normalized, and fed into convolutional layers before classification, demonstrating that neural networks can determine both particle type and quantity with high accuracy (Fig. 11.3).

11.5.12 Deep learning in modern microplastic detection

Recent advances in deep learning, particularly CNN-based systems, have revolutionized object detection across fields from healthcare to materials science (Fujiyoshi et al., 2019). These models can often surpass human performance when trained on sufficiently large and well-annotated datasets. For MP identification:

- CNNs have been used to recognize and localize MP microspheres based on images of scattered laser light.

- Automated MP detection devices have been designed using CNNs and customized optical setups (Grant-Jacob et al., 2018).
- Deep models can extract subtle spectral, spatial, and morphological cues that traditional algorithms often miss.

However, training such models requires large and diverse datasets an ongoing challenge in MP research, where standardized sample libraries are still being developed.

Given the small size and irregular shapes of many MP particles (Lee and Jhang., 2021), evaluated U-Net and Multi Res UNet architectures for segmentation and found both methods promising for accurate fiber identification.

***Deep Learning*:** Deep learning has rapidly emerged as a transformative tool in MP research, enabling new levels of precision and automation across environmental science, materials science, and computer vision. Its success relies on the ability of neural networks to learn complex patterns directly from raw images and spectra, reducing dependence on handcrafted features and traditional rule-based methods (Montesinos Lopez et al., 2022). One notable advancement is the hybrid deep learning framework proposed by (Lorenzo-Navarro et al., 2021), which integrates U-Net for particle segmentation with VGG16 for classification. In this two-stage design, U-Net first isolates individual MP particles within an image, accurately delineating particle boundaries and removing background interference. Each segmented particle is then passed to a refined CNN classifier (VGG16), yielding highly reliable polymer identification.

This hybrid approach outperforms both conventional computer vision and manual inspection, demonstrating excellent accuracy even when applied to images captured using standard laboratory equipment. Its scalability and minimal hardware requirements make it particularly valuable for laboratories lacking access to advanced imaging systems. As such, deep learning is reshaping MP analysis, offering standardized, reproducible, and high throughput workflows capable of supporting global monitoring efforts.

11.5.13 Integration of image analysis with AI and machine learning

Artificial intelligence (AI) and machine learning (ML) are increasingly intertwined with imaging technologies to strengthen the detection, classification, and quantification of MPs in environmental samples and consumer products. Advanced imaging techniques such as fluorescence microscopy paired with dyes like Nile Red enhance the visibility of MP particles, while AI automates the subsequent classification steps (Vitali et al., 2024). demonstrated this by combining automated image processing with a Random Forest classifier to analyze fluorescent MPs in bottled water. Their workflow significantly improved detection accuracy, reduced subjectivity, and enabled rapid, repeatable quantification suitable for routine laboratory analysis.

The integration of imaging and AI/ML pipelines represents a shift toward scalable, data-driven MP monitoring. These methods streamline labor-intensive tasks, minimize human error, and provide consistent results across operators and sample types. As environmental monitoring and food safety assessments increasingly rely on standardized procedures, AI enhanced image analysis is positioned to become an essential tool for MP surveillance and regulatory compliance.

11.5.14 Machine learning techniques for microplastic identification and quantification

Machine learning has become indispensable for MP research, offering powerful approaches for analyzing both labeled and unlabeled data. Across unsupervised and supervised learning domains, Machine Learning techniques enable particle grouping, spectral interpretation, and polymer classification with far greater efficiency and accuracy than traditional methods.

Unsupervised learning techniques are particularly useful for exploratory analysis:

- K-means clustering enables fast grouping of particles but assumes spherical clusters and is sensitive to outliers (Ikotun et al., 2023).
- Hierarchical cluster analysis (HCA) reveals nested cluster structures without requiring a predefined cluster number, though it is computationally intensive for large datasets (Zhao et al., 2024).
- HDBSCAN handles variable-density clusters well but requires careful parameter optimization (Tran et al., 2021).
- Principal component analysis (PCA) reduces data dimensionality and highlights spectral variance, yet it may miss nonlinear relationships (Fang et al., 2022).
- UMAP preserves complex data structure for visualization but provides limited interpretability (Vermeulen et al., 2021).

Supervised learning methods excel in polymer classification, toxicity prediction, and MP quantification:

- Support Vector Machines (SVM) offer high accuracy but scale poorly with large datasets (da Silva et al., 2020).
- K-Nearest Neighbors (KNN) is simple and effective but sensitive to feature selection and noise (Ertel, 2017).
- PLS-DA performs well for spectral analysis but may overfit without robust training sets (da Silva et al., 2020).
- Decision trees provide interpretability yet can overfit, while random forests mitigate this through ensemble design and offer strong robustness (Deepnarain et al., 2019).
- Neural networks and deep learning models excel at extracting complex features from images and spectra but require extensive datasets and are often criticized for limited transparency (Yang et al., 2019).

When integrated with high-resolution imaging systems and standardized spectral libraries, these ML approaches can dramatically advance MP detection, supporting everything from environmental monitoring to regulatory assessments.

11.6 Summary and conclusion

MP contamination in aquatic systems demands solutions that are preventive, adaptive, and systems wide. Circular economy strategies reduce plastic leakage at its sources, while engagement ensures societal alignment with sustainability goals. Policy interventions provide the regulatory backbone needed to scale action, and artificial intelligence offers the analytical capacity to guide and evaluate these efforts. Together, these approaches form a resilient framework for mitigating MP by aligning technological innovation with circular design and evidence-based governance, sustainable reduction in MP contamination are achievable. Further efforts should focus on strengthening cross-sector collaboration, expanding AI- enabled monitoring networks, and embedding circular economy principles into global plastic management strategies.

References

Adam, I., Walker, T.R., Bezerra, J.C., Clayton, A., 2020. Policies to reduce single– use plastic marine pollution in West Africa. Mar. Pol. 116, 103928.

Back, H.de M., Vargas, E.C., Estevam Alarcon, O., Pottmaier, D., 2022. Training and evaluating machine learning algorithms for ocean microplastics classification through vibrational spectroscopy. Chemosphere 287 (Pt 1), 131903. https://doi.org/10.1016/j.chemosphere.2021.131903.

Bertoldi, C., Lara, L.Z., Gomes, A.A., Fernandes, A.N., 2021. Microplastic abundance quantification via a computer-vision-based chemometrics-assisted approach. Microchem. J. (160), 105690. https://doi.org/10.1016/j.microc.2020.105690.

Bianco, V., Memmolo, P., Carcagnì, P., Merola, F., Paturzo, M., Distante, C., Ferraro, P., 2020. Microplastic identification via holographic imaging and machine learning. Adv. Intell. Syst. 2 (2). https://doi.org/10.1002/aisy.201900153.

Block, B., 2013. China Reports 66– Percent Drop in Plastic Bag Use. Eye on Earth. Worldwatch Institute. Retrieved November, 29.

Brandt, J., Mattsson, K., Hassellöv, M., 2021. Deep learning for reconstructing low-quality FTIR and Raman Spectra—A case study in microplastic analyses. Anal. Chem. 93 (49), 16360–16368. https://doi.org/10.1021/acs.analchem.1c02618.

Choi, E., Choi, Y., Lee, H., Kim, J.-W., Oh, H.B., 2024. Development of a machine-learning model for microplastic analysis in an FT-IR microscopy image. Bull. Kor. Chem. Soc. 45 (5), 472–481. https://doi.org/10.1002/bkcs.12835.

Convery, F., McDonnell, S., Ferreira, S., 2007. The most popular tax in Europe? Lessons from the Irish plastic bags levy. Environ. Resour. Econ. 38 (1), 1–11. https://doi.org/10.1007/s10640-006-9059-2.

Cornago, E., Borkey, P., Brown, A., 2021. Preventing single-use plastic waste: implications of different policy approaches. OECD Environ. Work. Pap (182).

da Silva, V.H., Murphy, F., Amigo, J.M., Stedmon, C., Strand, J., 2020. Classification and quantification of microplastics (ESCAPED_CHARACTER_LESS_THAN100 μm) using a focal plane array–fourier transform infrared imaging system and machine learning. Anal. Chem. 92, 13724–13733. https://doi.org/10.1021/acs.analchem.0c01324.

Deepnarain, N., Nasr, M., Kumari, S., Stenström, T.A., Reddy, P., Pillay, K., Bux, F., 2019. Decision tree for identification and prediction of filamentous bulking at full-scale activated sludge wastewater treatment plant. Proc. Safety Environ. Protect. 126, 25–34. https://doi.org/10.1016/j.psep.2019.02.023.

D' Amato, D., Droste, N., Allen, B., Kettunen, M., Lähtinen, K., Korhonen, J., Leskinen, P., Matthies, B.D., Toppinen, A., 2017. Green, circular, bio economy: a comparative analysis of sustainability avenues. J. Clean. Prod. 168, 716–734. https://doi.org/10.1016/j.jclepro.2017.09.053.

Enyoh, C.E., Wang, Q., 2024. Automated classification of undegraded and aged polyethylene terephthalate microplastics from ATR-FTIR spectroscopy using machine learning algorithms. J. Polym. Environ. 32 (9), 4143–4158. https://doi.org/10.1007/s10924-024-03199-4.

Ertel, W., 2017. Mach. Learn. Data Min. 175–243.

Excell, C., 2018. Legal Limits on Single-Use Plastics and Microplastics: A Global Review of National Laws and Regulations. United Nations Environment Programme (UNEP).

Fang, C., Luo, Y., Zhang, X., Zhang, H., Nolan, A., Naidu, R., 2022. Identification and visualisation of microplastics via PCA to decode Raman spectrum matrix towards imaging. Chemosphere 286 (Pt 2), 131736. https://doi.org/10.1016/j.chemosphere.2021.131736.

Fujiyoshi, H., Hirakawa, T., Yamashita, T., 2019. Deep learning-based image recognition for autonomous driving. IATSS Res. 43 (4), 244–252. https://doi.org/10.1016/j.iatssr.2019.11.008.

Geyer, R., 2020. In: Letcher, T.M. (Ed.), Chapter 2-Production, Use, and Fate of Synthetic Polymers. https://doi.org/10.1016/B978-0-12-817880-5.00002-5.

Giardino, M., Balestra, V., Janner, D., Bellopede, R., 2023. Automated method for routine microplastic detection and quantification. Sci. Total Environ. 859 (Pt 2), 160036. https://doi.org/10.1016/j.scitotenv.2022.160036.

Gillespie, I.M.M., Philp, J.C., 2013. Bioremediation, an environmental remediation technology for the bioeconomy. Trends Biotechnol. 31 (6), 329–332. https://doi.org/10.1016/j.tibtech.2013.01.015.

Grant-Jacob, J.A., Mackay, B.S., Baker, J.A.G., Heath, D.J., Xie, Y., Loxham, M., Eason, R.W., Mills, B., 2018. Real-time particle pollution sensing using machine learning. Opt. Exp. 26 (21), 27237–27246. https://doi.org/10.1364/OE.26.027237.

Hahn, A., Gerdts, G., Völker, C., Vincent, N., 2019. Using FTIRS as pre-screening method for detection of microplastic in bulk sediment samples. Sci. Total Environ. 689, 341–346. https://doi.org/10.1016/j.scitotenv.2019.06.227.

Han, X.-L., Jiang, N.-J., Hata, T., Choi, J., Du, Y.-J., Wang, Y.-J., 2023. Deep learning based approach for automated characterization of large marine microplastic particles. Mar. Environ. Res. 183, 105829. https://doi.org/10.1016/j.marenvres.2022.105829.

Hu, B., Dai, Y., Zhou, H., Sun, Y., Yu, H., Dai, Y., Wang, M., Ergu, D., Zhou, P., 2024. Using artificial intelligence to rapidly identify microplastics pollution and predict microplastics environmental behaviors. J. Hazard Mater. 474, 134865. https://doi.org/10.1016/j.jhazmat.2024.134865.

Huang, M., Han, K., Liu, W., Wang, Z., Liu, X., Guo, Q., 2024. Advancing microplastic surveillance through photoacoustic imaging and deep learning techniques. J. Hazard Mater. 470, 134188. https://doi.org/10.1016/j.jhazmat.2024.134188.

Hufnagl, B., Stibi, M., Martirosyan, H., Wilczek, U., Möller, J.N., Löder, M.G.J., Laforsch, C., Lohninger, H., 2022. Computer-assisted analysis of microplastics in environmental samples based on μFTIR imaging in combination with machine learning. Environ. Sci. Technol. Lett. 9 (1), 90–95. https://doi.org/10.1021/acs.estlett.1c00851.

Ikotun, A.M., Ezugwu, A.E., Abualigah, L., Abuhaija, B., Heming, J., 2023. K-means clustering algorithms: a comprehensive review, variants analysis, and advances in the era of big data. Inf. Sci. 622, 178–210. https://doi.org/10.1016/j.ins.2022.11.139.

Jentoft, S., Kristoffersen, T., 1989. Fishermen's Co-management: the case of the lofoten fishery. Human Organ. 48 (4), 355–365. https://doi.org/10.17730/humo.48.4.k045637kv290u886.

Jin, N., Song, Y., Ma, R., Li, J., Li, G., Zhang, D., 2022. Characterization and identification of microplastics using Raman spectroscopy coupled with multivariate analysis. Anal. Chim. Acta 1197, 339519. https://doi.org/10.1016/j.aca.2022.339519.

Kida, M., et al., 2024. Assessment of machine learning-based methods predictive suitability for migration pollutants from microplastics degradation. J. Hazard Mater. 461, 132565.

Kyvelou, S.S., Ierapetritis, D.G., Chiotinis, M., 2023. The future of fisheries comanagement in the context of the sustainable blue economy and the green deal: there is no green without blue. Sustainability 15 (10), 7784.

Lee, J., Jhang, C.,K., 2021. A systematic protocol of microplastics analysis from their identification to quantification in water environment: a comprehensive review. J. Hazard Mater. 403, 124049. https://doi.org/10.1016/j.jhazmat.2020.124049.

Lee, K.S., Chen, H.L., Ng, Y.S., Maul, T., Gibbins, C., Ting, K.N., Amer, M., Camara, M., 2022. U-Net skip-connection architectures for the automated counting of microplastics. Neural Comput. Appl. 34 (9), 7283–7297. https://doi.org/10.1007/s00521-021-06876-w.

Lee-Geiller, S., Kütting, G., 2021. From management to stewardship: a comparative case study of waste governance in New York City and Seoul metropolitan city. Resour. Conserv. Recycl. 164, 105110.

Liu, Y., Yao, W., Qin, F., Zhou, L., Zheng, Y., 2023. Spectral classification of large-scale blended (Micro)Plastics using FT-IR raw spectra and image-based machine learning. Environ. Sci. Technol. 57 (16), 6656–6663. https://doi.org/10.1021/acs.est.2c08952.

Lorenzo-Navarro, J., Castrillon-Santana, M., Sanchez-Nielsen, E., Zarco, B., Herrera, A., Martínez, I., Gomez, M., 2021. Deep learning approach for automatic microplastics counting and classification. Sci. Total Environ. 765, 142728. https://doi.org/10.1016/j.scitotenv.2020.142728.

Luo, Y., et al., 2023. Detection of microplastics and nanoplastics released from a kitchen blender using Raman imaging. J. Hazard Mater. 453, 131403.

Luo, Y., et al., 2024. Quantitative analysis of microplastics in water environments based on Raman spectroscopy and convolutional neural network. Sci. Total Environ. 926, 171925.

Maruf, M., Chang, Y.C., Yang, L., 2024. Managing institutional interlinkages for the protection of marine environment in the East Asian seas region and beyond: the case of marine plastic pollution. Ocean Coast Manag. 255. https://doi.org/10.1016/j.ocecoaman.2024.107232.

Masud, A.A., Islam, A.R.M.T., Mamun, A.A., Alam, G.M.M., Arabameri, A., Bari, A.B.M.M., Pal, S.C., Rakib, M.R.J., Senapathi, V., Bodrud-Doza, M., Idris, A.M., Malafaia, G., 2024. A new approach from public behavioral attitudes and perceptions towards microplastics: influencing factors, and policy proposals. Ocean Coastal Manag. 253. https://doi.org/10.1016/j.ocecoaman.2024.107144.

Milojevic, N., Cydzik– Kwiatkowska, A., 2021. Agricultural use of sewage sludge as a threat of microplastic (Mp) spread in the environment and the role of governance. Energies 14 (19), 6293.

Mohan, K., Lakshmanan, V.R., 2023. A critical review of the recent trends in source tracing of microplastics in the environment. Environ. Res. 239 (Pt 2), 117394. https://doi.org/10.1016/j.envres.2023.117394.

Montesinos Lopez, O.A., Montesinos Lopez, A., Crossa, J., 2022. Fundamentals of Artificial Neural Networks and Deep Learning. Multivariate Statistical Machine Learning Methods for Genomic Prediction. Springer International Publishing, Cham, pp. 379–425.

Ng, W., Minasny, B., McBratney, A., 2020. Convolutional neural network for soil microplastic contamination screening using infrared spectroscopy. Sci. Total Environ. 702, 134723. https://doi.org/10.1016/j.scitotenv.2019.134723.

Nijamdeen, T.W.G.F.M., Ratsimbazafy, H.A., Kodikara, K.A.S., Nijamdeen, T.W.G.F.A., Thahira, T., Peruzzo, S., Hugé, J., 2023. Mangrove management in Sri Lanka and stakeholder collaboration: a social network perspective. J. Environ. Manag. 330, 117116.

Nuamah, F., Tulashie, S.K., Debrah, J.S., Pèlèbè, R.O.E., 2023. Microplastics in the Gulf of Guinea: an analysis of concentrations and distribution in sediments, gills, and guts of fish collected off the coast of Ghana. Environ. Res. 234. https://doi.org/10.1016/j.envres.2023.116567.

Ogutu, M.O., Akor, J., Mulindwa, M.S., Heshima, O., Nsengimana, C., 2023. Implementing circular economy and sustainability policies in Rwanda: experiences of Rwandan manufacturers with the plastic ban policy. Frontiers in Sustainability 4. https://doi.org/10.3389/frsus.2023.1092107.

Onyena, A.P., Aniche, D.C., Ogbolu, B.O., Rakib, M.R.J., Uddin, J., Walker, T.R., 2021. Governance strategies for mitigating microplastic pollution in the marine environment: a review. Microplastics 1 (1), 15–46.

Park, S.Y., Kim, C.G., 2019. Biodegradation of micro-polyethylene particles by bacterial colonization of a mixed microbial consortium isolated from a landfill site. Chemosphere 222, 527–533. https://doi.org/10.1016/j.chemosphere.2019.01.159.

Pfau, S.F., Hagens, J.E., Dankbaar, B., Smits, A.J.M., 2014. Visions of sustainability in bioeconomy research. Sustainability 6 (3), 1222–1249. https://doi.org/10.3390/su6031222.

Phan, S., Luscombe, C.K., 2023. Recent trends in marine microplastic modeling and machine learning tools: potential for long-term microplastic monitoring. J. Appl. Phys. 133, 020701.

Poortinga, W., Whitmarsh, L., Suffolk, C., 2013. The introduction of a single-use carrier bag charge in Wales: attitude change and behavioural spillover effects. J. Environ. Psychol. 36, 240–247. https://doi.org/10.1016/j.jenvp.2013.09.001.

Post, C., Brülisauer, S., Waldschläger, K., Hug, W., Grüneis, L., Heyden, N., Schmor, S., Förderer, A., Reid, R., Reid, M., Bhartia, R., Nguyen, Q., Schüttrumpf, H., Amann, F., 2021. Application of laser-induced, deep UV Raman spectroscopy and artificial intelligence in real-time environmental monitoring—solutions and first results. Sensors 21 (11), 3911. https://doi.org/10.3390/s21113911.

Qian, N., Gao, X., Lang, X., Deng, H., Bratu, T.M., Chen, Q., Stapleton, P., Yan, B., Min, W., 2024. Rapid single-particle chemical imaging of nanoplastics by SRS microscopy. Proc. Natl. Acad. Sci. 121 (3). https://doi.org/10.1073/pnas.2300582121.

Qin, Y., Qiu, J., Tang, N., He, Y., Fan, L., 2024. Deep learning analysis for rapid detection and classification of household plastics based on Raman spectroscopy. Spectrochim. Acta Mol. Biomol. Spectrosc. 309, 123854. https://doi.org/10.1016/j.saa.2024.123854.

Qiu, Z., Zhao, S., Feng, X., He, Y., 2020. Transfer learning method for plastic pollution evaluation in soil using NIR sensor. Sci. Total Environ. 740, 140118. https://doi.org/10.1016/j.scitotenv.2020.140118.

Shah, M., et al., 2023. Enhancing marine conservation: YOLOv8-based underwater waste detection system. In: 2023 3rd International Conference on Innovative Mechanisms for Industry Applications (ICIMIA), pp. 1396–1404.

Shen, L., Worrell, E., Patel, M.K., 2010. Open-loop recycling: a LCA case study of PET bottle-to-fibre recycling. Resour. Conserv. Recycl. 55 (1), 34–52. https://doi.org/10.1016/j.resconrec.2010.06.014.

Shen, M., Zhang, Y., Zhu, Y., Song, B., Zeng, G., Hu, D., et al., 2019. Recent advances in toxicological research of nanoplastics in the environment: a review. Environ. Pollut. 252, 511–521. https://doi.org/10.1016/j.envpol.2019.05.102.

Shi, B., Patel, M., Yu, D., Yan, J., Li, Z., Petriw, D., Pruyn, T., Smyth, K., Passeport, E., Dwayne Miller, R.J., Howe, J.Y., 2022. Automatic quantification and classification of microplastics in scanning electron micrographs via deep learning. Sci. Total Environ. 825, 153903. https://doi.org/10.1016/j.scitotenv.2022.153903.

Sim, W., Song, S.W., Park, S., Jang, J.I., Kim, J.H., Cho, Y.-M., Kim, H.M., 2024. Unveiling microplastics with hyperspectral Raman imaging: from macroscale observations to real-world applications. J. Hazard Mater. 463, 132861. https://doi.org/10.1016/j.jhazmat.2023.132861.

Tan, A., Zhao, J., Zhao, Y., Li, X., Su, H., 2023. Determination of microplastics by FTIR spectroscopy based on quaternion parallel feature fusion and support vector machine. Chemometr. Intell. Lab. Syst. 243, 105018. https://doi.org/10.1016/j.chemolab.2023.105018.

Thar, A.S., et al., 2023. A comparative study of machine learning and deep learning models for microplastic classification using FTIR spectra. In: 2023 18th International Joint Symposium on Artificial Intelligence and Natural Language Processing (iSAINLP), pp. 1–6.

Tran, T. H., Cao, T. D., Tran, T. T. H., 2021. HDBSCAN eval. Perform. Hierarchical clust. Big Data 273–283.

Tran, H.-T., et al., 2023. Machine learning approaches for predicting microplastic pollution in peatland areas. Mar. Pollut. Bull. 194, 115417.

UNEP, 2019a. Innovative Pathways to Achieve Sustainable Consumption and Production (UNEP/EA.4/Res.1); UN Doc. United Nations, New York, NY, USA, pp. 1–5.

UNEP, 2019b. Marine Plastic Litter and Microplastics (UNEP/EA.4/Res.6); UN Doc. United Nations, New York, NY, USA, pp. 1–4.

United Nations, (UN). (2021). The sustainable development goals report. https://unstats.un.org/sdgs/report/2021/. (Accessed on 22 August 2021).

Usman, S., Abdull Razis, A.F., Shaari, K., Azmai, M.N.A., Saad, M.Z., Mat Isa, N., Nazarudin, M.F., 2022. The burden of microplastics pollution and contending policies and regulations. Int. J. Environ. Res. Publ. Health 19 (11). https://doi.org/10.3390/ijerph19116773.

Veerasingam, S., Ranjani, M., Venkatachalapathy, R., Bagaev, A., Mukhanov, V., Litvinyuk, D., Mugilarasan, M., Gurumoorthi, K., Guganathan, L., Aboobacker, V.M., Vethamony, P., 2021. Contributions of Fourier transform infrared spectroscopy in microplastic pollution research: a review. Crit. Rev. Environ. Sci. Technol. 51 (22), 2681–2743. https://doi.org/10.1080/10643389.2020.1807450.

Vermeulen, M., Smith, K., Eremin, K., Rayner, G., Walton, M., 2021. Application of uniform manifold approximation and projection (UMAP) in spectral imaging of artworks. Spectrochim. Acta Mol. Biomol. Spectrosc. 252, 119547. https://doi.org/10.1016/j.saa.2021.119547.

Vinay Kumar, B.N., Löschel, L.A., Imhof, H.K., Löder, M.G.J., 2021. Christian Laforsch, Analysis of microplastics of a broad size range in commercially important mussels by combining FTIR and Raman spectroscopy approaches. Environ. Pollut. 269, 116147. https://doi.org/10.1016/j.envpol.2020.116147.

Vitali, C., Peters, R.J.B., Janssen, H.-G., Undas, A.K., Munniks, S., Ruggeri, F.S., Nielen, M.W.F., 2024. Quantitative image analysis of microplastics in bottled water using artificial intelligence. Talanta 266, 124965. https://doi.org/10.1016/j.talanta.2023.124965.

Wagner, T.P., 2017. Reducing single-use plastic shopping bags in the USA. Waste Manag. 70, 3–12. https://doi.org/10.1016/j.wasman.2017.09.003.

Wander, L., Vianello, A., Vollertsen, J., Westad, F., Braun, U., Paul, A., 2020. Exploratory analysis of hyperspectral FTIR data obtained from environmental microplastics samples. Anal. Methods 12 (6), 781–791. https://doi.org/10.1039/c9ay02483b.

Watanabe, J.I., Shao, Y., Miura, N., 2019. Underwater and airborne monitoring of marine ecosystems and debris . SPIE, Japan. J. Appl. Remote Sens. 13 (4). https://doi.org/10.1117/1.JRS.13.044509.

Weisser, J., Beer, I., Hufnagl, B., Hofmann, T., Lohninger, H., Ivleva, N.P., Glas, K., 2021. From the well to the bottle: identifying sources of microplastics in mineral water. Water 13 (6), 841. https://doi.org/10.3390/w13060841.

World Health Organization (WHO). (2020). WHO calls for more research into microplastics and a crackdown on plastic pollution. Available online: https://www.who.int/news/item/22−08−2019−who−calls−for−more−research−into−microplastics−and−a−crackdown−on−plastic−pollution. (Accessed on 24 August 2020).

Xie, L., Luo, S., Liu, Y., Ruan, X., Gong, K., Ge, Q., Li, K., Valev, V.K., Liu, G., Zhang, L., 2023. Automatic identification of individual nanoplastics by Raman spectroscopy based on machine learning. Environ. Sci. Technol. 57 (46), 18203–18214. https://doi.org/10.1021/acs.est.3c03210.

Xue, Q., Wang, N., Yang, H., Yang, J., Bai, H., 2022. Detection of microplastics based on spatial heterodyne Raman spectroscopy. Spectrochim. Acta Mol. Biomol. Spectrosc. 283, 121712. https://doi.org/10.1016/j.saa.2022.121712.

Yan, X., Cao, Z., Murphy, A., Qiao, Y., 2022. An ensemble machine learning method for microplastics identification with FTIR spectrum. J. Environ. Chem. Eng. 10 (4), 108130. https://doi.org/10.1016/j.jece.2022.108130.

Yang, J., Xu, J., Zhang, X., Wu, C., Lin, T., Ying, Y., 2019. Deep learning for vibrational spectral analysis: recent progress and a practical guide. Anal. Chim. Acta 1081, 6–17. https://doi.org/10.1016/j.aca.2019.06.012.

Yang, Z., et al., 2023. Development of automated microplastic identification workflow for Raman micro-imaging and evaluation of the uncertainties during micro-imaging. Mar. Pollut. Bull. 193, 115200.

Zeng, G., Ma, Y., Du, M., Chen, T., Lin, L., Dai, M., Luo, H., Hu, L., Zhou, Q., Pan, X., 2024. Deep convolutional neural networks for aged microplastics identification by Fourier transform infrared spectra classification. Sci. Total Environ. 913, 169623. https://doi.org/10.1016/j.scitotenv.2023.169623.

Zhao, B., Richardson, R.E., You, F., 2024. Advancing microplastic analysis in the era of artificial intelligence: from current applications to the promise of generative AI. Nexus 1 (4), 100043. https://doi.org/10.1016/j.ynexs.2024.100043.

Zhu, Y., Yeung, C.H., Lam, E.Y., 2021. Digital holographic imaging and classification of microplastics using deep transfer learning. Appl. Opt. 60 (4), A38. https://doi.org/10.1364/AO.403366.

Index

Note: Page numbers followed by f indicate figures, t indicate tables.

N

O

P

Q

R

S

Printed and bound by CPI Group (UK) Ltd, Croydon, CR0 4YY
14/09/2026
02216410-0012